f, g, h, etc.	names of functions (p. 168)
$f(x)$, etc.	f of x or value of f at x (p. 169)
(a, b)	the ordered pair of numbers whose first component is a and whose second component is b (p. 172)
$\begin{vmatrix} a_1 & b_1 \\ a_2 & b_2 \end{vmatrix}$	second-order determinant (p. 227)
$\begin{vmatrix} a_1 & b_1 & c_1 \\ a_2 & b_2 & c_2 \\ a_3 & b_3 & c_3 \end{vmatrix}$	third-order determinant (p. 231)
$\log_b x$	logarithm to the base b of x (p. 254)
$\text{antilog}_b\, x$	antilogarithm to the base b of x (p. 263)
$A, B, C, \ldots$	point (p. 279)
$l_1, l_2, \ldots$	line (p. 279)
AB	ray, or half-line (p. 279)
AB or BA	line segment or length of line segment (p. 279)
$\angle ABC$, or $\angle B$ or $\angle 1$, $\angle 2$	angle or measure of angle (p. 279)
$\triangle ABC$	triangle with vertices A, B, and C (p. 287)
$\widehat{AB}$	arc of a circle with end points A and B (p. 296)
$\cong$	congruent to (p. 306)
$\sim$	similar to (p. 306)

Intermediate mathematics

Intermediate mathematics

IRVING DROOYAN / WILLIAM WOOTON

Los Angeles Pierce College

Wadsworth Publishing Company, Inc.

Belmont, California

L.C. Cat. Card No.: 78-154006

Printed in the United States of America

5 6 7 8 9 10—80 79 78 77

ISBN 0-534-00095-9

Preface

This textbook has a twofold purpose. It is designed for students preparing for occupational fields in which the mathematical topics ordinarily covered in a course in intermediate algebra are of primary concern. At the same time, the material is presented in enough detail so that those students wishing to go on to more advanced mathematics after completing this course will be prepared to do so.

Although complete coverage of the material would require two semester courses or three quarter courses, the last five chapters and Appendix are independent of each other. Any may be omitted for a shorter course.

The first seven chapters review material normally encountered in some form in a beginning algebra course. Here, however, the algebraic concepts are presented as a unified, related structure. The properties of the set of real numbers are explicitly listed in Chapter 1 and are used throughout Chapter 2 as a basis for the discussion of fundamental operations with polynomials. In Chapter 3, concerning operations with fractions, the fundamental principle of fractions is always invoked when writing equal fractions. Chapter 4 introduces linear equations and inequalities, and Chapters 5 and 6 lay the foundation for solving quadratic equations (Chapter 7) by acquainting the student with the properties of expressions involving rational-number exponents and radicals. The presentation of the material of Chapters 8 through 11 centers around the function concept, and heavy emphasis is placed on graphing. In particular, linear and quadratic functions and their graphs are explored in some detail in Chapters 8 and 9, and logarithms are developed in Chapter 11 from a consideration of the exponential function.

Chapters 12 and 13 provide a practical introduction to those parts of geometry and trigonometry that are most important in real-life applications. The Appendix offers a sound foundation in the use of the slide rule. Furthermore, the material on the slide rule is organized in a way that allows the instructor flexibility in determining the appropriate place to introduce the various units for a particular course.

Every effort has gone into making the material in this book teachable. The exercises following each section have been carefully graded, and many worked-out examples are built into the exercise sets. Answers, including graphs, are given for all odd-numbered exercises; and answers are provided for all review exercises, which conclude each chapter. A second color is used functionally to depict applications of fundamental mathematical principles. For easy reference, a list of mathematical symbols is printed on the inside covers.

We thank Charles C. Carico of Pierce College for his helpful comments and Alexander Kugushev, our editor, for his suggestions and encouragement in the preparation of this book.

Irving Drooyan
William Wooton

Contents

1

Properties of real numbers

2

Polynomials

3

Fractions

4

First-degree equations and inequalities: one variable

5

Exponents

6

Roots and radicals

7

Second-degree equations: one variable

8

Functions and graphs I

9

Functions and graphs II

10

Systems of equations

11

Exponential and logarithmic functions

12

Elements of geometry

13

Trigonometry

Appendix

Properties of real numbers

Algebra is frequently called a generalization of arithmetic, and that is the viewpoint maintained throughout this book. Our study is concerned, basically, with numbers and the properties associated with them. It is sometimes easy to lose sight of this fact in the midst of the seemingly complicated manipulations of symbols in which we engage, but we shall be better off if we strive to give meaning to everything we do in terms of numbers.

The names used for counting numbers (one, two, etc.) should, of course, be familiar to you. You should also be familiar with the symbols, called numerals (1, 2, etc.), used to represent them. In addition, you should know that such symbols as $\frac{1}{2}$, $\frac{3}{4}$, -3, $-\frac{2}{3}$, π, and $\sqrt{2}$ also represent numbers.

The point of this discussion is that the names and symbols used to talk about numbers are not the numbers themselves. There are places in algebra where confusing the two ideas—the number itself and the symbol used to represent it—can lead to misunderstanding. In this book, however, we propose to discuss such things as the number 2 or the number 5 or the number 8234. We do so simply as a matter of convenience. We do not wish to have always to say "the number represented by the numeral 2," because this intent should be clear from context.

1.1 Sets and symbolism—the real numbers

A **set** is simply a collection of some kind. It may be a collection of people, or books, or colors, or almost anything else; however, in algebra we are interested primarily in sets of numbers. Any one of the collection of things in a set is called a **member** or **element** of the set. For example, the counting numbers, 1, 2, 3, ..., are the elements of a set we call the set of **natural numbers**. Because there exists no last counting number, we refer to this set as an **infinite set**. A set whose elements can be arranged in some fashion and counted one by one until a last element is reached is called a **finite set**. For example, the set of numbers represented by the symbols on a die, 1, 2, 3, 4, 5, 6, is a finite set. The set containing no members is called the **empty set** or **null set**.

Sets are designated symbolically by means of capital letters—A, I, R, etc.—or by means of braces { } used in conjunction with words or symbols. Thus, when we write

$$\{\text{counting numbers less than } 7\}$$

or

$$1, 2, 3, 4, 5, 6\},$$

we mean the set whose members are the numbers 1, 2, 3, 4, 5, and 6. We say that two sets are equal if they have the same members; thus

$$\{2, 3, 4\} = \{\text{counting numbers between 1 and } 5\}.$$

We represent the empty set by the symbol $\emptyset$.

If every member of a given set A is also a member of another set B, we say that A is a **subset** of B. Thus $\{1, 2, 3\}$ is a subset of $\{1, 2, 3, 4, 5\}$. When discussing an individual element in a set, we generally denote this element by means of a lowercase letter, such as a, b, c, x, y, or z. Symbols used in this way are called **variables**. If the given set, called the **replacement set** of the variable, is a set of numbers (as it will be throughout this book), then the variable represents a number. That is, when we use the variable x, it is to be understood that x represents a number. A symbol used to denote the member of a set containing only one member is called a **constant**.

The set of numbers with which you first become acquainted is the set of natural numbers (the counting numbers),

$$\{1, 2, 3, \ldots\}.$$

You learn to add, subtract, multiply, and divide such numbers and to use them in simple quantitative problems. By the end of a first-year course in algebra, you are working with the set of real numbers. In particular, you should be familiar with the following sets:

1. The **natural numbers**, among whose elements are such numbers as 1, 2, 7, and 235.
2. The **integers**, whose elements consist of the natural numbers, their negatives, and zero. Among the integers are such numbers as -7, -3, 0, 5, and 11.
3. The **rational numbers**, whose elements are all those numbers that can be represented in the form a/b, where a and b are integers and b is not zero. Among the elements of the set of rational numbers are found such numbers as $-\frac{3}{4}$, $\frac{18}{27}$, 3, and -6.
4. The **irrational numbers**, whose elements are those numbers whose decimal representations are nonterminating, nonrepeating numerals. Among the elements of this set are such numbers as $\sqrt{2}$, π, and $-\sqrt{7}$. An irrational number cannot be represented in the form a/b where a and b are integers.
5. The **real numbers**, which is the set of all rational and all irrational numbers and which we shall designate by $\boldsymbol{R}$.

It is with the real numbers or subsets of the real numbers that we shall be most concerned in this book. The relationships between these sets are shown in Figure 1.1.

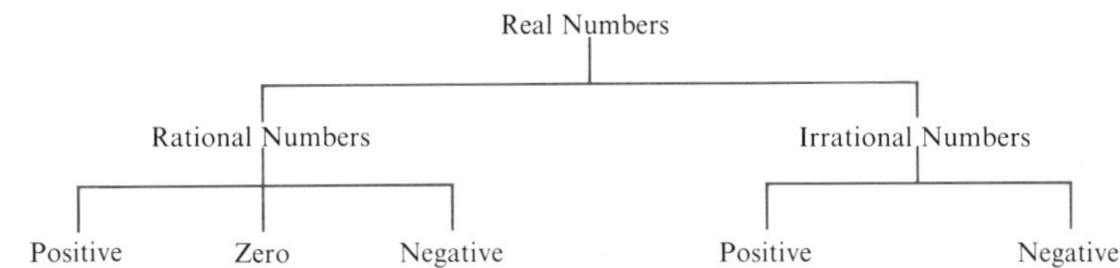

FIGURE **1.1**

EXERCISE 1.1

Specify each set by using numerals.

EXAMPLES

a. {natural numbers between 7 and 10} b. {natural-number multiples of 4}

SOLUTIONS

a. **{8, 9}**

b. **{4, 8, 12, ...}**

1. {first four natural numbers}
2. {first six natural numbers}
3. {odd natural numbers}
4. {even natural numbers}
5. {even natural numbers between 7 and 13}
6. {odd natural numbers between 4 and 10}
7. {first four natural-number multiples of 3}
8. {first four natural-number multiples of 5}

In Problems 9–14, let $A = \{6, -3, 3/5, \sqrt{7}, 0, 1/5, -\sqrt{3}, -5, -2/3, \sqrt{5}\}$.

EXAMPLES

a. What is the set whose members are all of the natural numbers contained in A?

b. What is the set whose members are all the negative integers contained in A?

SOLUTIONS

a. **{6}** b. **{− 3, − 5}**

9. What is the set whose members are all of the integers contained in A?
10. What is the set whose members are all of the rational numbers contained in A?
11. What is the set whose members are all of the irrational numbers contained in A?
12. What is the set whose members are all of the odd natural numbers contained in A?
13. What is the set whose members are all of the positive even integers contained in A?
14. What is the set whose members are all of the real numbers between 1 and 7 contained in A?
15. List all possible subsets of the set {1, 2, 3} that contain 2 members.
16. List all possible subsets of the set {0, 1, 2, 3} that contain 3 members.

In Problems 17–22, state whether the given set is finite or infinite.

17. The natural numbers whose numerals end in 2.
18. The people alive today.
19. The natural numbers whose numerals contain four digits.
20. The grains of sand in the world.

21. The odd natural numbers.
22. The rational numbers between 0 and 1.

Let a represent a member of the set A. In each of the following cases, state whether a is a variable or a constant.

23. $A = \{3, 4, 5, 6\}$
24. $A = \{5, 7\}$
25. $A = \{3\}$
26. $A = \{9\}$
27. $A = \{\text{integers between 7 and 9}\}$
28. $A = \{\text{natural numbers less than 2}\}$
29. $A = \{\text{real numbers greater than 2}\}$
30. $A = \{\text{rational numbers between 1 and 3}\}$

1.2 Axioms of equality and order

In mathematics when we make formal assumptions about numbers or their properties, we call the assumptions **axioms** or **postulates.**

The words *property*, *law*, and *principle* are sometimes used to denote assumptions, although these words may also be applied to certain consequences of axioms. In this book we shall use, in each situation, the word we believe to be the one most frequently encountered. The first such assumptions to be considered have to do with equality.

An **equality**, or an "is equal to" assertion involving the symbol $=$, is simply a mathematical statement that two symbols, or groups of symbols, are names for the same number. Thus, 3, $4 - 1$, and $2 + 1$ are all names for the same number; hence the equality

$$4 - 1 = 2 + 1$$

is a statement that $4 - 1$ and $2 + 1$ are different names for the same number.

We shall assume that the "is equal to" relationship has the following properties.

For all real numbers, a, b, and c:

$a = a$	**Reflexive law**
If $a = b$, then $b = a$.	**Symmetric law**
If $a = b$ and $b = c$, then $a = c$.	**Transitive law**
If $a = b$, then b may be substituted for a or a for b in any expression without altering the truth or falsity of the expression.	**Substitution law**

In writing an equality, we refer to the symbol or symbols to the left of the equals sign as the **left-hand member**, and those to the right as the **right-hand member** of the equality.

Because there exists a one-to-one correspondence between the real numbers and the points on a line (for each real number there corresponds one and only one point on the line, and vice versa), a line can be used to visualize relationships existing between real numbers. For example, to represent $\{1, 3, 5\}$ on a line, we simply scale a straight line in convenient units with increasing positive direction indicated by an arrowhead and indicate the required points with closed dots on the line, as shown in Figure 1.2. This geometric representation is called a **line graph**.

FIGURE **1.2**

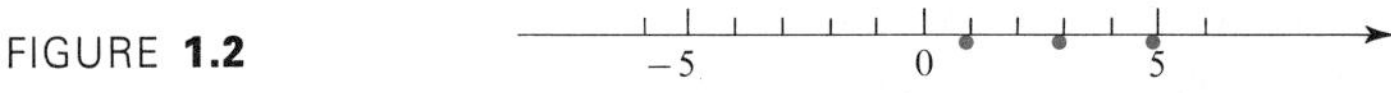

The real number corresponding to a point on a line graph is called the **coordinate** of the point, and the point is called the **graph** of the number. The point associated with 0 is called the **origin**.

A line graph can be used to separate the real numbers into three subsets—one set G whose elements are associated with the points on the line to the right of the origin, one set L whose elements are associated with the points on the line to the left of the origin, and one set whose only element is 0. The elements of the set G are called **positive numbers**, and in some cases we prefix their numerals with plus signs ($+1$, $+7$, $+\pi$) to denote the fact that they are positive. The elements of the set L are called **negative numbers**, and numerals representing such numbers are prefixed with minus signs (-2, $-\frac{1}{2}$, $-\pi$) to identify them as such. The number zero is neither positive nor negative; it serves to separate the positive and negative numbers. Since the signs $+$ and $-$ are used to identify positive and negative numbers, the real numbers are sometimes called **signed numbers**.

The addition of the positive real number c to a real number a can be visualized as the process of locating the point corresponding to a on a line graph, and then locating the sum $a + c$ by moving along the line c units to the right of a, as shown in Figure 1.3.

FIGURE **1.3**

We are now in a position to define the real number a as less than the real number b, or b as greater than a, if, for some positive real number c,

$$a + c = b.$$

This implies that, on a line graph, as in Figure 1.4, the point a will be found to the left of the point b. The number c is called the **difference** when a is

FIGURE **1.4**

subtracted from b. In addition to the symbol $=$ for equality, the following symbols are used in connection with the property of order:

$<$, read "is less than";
$\leq$, read "is less than or equal to";
$>$, read "is greater than";
$\geq$, read "is greater than or equal to."

The slant bar, /, is used in conjunction with an order symbol to indicate the word "not"; thus $\neq$ means "is not equal to" and $\nless$ means "is not less than." Furthermore $b > a$ has the same meaning as $a < b$; $b \geq a$ means the same as $a \leq b$; and so on.

Inequalities such as

$$1 < 2 \quad \text{and} \quad 3 < 5$$

are said to be in the *same sense*, because the left-hand member is less than the right-hand member in each case. Inequalities such as

$$1 < 2 \quad \text{and} \quad 5 > 3$$

are said to be of *opposite sense*, because in one case the left-hand member is less than the right-hand member but in the other case the left-hand member is greater than the right-hand member.

We adopt the following axioms with respect to the order of real numbers.

For all real numbers a, b, and c:

Exactly one of the following relationships holds: **Trichotomy law**

$$a < b, \qquad a = b, \qquad a > b.$$

If $a < b$ and $b < c$, then $a < c$. **Transitive law**

A line graph can be used to display an infinite set of points as well as a finite set, such as is the case in Figure 1.2. For example, Figure 1.5 is the

FIGURE **1.5**

graph of the set of all real numbers greater than or equal to 2 and less than 5. The closed dot on the left end of the graph indicates that the end point is a part of the graph, while the open dot on the right indicates that the end point is not in the graph.

EXERCISE 1.2

Each of the statements 1–18 is an application of one of the axioms of equality or order. Justify each statement by citing the appropriate axiom. (There may be more than one correct answer.)

EXAMPLES

a. If $3 = a$, then $a = 3$.

b. If $3 < a$ and $a = b$, then $3 < b$.

SOLUTIONS

a. **Symmetric law**

b. **Substitution law**

1. If $a = 2$, then $2 = a$.
2. If $a + 3 = b$, then $b = a + 3$.
3. For all b, $b + 2 = b + 2$.
4. For all x, $2x = 2x$.
5. If $x = 5$ and $x = y$, then $y = 5$.
6. If $a < 2$ and $2 < x$, then $a < x$.
7. If $x = 5$ and $y = x + 2$, then $y = 5 + 2$.
8. If $y = 6$ and $x + y = z$, then $x + 6 = z$.
9. If $3 < b$ and $b < c$, then $3 < c$.
10. For all a, $a + 5 = a + 5$.
11. If $x = y$ and $y = 6$, then $x = 6$.
12. For all x, $x - 3 = x - 3$.
13. For every number x, $x < 0$, $x = 0$, or $x > 0$.
14. If $x = 4$ and $y = x + 1$, then $y = 4 + 1$.
15. If $y = 3$ and $y + a = b$, then $3 + a = b$.
16. For every number b, $b < 0$, $b = 0$, or $b > 0$.
17. If $a = 2c$ and $c = 6$, then $a = 2 \cdot 6$.
18. If $-6 = x$, then $x = -6$.

Graph each set.

EXAMPLE

{integers between -2 and 3}

SOLUTION

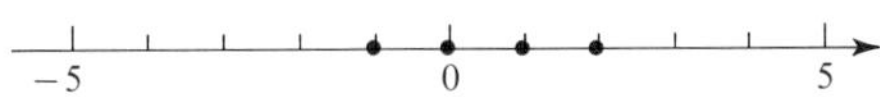

19. {natural numbers less than 6}
20. {natural numbers between 6 and 10}
21. {integers between −4 and 4}
22. {integers between −7 and 2}

EXAMPLE

{real numbers greater than −3 and less than or equal to 2}

SOLUTION

23. {real numbers greater than 5}
24. {real numbers greater than −3}
25. {real numbers less than or equal to 3}
26. {real numbers greater than 2 or less than −2}

Express each of the following relations by means of symbols.

EXAMPLES

a. 5 is greater than −7. b. x is between 5 and 8.

SOLUTIONS

a. $\mathbf{5 > -7}$ b. $\mathbf{5 < x < 8}$

27. 8 is greater than 2.
28. 2 is less than 7.
29. −4 is less than −3.
30. −4 is greater than −7.
31. x is not less than y.
32. a is not greater than b.
33. x is between 3 and 5.
34. x is between −2 and 7.

Replace each comma with an appropriate order symbol.

35. $-2, 5$ 36. $3, 4$ 37. $-7, -1$ 38. $0, -1$

39. $-5, -2$ 40. $\frac{-3}{2}, \frac{-3}{4}$ 41. $1\frac{1}{2}, \frac{3}{2}$ 42. $3, \frac{6}{2}$

Write an equivalent relation without using the slant bar.

43. $2 \ngeq 5$ 44. $-1 \nleq -2$ 45. $x \nless y$ 46. $x \ngtr z$

In Problems 47–50, write the relation in two ways—with the slant bar and without it.

47. Express "x is positive" by means of symbols.
48. Express "x is negative" by means of symbols.
49. Express "x is non-negative" by means of symbols.
50. Express "x is nonpositive" by means of symbols.

1.3 The properties of the real numbers

In addition to the equality and order axioms, we take as axioms the properties of the real numbers listed on page 11. Parentheses are used in some of the relations (and hereafter) to indicate that the symbols within the parentheses are to be viewed as a single entity.

The first and fourth axioms assert that the sum or product of any two real numbers is another real number and that each product and each sum is unique. When the performance of an operation on two elements of some set of numbers always results in another element of the same set, we say that the set is **closed** with respect to that operation. Thus, we assume that the set of real numbers is closed with respect to addition and multiplication.

The commutative laws state that the order in which we write two terms in a sum or two factors in a product does not alter the sum or product. Thus,

$$2 + 4 = 4 + 2$$

and

$$2 \cdot 4 = 4 \cdot 2.$$

The associative laws assert that three terms in a sum or three factors in a product may be associated in either of two ways; they therefore give meaning to $a + b + c$ and $a \cdot b \cdot c$. Thus,

$$(2 + 3) + 4 = 2 + (3 + 4)$$

and

$$(2 \cdot 3)4 = 2(3 \cdot 4).$$

PROPERTIES OF THE REAL NUMBERS

If a, b, and c are real numbers:

$a + b$ is a unique real number.	**Closure for addition**
$a + b = b + a$.	**Commutative law of addition**
$(a + b) + c = a + (b + c)$.	**Associative law of addition**
ab is a unique real number.	**Closure for multiplication**
$ab = ba$.	**Commutative law of multiplication**
$(ab)c = a(bc)$.	**Associative law of multiplication**
$a(b + c) = ab + ac$.	**Distributive law**
There exists a unique number 0 with the property $a + 0 = a$ and $0 + a = a$.	**Additive property of zero**
There exists a unique number 1 with the property $a \cdot 1 = a$ and $1 \cdot a = a$.	**Multiplicative property of one**
For each real number a, there exists a unique real number $-a$ (called the **negative** of a) with the property $a + (-a) = 0$ and $-a + a = 0$.	**Negative law**
For each real number a except zero, there exists a unique real number $1/a$ (called the **reciprocal** of a) with the property $a\left(\frac{1}{a}\right) = 1$ and $\left(\frac{1}{a}\right)a = 1$.	**Reciprocal law**

If both the associative and commutative laws are used, we can add any number of terms or multiply any number of factors in any order that we wish.

The distributive law gives us a means of relating the operations of addition and multiplication. Thus,

$$2(3+4) = 2(3) + 2(4).$$

If we use this law with the associative law, we can show that

$$a(b+c+d) = ab + ac + ad,$$
$$a(b+c+d+e) = ab + ac + ad + ae,$$

and so on. We refer to this latter property as the **generalized distributive law**. Furthermore, the commutative law of multiplication assures us that

$$a(b+c+d) = (b+c+d)a,$$
$$a(b+c+d+e) = (b+c+d+e)a,$$

and so forth.

The additive property of zero and the multiplicative property of one simply state some basic properties of the numbers zero and one. The sum of zero and any real number a is a, and the product of one and any real number a is a. The negative axiom asserts the existence of a negative for each real number and tells us that the sum of any real number and its negative is zero; for example,

$$3 + (-3) = 0.$$

In mathematics the word "unique" means that there is one and only one of the things being discussed. Since the number $-a$ is assumed to be unique, it follows that if $a + b = 0$, then $b = -a$ and $a = -b$; that is, if the sum of two numbers is zero, each is the negative of the other. For example, if $x + 3 = 0$, then x must equal -3, and if $-5 + y = 0$, then y must equal 5.

The reciprocal axiom guarantees the existence of a reciprocal for each nonzero number and tells us that the product of any real number and its reciprocal is 1; for example,

$$5 \cdot \frac{1}{5} = 1.$$

Since the number $1/a$ is also assumed to be unique, it follows that if $a \cdot b = 1$, then $b = 1/a$ and $a = 1/b$; that is, if the product of two numbers is one, each is the reciprocal of the other.

The following mathematical statements are consequences of the axioms above and will help simplify some later work. These properties and others that are developed in this text are listed on page 409 for convenient reference.

1. If $\boldsymbol{a = b}$, then

$$\boldsymbol{a + c = b + c} \quad \text{and} \quad \boldsymbol{c + a = c + b} \qquad \textbf{Addition law of equality}$$

2. If $\boldsymbol{a = b}$, then

$$\boldsymbol{ac = bc} \quad \text{and} \quad \boldsymbol{ca = cb} \qquad \textbf{Multiplication law of equality}$$

For example, if

$$x = y,$$

then

$$x + 2 = y + 2$$

and

$$2x = 2y.$$

3. For each real number a,

$$\boldsymbol{-(-a) = a} \qquad \textbf{Double-negative law}$$

For example,

$$-(-3) = 3 \quad \text{and} \quad -[-(-3)] = -3.$$

4. For each real number a,

$$\boldsymbol{a \cdot 0 = 0} \qquad \textbf{Zero-factor law}$$

For example,

$$2 \cdot 0 = 0 \quad \text{and} \quad -3 \cdot 0 = 0.$$

Because each real number is associated with a single point located a specific distance from the origin, we can associate a distance with each real number. However, $-a$ and a are each located the same distance from the origin, and if we wish to refer simply to this distance and not to its direction to the left or right of 0, we can use the notation $|a|$ (read "the absolute value of a"). Thus $|a|$ is always non-negative. That is,

$$|a| = a \quad \text{if } a \geq 0;$$
$$|a| = -a \quad \text{if } a < 0.$$

For example,

$$|3| = 3, \quad |-3| = 3, \quad \text{and} \quad |0| = 0.$$

EXERCISE 1.3

Justify each statement by citing the appropriate axiom or consequence of an axiom introduced in this section.

EXAMPLES

a. $2(3+1)=2\cdot 3+2\cdot 1$

b. $x+2=2+x$

SOLUTIONS

a. **Distributive law**

b. **Commutative law of addition**

1. $5+3=3+5$
2. $7+0=7$
3. $7\cdot \frac{1}{7}=1$
4. $5+(4+1)=(5+4)+1$
5. $3+(-3)=0$
6. $-(-8)=8$
7. If x and y are real numbers and $xy=z$, then z is a real number.
8. If a and b are real numbers and $a+b=c$, then c is a real number.
9. $(2\cdot 3)4=2(3\cdot 4)$
10. $3\cdot 1=3$
11. $3(2+5)=3\cdot 2+3\cdot 5$
12. $3(-2)=(-2)3$
13. $5+(-2)=(-2)+5$
14. $x\cdot 0=0\cdot x$
15. $(-4+3)+1=-4+(3+1)$
16. $2[1+(-5)]=2\cdot 1+2(-5)$
17. $-(-3)=3$
18. $a(b+c)=(b+c)a$
19. $(a+b)+c=c+(a+b)$
20. $a+(b+c)=a+(c+b)$
21. $a+(b+c)d=a+d(b+c)$
22. $a+(b+c)d=(b+c)d+a$
23. $a+c(b+d)=a+cb+cd$
24. $a[b+(c+d)]=ab+a(c+d)$
25. $(a+b)+[-(a+b)]=0$
26. $\frac{1}{3}(a+b)=\frac{1}{3}\cdot a+\frac{1}{3}\cdot b$
27. If a is negative, what is $-a$?
28. If $-a$ is positive, what is a?
29. If $-a$ is negative, what is a?
30. If a is positive, what is $-a$?

Write each real number without using absolute-value notation.

EXAMPLES

a. $|-8|$

b. $-|3|$

c. $-|-5|$

SOLUTIONS

a. **8**

b. **−3**

c. **−5**

31. $|4|$
32. $|-7|$
33. $-|-2|$
34. $-|6|$
35. $|0|$
36. $|-9|$

1.4 Sums and differences

The assumptions we have made regarding the real numbers permit us to formulate some convenient rules for performing operations on such numbers. The rules follow from the assumptions, but it is preferable to be able to manipulate the numbers in most cases of routine arithmetic without each time having to invoke the full logical machinery of the system. You should realize, however, that the rules developed here do not materialize out of thin air but are consequences of the assumptions we have made.

In this section we shall develop rules for the addition and subtraction of signed numbers. We can develop these rules by seeking relationships between the sums of pairs of signed numbers and the sums of pairs of positive numbers, relationships that are consequences of our assumptions. For example, consider the expression

$$(-7) + (-5).$$

We can relate this to the sum $7 + 5$ by observing that

$$[(-7) + (7)] + [(-5) + (5)] = 0.$$

It follows from the associative and commutative laws of addition that

$$[(7) + (5)] + [(-7) + (-5)] = 0,$$

and therefore $[(-7) + (-5)]$ must be the negative of $[(7) + (5)]$, since their sum is zero. Then

$$\begin{aligned}(-7) + (-5) &= -(7 + 5)\\ &= -12.\end{aligned}$$

By an identical argument, we can show that

$$(-a) + (-b) = -(a + b).$$

Thus:

To find the sum of two real numbers with like signs, add the absolute value of the numbers and prefix this sum with the sign of the numbers being added. The resulting real number is the sum.

For example,

$$5 + 3 = |5| + |3| = 8,$$

$$(-7) + (-8) = -(|-7| + |-8|) = -15,$$

$$(-2) + (-9) = -(|-2| + |-9) = -11.$$

This process is easily accomplished mentally.

Now what can we say about the sum of two numbers with unlike signs—say, $7 + (-5)$? We can begin by observing that

$$7 = 2 + 5,$$

and by the addition law,

$$7 + (-5) = (2 + 5) + (-5).$$

Now by the associative law of addition, we have

$$7 + (-5) = 2 + [5 + (-5)];$$

and the negative axiom tells us that $5 + (-5)$ is 0, so that

$$7 + (-5) = 2.$$

But 2 is just the difference between the absolute values of the numbers being added. In a similar way, we can show that this is always true by considering the sum $a + (-b)$ for both the case $a > b$ and the case $a < b$. These facts can be stated as follows:

To find the sum of two real numbers with unlike signs, find the difference of the absolute value of the numbers and prefix the difference with the sign on the number of greatest absolute value. This signed difference is the sum. If the numbers have the same absolute value, the sum is zero.

For example,

$$\begin{aligned} 5 + (-7) &= -(|-7| - |5|) = -2, \\ 8 + (-6) &= (|8| - |-6|) = 2, \\ (-12) + 10 &= -(|-12| - |10|) = -2, \\ (-5) + 5 &= (|-5| - |5|) = 0. \end{aligned}$$

Again, this process is easily accomplished mentally.

The difference of two real numbers is defined to be the real number c

$$a - b = c,$$

such that

$$b + c = a.$$

For example $7 - 4$ is the real number c such that $4 + c = 7$. In this case, obviously, c equals 3. Observe that $7 + (-4) = 3$; hence

$$7 - 4 = 7 + (-4).$$

Also, from the above definition of a difference, $4 - 7$ is the real number c such that $7 + c = 4$. In this case c equals -3. Observe that $4 + (-7) = -3$; hence

$$4 - 7 = 4 + (-7).$$

It can be shown that for all real numbers a and b,

$$a - b = a + (-b),$$

which fact can be stated as follows:

To subtract one real number b from another a, add the negative of b to a.

For example,

$$8 - (-3) = 8 + [-(-3)] = |8| + |3| = 11,$$
$$(-7) - (4) = (-7) + (-4) = -(|-7| + |-4|) = -11,$$
$$(-5) - (-2) = (-5) + [-(-2)] = -(|-5| - |2|) = -3.$$

For convenient reference, we shall call the relationship

$$\mathbf{a - b = a + (-b)}$$

the **alternate definition of a difference.**

We have now used the sign $-$ in three ways. It has been used to denote a negative number, such as -3; it has been used to designate the negative (in the sense of "opposite") of a number, such as $-(3)$ or $-(-3)$; and it has been used as a sign of operation to indicate the subtraction of one number from another.

Having seen that the difference $a - b$ is given by $a + (-b)$, we may consider the symbols $a - b$ as representing either the difference of a and b or, preferably, the sum of a and $(-b)$.

EXERCISE 1.4

Perform the operations indicated.

EXAMPLES

a. $3 + (-7)$ b. $(-3) - (-7)$ c. $(-2) + (-3) - (4)$

SOLUTIONS

a. **−4**

b. $(-3) + (7)$
4

c. $(-2) + (-3) + (-4)$
−9

1. $5 + 2$
2. $7 + (-2)$
3. $(-5) + (-1)$
4. $8 + (-5)$
5. $3 + 0$
6. $(-3) + 2$
7. $6 + (-10)$
8. $(-8) + (-12)$
9. $(5) + (-5)$
10. $(-8) + 0$
11. $8 + (-3)$
12. $-10 + (10)$

13. $8 - 3$	14. $3 - 5$	15. $25 - (-2)$
16. $-16 - 8$	17. $-18 - (-12)$	18. $22 - 29$
19. $7 - (-7)$	20. $-18 - (-18)$	21. $(-12) - (-2)$
22. $-4 - (-6)$	23. $0 - (-5)$	24. $-8 - 0$
25. $3 + 2 - 4$	26. $8 - 2 + 1$	27. $8 - (-2) + (-3)$
28. $(-3) + 4 - 7$	29. $(5) - (-2) - 1$	30. $(-3) - (5) - 6$
31. $8 + (0) - (-1)$	32. $8 + 0 - 8$	33. $2 - (2) - (-5)$
34. $25 + 18 - 43$	35. $8 - 8 + 6$	36. $8 + 3 - 5$

Add.

37.	38.	39.	40.
215	−135	177	272
−342	211	−23	408
87	46	−185	−391
−23	−317	28	−310

41.	42.	43.	44.
2.03	−5.280	26.1	0.021
−0.17	−6.004	147.3	−1.803
2.38	1.982	−21.9	3.200
−3.19	9.213	−86.2	−1.810

Subtract.

45.	46.	47.	48.
247	−623	485	−385
115	412	−271	−415

49.	50.	51.	52.
3.18	−0.82	81.70	−8.130
2.19	1.63	−6.92	−0.617

1.5 Products and quotients

It is assumed that the multiplication of two positive real numbers always results in a product that is a positive real number. Thus, if a and b are positive,

$$a \cdot b = ab,$$

where ab is positive. What can be said about the product of a positive and a negative number? Let us begin by considering the equality

$$3 + (-3) = 0.$$

By the multiplication law,

$$2[3 + (-3)] = 2 \cdot 0;$$

and the distributive law and the zero-factor law permit us to write

$$2 \cdot 3 + 2(-3) = 0.$$

Now, since $2(-3)$ must be a number that adds to $2 \cdot 3$, or 6, to yield zero, it follows that $2(-3)$ is the negative of $2 \cdot 3$, and

$$2(-3) = -(2 \cdot 3) = -6.$$

Thus the product of two real numbers with unlike signs is negative. This argument can be developed for the product of any two real numbers with unlike signs. Thus,

$$\boldsymbol{a(-b) = -(ab) = -ab},$$

or, alternatively,

$$(-b)(a) = -(ab) = -ab.$$

Furthermore, by a similar argument, we can show that

$$(-a)(-b) = -(-ab),$$

or, from the double-negative law, that

$$\boldsymbol{(-a)(-b) = ab}.$$

Both of the foregoing facts can be summarized as follows:

To find the product of two real numbers, multiply the absolute values of the numbers. If the numbers have like signs, the product is positive; if the numbers have unlike signs, the product is negative.

For example,

$$(3)(-2) = -6, \qquad (-3)(-2) = 6,$$
$$(-2)(3) = -6, \qquad (3)(2) = 6.$$

We define the quotient of two real numbers a and b to be the number q,

$$\frac{a}{b} = q \quad (b \neq 0),$$

such that

$$bq = a.$$

In general, a fraction bar is used to indicate that one algebraic expression is to be divided by another; the dividend is the numerator and the divisor is the denominator. Note that the denominator is restricted to nonzero numbers, for, if b is zero and a is not zero, then there exists no q such that

$$0 \cdot q = a.$$

Again, if b is zero and a is zero, then for *any* q

$$0 \cdot q = 0,$$

and the quotient is not unique. Thus we can state:

Division by zero is not defined.

On the other hand, zero divided by any nonzero real number b is 0. That is, in the quotient

$$\frac{0}{b} = q,$$

q must equal zero, because if $b \neq 0$, $b \cdot q = 0$ only if $q = 0$.

Since the quotient of two numbers a/b is a number q such that $bq = a$, the sign of the quotient of two signed numbers must be consistent with the rules of signs for the product of two signed numbers. Therefore, for $a, b, q > 0$,

$$\frac{+a}{+b} = +q, \qquad \frac{-a}{-b} = +q,$$

$$\frac{+a}{-b} = -q, \qquad \frac{-a}{+b} = -q.$$

Thus:

To divide one real number by another nonzero real number, find the quotient of the absolute values of the numbers. If the numbers have like signs, the quotient is positive; if the numbers have unlike signs, the quotient is negative.

It is important to recognize that a/b and $a(1/b)$ represent the same number. To see that this is true, assume that

$$a\left(\frac{1}{b}\right) = q \quad (b \neq 0). \tag{1}$$

Then multiply each member of this equality by b to obtain

$$a\left(\frac{1}{b}\right)b = qb; \tag{2}$$

since $(1/b)b = 1$, Equation (2) becomes

$$a = qb.$$

But, by definition, $a = qb$ implies that $q = a/b$; therefore, Equation (1) can be written

$$\boldsymbol{\frac{a}{b} = a\left(\frac{1}{b}\right)}.$$

For convenient reference, we shall call this relationship the **alternate definition of a quotient**.

EXERCISE 1.5

Multiply.

EXAMPLES

a. $(-3)(2)$ b. $(-7)(-5)$ c. $(-2)(-3)(-5)$

SOLUTIONS

a. **−6** b. **35** c. $6(-5)$
−30

1. $5(-3)$
2. $(-4)(2)$
3. $(-8)(-3)$
4. $(4)(2)$
5. $(-6)(7)$
6. $(-3)(-3)$
7. $(-15)(-3)$
8. $(7)(9)$
9. $(-11)(0)$
10. $(3)(0)$
11. $(-8)(-7)$
12. $(2)(-8)$
13. $(2)(3)(5)$
14. $(-8)(1)(2)$
15. $(-6)(-2)(-3)$
16. $(8)(-1)(5)$
17. $(3)(-2)(0)$
18. $(-75)(0)(-5)$
19. $(-4)(4)(4)$
20. $(-3)(3)(-3)$
21. $(-2)(-2)(-2)$
22. $(5)(10)(15)$
23. $(-2)(-3)(-4)$
24. $(-1)(-10)(-5)$
25. $\begin{array}{r} 271 \\ \times 13 \\ \hline \end{array}$
26. $\begin{array}{r} -183 \\ \times 82 \\ \hline \end{array}$
27. $\begin{array}{r} 386 \\ \times(-43) \\ \hline \end{array}$
28. $\begin{array}{r} -1422 \\ \times(-65) \\ \hline \end{array}$
29. $\begin{array}{r} 82.3 \\ \times(-0.5) \\ \hline \end{array}$
30. $\begin{array}{r} -68.4 \\ \times 2.7 \\ \hline \end{array}$
31. $\begin{array}{r} 0.283 \\ \times(-8.7) \\ \hline \end{array}$
32. $\begin{array}{r} -81.63 \\ \times(-2.9) \\ \hline \end{array}$

Divide.

EXAMPLES

a. $\dfrac{-18}{9}$ b. $\dfrac{-27}{-3}$ c. $\dfrac{6}{-2}$

SOLUTIONS

a. **−2** b. **9** c. **−3**

33. $\dfrac{18}{2}$
34. $\dfrac{16}{4}$
35. $\dfrac{20}{-5}$
36. $\dfrac{-40}{10}$
37. $\dfrac{-12}{-2}$
38. $\dfrac{27}{-9}$
39. $\dfrac{0}{-7}$
40. $\dfrac{-40}{-4}$

41. $\frac{-32}{0}$ 42. $\frac{0}{27}$ 43. $\frac{-15}{-3}$ 44. $\frac{-63}{0}$

45. $23\overline{)1886}$ 46. $-47\overline{)1457}$ 47. $17\overline{)-3094}$

48. $-62\overline{)-2914}$ 49. $9.1\overline{)25.48}$ 50. $-4.7\overline{)1.974}$

51. $0.62\overline{)-0.2108}$ 52. $-13\overline{)-2.340}$

From the definition of a quotient, rewrite each of the following expressions in the form $a = bq$.

EXAMPLES

a. $\frac{18}{-3} = -6$ b. $\frac{-27}{-3} = 9$ c. $\frac{-14}{2} = -7$

SOLUTIONS

a. $\mathbf{18 = (-3)(-6)}$ b. $\mathbf{-27 = (-3)(9)}$ c. $\mathbf{-14 = 2(-7)}$

53. $\frac{-32}{2} = -16$ 54. $\frac{-14}{7} = -2$ 55. $\frac{24}{-3} = -8$

56. $\frac{36}{9} = 4$ 57. $\frac{56}{-7} = -8$ 58. $\frac{-54}{-9} = 6$

Rewrite each of the following quotients as a product in which one factor is the reciprocal of a natural number.

EXAMPLES

a. $\frac{3}{4}$ b. $\frac{16}{5}$ c. $\frac{7}{9}$

SOLUTIONS

a. $\mathbf{3\left(\frac{1}{4}\right)}$ b. $\mathbf{16\left(\frac{1}{5}\right)}$ c. $\mathbf{7\left(\frac{1}{9}\right)}$

59. $\frac{5}{7}$ 60. $\frac{23}{5}$ 61. $\frac{2}{9}$ 62. $\frac{5}{12}$

63. $\frac{13}{11}$ 64. $\frac{3}{13}$ 65. $\frac{7}{10}$ 66. $\frac{3}{1000}$

Rewrite each of the following products as a quotient.

EXAMPLES

a. $4\left(\frac{1}{3}\right)$ b. $3\left(\frac{1}{100}\right)$ c. $\left(\frac{1}{10}\right)9$

SOLUTIONS

a. $\frac{4}{3}$ b. $\frac{3}{100}$ c. $\frac{9}{10}$

67. $3\left(\frac{1}{2}\right)$
68. $8\left(\frac{1}{3}\right)$
69. $\left(\frac{1}{7}\right)2$
70. $\left(\frac{1}{100}\right)3$
71. $\left(\frac{1}{8}\right)5$
72. $\left(\frac{1}{7}\right)6$
73. $9\left(\frac{1}{2}\right)$
74. $7\left(\frac{1}{10{,}000}\right)$

1.6 Prime factors

In discussing the product abc, we refer to the numbers a, b, and c as **factors** of the product. Thus 2 and 3 are factors of 6; 5 and 1 are factors of 5. If a natural number greater than 1 has no factors that are natural numbers other than itself and 1, it is said to be a **prime number**. Thus, 2, 3, 5, 7, 11, etc., are prime numbers. A natural number greater than 1 that is not a prime number is said to be a **composite number**. Thus 4, 6, 8, 9, 10, etc., are composite numbers. When a composite number is exhibited as a product of prime factors only, it is said to be **completely factored**. For example, although 30 may be factored into (5)(6), (10)(3), (15)(2), or (1)(30), if we continue the factorization we arrive at the set of prime factors 2, 3, and 5 in each case. Although it is unnecessary to develop the argument here, it is a fact that, except for order, *each composite number has one and only one prime factorization.*

Notice again that the words "composite" and "prime" are used in reference to natural numbers only. Integers, rational numbers, and irrational numbers are not referred to as either prime or composite. Any negative integer can, however, be expressed as the product $(-1)a$, where a is a natural number. Hence, if we refer to the completely factored form of a negative integer, we refer to the product of (-1) and the prime factors of the associated natural number.

It is evident that there are infinitely many sets of factors for a natural number if real numbers are considered as factors. Thus 30 can also be expressed as $(1/2)(60)$, $(1/3)(90)$, $4(15/2)$, etc., where the possibilities are unlimited.

EXERCISE 1.6

Express each of the following integers in completely factored form. If the integer is a prime number, so state.

EXAMPLES

a. 36 b. 29 c. -51

SOLUTIONS

a. $\mathbf{2 \cdot 2 \cdot 3 \cdot 3}$ b. **Prime** c. $\mathbf{-1 \cdot 3 \cdot 17}$

1. 6	2. 22	3. 25	4. 81
5. 18	6. 24	7. 17	8. 23
9. 56	10. 65	11. -38	12. -47
13. 20	14. 29	15. 106	16. 107
17. -200	18. -300	19. 400	20. 600
21. 205	22. 206	23. -207	24. -209
25. 240	26. 640	27. 244	28. 696

Chapter review

[1.1] 1. Specify {first three natural-number multiples of 6} by using numerals.

2. What is the set whose members are all the integers contained in

$$\left\{-3, -\frac{5}{2}, 0, \sqrt{2}, 5, \frac{11}{2}\right\}?$$

[1.2] Justify each of the statements 3–5, by citing the appropriate axiom.

3. If $x < 5$ and $5 < y$, then $x < y$.
4. If $x = y + 1$ and $y = 2$, then $x = 2 + 1$.
5. If $y = x + 2$, then $x + 2 = y$.

6. Graph {integers between -3 and 5}.
7. Graph {real numbers less than or equal to -2}.
8. Write $x \not< 0$ without using the slant bar.

[1.3] Justify each of the statements 9–12, by citing the appropriate axiom.

9. $(x + 3) + y = x + (3 + y)$
10. $2(x + 1) = 2 \cdot x + 2 \cdot 1$
11. $(x + 2)y = y(x + 2)$
12. $3x + (-3x) = 0$

13. If $-x$ is positive, what is x?

14. Write $-|-3|$ without using absolute-value notation.

[1.4] In problems 15–16, perform the indicated operations.

15. $2-7+3$

16. $4-(-3)-2$

17. Add: $\begin{array}{r} 218 \\ -617 \\ 83 \\ \hline \end{array}$

18. Subtract: $\begin{array}{r} -81.2 \\ 63.4 \\ \hline \end{array}$

In Problems 19–20, rewrite each difference as a sum.

19. $4-7$

20. $12-3$

[1.5] In Problems 21–24, perform the indicated operations.

21. $2(-3)(-4)$

22. $\begin{array}{r} 42.1 \\ \times(-3.1) \\ \hline \end{array}$

23. $\dfrac{-28}{-4}$

24. $-12\overline{)25.2}$

25. Rewrite the quotient $\frac{3}{5}$ as a product.

26. Rewrite the product $3\left(\dfrac{1}{100}\right)$ as a quotient.

[1.6] Express each of the following in completely factored form.

27. 68

28. 120

29. 91

30. -425

Polynomials

2.1 Definitions

As we have seen, the sum of two numbers is expressed by writing $a + b$, and their product, by writing ab. In some cases, as in some places in Chapter 1, a centered dot (·) is used to indicate multiplication—for example, $a \cdot b$, $x \cdot y$, and $2 \cdot 3$. In other cases, parentheses are written around one or both of the symbols, as (2)(3), 2(3), and $(x)(x)$. If the factors in a product are identical, the number of factors is indicated by means of an **exponent**. An exponent is a small numeral written at the upper right of a given numeral to indicate the number of times the given numeral represents a factor in the product; thus

$$x \cdot x = x^2, \quad y \cdot y \cdot y = y^3, \quad \text{and} \quad 3 \cdot 3 \cdot 3 \cdot 3 \cdot 3 = 3^5.$$

In general, if a is a real number and m is a natural number,

$$\boldsymbol{a^m = a \cdot a \cdot a \cdot \cdots \cdot a} \quad \textbf{(}\boldsymbol{m}\textbf{ factors.)}$$

If no exponent appears, the exponent 1 is intended—that is, $a = a^1$. The symbol to which the exponent is attached is called the **base**, and the product is said to be a **power** of the base.

Any collection of numerals, variables, and signs of operation is called an **expression**. In an expression of the form $A + B + C + \cdots$, A, B, and C, are called **terms** of the expression. Any factor or group of factors in a term is said to be the **coefficient** of the remaining factors in the term. Thus, in the term $3xyz$, $3x$ is the coefficient of yz, y is the coefficient of $3xz$, 3 is the coefficient of xyz, and so on. Hereafter, the word "coefficient" will refer to the numeral portion of the term, unless otherwise indicated. For example, the coefficient of the term $4xy$ is 4; the coefficient of $-2a^2b$ is -2. If no numeral coefficient appears in a term, the coefficient is understood to be 1. Thus x is viewed as $(1)x$.

An expression in which the operations involved consist solely of addition, subtraction, multiplication, and division and in which all variables occur as natural-number powers only, is called a **rational expression**. Thus,

$$5, \quad \frac{x+y}{2}, \quad x - \frac{1}{x}, \quad \frac{3x+2y}{x-y}, \quad \text{and} \quad x^2 + 2x + 1$$

are rational expressions. Any rational expression in which no variable occurs in a denominator is called a **polynomial**. Thus,

$$x^2, \quad 5, \quad 3x^2 - 2x + 1, \quad \frac{1}{5}x - 2, \quad \text{and} \quad x^2y - 2x + y$$

are polynomials. A polynomial consisting of only one term is called a **monomial**. If the polynomial contains two or three terms, we refer to it as a **binomial** or **trinomial**, respectively. For example,

$$3, \quad x^2y, \quad \text{and} \quad 5x^3 \quad \text{are monomials,}$$

$$x + y, \quad 2x^2 - 3x, \quad \text{and} \quad 4x + 3y \quad \text{are binomials,}$$

$$4x + 3y + 2z, \quad x^2 + 2x + 1, \quad \text{and} \quad 5xy + 2x - 3y \quad \text{are trinomials;}$$

and each of these examples is a polynomial.

The **degree** of a monomial is given by the exponent on the variable in the monomial. Thus $3x^4$ is of fourth degree, and $7x^5$ is of fifth degree. If the monomial contains more than one variable, the degree is given by the sum of the exponents on the variables. Thus the monomial $3x^2y^3z$ is of sixth degree in x, y, and z. It can also be described as being of second degree in x, or third degree in y, or of first degree in z. The degree of a polynomial is the same as the degree of its term of largest degree;

$$3x^2 + 2x + 1 \text{ is a second-degree polynomial,}$$

$$x^5 - x - 1 \text{ is a fifth-degree polynomial,}$$

$$2x - 3 \text{ is a first-degree polynomial.}$$

EXERCISE 2.1

Write an equal expression using exponents.

EXAMPLES

a. $xx + xyyy$ b. $aac - abbbcc$ c. $2r - rrrs$

SOLUTIONS

a. $\mathbf{x^2 + xy^3}$ b. $\mathbf{a^2c - ab^3c^2}$ c. $\mathbf{2r - r^3s}$

1. $yyyy$
2. xyy
3. $2aab$
4. $3abbb$
5. $2xxyyy$
6. $xyyyyz$
7. $rr + rsss$
8. $2abb - aab$
9. $xx + yyyy$
10. $2bb - 3aab$
11. $3aac - abbcc$
12. $2aaab - 2bbcc$

Write an equal expression without exponents.

EXAMPLES

a. $-x^3(-y)^2$ b. $2x^3 - (2x)^3$ c. $a^2 - a(-3b^2)$

SOLUTIONS

a. $\mathbf{-xxx(-y)(-y)}$ b. $\mathbf{2xxx - (2x)(2x)(2x)}$ c. $\mathbf{aa - a(-3)bb}$

13. x^2y^4
14. x^3y
15. $2ab^3$
16. $3ab^5$
17. $7r^3s^2t$
18. $2rs^3t^2$
19. $(-x)^3$
20. $-x^3$
21. $-y^2x$
22. $(-y)^2x$
23. $-(x^3)(-y)^2$
24. $-(-x)^2y^3$
25. $r^2st + s^2t$
26. $rs^2t - rt^2$
27. $2x^3 - xy^2$
28. $(2x)^3 - (xy)$
29. $ab^2 - ab^4$
30. $(ab)^2 - (ab)^4$

Identify the given polynomial as a monomial, binomial, or trinomial and determine the degree of the polynomial.

EXAMPLES

a. $2x^3 - x$ b. $x^2y^3 + y^4 + x^2$

SOLUTIONS

a. **Binomial; degree 3 in x**

b. **Trinomial; degree 5 in x and y; degree 2 in x; degree 4 in y**

31. $x^3 + 2x^2$
32. x^3
33. $3y^2 + 6y + 4$
34. $7y^3 - y$
35. $4a^3$
36. $a^3 - 2a + 1$
37. $x^3y + xy^2$
38. $x^3 - x^2y^2 + y^4$
39. $\frac{4}{3}\pi r^3$
40. $\frac{1}{3}\pi r^2 h$
41. $s^2 - \pi r^2$
42. $2\pi rh + \pi r^2$

How many terms are there in each of the following expressions as written?

43. $a(b + c)$
44. $(x + y)^2 + 2$
45. $\frac{x + y}{3} - 2x$
46. $3 - \frac{2x + y^2}{x}$
47. $IR_1 + IR_2$
48. $I(R_1 + R_2)$
49. $\frac{R_1}{1} + \frac{1}{R_2}$
50. $\frac{1}{R_1 + R_2}$

2.2 Sums and differences

By the symmetric law of equality, the distributive law can be written in the form

$$ab + ac = a(b + c),$$

and by the commutative law, it can be written

$$ab + ac = (b + c)a.$$

If we now define **like terms** to be terms that may differ only in their numeral coefficients, we can state:

The sum of given like terms is another like term whose coefficient is the sum of the coefficients of the given terms.

For example,

$$\begin{aligned} 3x + 2x &= (3 + 2)x \\ &= 5x, \\ 3y + 2y + 5y &= (3 + 2 + 5)y \\ &= 10y, \\ 2x^2y + 6x^2y + x^2y &= (2 + 6 + 1)x^2y \\ &= 9x^2y. \end{aligned}$$

Because $a - b$ is equal to $a + (-b)$, we shall view the signs in any polynomial as signs denoting positive or negative coefficients, and the operation involved shall be understood to be addition. Thus

$$\begin{aligned} 3x - 5x + 4x &= (3x) + (-5x) + (4x) \\ &= (3 - 5 + 4)x \\ &= 2x. \end{aligned}$$

Grouping devices such as parentheses have been used here to indicate that various expressions are to be viewed as a single number. The expression

$$3x + (2x + 5x)$$

represents the addition of $3x$ to the sum of $2x$ and $5x$, while the expression

$$(3x + 2x) + 5x$$

represents the addition of $5x$ to the sum of $3x$ and $2x$. But the associative law asserts that these expressions are identical. Hence, the order in which we group terms in expressions of addition is immaterial. Thus,

$$3x + 2x + 5x$$

is equal to either of the grouped expressions above.

Similarly,

$$\begin{aligned} 2y + (3y - 2y + y) &= 2y + 3y - 2y + y \\ &= 4y \end{aligned}$$

and

$$\begin{aligned} (3x^2 + 2x + 1) + (2x^2 - x + 3) &= 3x^2 + 2x + 1 + 2x^2 - x + 3 \\ &= 5x^2 + x + 4. \end{aligned}$$

Because $-(b + c) = -b - c$, an expression such as

$$a - (b + c),$$

in which a set of parentheses is preceded by a negative sign, can be written as

$$a - b - c,$$

and then simplified if possible. For example,

$$\begin{aligned} (x^2 + 2x) - (2x^2 - 3x + 2) &= x^2 + 2x - 2x^2 + 3x - 2 \\ &= -x^2 + 5x - 2. \end{aligned}$$

In any expression where grouping devices are nested—that is, where groups occur within groups—you can avoid a great deal of difficulty by removing the inner devices first and working outward. Thus

$$\begin{aligned} 3x - [2 - (3x + 1)] &= 3x - [2 - 3x - 1] \\ &= 3x - 2 + 3x + 1 \\ &= 6x - 1. \end{aligned}$$

EXERCISE 2.2

Simplify.

EXAMPLES

a. $2x - 3x + 7x$ b. $(a - a^2) + (5a - 3a^2)$ c. $(3a^2 - b) + (2b - a^2)$

SOLUTIONS

a. $\mathbf{6x}$

b. $a - a^2 + 5a - 3a^2$
$\mathbf{-4a^2 + 6a}$

c. $3a^2 - b + 2b - a^2$
$\mathbf{2a^2 + b}$

1. $5a + 4a$ 2. $3a^2 - 5a^2$ 3. $5x - 2x$
4. $-5x - 8x$ 5. $a^4b - 3a^4b$ 6. $-2ab - 3ab$
7. $-xy^2 - xy^2$ 8. $4xy + 2xy$ 9. $2a - a - 4a$
10 $3a^2 - a^2 + a^2$ 11. $6c + 3c - c$ 12. $-3c - 2c + 15c$
13. $3c + 2d - c$ 14. $4a + 3a - b$ 15. $b - 2a + b$
16. $b^2 - 5b^2 + a^2$ 17. $2xy + x^2y + xy$ 18. $xy^2 + x^2y + 3x^2y$
19. $2x^2 + 3x - 5x + 6x^3 + x^2$ 20. $y^2 - 3xy + 2y - 2xy + 3y^2$
21. $x + (2x + 3y)$ 22. $4y + (2x - 3y)$
23. $(a^2 + a) + (a^2 - a)$ 24. $(a^2 - b) + (b - a^2)$
25. $(3R + 2S) + (2R - S)$ 26. $5S + (R - 2S)$
27. $(3R^2 + S) + (S - R^2)$ 28. $(S^2 + RS) + (2RS - S^2)$
29. $3\pi R^2 + (\pi R^2 h - \pi R^2)$ 30. $2S^2 + (\pi R^2 + S^2) - S^2$

Add the bottom polynomial to the top polynomial.

31. $\begin{array}{r} 2z^2 - 3z + 5 \\ -z^2 + z - 3 \\ \hline \end{array}$

32. $\begin{array}{r} 4t^3 - 2t^2 + 3t \\ -2t^3 + 5t^2 + 3t \\ \hline \end{array}$

33. $\begin{array}{r} c^2d^2 - 3c^2d + 2cd^2 \\ 3c^2d - 2cd^2 + 5 \\ \hline \end{array}$

34. $\begin{array}{r} n^3 - 5n^2 + 2n \\ - 2n^2 - n + 2 \\ \hline \end{array}$

35. $\begin{array}{r} 3x^2 \qquad + 5 \\ x^2 - 7x + 3 \\ \hline \end{array}$

36. $\begin{array}{r} t^2 + 2ts - s^2 \\ 3t^2 \qquad - s^2 \\ \hline \end{array}$

Simplify.

EXAMPLES

a. $x - (x - 2)$ b. $(3x - z) - (x + y)$ c. $-(a + b) - (a - 2b)$

Solutions overleaf.

SOLUTIONS

a. $x - x + 2$
$\mathbf{2}$

b. $3x - z - x - y$
$\mathbf{2x - z - y}$

c. $-a - b - a + 2b$
$\mathbf{-2a + b}$

37. $(3x + 2) - (-x + 3)$
38. $(x + 2) - (2x - 3)$
39. $(x^2 + 3x - 2) - (2x^2 - 1)$
40. $(2x^2 - x + 4) - (x^2 - x)$
41. $(a - b + c) - (a + b + c)$
42. $(a - b + c) - (a + 3b + c)$
43. Subtract $x^2 - 3x$ from $2x^2 + x - 2$.
44. Subtract $x^2 - 3x + 1$ from $x - 1$.
45. Subtract $a + b - 2c$ from $2a - 3$.
46. Subtract $a^2 + 2a$ from 0.
47. Subtract $x^2 - 2x$ from the sum of $x^2 + 2x$ and $2x^2 - x + 2$.
48. Subtract $2y^2 - y + 1$ from the sum of $y + 2$ and $y^2 - 4y + 3$.

Subtract the bottom polynomial from the top polynomial.

49. $\begin{array}{l} 3d^2 + 7d - 5 \\ \underline{\;\, d^2 - 2d - 6} \end{array}$

50. $\begin{array}{l} \;\;\,8n^3 - 3n + 2 \\ \underline{-2n^3 - 5n + 2} \end{array}$

51. $\begin{array}{r} 4t^3 + 3t^2 - 2t \qquad \\ \underline{5t^2 + \;\, t - 5} \end{array}$

52. $\begin{array}{r} 10k^3 - 3k^2 - 2k \qquad \\ \underline{5k^2 - 7k - 3} \end{array}$

53. $\begin{array}{l} \;\;m^2 \qquad\quad - 8 \\ \underline{2m^2 - 2m + 6} \end{array}$

54. $\begin{array}{l} \;\;\,x^3 + 3x^2y - 2xy^2 \\ \underline{-3x^3 \qquad\quad + \;\, xy^2} \end{array}$

Simplify.

55. $y - [2y + (y + 1)]$
56. $3a + [2a - (a + 4)]$
57. $3 - [2x - (x + 1) + 2]$
58. $5 - [3y + (y - 4) - 1]$
59. $(3A + 2) - [A + (2 + A) + 1]$
60. $(A - 3) + [2A - (3 + A) - 2]$
61. $[A - (2B + C)] - [2A + (B - C)]$
62. $[3A - (B - 2C)] - [4A - (2B + C)]$

2.3 Products

Consider the product

$$a^m a^n,$$

where m and n are natural numbers. Since

$$a^m = a\,a\,a \cdot \cdots \cdot a \text{ (} m \text{ factors),}$$

and

$$a^n = aaa \cdot \cdots \cdot a \ (n \text{ factors}),$$

it follows that

$$\begin{aligned} a^m a^n &= (\overbrace{aaa \cdot \cdots \cdot a}^{m \text{ factors}})(\overbrace{aaa \cdot \cdots \cdot a}^{n \text{ factors}}) \\ &= (\overbrace{aaa \cdot \cdots \cdot a}^{m+n \text{ factors}}), \end{aligned}$$

$$\boldsymbol{a^m a^n = a^{m+n}} \tag{1}$$

Equation (1) is referred to as the **first law of exponents.** Thus we can multiply two natural-number powers of the same base simply by adding the exponents and writing the sum as an exponent on the same base. For example,

$$\begin{aligned} x^2x^3 &= x^5, \\ xx^3x^4 &= x^8, \\ y^3y^4y^2 &= y^9. \end{aligned}$$

In multiplying two monomials, say

$$(3x^2y)(2xy^2),$$

the commutative and associative laws are used; and we write

$$3 \cdot 2 \cdot x^2 \cdot x \cdot y \cdot y^2,$$

which simplifies to

$$6x^3y^3.$$

The generalized distributive law gives meaning to the product of a monomial and a polynomial containing more than one term. For example,

$$\begin{aligned} 3x(x + y + z) &= 3x \cdot x + 3x \cdot y + 3x \cdot z \\ &= 3x^2 + 3xy + 3xz. \end{aligned}$$

The distributive law can be applied successively to multiply polynomials containing more than one term. For example,

$$\begin{aligned} (3x + 2y)(x - y) &= 3x(x - y) + 2y(x - y) \\ &= 3x^2 - 3xy + 2xy - 2y^2 \\ &= 3x^2 - xy - 2y^2. \end{aligned}$$

The following list of binomial products represents types so frequently encountered hereafter that you should learn to recognize them on sight:

$$(x + a)(x + b) = x^2 + (a + b)x + ab,$$

$$(x + a)^2 = x^2 + 2ax + a^2,$$

$$(x + a)(x - a) = x^2 - a^2,$$

$$(ax + by)(cx + dy) = acx^2 + (ad + bc)xy + bdy^2.$$

EXERCISE 2.3

Multiply.

EXAMPLES

a. $(y^3)(y^4)$ b. $(-2x^2y)(-5xy^3)$ c. $(-2x^2)(3xy)(y^2)$

SOLUTIONS

a. $\mathbf{y^7}$ b. $\mathbf{10x^3y^4}$ c. $\mathbf{-6x^3y^3}$

1. $(2x^2)(3x)$
2. $(3x)(x^2y)$
3. $(4a)(-5a)(a^2)$
4. $(2)(-3b)(4ab)$
5. $(-2x)(xy)(y^2)$
6. $(3x^2)(2xy)(y^3)$
7. $(-a)(-b)(c^2)$
8. $(a^3)(-2ab^2)(-b^3)$
9. $(-x)(-xy^2)(-y^2)$
10. $(-3x^2y)(-xy)(-2xy^2)$
11. $(-xyz)(-2xy)(-4yz^2)$
12. $(-5xz^2)(-2x^2y)(-2y^2z)$

EXAMPLES

a. $2(x^2 - x - 1)$ b. $-(a + b - c)$ c. $2a(a^2 - 3a)$

SOLUTIONS

a. $\mathbf{2x^2 - 2x - 2}$

b. $-1(a + b - c)$
$\mathbf{-a - b + c}$

c. $2a \cdot a^2 - 2a \cdot 3a$
$\mathbf{2a^3 - 6a^2}$

13. $3(x + 1)$
14. $-2(x^2 - x)$
15. $2(l + w)$
16. $-3a^2(a + a^2)$
17. $-(x^2 + x - 2)$
18. $-(x^2 - x + 5)$
19. $abc(a - b + c)$
20. $ab(2a - b + 3c)$
21. $ab(a^2 - ab + b^2)$
22. $2ab(ab^2 - ab + a^2b)$
23. $2r(2h + \pi r - r)$
24. $2L(T - 3P + 2)$

EXAMPLES

a. $(x-3)(x+5)$ b. $(x+3)^2$ c. $(2x-3y)^2$

SOLUTIONS

a. $x^2+5x-3x-15$
$\mathbf{x^2+2x-15}$

b. $(x+3)(x+3)$
$x^2+3x+3x+9$
$\mathbf{x^2+6x+9}$

c. $(2x-3y)(2x-3y)$
$4x^2-6xy-6xy+9y^2$
$\mathbf{4x^2-12xy+9y^2}$

25. $(x+2)(x+5)$
26. $(x-3)(x+2)$
27. $(x-1)(x+3)$
28. $(x-5)(x-7)$
29. $(y-3)^2$
30. $(x+2)^2$
31. $(x-2a)(x+a)$
32. $(x-3a)(x+2a)$
33. $(x+y)^2$
34. $(x-2y)^2$
35. $(x-y)(x+y)$
36. $(x-3y)(x+3y)$
37. $(5x+1)(2x+3)$
38. $(2x+3)(x-5)$
39. $(3a-1)^2$
40. $(2a+3)^2$
41. $(2x-5)(2x+5)$
42. $(3x-2)(3x+2)$
43. $(2x-a)(x+2a)$
44. $(3x+a)(x+5a)$
45. $(E_1+1)(E_1-1)$
46. $(V_1+V)(V_1-V)$
47. $(E_1+E_2)^2$
48. $(V_1-V_2)^2$

EXAMPLES

a. $3(x-1)(x+2)$ b. $a(a-b)(c-d)$

SOLUTIONS

a. $3(x^2+x-2)$
$\mathbf{3x^2+3x-6}$

b. $a(ac-ad-bc+bd)$
$\mathbf{a^2c-a^2d-abc+abd}$

49. $2(x+1)(x+3)$
50. $3(x-1)(x+2)$
51. $x(x-2)(x+3)$
52. $x(x+1)(x+3)$
53. $3(a+2)^2$
54. $4(a-5)^2$
55. $-3(2x-1)(x+2)$
56. $-(2x-3)(4x+5)$
57. $-2x(x+3)^2$
58. $-2x(x-4)^2$
59. $3x^2(x-2)(3x-5)$
60. $2x^2(2x+1)(x-3)$
61. $(a-2b)(a+2b)$
62. $(3a-5b)(3a+5b)$
63. $(x+a)(y+b)$
64. $(x-b)(y+2a)$
65. $(a-b)(a^2+ab+b^2)$
66. $(a+b)(a^2-ab+b^2)$

Simplify.

EXAMPLES

a. $2x-[x+3(x-1)-5]$ b. $a(a-[3-(a+1)]+4)$

Solutions overleaf.

SOLUTIONS

a. $2x - (x + 3x - 3 - 5)$
$2x - x - 3x + 3 + 5$
$\mathbf{-2x + 8}$

b. $a(a - [3 - a - 1] + 4)$
$a(a - 3 + a + 1 + 4)$
$a(2a + 2)$
$\mathbf{2a^2 + 2a}$

67. $3[x + (2x + 3)]$
68. $2y[y - (3y + 2)]$
69. $2[a - (a - 1) + 2]$
70. $3[2a - (a + 1) + 3]$
71. $a[a - (2a + 3) - (a - 1)]$
72. $-2a[3a + (a - 3) - (2a + 1)]$
73. $-[a - 3(a + 1) - (2a + 1)]$
74. $-[(a + 1) - 2(3a - 1) + 4]$
75. $2(a - [a - 2(a + 1) + 1] + 1)$
76. $-(4 - [3 - 2(a - 1) + a] + a)$
77. $-x(x - 3[2x - 3(x + 1)] + 2)$
78. $x(4 - 2[3 - 4(x + 1)] - x)$
79. $2x(x + 3[2(2x - 1) - (x - 1)] + 5)$
80. $-x(4 - 2[(x + 1) - 3(x + 2)] - x)$

2.4 Quotients

Consider the quotient

$$\frac{a^m}{a^n} \quad (a \neq 0),$$

where m and n are natural numbers and $m > n$. We can write

$$\begin{aligned} \frac{a^m}{a^n} &= a^m\left(\frac{1}{a^n}\right) \\ &= a^{m-n} \cdot a^n\left(\frac{1}{a^n}\right) \\ &= a^{m-n}\left(a^n \cdot \frac{1}{a^n}\right); \end{aligned}$$

$$\mathbf{\frac{a^m}{a^n} = a^{m-n} \quad (a \neq 0).} \tag{1}$$

Equation (1) is called the **second law of exponents**. For example, for $x, y, z \neq 0$,

$$\begin{aligned} \frac{x^5}{x^2} &= x^{5-2} \\ &= x^3, \end{aligned}$$

$$\begin{aligned} \frac{6x^3y^2}{3x^2y} &= \left(\frac{6}{3}\right)(x^{3-2}y^{2-1}) \\ &= 2xy, \end{aligned}$$

and

$$\frac{18x^3y^2z^4}{6x^2yz^2} = \left(\frac{18}{6}\right)(x^{3-2}y^{2-1}z^{4-2})$$
$$= 3xyz^2.$$

EXERCISE 2.4

Divide.

EXAMPLES

a. $\dfrac{6x^2y^3}{-2xy}$ b. $\dfrac{a^3b^5c^2}{ab^2c}$ c. $\dfrac{4(x+2t)^2}{2(x+2t)}$

SOLUTIONS

a. $\left(\dfrac{6}{-2}\right)x^{2-1}y^{3-1}$
$-3xy^2$

b. $a^{3-1}b^{5-2}c^{2-1}$
a^2b^3c

c. $\left(\dfrac{4}{2}\right)(x+2t)^{2-1}$
$2(x+2t)$

1. $\dfrac{x^3y^2}{xy}$ 2. $\dfrac{a^3b^2}{ab}$ 3. $\dfrac{6a^3b}{3a}$ 4. $\dfrac{12a^2c^2}{3ac}$

5. $\dfrac{7a^2b}{-a}$ 6. $\dfrac{-a^5}{a^2}$ 7. $\dfrac{x^3y^2}{xy}$ 8. $\dfrac{x^2y^3z^4}{xyz}$

9. $\dfrac{(x-2)^3}{(x-2)^2}$ 10. $\dfrac{(a-b)^2}{a-b}$ 11. $\dfrac{-6x^2y^2z^4}{-3xz^2}$ 12. $\dfrac{a^2b^5c^3}{-abc^2}$

13. $\dfrac{x^3(x+y)^2}{x(x+y)}$ 14. $\dfrac{y^2(x-2y)^3}{y(x-2y)}$ 15. $\dfrac{6y^5(2x+y)^3}{-3y^2(2x+y)^2}$

16. $\dfrac{-12x^3(x-y)^2}{-2x(x-y)}$ 17. $\dfrac{(x+y)^3(x-y)^2}{x+y}$ 18. $\dfrac{(2x-y)(x-2y)^3}{x-2y}$

19. $\dfrac{(y-z)^3(y+2z)^4}{(y-z)^2(y+2z)}$ 20. $\dfrac{(y+3z)^2(y-z)^3}{(y+3z)(y-z)^2}$

21. $\dfrac{R^3(E_1+E_2)^2}{R}$ 22. $\dfrac{R^3(E_1+E_2)^2}{E_1+E_2}$

23. $\dfrac{(E_1+E_2)^2(E_1-E_2)^3}{E_1-E_2}$ 24. $\dfrac{(E_1+E_2)^2(E_1-E_2)^3}{E_1+E_2}$

25. Assuming we could apply the second law of exponents in rewriting x^n/x^n, we would obtain x^{n-n} or x^0. What meaning must we assign to the symbol x^0 so that this result is consistent with the definition of a quotient? What restriction must be placed on x in this case?

2.5 Factoring monomials from polynomials

The distributive law in the form

$$ax + bx + cx + dx = x(a + b + c + d)$$

furnishes a means of writing a polynomial as a single term comprised of two or more factors. This process is called **factoring**. Thus, by the distributive law,

$$3x^2 + 6x = 3x(x + 2).$$

Of course, we can also write

$$3x^2 + 6x = 3(x^2 + 2x) \quad \text{or} \quad 3x^2 + 6x = 3x^2\left(1 + \frac{2}{x}\right)$$

or any other of an infinite number of such expressions. We are, however, primarily interested in factoring a polynomial into a unique (except for signs and order of factors) form referred to as the **completely factored form.** A polynomial with integral coefficients is in completely factored from if:

1. it is written as a product of polynomials with integral coefficients;
2. no polynomial—other than a monomial—in the factored form contains a polynomial factor with integral coefficients.

Common monomial factors can be factored from a polynomial by first identifying such common factors and then writing the resultant factored expression. For example, observe that the polynomial

$$6x^3 + 9x^2 - 3x$$

contains the monomial $3x$ as a factor of each term. We therefore write

$$6x^3 + 9x^2 - 3x = 3x(\qquad\qquad)$$

and insert within the parentheses the appropriate polynomial factor. This factor can be determined by inspection. We ask ourselves for the monomials that multiply $3x$ to yield $6x^3$, $9x^2$, and $-3x$ in turn. The final result appears

$$6x^3 + 9x^2 - 3x = 3x(2x^2 + 3x - 1).$$

EXERCISE 2.5

Determine the missing factor.

EXAMPLES

a. $x^5 = x^2(?)$ b. $x^3y^7 = xy^3(?)$

SOLUTIONS

a. $x^5 = x^2(\mathbf{x^3})$ b. $x^3y^7 = xy^3(\mathbf{x^2y^4})$

1. $x^6 = x^4(?)$
2. $y^3 = y^2(?)$
3. $xy^4 = xy^2(?)$
4. $x^4y^3 = x^2y^2(?)$
5. $6x^3y^5 = 2x^3y(?)$
6. $12xy^6 = 4xy^2(?)$

Factor completely.

EXAMPLES

a. $6x - 18$ b. $18x^2y - 24xy^2$ c. $F_1t + F_2t$

SOLUTIONS

a. $6(? - ?)$
$6(x - 3)$

b. $6xy(? - ?)$
$6xy(3x - 4y)$

c. $t(? + ?)$
$t(F_1 + F_2)$

7. $2x + 6$
8. $3x - 9$
9. $4x^2 + 8x$
10. $3x^2 + 6x$
11. $3x^2 - 6x - 3$
12. $x^3 - x^2 + x$
13. $24a^2 + 12a - 6$
14. $15a^3 + 18a^2 - 3a$
15. $2x^4 - 4x^2 + 8x$
16. $6x^3 + 2x^2 - 2x$
17. $xy^2 + x^2y$
18. $2x^3y^2 - 4xy^2$
19. $IR_1 + IR_2$
20. $I^2R_1 + IR_2$
21. $\pi R^2 + 2\pi R$
22. $4\pi R^2 - \pi R$
23. $R_0 + R_0 at$
24. $2R_1{}^2 - R_1t$

Supply the missing factors and hence, in each case, write the expression on the left in completely factored form.

EXAMPLES

a. $-5x + 10 = -5(\ \ ?\ \ )$ b. $-IR^2 - IR = -IR(\ \ ?\ \ )$

SOLUTIONS

a. **$-5(x - 2)$** b. **$-IR(R + 1)$**

25. $-2x + 2 = -2(\ \ ?\ \ )$
26. $-6x - 9 = -3(\ \ ?\ \ )$
27. $-ab - ac = ?(b + c)$
28. $-a^2 + ab = ?(a - b)$
29. $-xy - x^2y = -xy(\ \ ?\ \ )$
30. $-x^3y + y^2x = -xy(\ \ ?\ \ )$
31. $-x + x^2 - x^3 = -x(\ \ ?\ \ )$
32. $-y + 2xy + xy^2 = -y(\ \ ?\ \ )$
33. $-2ER_1 + ER_2 = -E(\ \ ?\ \ )$
34. $-IR - IR^2 = -IR(\ \ ?\ \ )$
35. $-V_0 - 0.00365V_0T = -V_0(\ \ ?\ \ )$
36. $-0.2E^2 + EQ = -E(\ \ ?\ \ )$

2.6 Factoring quadratic polynomials

One very common type of factoring is that involving quadratic (second-degree) binomials or trinomials. From Section 2.3, recall that

(1) $(x + a)(x + b) = x^2 + (a + b)x + ab$;
(2) $(x + a)^2 = x^2 + 2ax + a^2$;
(3) $(x + a)(x - a) = x^2 - a^2$;
(4) $(ax + by)(cx + dy) = acx^2 + (ad + bc)xy + bdy^2$.

These four forms are those most commonly encountered in the chapters that follow. In this section, we are interested in viewing these relationships from right to left—that is, from polynomial to factored form. Again, we shall require integral coefficients and positive integral exponents on the variables.

As an example of an application of form (1), consider the trinomial

$$x^2 + 6x - 16.$$

We desire, if possible, to find two binomial factors of the form

$$(x + a)(x + b)$$

whose product is the given trinomial. We see from form (1) that a and b are two integers such that $a + b = 6$ and $ab = -16$; that is, their sum must be the coefficient of the linear term $6x$ and their product must be -16. By inspection, or by trial and error, we determine that the two numbers are 8 and -2, so that

$$(x^2 + 6x - 16) = (x + 8)(x - 2).$$

Form (2) is simply a special case of (1), the square of a binomial. Thus

$$\begin{aligned} x^2 + 8x + 16 &= (x + 4)(x + 4) \\ &= (x + 4)^2. \end{aligned}$$

Form (3) is another special case of (1), in which the coefficient of the first-degree term in x is zero. For example,

$$x^2 - 25 = (x - 5)(x + 5).$$

In particular, form (3) states that the difference of the squares of two numbers is equal to the product of the sum and the difference of the two numbers.

Form (4) is a generalization of (1)—that is, in (4) we are confronted with a quadratic trinomial where the coefficient of the term of second degree in x is other than 1. We illustrate the factoring of such a trinomial by example:

$$8x^2 - 9 - 21x$$

1. Write in decreasing powers of x.

$$8x^2 - 21x - 9$$

2. Consider possible combinations of first-degree factors of the first term.

$$\begin{array}{l}(8x \quad)(x \quad) \\ (4x \quad)(2x \quad)\end{array}$$

3. Consider combinations of the factors of the last term.

$$\begin{array}{l}(8x \quad 9)(x \quad 1) \\ (8x \quad 1)(x \quad 9) \\ (8x \quad 3)(x \quad 3) \\ (4x \quad 9)(2x \quad 1) \\ (4x \quad 1)(2x \quad 9) \\ (4x \quad 3)(2x \quad 3)\end{array}$$

(inner products bracketed: 3; outer products bracketed: 2)

4. Select the combination(s) of products (2) and (3) whose sum(s) could be the second term ($-21x$).

$$(8x \quad 3)(x \quad 3)$$

5. Insert the proper signs.

$$(8x + 3)(x - 3)$$

This process can normally be accomplished mentally; it is written out in detail here for the purposes of illustration only.

If a polynomial of more than one term contains a common monomial factor in each of its terms, this monomial should be factored from the polynomial before seeking other factors. Thus

$$\begin{aligned} 32x^2 - 84x - 36 &= 4(8x^2 - 21x - 9) \\ &= 4(8x + 3)(x - 3). \end{aligned}$$

EXERCISE 2.6

Factor completely.

EXAMPLES

a. $x^2 - 2x - 3$ b. $x^2 - 4$ c. $x^2 - 9y^2$

SOLUTIONS

a. $\mathbf{(x-3))(x+1)}$

b. $\mathbf{(x-2)(x+2)}$

c. $x^2 - (3y)^2$
$\mathbf{(x-3y)(x+3y)}$

1. $x^2 + 8x + 12$
2. $x^2 + 5x + 6$
3. $y^2 - 8y + 15$
4. $y^2 - 10y + 21$
5. $x^2 - x - 6$
6. $y^2 - 2y - 15$
7. $x^2 - 2x - 3$
8. $x^2 + 5x - 14$
9. $y^2 + 3y - 4$
10. $y^2 - 2y - 24$
11. $x^2 - x - 42$
12. $x^2 + 2x - 63$
13. $x^2 - 2xy + y^2$
14. $x^2 - 2xy - 3y^2$
15. $x^2 + 6xy + 5y^2$
16. $x^2 - 9xy + 20y^2$
17. $x^2 - 1$
18. $x^2 - 25$
19. $a^2 - 9$
20. $a^2 - 36$
21. $4 - b^2$
22. $9 - a^2$
23. $(ab)^2 - 1$
24. $4 - (ab)^2$
25. $a^2 - (2xy)^2$
26. $(xy)^2 - b^2$
27. $9x^2 - y^2$
28. $x^2 - 4y^2$
29. $x^4 - 9$
30. $25 - b^2$

EXAMPLES

a. $5x^2 - 9x - 2$ b. $9x^2 - 16y^2$

SOLUTIONS

a. $\mathbf{(5x+1)(x-2)}$

b. $(3x)^2 - (4y)^2$
$\mathbf{(3x-4y)(3x+4y)}$

31. $3x^2 + 4x + 1$
32. $4a^2 - 5a + 1$
33. $9x^2 - 21x - 8$
34. $10x^2 - 3x - 18$
35. $3x^2 - 7ax + 2a^2$
36. $9x^2 + 9ax - 10a^2$
37. $9x^2 - y^2$
38. $4x^2 - 9y^2$
39. $4x^2 + 12x + 9$
40. $4y^2 + 4y + 1$
41. $4x^2y^2 - 25$
42. $64x^2y^2 - 1$

EXAMPLES

a. $4a^3 - 5a^2 + a$ b. $8x^5 - 2x^3$

SOLUTIONS

a. $a(4a^2 - 5a + 1)$
$a(4a - 1)(a - 1)$

b. $2x^3(4x^2 - 1)$
$2x^3(2x - 1)(2x + 1)$

43. $3x^2 + 12x + 12$
44. $2x^2 + 6x - 20$
45. $2a^3 - 8a^2 - 10a$
46. $2a^3 + 15a^2 + 7a$
47. $4a^2 - 8ab + 4b^2$
48. $20a^2 + 60ab + 45b^2$
49. $4x^2y - 36y$
50. $x^2 - 4x^2y^2$
51. $12x - x^2 - x^3$
52. $x^2 - 2x^3 + x^4$
53. $x^4y^2 - x^2y^2$
54. $x^3y - xy^3$

EXAMPLES

a. $x^4 + 2x^2 + 1$

b. $x^4 - 3x^2 - 4$

SOLUTIONS

a. **$(x^2 + 1)^2$**

b. $(x^2 - 4)(x^2 + 1)$
$(x - 2)(x + 2)(x^2 + 1)$

55. $y^4 + 3y^2 + 2$
56. $a^4 + 5a^2 + 6$
57. $3x^4 + 7x^2 + 2$
58. $4x^4 - 11x^2 - 3$
59. $x^4 + 3x^2 - 4$
60. $x^4 - 6x^2 - 27$
61. $x^4 - 5x^2 + 4$
62. $y^4 - 13y^2 + 36$
63. $2a^4 - a^2 - 1$
64. $3x^4 - 11x^2 - 4$
65. $x^4 + a^2x^2 - 2a^4$
66. $4x^4 - 33a^2x^2 - 27a^4$
67. For what value(s) of the variable is $x(x - 2)$ equal to 0?
68. For what value(s) of the variable is $(x - 3)(x + 2)$ equal to 0?
69. For what value(s) of the variable is $x^2 - 3x + 2$ equal to 0? *Hint:* Factor the polynomial.
70. For what value(s) of the variable is $x^3 + x^2 - 12x$ equal to 0?

2.7 Factoring other polynomials

There are a few other polynomials that occur frequently enough to justify a study of their factorization. In particular, the forms

(1) $(a + b)(x + y) = ax + ay + bx + by$,
(2) $(x + a)(x^2 - ax + a^2) = x^3 + a^3$, and
(3) $(x - a)(x^2 + ax + a^2) = x^3 - a^3$

are often encountered in one or another area of mathematics. We are again interested in viewing these relationships from right to left. Expressions such as the right member of form (1) are factorable by grouping. For example, to factor

$$3x^2y + 2y + 3xy^2 + 2x,$$

we rewrite it in the form

$$3x^2y + 2x + 3xy^2 + 2y$$

and factor the common monomials x and y from the first group of two terms and the second group of two terms, respectively, obtaining

$$x(3xy + 2) + y(3xy + 2).$$

If now we factor the common binomial $(3xy + 2)$ from each term, we have

$$(3xy + 2)(x + y).$$

The application of forms (2) and (3) are direct. Thus, as an example of form (2),

$$\begin{aligned} 8a^3 + b^3 &= (2a)^3 + b^3 \\ &= (2a + b)[(2a)^2 - 2ab + b^2] \\ &= (2a + b)[4a^2 - 2ab + b^2]. \end{aligned}$$

EXERCISE 2.7

Factor.

EXAMPLES

a. $yb - ya + xb - xa$

b. $x^2 + xb - ax - ab$

SOLUTIONS

a. $y(b - a) + x(b - a)$
$(b - a)(y + x)$

b. $x(x + b) - a(x + b)$
$(x + b)(x - a)$

1. $ax^2 + x + ax + 1$
2. $5a + ab + 5b + b^2$
3. $ax^2 + x + a^2x + a$
4. $a + ab + b + b^2$
5. $x^2 + ax + xy + ay$
6. $x^3 - x^2y + xy - y^2$
7. $3ab - cb - 3ad + cd$
8. $1 - x - y + xy$
9. $3x + y - 6x^2 - 2xy$
10. $5xz - 5yz - x + y$
11. $a^3 + 2ab^2 - 4b^3 - 2a^2b$
12. $6x^3 - 4x^2 + 3x - 2$

EXAMPLES

a. $x^3 + 8$

b. $8x^3 - y^3$

SOLUTIONS

a. $x^3 + (2)^3$
$(x + 2)(x^2 - 2x + 2^2)$
$\mathbf{(x + 2)(x^2 - 2x + 4)}$

b. $(2x)^3 - y^3$
$(2x - y)[(2x)^2 + 2xy + y^2]$
$\mathbf{(2x - y)(4x^2 + 2xy + y^2)}$

13. $x^3 - 1$
14. $y^3 + 27$
15. $(2x)^3 + y^3$
16. $y^3 - (3x)^3$
17. $a^3 - 8b^3$
18. $8a^3 + b^3$
19. $(xy)^3 - 1$
20. $8 + x^3y^3$
21. $27a^3 + 64b^3$
22. $8a^3 - 125b^3$
23. $x^3 + (x - y)^3$
24. $(x + y)^3 - z^3$
25. Show that $ac - ad + bd - bc$ can be factored both as $(a - b)(c - d)$ and as $(b - a)(d - c)$.
26. Show that $a^2 - b^2 - c^2 + 2bc$ can be factored as $(a - b + c)(a + b - c)$.

2.8 Numerical evaluation

If x represents a real number, then so does the expression

$$x^2 + 2x + 1.$$

We can find the value of the expression when x represents 2 by replacing each x in the expression with 2 and performing the indicated operations. Thus

$$(2)^2 + 2(2) + 1 = 9.$$

This process is referred to as **numerical evaluation.**

Now consider the expression

$$y + yx^2.$$

By specifying that y represents 2 and x represents 3, we have

$$2 + 2(3)^2 = 20.$$

To prevent possible misinterpretations of expressions in numerical evaluations we should always perform operations in the following order:

1. Perform any operations inside grouping devices. (Consider fraction bars as grouping devices.)
2. Compute powers.
3. Perform all other multiplication operations and division operations in the order in which they occur from left to right.
4. Perform the remaining additions and subtractions in the order in which they occur from left to right.

EXERCISE 2.8

Simplify.

EXAMPLES

a. $\frac{3^2+3}{6}-\frac{2^3-4}{4}$

b. $\frac{4(3-1)^2}{2^3}+\frac{(-2)^3\cdot 3}{-3\cdot 2^2}$

SOLUTIONS

a. $\frac{9+3}{6}-\frac{8-4}{4}$

$\frac{12}{6}-\frac{4}{4}$

$2-1$

1

b. $\frac{4(2)^2}{8}+\frac{(-8)(3)}{-3\cdot 4}$

$\frac{4(4)}{8}+\frac{-24}{-12}$

$2+2$

4

1. $(-3)^2$
2. -3^2
3. 4^2-3^2
4. $4^2-(-3)^2$
5. $\frac{6\cdot 2}{4}+5$
6. $\frac{(3)(6)}{9}-\frac{4\cdot 8}{16}$
7. $6+\frac{8(2+1)}{6}$
8. $3+\frac{3(3-6)}{9}-5$
9. $\frac{5^2+3}{2}+\frac{3^2-1}{2}$
10. $\frac{3^2-2^2}{5}-(2\cdot 3)^2$
11. $\frac{(3)^2-2}{7}+\frac{(-3)^2-2^2}{5}$
12. $\frac{4(-2)^2-(-1)^2}{3\cdot 5}$

Given $x=-2$, evaluate each expression.

EXAMPLES

a. x^2-2x-1

b. $-(3x)^2+x$

SOLUTIONS

a. $(-2)^2-2(-2)-1$

$4+4-1$

7

b. $-[3(-2)]^2+(-2)$

$-[-6]^2-2$

$-36-2$

−38

13. $2x$
14. x^2+1
15. x^2+3x+1
16. $1-2x-3x^2$
17. $(3x)^2+3x^2$
18. $-(x)^2+x^3$

19. $(2x-4)^2 - x^2$
20. $(3x-1)^2 + 2x^2$
21. $-[(x-1)-(x+3)]$
22. $-[(x-1)-(x-1)^2]$
23. $-[(x+1)^2-(x-2)^2]$
24. $-[(2x-1)^2-x^3]$

Given $x = 3$ and $y = -2$, evaluate each expression.

25. $2x + 3y$
26. $x^2 - y$
27. $(2x - y)^2$
28. $(x + 3y)^2$
29. $x^2 - y^3$
30. $(2x)^2 - xy$
31. $(3x)^2 - (-2y)^3$
32. $(x-y)^2 - y^2$
33. $3(x-y)^2(x+y)$
34. $(y-y^2)^2 + (x-x^2)^2$
35. $-[(x+y)^2-(x-y)^2]$
36. $x[(y-x)-(x-y)^2]$

Evaluate each of the following expressions for the given values of the variables.

37. $\dfrac{R+r}{r}$; $R = 12$ and $r = 2$
38. $\dfrac{E-e}{R}$; $E = 18$, $e = 2$, and $R = 4$
39. $\dfrac{5(F-32)}{9}$; $F = 212$
40. $P + Prt$; $P = 1000$, $r = .04$, and $t = 2$
41. $\dfrac{rl-a}{r-1}$; $r = 2$, $l = 12$, and $a = 4$
42. $\dfrac{Mv^2}{g}$; $M = 64$, $v = 2$, and $g = 32$
43. $V_0(1 + 0.00365T)$; $V_0 = 3$ and $T = 1000$
44. $R_0(1 + at)$; $R_0 = 2.5$, $a = .05$, and $t = 20$

Chapter review

[2.1] Write an equal expression using exponents.

1. $2xyy - 3xxy$
2. $x - xyy$
3. $xxy + yyyyy$
4. What is the degree of the polynomial $4x^2 - x + 1$?

[2.2] Simplify.

5. $3b + (a - 2b)$
6. $(a - 2b) - (a - b + 2c)$
7. $x - (x + y) - y$
8. $2x - [x - (x - 1) + 2]$

[2.3] Multiply.

9. $(-x)(2y)(-xy)$

10. $2x(x^2 - 3x)$

11. $(x - 3y)(x + y)$

12. $3[(3x - 2)(x + 1) - (x - 2)^2]$

[2.4] Divide.

13. $\dfrac{24x^2y^3}{-3xy}$

14. $\dfrac{(x - 2)(x + 3)^2}{x + 3}$

[2.5] Factor.

15. $4x - 12$

16. $2x^3 - 4x^2 + 6x$

[2.6] Factor completely.

17. $x^2 - 7x + 12$

18. $2y^2 + 5y - 3$

19. $x^2 - 25y^2$

20. $2x^3 + 3x^2 - 2x$

[2.7] Factor.

21. $2xy + 2x^2 + y + x$

22. $x^3 - 8a^3$

[2.8] Evaluate each expression.

23. $2x^2 - x + 1; \quad x = -3$

24. $(2x - y)^2 - x^2; \quad x = -2$ and $y = 3$

25. $V_0(1 - 0.00214T); \quad V_0 = 4$ and $T = 1000$

Fractions

An **algebraic fraction** (or simply a **fraction**) is a symbol indicating the quotient of two expressions;

$$\frac{3}{x-2}, \quad \frac{4x+1}{x}, \quad \text{and} \quad \frac{4x}{x+3}$$

are examples of algebraic fractions. For each replacement of the variable(s) for which the numerator and denominator represent real numbers and for which the denominator is not zero, the fraction represents a real number. The variable(s) in the denominator must be restricted from values for which the denominator vanishes (is equal to zero), because division by zero is not defined. A fraction is said to be **undefined** for any such value(s) of the variable(s).

From the alternate definition of a quotient, $a/b = a(1/b)$. Therefore, if

$$\frac{a}{b} = \frac{c}{d} \quad (b, d \neq 0),$$

—that is, if two fractions represent the same number—it follows that

$$bd\left(\frac{a}{b}\right) = bd\left(\frac{c}{d}\right),$$

$$bda\left(\frac{1}{b}\right) = bdc\left(\frac{1}{d}\right),$$

$$adb\left(\frac{1}{b}\right) = bcd\left(\frac{1}{d}\right);$$

and since $b(1/b)$ is equal to 1 and $d(1/d)$ is equal to 1, we have

$$ad(1) = bc(1),$$
$$ad = bc.$$

Since each step in this argument is reversible, we can say that if $b, d \neq 0$, then

$$\boldsymbol{\frac{a}{b} = \frac{c}{d}} \quad \textbf{if and only if} \quad \boldsymbol{ad = bc}. \tag{1}$$

This establishes a means by which we can identify fractions that represent the same number. For example,

$$\frac{2}{3} = \frac{4}{6} \quad \text{because} \quad 2 \cdot 6 = 3 \cdot 4;$$

$$\frac{6}{9} = \frac{8}{12} \quad \text{because} \quad 6 \cdot 12 = 9 \cdot 8;$$

$$\frac{2}{3} \neq \frac{3}{4} \quad \text{because} \quad 2 \cdot 4 \neq 3 \cdot 3.$$

Many times in algebra, for one reason or another, we wish to replace a given fraction with an equal fraction. We can write any number of fractions equal to a given fraction. For example, $4/6$, $6/9$, $8/12$ are all equal to $2/3$. We can go from a given fraction to an equal fraction by means of the relationship

$$\boldsymbol{\frac{a}{b} = \frac{ac}{bc}},$$

where $b, c \neq 0$. This is generally called the **fundamental principle of fractions.** This principle asserts that an equal fraction is obtained if the numerator and the denominator of a fraction are each multiplied by the same nonzero number. The validity of the fundamental principle follows from (1) above; that is, for $b, c \neq 0$,

$$\frac{a}{b} = \frac{ac}{bc}$$

because

$$a(bc) = b(ac).$$

After the product of two fractions is considered in Section 3.6, we can interpret the fundamental principle of fractions as an application of the multiplicative property of 1, because for all $c \neq 0$, $c/c = 1$. Thus, if $b, c \neq 0$,

$$\begin{aligned}\frac{ac}{bc} &= \frac{a}{b} \cdot \frac{c}{c} \\ &= \frac{a}{b} \cdot 1 \\ &= \frac{a}{b}.\end{aligned}$$

3.1 Signs of fractions

There are three signs associated with a fraction: a sign for the numerator, a sign for the denominator, and a sign for the fraction itself. If any two of the three signs are changed, the resulting fraction is equal to the original fraction. For example,

$$\frac{-a}{-b} = -\frac{-a}{b} = -\frac{a}{-b} = \frac{a}{b}$$

and

$$-\frac{-a}{-b} = -\frac{a}{b} = \frac{a}{-b} = \frac{-a}{b}.$$

The forms a/b and $-a/b$, in which the sign of the fraction and the sign of the denominator are both positive, are generally the most convenient representations and will be referred to as **standard forms**. Thus,

$$\frac{-3}{5}, \quad \frac{3}{5}, \quad \text{and} \quad \frac{7}{10}$$

are in standard form, while

$$\frac{3}{-5}, \quad -\frac{-3}{5}, \quad \text{and} \quad -\frac{7}{10}$$

are not.

If the denominator of the fraction is an expression containing more than one term, there are two possibilities for standard form. Since $-(a-b) = b-a$, we have

$$\frac{-b}{a-b} = \frac{-b}{-(-a+b)}$$
$$= \frac{b}{b-a},$$

and either $\frac{-b}{a-b}$ or $\frac{b}{b-a}$ may be taken as standard form, as convenience dictates.

In the exercise following and in all succeeding exercises we shall agree that no variable in a denominator will assume a value for which the denominator vanishes (is equal to 0*).*

EXERCISE 3.1

Write each fraction in standard form.

EXAMPLES

a. $\frac{3}{-4}$ b. $-\frac{5}{-y}$ c. $-\frac{a}{a-2}$

SOLUTIONS

a. $\mathbf{\frac{-3}{4}}$ b. $\mathbf{\frac{5}{y}}$ c. $\mathbf{\frac{-a}{a-2}}$ or $\mathbf{\frac{a}{2-a}}$

1. $-\frac{3}{5}$
2. $-\frac{-5}{-6}$
3. $\frac{-4}{-7}$
4. $\frac{3}{-8}$
5. $\frac{4x}{-y}$
6. $\frac{2y}{-3z}$
7. $-\frac{-2}{x}$
8. $-\frac{-6}{-xy}$
9. $-\frac{-3x}{-2y^2}$
10. $\frac{8xy^2}{-3z}$
11. $-\frac{2}{x-y}$
12. $-\frac{x}{x-y}$
13. $-\frac{R-S}{3}$
14. $\frac{R_1-R_2}{-2}$
15. $-\frac{L+C}{-L}$
16. $-\frac{T-T_1}{-T_1}$

Write each fraction on the left as an equal fraction in standard form, using the denominator shown on the right.

EXAMPLE

$\frac{-1}{2-x}, \frac{?}{x-2}$

SOLUTION

Express $2-x$ as $-(-2+x)=-(x-2)$.

$$\frac{-1}{2-x}=\frac{-1}{-(x-2)}$$
$$=\mathbf{\frac{1}{x-2}}$$

17. $\frac{-4}{3-y}, \frac{?}{y-3}$

18. $\frac{-3}{2-x}, \frac{?}{x-2}$

19. $\frac{1}{x-z}, \frac{?}{z-x}$

20. $\frac{6}{z-y}, \frac{?}{y-z}$

21. $\frac{x-2}{3-x}, \frac{?}{x-3}$

22. $-\frac{2x-5}{3-y}, \frac{?}{y-3}$

23. $\frac{R-2}{S-R}, \frac{?}{R-S}$

24. $\frac{T_1-3}{T_2-T_1}, \frac{?}{T_1-T_2}$

25. $-\frac{2}{R_1-R_2}, \frac{?}{R_2-R_1}$

26. $-\frac{3}{M-M_0}, \frac{?}{M_0-M}$

27. $\frac{x_2-x_1}{y_2-y_1}, \frac{?}{y_1-y_2}$

28. $-\frac{x_2-x_1}{y_2-y_1}, \frac{?}{y_1-y_2}$

3.2 Changing fractions to lowest terms I

A fraction is said to be **in lowest terms** when the numerator and denominator do not contain certain types of factors in common. The arithmetic fraction a/b, where a and b are integers and $b \neq 0$, is in lowest terms providing a and b are relatively prime—that is, providing they contain no common integral factors other than 1 and -1. If numerator and denominator of a fraction are polynomials with integral coefficients, then the fraction is said to be in lowest terms if the numerator and denominator do not contain a common polynomial factor with integral coefficients.

To express a given fraction in lowest terms (called *reducing* the fraction), we can factor the numerator and denominator completely, and then apply the fundamental principle of fractions. For example,

$$\frac{8x^3y}{6x^2y^3} = \frac{2 \cdot 2 \cdot 2 \cdot x \cdot x \cdot x \cdot y}{2 \cdot 3 \cdot x \cdot x \cdot y \cdot y \cdot y}$$

$$= \frac{2 \cdot 2 \cdot x}{3 \cdot y \cdot y}\left(\frac{2 \cdot x \cdot x \cdot y}{2 \cdot x \cdot x \cdot y}\right)$$

$$= \frac{2 \cdot 2 \cdot x}{3 \cdot y \cdot y}(1)$$

$$= \frac{4x}{3y^2}.$$

Diagonal lines are sometimes used to abbreviate this procedure. For example, instead of writing

$$\frac{y}{y^2} = \frac{1}{y}\left(\frac{y}{y}\right) = \frac{1}{y}(1) = \frac{1}{y},$$

we can write

$$\frac{y}{y^2} = \frac{\overset{1}{\cancel{y}}}{\underset{y}{\cancel{y^2}}} = \frac{1}{y},$$

using the diagonal lines instead of writing the factor y/y. Reducing a fraction to lowest terms should be accomplished mentally whenever possible.

The division of a polynomial containing more than one term by a monomial may also be considered a special case of changing a fraction to lowest terms, provided the monomial is contained as a factor in each term in the polynomial. For example,

$$\frac{9x^3 - 6x^2 + 3x}{3x} = \frac{\overset{1}{\cancel{3x}}(3x^2 - 2x + 1)}{\underset{1}{\cancel{3x}}}$$

$$= 3x^2 - 2x + 1.$$

Alternatively, we may utilize an approach based on the representation of the sum of two or more fractions, which is developed at length in Section 3.5; namely,

$$\frac{a}{g} + \frac{b}{g} + \frac{c}{g} = \frac{a + b + c}{g};$$

and by the symmetric law of equality,

$$\frac{a+b+c}{g} = \frac{a}{g} + \frac{b}{g} + \frac{c}{g}.$$

As an illustration,

$$\frac{9x^3 - 6x^2 + 3x}{3x} = \frac{\overset{3\cdot x^2}{\cancel{9x^3}}}{\underset{1\cdot 1}{\cancel{3x}}} - \frac{\overset{2\cdot x}{\cancel{6x^2}}}{\underset{1\cdot 1}{\cancel{3x}}} + \frac{\overset{1}{\cancel{3x}}}{\underset{1}{\cancel{3x}}}$$

$$= 3x^2 - 2x + 1.$$

This second approach has the advantage of being applicable even though the denominator is not a factor of the numerator, but in this case it cannot be looked upon as a matter of reducing a fraction to lowest terms.

EXERCISE 3.2

Reduce to lowest terms.

EXAMPLES

a. $\frac{16}{24}$ b. $\frac{12xy^2}{8x^2y}$ c. $\frac{2y + xy}{2y}$

SOLUTIONS

a. $\frac{\overset{2}{\cancel{16}}}{\underset{3}{\cancel{24}}}$

$\mathbf{\frac{2}{3}}$

b. $\frac{\overset{3\cdot 1\cdot y}{\cancel{12xy^2}}}{\underset{2\cdot x\cdot 1}{\cancel{8x^2y}}}$

$\mathbf{\frac{3y}{2x}}$

c. $\frac{\overset{1}{\cancel{y}}(2 + x)}{\underset{1}{\cancel{2y}}}$

$\mathbf{\frac{2 + x}{2}}$

1. $\frac{6}{8}$ 2. $\frac{8}{30}$ 3. $\frac{20}{12}$ 4. $\frac{24}{18}$

5. $\frac{-20}{44}$ 6. $\frac{-36}{48}$ 7. $\frac{3x^2}{12x^5}$ 8. $\frac{9x^3y^4}{36x^5y^2}$

9. $\frac{2x^2y^3}{6x^3y}$ 10. $\frac{14x^2y^3}{7x^2y}$ 11. $\frac{a^2bc}{-ab^2c}$ 12. $\frac{32a^4b^2y}{-16a^3b^2y^2}$

13. $\dfrac{-2abc^2}{-6a^3b^3c^3}$ 14. $\dfrac{-3ab^2c^2}{-12a^2b^2c^2}$ 15. $\dfrac{-18a^3bc}{4ab^2c}$ 16. $\dfrac{2ab^3c^2}{-12a^3bc}$

Divide. Work two ways when possible.

EXAMPLES

a. $\dfrac{2y^3 - 6y^2 + 10y}{2y}$

b. $\dfrac{4x^2 - 2x + 1}{2x}$

SOLUTIONS

a. $\dfrac{\overset{1}{\cancel{2y}}(y^2 - 3y + 5)}{\underset{1}{\cancel{2y}}}$ $\quad \dfrac{\overset{1\cdot y^2}{\cancel{2y^3}}}{\underset{1\cdot 1}{\cancel{2y}}} - \dfrac{\overset{3\cdot y}{\cancel{6y^2}}}{\underset{1\cdot 1}{\cancel{2y}}} + \dfrac{\overset{5\cdot 1}{\cancel{10y}}}{\underset{1\cdot 1}{\cancel{2y}}}$

$\mathbf{y^2 - 3y + 5}$

b. $\dfrac{4x^2}{2x} - \dfrac{2x}{2x} + \dfrac{1}{2x}$

$\mathbf{2x - 1 + \dfrac{1}{2x}}$

17. $\dfrac{4x - 6}{2}$ 18. $\dfrac{6x - 9}{3}$ 19. $\dfrac{ay - a}{a}$

20. $\dfrac{bx^2 - bx}{bx}$ 21. $\dfrac{R^2 - R}{2R}$ 22. $\dfrac{2R^2 + 4R}{4R}$

23. $\dfrac{4P^2 + 6P}{4P}$ 24. $\dfrac{6P - 2P^3}{4P}$

25. $\dfrac{a^3 - 3a^2 + 2a}{a}$ 26. $\dfrac{3x^3 - 6x^2 + 3x}{-3x}$

27. $\dfrac{8a^2 + 4a + 1}{2}$ 28. $\dfrac{12x^3 - 8x^2 + 3x}{4x}$

29. $\dfrac{2x^3 - 4x^2 - 3x}{2x}$ 30. $\dfrac{8a^2 - 4ax^2 + ax}{2ax}$

31. $\dfrac{4R_1 + 6R_2 + 8}{2}$ 32. $\dfrac{2R_1 - 6R_2 + 8}{2}$

33. $\dfrac{R^2R_1 + RR_2 - R}{R}$ 34. $\dfrac{2R^2R_1 - 4RR_2 - 2R}{4R}$

3.3 Changing fractions to lowest terms II

If the divisor is contained as a factor in the dividend, the division of one polynomial by another polynomial, where each contains more than one term, may also be considered as an example of reducing a fraction to lowest terms. Thus to devide $2x^2 + x - 15$ by $x + 3$ we can write

$$\frac{2x^2 + x - 15}{x + 3} = \frac{(2x - 5)\overset{1}{\cancel{(x + 3)}}}{\underset{1}{\cancel{(x + 3)}}}.$$

$$= 2x - 5.$$

As always, the process is not valid if the divisor is zero. In this example, it is required that $x \neq -3$.

The preceding result may also be obtained through a method similar to the long division process used in arithmetic. For example consider the quotient:

$$x + 3\overline{)2x^2 + x - 15}$$

Divide $2x^2$ by x to obtain first term of quotient, $2x$. Subtract product of $2x$ and $x + 3$ from $2x^2 + x$, and "bring down" -15.

$$\begin{array}{r} 2x \phantom{{}+x-15} \\ x + 3\overline{)2x^2 + x - 15} \\ \underline{2x^2 + 6x} \phantom{{}-15} \\ -5x - 15 \end{array}$$

Divide $-5x$ by x to obtain second term of quotient, -5. Subtract product of -5 and $x + 3$ from $-5x - 15$. The remainder is 0.

$$\begin{array}{r} 2x - 5 \phantom{{}-15} \\ x + 3\overline{)2x^2 + x - 15} \\ \underline{2x^2 + 6x} \phantom{{}-15} \\ -5x - 15 \\ \underline{-5x - 15} \\ 0 \end{array}$$

This latter procedure is most useful when the divisor is not a factor of the dividend. If such is the case, the division process will produce a remainder that may be expressed by a fraction. For example, in dividing

$$(x^2 + 2x + 2) \text{ by } (x + 1) \quad \text{for } x \neq -1,$$

we obtain

$$\begin{array}{r} x + 1 \phantom{{}+2} \\ x + 1\overline{)x^2 + 2x + 2} \\ \underline{x^2 + x} \phantom{{}+2} \\ x + 2 \\ \underline{x + 1} \\ 1 \end{array}$$

and the result may be expressed as

$$x + 1 + \frac{1}{x + 1}.$$

EXERCISE 3.3

Divide. Work two ways when possible.

EXAMPLE

$$\frac{y^2+y-6}{y-2}$$

SOLUTION

$$\frac{(y+3)\overset{1}{\cancel{(y-2)}}}{\underset{1}{\cancel{(y-2)}}}$$

$y+3$

$$\begin{array}{r} y+3 \\ y-2\overline{)y^2+y-6} \\ \underline{y^2-2y} \\ 3y-6 \\ \underline{3y-6} \end{array}$$

1. $\frac{y^2+5y-14}{y-2}$
2. $\frac{x^2+5x+6}{x+3}$
3. $\frac{y^2-y-2}{y+1}$
4. $\frac{x^2-6x+5}{x-5}$
5. $\frac{x^2+4x+4}{x+2}$
6. $\frac{x^2-6x+9}{x-3}$
7. $\frac{y^2-5y+4}{y-4}$
8. $\frac{y^2-y-6}{y-3}$
9. $\frac{R^2-5R+6}{R-2}$
10. $\frac{R^2-4R+3}{R-3}$
11. $\frac{S^2+2S-35}{S-5}$
12. $\frac{S^2-5S-6}{S-6}$
13. $\frac{4y^2+12y+5}{2y+1}$
14. $\frac{2x^2+13x-7}{2x-1}$
15. $\frac{4y^2-4y-5}{2y+1}$
16. $\frac{2x^2-3x-15}{2x+5}$
17. $\frac{3x^2+5x+2}{3x+2}$
18. $\frac{3x^2-8x+4}{3x-2}$
19. $\frac{4R^2+7R+2}{4R-1}$
20. $\frac{4R^2-11R+3}{4R+1}$
21. $\frac{6S^2+S-2}{3S+2}$
22. $\frac{6S^2+5S-6}{3S-2}$
23. $\frac{6P^2-P-1}{3P+1}$
24. $\frac{3P^2+8P+5}{3P+5}$

EXAMPLE

$$\frac{x^2-9}{x-3}$$

SOLUTION

$$\frac{\overset{1}{\cancel{(x-3)}}(x+3)}{\underset{1}{\cancel{(x-3)}}}$$

$x+3$

Insert $0x$ in numerator in the division process.

$$\begin{array}{r|l} & x + 3 \\ x-3 & x^2 + 0x - 9 \\ & x^2 - 3x \\ \hline & \quad 3x - 9 \\ & \quad 3x - 9 \\ \hline \end{array}$$

25. $\dfrac{x^2 - 36}{x + 6}$ 26. $\dfrac{x^2 - 25}{x - 5}$ 27. $\dfrac{4y^2 - 9}{2y - 3}$

28. $\dfrac{9y^2 - 1}{3y + 1}$ 29. $\dfrac{25R^2 - 16}{5R - 4}$ 30. $\dfrac{16S^2 - 9}{4S + 3}$

3.4 Building fractions

In the preceding section, we changed fractions to equal fractions in lowest terms by applying the fundamental principle in the form

$$\frac{ac}{bc} = \frac{a}{b}\left(\frac{c}{c}\right) = \frac{a}{b}.$$

In this section, we shall change fractions to equal fractions by applying the fundamental principle in the form

$$\frac{a}{b} = \frac{a}{b}\left(\frac{c}{c}\right) = \frac{ac}{bc}.$$

For example, $\frac{1}{2}$ can be changed to an equal fraction with a denominator of 8 by multiplying the numerator and the denominator by 4. Thus

$$\frac{1}{2} = \frac{1}{2} \cdot \frac{4}{4} = \frac{4}{8}.$$

This process is called "building a fraction," and the number 4 is said to be a building factor.

In general, to build a fraction a/b to an equal fraction with bc as a denominator (i.e., $a/b = ?/bc$), you can usually determine the building factor c by inspection and then multiply the numerator and the denominator of the original fraction by this building factor. If the building factor cannot be obtained by inspection, the desired denominator (bc) can be divided by the denominator of the given fraction (b) to determine the building factor (c).

EXERCISE 3.4

Express each of the given fractions as an equal fraction with the given denominator.

EXAMPLE

$\frac{3}{4xy}; \frac{?}{8x^2y^2}$

SOLUTION

Obtain building factor.

$$(8x^2y^2 \div 4xy = 2xy)$$

Multiply numerator and denominator of given fraction by building factor $(2xy)$.

$$\frac{3}{4xy} \cdot \frac{(2xy)}{(2xy)}$$

$$\mathbf{\frac{6xy}{8x^2y^2}}$$

1. $\frac{3}{4}; \frac{?}{12}$
2. $\frac{1}{5}; \frac{?}{10}$
3. $\frac{16}{-5}; \frac{?}{15}$
4. $\frac{7}{-6}; \frac{?}{18}$
5. $4; \frac{?}{12}$
6. $6; \frac{?}{3}$
7. $\frac{2}{6x}; \frac{?}{12x}$
8. $\frac{5}{3y}; \frac{?}{18y}$
9. $\frac{-a^2}{b^2}; \frac{?}{3b^3}$
10. $\frac{-a^2}{b^2}; \frac{?}{6b^3}$
11. $y; \frac{?}{xy}$
12. $x; \frac{?}{x^2y}$
13. $\frac{1}{3}; \frac{?}{3(x+y)}$
14. $\frac{1}{5}; \frac{?}{10(x-y)}$
15. $\frac{x-1}{3}; \frac{?}{9(x+1)}$
16. $\frac{x-2}{2}; \frac{?}{6(x+1)}$
17. $3a; \frac{?}{9(a+1)}$
18. $5a^2; \frac{?}{3(a+2)}$

EXAMPLE

$\frac{3}{2a-2b}; \frac{?}{4a^2-4b^2}$

SOLUTION

Factor denominators.

$$\frac{3}{2(a-b)};\ \frac{}{4(a-b)(a+b)}$$

Obtain building factor.

$$[4(a-b)(a+b) \div 2(a-b) = 2(a+b)]$$

Multiply numerator and denominator of given fraction by building factor $2(a+b)$

$$\frac{3}{2(a-b)} \cdot \frac{2(a+b)}{2(a+b)}$$

$$\frac{6(a+b)}{4a^2-4b^2}$$

19. $\dfrac{3}{a-b};\ \dfrac{?}{a^2-b^2}$

20. $\dfrac{5}{2a+b};\ \dfrac{?}{4a^2-b^2}$

21. $\dfrac{3x}{y+2};\ \dfrac{?}{y^2-y-6}$

22. $\dfrac{5x}{x+3};\ \dfrac{?}{x^2+x-6}$

23. $\dfrac{-2}{x+1};\ \dfrac{?}{x^2+3x+2}$

24. $\dfrac{-3}{y+2};\ \dfrac{?}{y^2+3y+2}$

25. $\dfrac{-3}{2a-2};\ \dfrac{?}{6a^2-24a+18}$

26. $\dfrac{-2}{3a-6};\ \dfrac{?}{3a^2-3a-6}$

27. $\dfrac{2}{a-b};\ \dfrac{?}{b^2-a^2}$

28. $\dfrac{7}{x-y};\ \dfrac{?}{y^2-x^2}$

29. $\dfrac{x}{2-x};\ \dfrac{?}{x^2-3x+2}$

30. $\dfrac{y}{3-2y};\ \dfrac{?}{2y^2-y-3}$

3.5 Sums and differences of fractions

We can give meaning to the sum of fractions with like denominators, say

$$\frac{a}{c} + \frac{b}{c},$$

by observing that the alternative definition of a quotient enables us to write this sum as

$$a\left(\frac{1}{c}\right) + b\left(\frac{1}{c}\right).$$

By the distributive law, we have

$$(a + b)\left(\frac{1}{c}\right).$$

from which we get

$$\frac{a}{c} + \frac{b}{c} = \frac{a + b}{c}.$$

This result suggests the following definition:

The sum of two or more fractions with the same denominator is a fraction with the same denominator whose numerator is the sum of the numerators of the fractions being added.

For example,

$$\frac{3x}{5} + \frac{x}{5} = \frac{3x + x}{5}$$

$$= \frac{4x}{5}.$$

The difference of two fractions

$$\frac{a}{c} - \frac{b}{c}$$

can be viewed as the sum

$$\frac{a}{c} + \left(\frac{-b}{c}\right),$$

from which we get

$$\frac{a}{c} - \frac{b}{c} = \frac{a - b}{c}.$$

For example,

$$\frac{5xy}{7} - \frac{3xy}{7} = \frac{5xy - 3xy}{7}$$

$$= \frac{2xy}{7}.$$

If fractions to be added or subtracted have unlike denominators, we can build them to equal fractions having common denominators and then add. The **least common denominator** (L.C.D.) of two or more fractions is the **least common multiple** (L.C.M.) of the denominators of the fractions.

The **least common multiple** (L.C.M.) of two or more natural numbers is the smallest natural number that is exactly divisible by each of the given numbers.

For example, 24 is the L.C.M. of 3 and 8, because 24 is the smallest natural number each will divide without remainder. To find the L.C.M. of a set of natural numbers, we

1. express each number in prime-factor form;
2. write as factors of a product each *different* prime factor occurring in any of the numbers, writing each factor the greatest number of times it occurs in any one of the given numbers.

For example, the L.C.M. of 12, 15, and 18 is found in the following manner:

1.
$$\begin{array}{ccc} 12 & 15 & 18 \\ 2\cdot 2\cdot 3 & 3\cdot 5 & 3\cdot 3\cdot 2 \end{array}$$
2. The L.C.M. is $2^2 \cdot 3^2 \cdot 5$, or 180.

The notion of a least number among several polynomial expressions is, in general, meaningless. However, we can define the least common multiple of a set of polynomials in a manner analogous to that described above—with the polynomial of lowest degree yielding a polynomial quotient upon division by each of the given polynomials.

We can find the L.C.M. of a set of polynomials with integral coefficients in a manner comparable to that used with a set of natural numbers. The L.C.M. of x^2, $x^2 - 9$, and $x^3 - x^2 - 6x$ is found in the following manner:

1.
$$\begin{array}{ccc} x^2 & x^2 - 9 & x^3 - x^2 - 6x \\ x\cdot x & (x-3)(x+3) & x(x-3)(x+2) \end{array}$$
2. The L.C.M. is $x^2(x - 3)(x + 3)(x + 2)$.

Since polynomials are easier to work with in factored form, it is usually advantageous to leave the L.C.M. of a set of polynomials in a factored form rather than actually carrying out the indicated multiplication.

A sum of fractions with unlike denominators can be expressed as a sum of fractions with like denominators by using the least common denominator (L.C.D.) of the fractions. Thus,

$$\frac{a}{b} + \frac{c}{d} = \frac{a}{b}\left(\frac{d}{d}\right) + \frac{c}{d}\left(\frac{b}{b}\right),$$

from which we obtain

$$\boldsymbol{\frac{a}{b} + \frac{c}{d} = \frac{ad + cb}{bd}}.$$

For example, the L.C.D. of the fractions in the sum

$$\frac{5}{12} + \frac{7}{9}$$

can be found by the method described above. In this case

$$12 = 2 \cdot 2 \cdot 3 \quad \text{and} \quad 9 = 3 \cdot 3.$$

Hence, the L.C.M. is $2 \cdot 2 \cdot 3 \cdot 3 = 36$, and the sum can be expressed as

$$\frac{5}{12}\left(\frac{3}{3}\right) + \frac{7}{9}\left(\frac{4}{4}\right) = \frac{15}{36} + \frac{28}{36}$$

$$= \frac{43}{36}.$$

As before, the difference of two fractions

$$\frac{a}{b} - \frac{c}{d}$$

can be viewed as the sum

$$\frac{a}{b} + \frac{-c}{d},$$

and written

$$\frac{a}{b}\left(\frac{d}{d}\right) + \frac{-c}{d}\left(\frac{b}{b}\right) = \frac{ad - cb}{bd}.$$

Thus,

$$\mathbf{\frac{a}{b} - \frac{c}{d} = \frac{ad - cd}{bd}}.$$

For example, the L.C.D. of the fractions in the difference

$$\frac{4}{x} - \frac{2}{3y}$$

is $3xy$. Hence, each fraction can be written as an equal fraction with denominator $3xy$ and the difference expressed as the sum

$$\frac{4}{x}\left(\frac{3y}{3y}\right) + \frac{-2}{3y}\left(\frac{x}{x}\right) = \frac{12y}{3xy} + \frac{-2x}{3xy}$$

$$= \frac{12y - 2x}{3xy}$$

EXERCISE 3.5

Write each expression as a single fraction.

EXAMPLES

a. $\frac{a}{3x} + \frac{b}{3x}$

b. $\frac{a+1}{b} - \frac{a-1}{b}$

SOLUTIONS

Add numerators.

a. $\dfrac{\mathbf{a+b}}{\mathbf{3x}}$

Write each fraction in standard form.

b. $\dfrac{a+1}{b}+\dfrac{-(a-1)}{b}$

Add numerators and simplify.

$\dfrac{a+1-(a-1)}{b}$

$\dfrac{a+1-a+1}{b}$

$\dfrac{\mathbf{2}}{\mathbf{b}}$

1. $\dfrac{7}{11}+\dfrac{2}{11}$
2. $\dfrac{3}{5}+\dfrac{6}{5}$
3. $\dfrac{3}{7}+\dfrac{5}{7}+\dfrac{2}{7}$
4. $\dfrac{3}{11}-\dfrac{2}{11}-\dfrac{5}{11}$
5. $\dfrac{x}{2}-\dfrac{3}{2}$
6. $\dfrac{y}{7}-\dfrac{5}{7}$
7. $\dfrac{a}{6}+\dfrac{b}{6}-\dfrac{c}{6}$
8. $\dfrac{x}{3}-\dfrac{2y}{3}+\dfrac{z}{3}$
9. $\dfrac{7}{3x}-\dfrac{5}{3x}$
10. $\dfrac{y}{3z}-\dfrac{2}{3z}$
11. $\dfrac{x-1}{2y}+\dfrac{x}{2y}$
12. $\dfrac{y+1}{b}+\dfrac{y-1}{b}$
13. $\dfrac{2a-b}{3x}+\dfrac{a-b}{3x}+\dfrac{a+b}{3x}$
14. $\dfrac{3x-1}{4y}-\dfrac{2-x}{4y}-\dfrac{x-3}{4y}$
15. $\dfrac{3}{x+2y}-\dfrac{x+3}{x+2y}-\dfrac{x-1}{x+2y}$
16. $\dfrac{2}{a-3b}-\dfrac{b-2}{a-3b}+\dfrac{b}{a-3b}$
17. $\dfrac{a+1}{a^2-2a+1}+\dfrac{5-3a}{a^2-2a+1}$
18. $\dfrac{x+4}{x^2-x+2}+\dfrac{2x-3}{x^2-x+2}$

Find the least common multiple.

EXAMPLES

a. 24, 30, 20

b. $2a$, $4b$, $6ab^2$

SOLUTIONS

a. $2\cdot 2\cdot 2\cdot 3$, $2\cdot 3\cdot 5$, $2\cdot 2\cdot 5$
$\mathbf{2^3\cdot 3\cdot 5}$, or **120**

b. $2\cdot a$, $2\cdot 2\cdot b$, $2\cdot 3\cdot a\cdot b\cdot b$
$\mathbf{2^2\cdot 3ab^2}$, or $\mathbf{12ab^2}$

19. 4, 6, 10
20. 3, 4, 5
21. 6, 8, 15
22. 4, 15, 18
23. 14, 21, 36
24. 4, 11, 22
25. $2ab, 6b^2$
26. $12xy, 24x^3y^2$
27. $6xy, 8x^2, 3xy^2$
28. $7x, 8y, 6z$
29. $a - b, a(a - b)^2$
30. $6(x + y)^2, 4x(x + y)$

EXAMPLES

a. $2a - 2, \quad a - 1$

b. $x^2 - 1, \quad 2(x - 1)^2$

SOLUTIONS

a. $2(a - 1), \quad (a - 1)$
$2(a - 1)$

b. $(x - 1)(x + 1), \quad 2(x - 1)(x - 1)$
$2(x - 1)^2(x + 1)$

31. $a^2 - b^2, a - b$
32. $x + 2, x^2 - 4$
33. $a^2 + 5a + 4, (a + 1)^2$
34. $x^2 - 3x + 2, (x - 1)^2$
35. $x^2 + 3x - 4, (x - 1)^2$
36. $x^2 - x - 2, x^2 - 4x + 4$
37. $x^2 - x, (x - 1)^3$
38. $y^2 + 2y, (y + 2)^2$

Write each expression as a single fraction.

EXAMPLE

$$\frac{2}{3x} - \frac{5}{6x}$$

SOLUTION

Write in standard form and build each fraction to a fraction with denominator $6x$.

$$\frac{(2)2}{(2)3x} + \frac{-5}{6x}$$

$$\frac{4}{6x} + \frac{-5}{6x}$$

$$\mathbf{\frac{-1}{6x}}$$

39. $\frac{3}{8} + \frac{1}{4}$
40. $\frac{2}{3} + \frac{1}{4}$
41. $\frac{3}{5} - \frac{1}{7}$
42. $\frac{5}{6} - \frac{2}{5}$
43. $\frac{2}{ax} + \frac{2}{x}$
44. $\frac{3}{by} + \frac{2}{b}$

45. $\dfrac{5}{x^2} - \dfrac{3}{y}$
46. $\dfrac{7}{y^2} - \dfrac{2}{x^3}$
47. $\dfrac{1}{R_1} + \dfrac{1}{R_2}$
48. $\dfrac{1}{f_1} - \dfrac{1}{f_2}$
49. $\dfrac{a-2}{6} - \dfrac{a+1}{3}$
50. $\dfrac{x-3}{4} + \dfrac{5-x}{10}$
51. $\dfrac{2x-y}{2y} + \dfrac{x+y}{x}$
52. $\dfrac{3a+2b}{3b} - \dfrac{a+2b}{6a}$
53. $\dfrac{R+3S}{3S} - \dfrac{R-S}{3R}$
54. $\dfrac{2R-S}{3S} - \dfrac{R+S}{6S}$
55. $\dfrac{2R_1+R_2}{R_1} + \dfrac{R_1-R_2}{2R_1}$
56. $\dfrac{R_1-3R_2}{4R_2} + \dfrac{R_1+2R_2}{3R_2}$

EXAMPLE

$$\frac{5}{a^2-9} - \frac{1}{a-3}$$

SOLUTION

Factor denominators and write in standard form.

$$\frac{5}{(a-3)(a+3)} + \frac{-1}{(a-3)}$$

Build each fraction to a fraction with denominator $(a-3)(a+3)$

$$\frac{5}{(a-3)(a+3)} + \frac{-1(a+3)}{(a-3)(a+3)}$$

Combine fractions and simplify.

$$\frac{5-(a+3)}{(a-3)(a+3)} = \mathbf{\frac{2-a}{(a-3)(a+3)}}$$

57. $\dfrac{7}{5x-10} - \dfrac{5}{3x-6}$
58. $\dfrac{2}{y+2} - \dfrac{3}{y+3}$
59. $\dfrac{5}{2x-6} - \dfrac{3}{x+3}$
60. $\dfrac{1}{2x+1} - \dfrac{3}{x-2}$
61. $\dfrac{a}{3x+2} - \dfrac{a}{x-1}$
62. $\dfrac{a+1}{a+2} - \dfrac{a+2}{a+3}$
63. $\dfrac{5x-y}{3x+y} - \dfrac{6x-5y}{2x-y}$
64. $\dfrac{x+2y}{2x-y} - \dfrac{2x+y}{x-2y}$
65. $\dfrac{R-2S}{R+S} - \dfrac{2R-S}{R-S}$
66. $\dfrac{1}{R^2-1} - \dfrac{1}{(R-1)^2}$
67. $\dfrac{S}{S^2-16} - \dfrac{S+1}{S^2-5S+4}$
68. $\dfrac{8}{R^2-4S^2} - \dfrac{2}{(R-2S)^2}$

3.6 Products and quotients of fractions

We wish to define both the product and quotient of two fractions to be consistent with the assumptions and definitions we have made with respect to real numbers. First, let us examine the product

$$\frac{a}{b} \cdot \frac{c}{d} = p \qquad (1)$$

and inquire as to the nature of p. We have

$$\frac{a}{b} \cdot \frac{c}{d} = p \qquad \text{(given)}$$

$$bd\left(\frac{a}{b}\right)\left(\frac{c}{d}\right) = bdp \qquad \text{(multiplication law)}$$

$$(b)(d)(a)\left(\frac{1}{b}\right)(c)\left(\frac{1}{d}\right) = bdp \qquad \text{(alternate definition of a quotient)}$$

$$(a)(c)(b)\left(\frac{1}{b}\right)(d)\left(\frac{1}{d}\right) = bdp \qquad \text{(associative and commutative laws of multiplication)}$$

$$(a)(c)(1)(1) = bdp \qquad \text{(reciprocal property)}$$

$$ac = bdp \qquad \text{(multiplicative property of 1)}$$

$$p = \frac{ac}{bd} \qquad \text{(definition of a quotient)}$$

Hence, by substituting $\frac{ac}{bd}$ for p in (1) above, we have

$$\mathbf{\frac{a}{b} \cdot \frac{c}{d} = \frac{ac}{bd}}.$$

Therefore, if the product of two fractions is to have validity in the area of real numbers, we must make the definition:

The product of two fractions is the fraction whose numerator is the product of the numerators and whose denominator is the product of the denominators of the two fractions.

For example,

$$\frac{6x^2}{y} \cdot \frac{xy}{2} = \frac{\overset{3}{\cancel{6}}x^3\overset{1}{\cancel{y}}}{\underset{1}{\cancel{2}} \cdot \underset{1}{\cancel{y}}}$$

$$= 3x^3.$$

If any of the fractions or any of the factors of the numerators or denominators of the fractions have negative signs attached, it is advisable to proceed as if all of the signs were positive and then attach the appropriate sign to the product. If there is an even number of negative signs involved, the product has a positive sign; if there is an odd number of negative signs involved, the product has a negative sign. For example,

$$\frac{-4(y-x)}{y}\cdot\frac{y^2}{x(y-x)}=\frac{-4\overset{1}{\cancel{(y-x)}}}{\underset{1}{\cancel{y}}}\cdot\frac{\overset{y}{\cancel{y^2}}}{x\underset{1}{\cancel{(y-x)}}}$$

$$=\frac{-4y}{x}$$

When dividing one fraction by another,

$$\frac{a}{b}\div\frac{c}{d}=q \tag{2}$$

we seek the quotient q such that

$$\left(\frac{c}{d}\right)q=\frac{a}{b}\quad\text{where } b,\ c,\ \text{and } d\neq 0.$$

To obtain q in terms of the other variables, we use the multiplication law and multiply each member of this equality by d/c,

$$\left(\frac{d}{c}\right)\left(\frac{c}{d}\right)q=\left(\frac{d}{c}\right)\left(\frac{a}{b}\right),$$

from which

$$q=\left(\frac{d}{c}\right)\left(\frac{a}{b}\right)=\frac{da}{cb}.$$

Hence, by substituting $\frac{da}{cb}$, or $\frac{ad}{bc}$ for q in (2) above, we have

$$\boldsymbol{\frac{a}{b}\div\frac{c}{d}=\frac{ad}{bc}}.$$

Thus:

To divide one fraction by another multiply the dividend by the reciprocal of the divisor.

This is the technique one learns in arithmetic—"invert the divisor and multiply."

EXERCISE 3.6

Multiply.

EXAMPLES

a. $\dfrac{3c^2}{5ab} \cdot \dfrac{10a^2b}{9c}$

b. $\dfrac{4a^2 - 1}{a^2 - 4} \cdot \dfrac{a^2 + 2a}{4a + 2}$

SOLUTIONS

a. $\dfrac{\overset{1 \cdot c}{\cancel{3c^2}}}{\underset{1 \cdot 1 \cdot 1}{\cancel{5ab}}} \cdot \dfrac{\overset{2 \cdot a \cdot 1}{\cancel{10a^2b}}}{\underset{3 \cdot 1}{\cancel{9c}}}$

$\mathbf{\dfrac{2ac}{3}}$

b. $\dfrac{(2a - 1)\overset{1}{\cancel{(2a+1)}}}{(a - 2)\underset{1}{\cancel{(a+2)}}} \cdot \dfrac{a\overset{1}{\cancel{(a+2)}}}{2\underset{1}{\cancel{(2a+1)}}}$

$\mathbf{\dfrac{a(2a - 1)}{2(a - 2)}}$

1. $\dfrac{16}{38} \cdot \dfrac{19}{12}$

2. $\dfrac{4}{15} \cdot \dfrac{3}{16}$

3. $\dfrac{21}{4} \cdot \dfrac{2}{15}$

4. $\dfrac{7}{8} \cdot \dfrac{48}{64}$

5. $\dfrac{24}{3} \cdot \dfrac{20}{36} \cdot \dfrac{3}{4}$

6. $\dfrac{3}{10} \cdot \dfrac{16}{27} \cdot \dfrac{30}{36}$

7. $\dfrac{12a^2b}{5c} \cdot \dfrac{10b^2c}{24a^3b}$

8. $\dfrac{a^2}{bc} \cdot \dfrac{3b^3c}{4a}$

9. $\dfrac{-2ab}{7c} \cdot \dfrac{3c^2}{4a^3} \cdot \dfrac{-6a}{15b^2}$

10. $\dfrac{10x}{12y} \cdot \dfrac{3x^2z}{5x^3z} \cdot \dfrac{6y^2x}{3yz}$

11. $5a^2b^2 \cdot \dfrac{1}{a^3b^3}$

12. $15x^2y \cdot \dfrac{3}{45xy^2}$

13. $\dfrac{5x + 25}{2x} \cdot \dfrac{4x}{2x + 10}$

14. $\dfrac{3y}{4xy - 6y} \cdot \dfrac{2x - 3y}{12x}$

15. $\dfrac{4a^2 - 1}{a^2 - 16} \cdot \dfrac{a^2 - 4a}{2a + 1}$

16. $\dfrac{9x^2 - 25}{2x - 2} \cdot \dfrac{x^2 - 1}{6x - 10}$

17. $\dfrac{R^2 - R - 20}{R^2 + 7R + 12} \cdot \dfrac{(R + 3)^2}{(R - 5)^2}$

18. $\dfrac{4R^2 + 8R + 3}{2R^2 - 5R + 3} \cdot \dfrac{6R^2 - 9R}{1 - 4R^2}$

19. $\dfrac{7S + 14}{14S - 28} \cdot \dfrac{4 - 2S}{S + 2} \cdot \dfrac{S - 3}{S + 1}$

20. $\dfrac{5S^2 - 5S}{3} \cdot \dfrac{S^2 - 9S - 10}{4S - 40} \cdot \dfrac{S^2}{2 - 2S^2}$

Divide.

EXAMPLES

a. $\dfrac{a^2b}{c} \div \dfrac{ab^3}{c^2}$

b. $\dfrac{3a^2 - 3}{2a + 2} \div \dfrac{3a - 3}{2}$

SOLUTIONS

a. $\dfrac{\overset{a \cdot 1}{\cancel{a^2 b}}}{\underset{1}{\cancel{c}}} \cdot \dfrac{\overset{c}{\cancel{c^2}}}{\underset{1 \cdot b^2}{\cancel{ab^3}}}$

$\dfrac{\mathbf{ac}}{\mathbf{b^2}}$

b. $\dfrac{\overset{1}{\cancel{3}}\overset{1}{\cancel{(a-1)}}\overset{1}{\cancel{(a+1)}}}{\underset{1}{\cancel{2}}\underset{1}{\cancel{(a+1)}}} \cdot \dfrac{\overset{1}{\cancel{2}}}{\underset{1}{\cancel{3}}\underset{1}{\cancel{(a-1)}}}$

1

21. $\dfrac{3}{4} \div \dfrac{9}{16}$

22. $\dfrac{2}{3} \div \dfrac{9}{15}$

23. $\dfrac{3}{5} \div \dfrac{6}{25}$

24. $\dfrac{4}{7} \div \dfrac{3}{14}$

25. $\dfrac{xy}{a^2b} \div \dfrac{x^3y^2}{ab}$

26. $\dfrac{9ab^3}{x} \div \dfrac{3}{2x^3}$

27. $\dfrac{-28x^2y^3}{15a^2b} \div \dfrac{-21x^2y^2}{35ab}$

28. $\dfrac{24a^3b}{-6xy^2} \div \dfrac{3a^2b}{12x}$

29. $\dfrac{x^2 - xy}{xy} \div \dfrac{2x - 2y}{xy}$

30. $\dfrac{r^2 + 2rs}{r} \div \dfrac{r + 2s}{s}$

31. $\dfrac{12 - 6x}{3x + 9} \div \dfrac{4x - 8}{-(5x + 15)}$

32. $\dfrac{18 - 4y}{3y + 2} \div \dfrac{6y - 27}{-(6y + 4)}$

33. $\dfrac{P^2 - P - 6}{P^2 + 2P - 15} \div \dfrac{P^2 - 4}{P^2 - 25}$

34. $\dfrac{M^2 + M - 2}{M^2 + 2M - 3} \div \dfrac{M^2 + 7M + 10}{M^2 - 2M - 15}$

35. $\dfrac{N^2 - N}{N^2 - 2N - 3} \cdot \dfrac{N^2 + 2N + 1}{N^2 + 4N} \div \dfrac{N^2 - 3N - 4}{N^2 - 16}$

36. $\dfrac{S^2 - 4S + 3}{S^2} \cdot \dfrac{S^2 + S}{S^2 - 6S + 9} \div \dfrac{S^2 - 2S - 3}{S^2 - S - 6}$

3.7 Complex fractions

A fraction that contains a fraction or fractions in either the numerator or denominator or both is called a **complex fraction**.

In simplifying complex fractions, it is helpful to remember that fraction bars serve as grouping devices in the same sense as do parentheses or brackets. Thus the complex fraction

$$\frac{x + \dfrac{3}{4}}{x - \dfrac{1}{2}}$$

means

$$\left(x + \frac{3}{4}\right) \div \left(x - \frac{1}{2}\right).$$

Complex fractions may be simplified in either of two ways. In simple examples, it is easier to apply the fundamental principle of fractions and multiply the numerator and the denominator of the complex fraction by the L.C.D. of all of the fractions appearing therein. Thus, in the preceding example, applying the fundamental principle of fractions and the distributive law, we have

$$\frac{4}{4}\cdot\frac{\left(x+\frac{3}{4}\right)}{\left(x-\frac{1}{2}\right)}=\frac{4x+3}{4x-2}.$$

Alternatively, this fraction may be simplified by first representing the numerator as a single fraction and the denominator as a single fraction, and then inverting the denominator and multiplying. Thus

$$\frac{x+\frac{3}{4}}{x-\frac{1}{2}}=\frac{\frac{(4x+3)}{4}}{\frac{(2x-1)}{2}}$$

$$=\frac{4x+3}{\overset{}{\cancel{4}}_{2}}\cdot\frac{\overset{1}{\cancel{2}}}{2x-1}$$

$$=\frac{4x+3}{2(2x-1)}.$$

The second method is generally more convenient in complicated situations.

EXERCISE 3.7

Simplify.

EXAMPLE

$$\frac{\frac{3}{a}-\frac{1}{2a}}{\frac{1}{3a}+\frac{5}{6a}}$$

SOLUTION

Find the L.C.D. for all fractions in numerator and denominator ($6a$); multiply each term in the numerator and each term in the denominator by $6a$ and simplify.

$$\frac{(6a)\dfrac{3}{a} - (6a)\dfrac{1}{2a}}{(6a)\dfrac{1}{3a} + (6a)\dfrac{5}{6a}}$$

$$\frac{18-3}{2+5}$$

$$\mathbf{\frac{15}{7}}$$

1. $\dfrac{\dfrac{3}{7}}{\dfrac{2}{7}}$ 2. $\dfrac{\dfrac{3}{5}}{\dfrac{1}{10}}$ 3. $\dfrac{\dfrac{2}{5}}{\dfrac{7}{10}}$ 4. $\dfrac{\dfrac{5}{2}}{\dfrac{21}{4}}$

5. $\dfrac{\dfrac{b}{c}}{\dfrac{b^2}{a}}$ 6. $\dfrac{\dfrac{5x}{6y}}{\dfrac{4x}{5y}}$ 7. $\dfrac{\dfrac{2x}{5y}}{\dfrac{3x}{10y^2}}$ 8. $\dfrac{\dfrac{3ab}{4}}{\dfrac{3b}{8a^2}}$

9. $\dfrac{\dfrac{3}{4}}{4-\dfrac{1}{4}}$ 10. $\dfrac{3-\dfrac{1}{3}}{4+\dfrac{2}{3}}$ 11. $\dfrac{1-\dfrac{2}{3}}{3+\dfrac{1}{3}}$ 12. $\dfrac{\dfrac{1}{2}+\dfrac{3}{4}}{\dfrac{1}{2}-\dfrac{3}{4}}$

13. $\dfrac{3+\dfrac{2}{r}}{5+\dfrac{1}{r}}$ 14. $\dfrac{3-\dfrac{2}{t}}{2+\dfrac{3}{t}}$ 15. $\dfrac{R_1-\dfrac{1}{R_2}}{R_1+\dfrac{1}{R_2}}$ 16. $\dfrac{T_2+\dfrac{T_2}{T_1}}{T_2-\dfrac{T_2}{T_1}}$

EXAMPLE

$$\frac{a}{a+\dfrac{3}{3+\dfrac{1}{2}}}$$

SOLUTION

Write $3+\dfrac{1}{2}$ as $\dfrac{3\cdot 2}{1\cdot 2}+\dfrac{1}{2}$, or $\dfrac{7}{2}$.

Solution continued overleaf

$$\frac{a}{\dfrac{a+\dfrac{3}{7}}{2}}$$

Write $\dfrac{\frac{3}{7}}{2}$ as $3 \cdot \frac{2}{7}$ or $\frac{6}{7}$.

$$\frac{a}{a+\dfrac{6}{7}}$$

Write $a+\frac{6}{7}$ as $\frac{a \cdot 7}{1 \cdot 7}+\frac{6}{7}$, or $\frac{7a+6}{7}$.

$$\frac{a}{\dfrac{7a+6}{7}}$$

Simplify by inverting divisor $\frac{7a+6}{7}$ and multiplying to obtain $a \cdot \frac{7}{7a+6}$.

$$\boldsymbol{\frac{7a}{7a+6}}$$

17. $2-\dfrac{2}{3+\dfrac{1}{2}}$

18. $1-\dfrac{1}{1-\dfrac{1}{3}}$

19. $a-\dfrac{a}{a+\dfrac{1}{4}}$

20. $x-\dfrac{x}{1-\dfrac{x}{1-x}}$

21. $1-\dfrac{1}{1-\dfrac{1}{y-2}}$

22. $2y+\dfrac{3}{3-\dfrac{2y}{y-1}}$

23. $\dfrac{1+\dfrac{1}{1-\frac{a}{b}}}{1-\dfrac{3}{1-\frac{a}{b}}}$

24. $\dfrac{1-\dfrac{1}{\frac{a}{b}+2}}{1+\dfrac{3}{\frac{a}{2b}+1}}$

25. $\dfrac{1}{1+\dfrac{R_1}{R_1+R_2}}$

26. $\dfrac{1}{1-\dfrac{R_2}{R_1-R_2}}$

27. $\dfrac{1}{\dfrac{1}{T_1}+\dfrac{1}{T_2}}$

28. $\dfrac{T_2-T_1}{\dfrac{1}{T_1}-\dfrac{1}{T_2}}$

Chapter review

[3.1] Write each fraction in standard form.

1. $-\frac{3x}{2}$

2. $\frac{x-y}{-x}$

[3.2–3.3] Reduce to lowest terms.

3. $\frac{4x^2y^3}{6x^3y}$

4. $\frac{x+xy}{x}$

5. $\frac{2x^2-5x-3}{x-3}$

6. $\frac{4T^2-9}{2T-3}$

[3.4–3.5] Add.

7. $\frac{a}{b}+\frac{2a}{3b}$

8. $\frac{b+2}{b^2-1}-\frac{b-2}{b-1}$

9. $\frac{2}{R-S}-\frac{3}{S-R}$

10. $P+\frac{PQ}{P-Q}$

[3.6] Multiply.

11. $\frac{x^2y}{6}\cdot\frac{3}{xy^3}$

12. $\frac{3x}{2x-xy}\cdot\frac{x^2-y^2}{3x+3y}$

13. $\frac{x-y}{a}\cdot\frac{3a^2}{6x-6y}$

14. $\frac{x-3}{x^2-1}\cdot\frac{x^2+2x-3}{x^2+6x+9}$

Divide.

15. $\frac{xy^2}{z}\div\frac{x^2y}{z}$

16. $\frac{a^2}{b-a}\div\frac{a^3-a^2}{a-b}$

17. $R\div\frac{R^3}{S}$

18. $\frac{E^3}{R^2}\div E^3$

[3.7] Simplify.

19. $\dfrac{2-\frac{1}{3}}{3+\frac{1}{6}}$

20. $1-\dfrac{y}{y-\frac{1}{y}}$

First-degree equations and inequalities: one variable

4.1 Equations

Word sentences, such as " The sum of five and two is seven " or " The sum of five and two is nine," can be labeled true or false. Such sentences are called **statements.** Other word sentences, such as " He is six feet tall " or " It is less than three feet in length," cannot be labeled true or false, because the words " he " and " it " do not specify any particular person or object. Such sentences are called **open sentences.**

We can draw a useful analogy between word sentences and the symbolic sentences of mathematics. The symbolic sentences

$$5 + 2 = 7 \tag{1}$$

and

$$5 + 2 = 9 \tag{2}$$

are statements, because we can determine by inspection that (1) is true and (2) is false. The symbolic sentence

$$x + 3 = 5 \tag{3}$$

is an open sentence, because we cannot make a judgment about the truth or falsity thereof until the variable x has been replaced with an element from its replacement set. Symbolic sentences involving only equality relationships, whether statements or open sentences, are called **equations.**

Equations have certain advantages over word sentences when we are discussing numbers. Since symbols are used in place of whole groups of words, equations are much more concise: also, mechanical transformations of equations from one form to another are possible—transformations that enable us to generate a logical sequence of equivalent equations. Such transformations, and the logical assumptions upon which they rest, will be discussed in Section 4.2.

If when we replace the variable in an equation with a numeral the resulting statement is true, the number represented by the numeral is called a **solution** or **root** of the equation and is said to *satisfy* the equation. The set of all numbers that satisfy the equation is called the **solution set.** For example, if we substitute 2 for x in (3), we obtain the true statement

$$2 + 3 = 5$$

and 2 is a solution of the equation.

In discussing equations, a variable is frequently referred to as an **unknown,** although, again, its function is simply to represent an unspecified element of some set of numbers. The replacement set of the variable will be assumed to be the real numbers unless otherwise specified. In practical situations, however, it is always well to have a meaningful replacement set in mind. If we were investigating a situation involving the height of a man, we certainly would not consider negative numbers in the replacement set for any variable representing his height.

One major type of equation with which we are concerned in this book is the polynomial equation in one variable. This is an equation in which one member is zero and the other member is a polynomial in simplest form. The equations

$$x^2 + 2x + 1 = 0, \tag{4}$$

$$2y - 3 = 0, \tag{5}$$

$$-z^4 + z^3 + 1 = 0, \tag{6}$$

are polynomial equations in x, y, and z, respectively; (4) is of second degree, (5) is of first degree, and (6) is of fourth degree. In this chapter we shall study first-degree polynomial equations in one variable.

A **conditional equation** is an equation that is false for at least one element in the replacement set of the variable. Thus, if x represents a real number,

$$x + 2 = 5 \tag{7}$$

places a condition on the variable x that is not met by all real numbers and, hence, is a conditional equation. Equations such as

$$a + 0 = a \quad \text{and} \quad a(b + c) = ab + ac$$

that are true for all values of the variable(s) are called **identities.** Equations in which one variable, such as d, is expressed in terms of one or more other variables, such as r and t, are sometimes called **formulas**. For example,

$$d = rt$$

is a formula relating the distance traveled with a constant rate and a period of time. Formulas are very important in many practical applications of mathematics.

EXERCISE 4.1

Determine whether the equation is satisfied by the given number.

EXAMPLE

$3y + 6 = 4y - 4$, by 10

SOLUTION

Substitute 10 for y and simplify each member.
Does $3(10) + 6 = 4(10) - 4$?
Does $30 + 6 = 40 - 4$?
Does $36 = 36$?
Yes. 10 is a solution.

1. $x - 3 = 7$, by 4
2. $2x - 6 = 3$, by 4
3. $2x + 6 = 3x - 5$, by 3
4. $5x - 1 = 2x + 2$, by 1
5. $3x + 2 = 8 + x$, by 3
6. $3x - 5 = 2x + 7$, by -1
7. $3 = 6x + 3(x - 2)$, by -1
8. $6 - 2x = 6(2x + 1)$, by 0
9. $0 = 6x - 24$, by 4
10. $0 = 7x + 7$, by -1
11. $\frac{1}{4}x - 3 = x + 2$, by 8
12. $\frac{1}{3}x - 2x = -12 + \frac{1}{3}x$, by 6
13. $\frac{2x + 7}{5} + 3 = 0$, by 4
14. $\frac{3(2x + 5)}{9} - x = 0$, by 5
15. $\frac{3}{3 + 1} = 1 - \frac{2}{x + 1}$, by $\frac{1}{2}$
16. $\frac{x - 3}{2} + \frac{5}{4} = 1$, by $\frac{3}{2}$
17. $1 + \frac{1}{1 + \frac{1}{R}} = \frac{7}{4}$, by 3
18. $\frac{1}{\frac{1}{R} + \frac{1}{2R}} = \frac{4}{3}$, by 2
19. $R + \frac{R}{1 + \frac{1}{R}} = \frac{5}{12}$, by $\frac{1}{3}$
20. $R - \frac{1}{R - \frac{1}{R}} = \frac{28}{15}$, by $\frac{2}{3}$

4.2 Conditional equations

In this section, we are concerned with procedures for finding solution sets of first-degree conditional equations. We shall sometimes refer to this process as "solving the equation."

We define **equivalent equations** to be equations that have identical solution sets. Thus

$$2x + 1 = x + 4,$$

$$2x = x + 3,$$

$$x = 3$$

are equivalent equations, because $\{3\}$ is the solution set of each.

Since equations whose solution sets contain at least one member assert that, for some or all values of the variable, the left and right members are names for the same number, and since for each value of the variable a polynomial represents a number, we have the following facts:

1. *The addition or subtraction of the same polynomial to or from each member of an equation produces an equivalent equation.*
2. *The multiplication or division of each member of an equation by the same polynomial representing a nonzero number produces an equivalent equation.*

The application of these properties permits us to transform an equation whose solution set may not be obvious, through a series of equivalent equations until we reach an equation that has an obvious solution. For example, consider the equation

$$2x + 1 = x + 4.$$

We can add $-x - 1$ to each member to obtain the equivalent equations

$$2x + 1 + (-x - 1) = x + 4 + (-x - 1),$$

$$x = 3,$$

where the solution 3 is obvious. Any application of statement 1 or 2 above is called an **elementary transformation.** An elementary transformation *always* results in an equivalent equation. Care must be exercised in the application of the second statement, for we have specifically excluded multiplication or division by zero.

We can always ascertain whether what we think is a solution of an equation is such in reality by substituting the suggested solution in the original equation and verifying that the resulting statement is true. If each of the equations in a sequence is obtained by means of an elementary transformation, the sole purpose for such checking is to detect arithmetic errors. We shall dispense with checking solution sets in the examples that follow except in cases where we apply what may be a nonelementary transformation—that is, where we multiply or divide by an expression containing a variable.

Throughout this section, "equation" has been used in its general sense—that of an open sentence. We are primarily concerned at this time with first-degree equations. Any equation that can be written in the form

$$ax + b = 0 \quad (a \neq 0) \tag{3}$$

through elementary transformation is called a first-degree equation in one variable. We can show that such an equation always has one and only one solution by first multiplying both members by $1/a$ to obtain

$$x + \frac{b}{a} = 0$$

and then noting that, by the negative axiom,

$$x = -\frac{b}{a}.$$

Hence this *one* solution, $-b/a$, constitutes the one member of the solution set of the original first-degree equation in one variable.

EXERCISE 4.2

Solve.

EXAMPLE

$4x + 3 = 2x - 7$

SOLUTION

Add $-2x - 3$ to each member.

$$2x = -10$$

Divide each member by 2.

$$x = -5$$

The solution set is **{−5}**.

1. $x + 2 = 5$
2. $3 = x - 4$
3. $4 - x = x + 2$
4. $2 - y = 4 + 3y$
5. $2x = 3 + x$
6. $3x + 2 = 4x - 6 + x$

7. $6x - 2 = 3x + 10$ 8. $3x - 1 = 2x - 1$ 9. $3x + 3 = x + 3$
10. $3x - 5 = 0$ 11. $0 = 6x + 4$ 12. $3x - 2 = x - 5$

EXAMPLE

$6(x - 2) - (x + 1) = 2$

SOLUTION

Apply distributive law.

$$6x - 12 - x - 1 = 2$$

Combine like terms.

$$5x - 13 = 2$$

Add 13 to each member.

$$5x = 15$$

Divide each member by 5.

$$x = 3$$

The solution set is **{3}**.

13. $2(x + 5) = 16$ 14. $3(x + 1) = 3$ 15. $6 = 2(2x - 1)$
16. $3 = 4(2x + 1)$ 17. $x - (8 - x) = 2$ 18. $2x - (3 - x) = 0$
19. $3(7 - 2x) = 30 - 7(x + 1)$ 20. $6x - 2(2x + 5) = 6(5 + x)$
21. $-3[x - (2x + 3) - 2x] = -9$ 22. $5[2 + 3(x - 2)] + 20 = 0$
23. $-2[x - (x - 1)] = -3(x + 1)$ 24. $3[2x - (x + 2)] = -3(3 - 2x)$

EXAMPLE

$4 - (x - 1)(x + 2) = 8 - x^2$

SOLUTION

Apply distributive law.

$$4 - (x^2 + x - 2) = 8 - x^2$$
$$4 - x^2 - x + 2 = 8 - x^2$$

Add $x^2 - 6$ to each member.

$$-x = 2$$

Divide each member by -1.

$$x = -2$$

The solution set is **{−2}**.

25. $(x - 1)(x - 1) = x^2 - 11$ 26. $(2x + 1)(x - 3) = (x - 2)(2x + 1)$

27. $(x - 2)^2 = x^2 - 8$

28. $(x - 1)^2 = x^2 - 15$

29. $4 - (x - 3)(x + 2) = 10 - x^2$

30. $0 = x^2 + (2 - x)(5 + x)$

EXAMPLE

$\frac{x}{3} + 4 = x - 2$

SOLUTION

Multiply each member by 3.

$$3\left(\frac{x}{3} + 4\right) = 3(x - 2)$$

$$x + 12 = 3x - 6$$

Add $-x + 6$ to each member.

$$18 = 2x$$

Divide each member by 2.

$$9 = x$$

The solution set is **{9}**.

31. $\frac{x}{3} = 2$

32. $\frac{2x}{3} = -4$

33. $\frac{5x - x}{6} = 4$

34. $7 - \frac{x}{3} = x - 1$

35. $6 - \frac{x}{2} = x$

36. $\frac{2x}{3} - 3 = x - 5$

EXAMPLE

$\frac{2x}{3} - x = \frac{5}{2}$

SOLUTION

Determine the L.C.D.

$$\text{L.C.D.} = 3 \cdot 2 = 6$$

Multiply each member of the equation by 6.

$$6\left(\frac{2x}{3}\right) - 6(x) = 6\left(\frac{5}{2}\right)$$

$$4x - 6x = 15$$

Combine like terms.

$$-2x = 15$$

Divide each member by -2.

$$x = -\frac{15}{2}$$

The solution set is $\left\{-\frac{15}{2}\right\}$.

37. $\frac{5x}{2} - 1 = x - \frac{1}{2}$

38. $1 + \frac{x}{9} = \frac{4}{3}$

39. $\frac{x}{5} - \frac{x}{2} = 9$

40. $\frac{y}{2} + \frac{y}{3} - \frac{y}{4} = 7$

41. $\frac{2x - 1}{5} = \frac{x + 1}{2}$

42. $\frac{2x}{3} - \frac{2x + 5}{6} = \frac{1}{2}$

EXAMPLE

$\frac{2}{3} = 6 - \frac{x + 10}{x - 3}$

SOLUTION

Determine the L.C.D.

$$(3)(x - 3)$$

Multiply each member of the equation by $(3)(x - 3)$.

$$(3)(x - 3)\frac{2}{3} = (3)(x - 3)6 - (3)(x - 3)\frac{x + 10}{x - 3}$$

$$2(x - 3) = 18(x - 3) - 3(x + 10)$$

Rewrite without parentheses.

$$2x - 6 = 18x - 54 - 3x - 30$$

Combine like terms and add $-15x + 6$ to each member.

$$-13x = -78$$

Divide each member by -13.

$$x = 6$$

The solution set is $\{6\}$.

43. $\frac{3}{5} = \frac{x}{x + 2}$

44. $\frac{2}{x - 2} = \frac{9}{x + 12}$

45. $\frac{x}{x - 2} = \frac{2}{x - 2} + 1$

46. $\frac{5}{x - 3} = \frac{x + 2}{x - 3} - 1$

47. $\frac{2}{y + 1} + \frac{1}{3y + 3} = \frac{1}{6}$

48. $\frac{y}{y + 2} - \frac{3}{y - 2} = \frac{y^2 + 8}{y^2 - 4}$

Solve each formula for the specified variable.

49. $E = IR$; I, given that $E = 2.22$ and $R = 0.60$.

50. $F = \frac{9}{5}C + 32$; C, given that $F = 176$.

51. $s - \frac{1}{2}gt^2 = 0$; s, given that $g = 32.2$ and $t = 4$.

52. $S = \frac{a}{1 - r}$; r, given that $a = 6$ and $S = 9$.

53. $A = \frac{h}{2}(b + c)$; b, given that $A = 240$, $h = 20$, and $c = 12$.

54. $A = \pi r^2 + 2\pi rh$; h, given that $A = 240$ and $r = 3$. (Use $\pi = 3.14$.)

4.3 Solving equations for specified symbols

An equation containing more than one variable, or containing symbols such as a, b, and c, representing constants, can be solved for one of the symbols in terms of the remaining symbols by using the methods developed in the preceding section. In general, we apply elementary transformations until we obtain the desired symbol by itself as one member of an equation.

EXERCISE 4.3

Solve for x, y, or z. Leave the results in the form of an equation equivalent to the given equation.

EXAMPLE

$5by - 2a = 2ay$

SOLUTION

Add $2a - 2ay$ to each member.

$$5by - 2ay = 2a$$

Factor (or combine like terms).

$$y(5b - 2a) = 2a$$

Divide each member by $5b - 2a$.

$$\mathbf{y = \frac{2a}{5b - 2a}}$$

1. $b = x - c + a$
2. $cz + b = 0$
3. $\frac{a}{b}y + c = 0$
4. $\frac{ax}{b} - c = a$
5. $3y - 3b = y - b$
6. $3x + 6a = x + b$
7. $cx = c - x$
8. $ax + b = x$
9. $a(a - x) = b(b - x)$
10. $4z - 3(z - b) = 8b$
11. $(x - 2)(a + 3) = a$
12. $(y - 4)(b + 3) = 2b$
13. $\frac{1}{x} + \frac{1}{a} = 6$
14. $\frac{b}{2y} - \frac{1}{3} = \frac{b}{3y}$
15. $\frac{1}{a} + \frac{1}{b} = \frac{1}{x}$
16. $\frac{x}{a} - \frac{a}{b} = \frac{b}{c}$
17. $\frac{a-2}{b} + \frac{3}{2b} = \frac{2}{x}$
18. $\frac{2x + 4a}{3a} = \frac{3x + 4a}{2a}$

Solve.

EXAMPLE

$v = k + gt$, for g

SOLUTION

Add $-k$ to each member.

$$v - k = gt$$

Divide each member by t.

$$\frac{v-k}{t} = g$$

$$\mathbf{g = \frac{v-k}{t}}$$

19. $v = k + gt$, for k
20. $E = mc$, for m
21. $f = ma$, for m
22. $E = IR$, for R
23. $pv = K$, for v
24. $W = I^2R$, for R
25. $I = prt$, for p
26. $V = lwh$, for h
27. $s = \frac{1}{2}gt^2$, for g
28. $y = \frac{k}{z}$, for z
29. $v = k + gt$, for t
30. $p = 2l + 2w$, for l
31. $A = \frac{h}{2}(b + c)$, for c

32. $S = 2r(r + h)$, for h

33. $S = 3\pi d + 5\pi D$, for d

34. $A = 2\pi rh + 2\pi r^2$, for h

35. $S = \dfrac{a}{1 - r}$, for r

36. $\dfrac{1}{r} = \dfrac{1}{r_1} + \dfrac{1}{r_2}$, for r

4.4 Inequalities

Open sentences of the form

$$x + 3 \geq 10, \tag{1}$$

$$\frac{-2y - 3}{3} < 5, \tag{2}$$

and so forth, are called **inequalities.** For appropriate values of the variable, one member of an inequality represents a real number that is less than ($<$), less than or equal to ($\leq$), greater than or equal to ($\geq$), or greater than ($>$) the real number represented by the other member.

Any element of the replacement set of the variable for which an inequality is true is called a **solution,** and the set of all solutions to an inquality is called the **solution set** of the inequality.

As with equations, we shall solve a given inequality by generating a series of equivalent inequalities (inequalities having the same solution set) until we arrive at one whose solution set is obvious. To do this we shall need some methods of obtaining equivalent inequalities.

We note that

$$2 < 3,$$

and that

$$2 + 5 < 3 + 5,$$

and

$$2 - 5 < 3 - 5.$$

That is, the addition or subtraction of 5 produces an inequality in the same sense as the original inequality. The line graphs in Figure 4.1 make these statements clearer.

Line graphs a and b demonstrate that either the addition of 5 or the subtraction of 5 from each member of $2 < 3$ simply shifts the members the same number of units to the right or left on the number line, with the order of the members left unchanged. This will be the case for the addition or

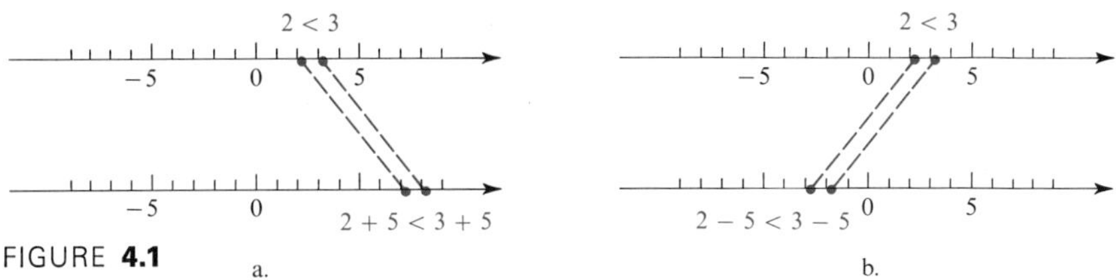

FIGURE **4.1**

subtraction of any real number to or from each member of an inequality. Since for any real value of the variable involved a polynomial represents a real number, we have the following property of inequalities:

1. *The addition or subtraction of the same polynomial to or from each member of an inequality produces an equivalent inequality in the same sense.*

Next, if we multiply each member of

$$2 < 3$$

by 2, we have

$$4 < 6,$$

where the products form an inequality in the same sense. If, however, we multiply each member of

$$2 < 3$$

by -2, we have

$$-4 > -6,$$

where the inequality is in the opposite sense. The line graphs in Figure 4.2 show why this is so. Multiplying each member of $2 < 3$ by 2 simply moves each member out twice as far in a positive direction, as shown in a. Multiplying by -2, however, doubles the absolute value of each member, but the products are negative and hence are reflected through 0, and the sense of the inequality is reversed as shown in b.

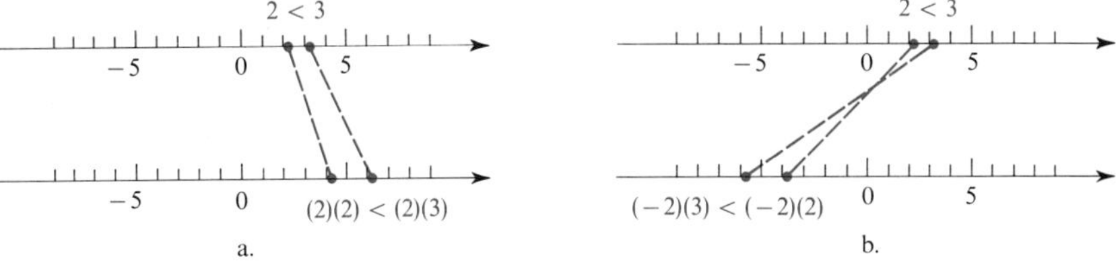

FIGURE **4.2**

A similar consideration in the case of division leads us to assert the following additional properties of inequalities:

2. *If each member of an inequality is multiplied or divided by the same positive number, the result is an equivalent inequality in the same sense.*
3. *If each member of an inequality is multiplied or divided by the same negative number, the result is an equivalent inequality in the opposite sense.*

In properties 2 and 3, the word "number" rather than "polynomial" is used in referring to multipliers and divisors. This is done because we must distinguish between the positive and negative cases, and we cannot, in general, refer to a polynomial as being either positive or negative. This does not, however, deter us from using polynomial multipliers or divisors so long as we are careful to distinguish between those values of any variable or variables involved for which property 2 is applicable and those for which property 3 is applicable—that is, so long as we consider separately those values of the variable(s) for which the polynomial represents a positive number and those for which it represents a negative number. Note that neither property permits multiplying or dividing by zero, and variables in polynomial multipliers and divisors are restricted from values for which the polynomial vanishes. The result of applying any of our three inequality properties is called an elementary transformation.

These properties can be applied to solve first-degree inequalities in the same way the equality statements are applied to solve first-degree equations. As an example, let us find the solution set of

$$\frac{x-3}{4} < 2.$$

By property (2), we can multiply each member by 4 to obtain

$$x - 3 < 8$$

By property (1), we can add 3 to each member, giving us

$$x < 11,$$

where the solution set is obvious. This solution set can be graphed on a line graph, as shown in Figure 4.3, where the heavy line shows points whose coordinates are in the solution set.

FIGURE **4.3**

There is a set notation that is frequently useful in discussing solution sets, particularly solution sets of inequalities. We may write

$$S = \{x \mid x < 11\}$$

for the solution set in the previous example, where the braces indicate that we are discussing a set, and the vertical line is read "such that." This symbolism is read, "S is the set of all x such that x is less than 11."

Inequalities sometimes appear in the form

$$-6 < 3x \leq 15, \tag{3}$$

where an expression is bracketed between two inequality symbols. The solution set to such an inequality is obtained in the same manner as that of any other inequality. In (3) above, each expression may be divided by 3 to obtain

$$-2 < x \leq 5.$$

The solution set,

$$S = \{x \mid -2 < x \leq 5\},$$

is illustrated on a line graph in Figure 4.4.

FIGURE **4.4**

EXERCISE 4.4

Solve each inequality and represent the solution set on a line graph.

EXAMPLE

$$\frac{x+4}{3} \leq 6 + x$$

SOLUTION

Multiply each member by 3.

$$x + 4 \leq 18 + 3x$$

Add $-4 - 3x$ to each member.

$$-2x \leq 14$$

Solution continued overleaf.

Divide each member by -2 and reverse the sense of the inequality.

$$x \geq -7$$

The solution set is $\{x \mid x \geq -7\}$.

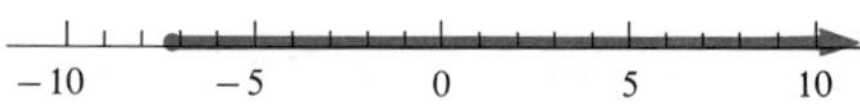

1. $3x < 6$
2. $2x > 8$
3. $x + 7 > 8$
4. $x - 5 \leq 7$
5. $-3x > 12$
6. $-4x < 20$
7. $2x - 3 < 4$
8. $3x + 2 > 8$
9. $3 - 2x > 12$
10. $4 - 3x < 13$
11. $2 - x < -3$
12. $3 - x > -4$
13. $3x - 2 \geq 1 + 2x$
14. $2x + 3 \leq x - 1$
15. $\dfrac{2x - 6}{3} > 0$
16. $\dfrac{2x - 3}{2} \leq 0$
17. $\dfrac{5x - 7x}{3} \geq 0$
18. $\dfrac{x - 3x}{5} \leq 6$
19. $\dfrac{R + 2}{3} < \dfrac{R - 2}{4}$
20. $\dfrac{2R}{5} > R - \dfrac{1}{2}$
21. $\dfrac{T - 1}{3} \geq 2 + \dfrac{T}{2}$
22. $\dfrac{2 - T}{5} \leq -\dfrac{8 + T}{3}$
23. $\dfrac{P - 2P}{4} < \dfrac{2P}{3} + \dfrac{1}{2}$
24. $\dfrac{P}{2} - \dfrac{2P}{3} > \dfrac{P}{4}$

4.5 Word problems

Word problems state relationships between numbers. The problem may be explicitly concerned with numbers, or it may be concerned with numerical measures of physical quantities. In either event, we seek a number or numbers for which the stated relationship holds. To express the quantitative ideas symbolically in the form of an equation is the most difficult part of solving word problems. Although, unfortunately, there is no single means available to do this, the following suggestions are frequently helpful:

1. Determine the quantities asked for and represent them as both symbols and word phrases. Since at this time we are using one variable only, all quantities asked for should be represented in terms of this variable.
2. Where applicable, draw a sketch and label all known quantities thereon; label the unknown quantities in terms of symbols.
3. Find in the problem a quantity that can be represented in two different ways and write this representation as an equation. The equation may derive from:
   a. the problem itself, which may state a relationship explicitly; for example, "What number added to 4 gives 7?" produces the equation $x + 4 = 7$;

b. formulas or relationships that are part of our general mathematical background; for example, $A = \pi r^2$, $d = rt$, etc.

4. Solve the resulting equation.
5. Check the results against the original problem. It is not sufficient to check the result in the equation, because the equation itself may be in error.

The key to working word problems is an ability to understand what you read. If you do not understand the problem, you cannot set up the requisite equation. You should always read a problem through carefully, because a hasty glance often leads to an erroneous interpretation of what is being said.

At this stage of your mathematical development, as much or more attention should be given to setting up equations as is given to their solution. Although some of the problems in this book can be solved without recourse to algebra, the practice in setting up the equations involved will help when more difficult problems are encountered.

EXERCISE 4.5

(a) Set up an equation for each problem.
(b) Solve.

General Problems

EXAMPLE

The first stage of a rocket burns 28 seconds longer than the second stage. If the total burning time for both stages is 152 seconds, how long does each stage burn?

SOLUTION

(a) Represent the unknown quantities using symbols. Let t represent a burning time of the first stage; then $t - 28$ represents the burning time of the second stage. Write an equation relating the unknown quantities.

$$\mathbf{t + (t - 28) = 152}$$

(b) Solve the equation.

$$\begin{aligned} t + t - 28 &= 152 \\ 2t - 28 &= 152 \\ 2t &= 180 \\ t &= 90 \\ t - 28 &= 62 \end{aligned}$$

Therefore, the first stage burns **90 seconds** and the second stage **62 seconds.**

Check: Does $90 + 62 = 152$? Yes.

1. One number is 2 less than three times another. Find the numbers if their sum is 30.
2. If $\frac{3}{5}$ of a number is 8 less than the number, what is the number?
3. A 40-foot rope is cut into two pieces, one of which is 2 feet shorter than six times the length of the other. How long is each piece of rope?
4. In an election, 7179 votes were cast for two candidates. If 6 votes had switched from the winner to the loser, the loser would have won by a single vote. How many votes were cast for each candidate?
5. One side of a long-playing record contains 5 songs, with 15-second intervals between songs. If all of the songs take an equal amount of time to play, and if the entire side takes 26 minutes to play, how many minutes does each song last?
6. In a 30-minute T.V. program, the entertainment portion lasts 2 minutes longer than three times the duration of the commercial portion. How much time is devoted to commercials during the 30 minutes?

Consecutive-Integer Problems

Solving consecutive-integer problems ordinarily involves using the fact that successive integers differ by 1; that is, successive integers can be represented by x, $x + 1$, $x + 2$, ..., etc. Of course, successive even integers, or successive odd integers, are thus represented by x, $x + 2$, $x + 4$, ..., etc.

EXAMPLE

The sum of three consecutive integers is 21 greater than 2 times the smallest of the three integers. Find the integers.

SOLUTION

(a) Represent the integers using symbols.
Let x represent an integer; then, $x + 1$ and $x + 2$ represent the next two consecutive integers. Write an equation representing the conditions given in the problem.

$$\mathbf{(x) + (x + 1) + (x + 2) = 2x + 21}$$

(b) Solve the equation.

$$\begin{aligned} 3x + 3 &= 2x + 21 \\ x &= 18 \\ x + 1 &= 19 \\ x + 2 &= 20 \end{aligned}$$

Therefore, the integers are **18**, **19**, and **20**.

Check: Does $18 + 19 + 20 = 2 \cdot 18 + 21$? Yes.

7. Find three consecutive integers whose sum is 42.
8. The sum of two consecutive integers is 8 less than three times the lesser integer. Find the integers.
9. If $\frac{1}{3}$ of an integer is added to $\frac{1}{2}$ the next consecutive integer, the sum is 33. Find the integers.
10. If $\frac{1}{2}$ of an integer is added to $\frac{1}{5}$ the next consecutive integer, the sum is 17. Find the integers.
11. The sum of three consecutive odd integers is 63. Find the integers.
12. If the lesser of two consecutive even integers is multiplied by -4, the product is 46 more than the sum of the integers. Find the integers.

Age Problems

Problems involving ages are similar to integer problems, because, ordinarily, ages are considered in whole numbers of years. In solving age problems, it is often necessary to express a person's (or thing's) age at various times in terms of his (or its) age now. The table below illustrates a few such cases.

If age today is	then age five years ago was	and age ten years from today will be
x	$x-5$	$x+10$
$y-3$	$y-8$	$y+7$
$z+2$	$z-3$	$z+12$

EXAMPLE

A man is presently nine times as old as his son. In 9 years, he will be only three times as old as his son. How old is each now?

SOLUTION

(a) Let $x =$ son's age now; then $9x =$ father's age now. In 9 years, $x + 9 =$ son's age, and $9x + 9 =$ father's age. Write an equation relating the ages 9 years hence.

$$\mathbf{9x + 9 = 3(x + 9)}$$

(b) Solve the equation.

$$9x + 9 = 3x + 27$$
$$6x = 18$$
$$x = 3$$
$$9x = 27$$

Therefore, today, the son's age is **3 years** and the father's age is **27 years**.

Check: Does $27 = 9 \cdot 3$? Yes.

Does $27 + 9 = 3(3 + 9)$? $36 = 36$? Yes.

13. A boy is three times as old as his sister. In 4 years, he will be twice as old as his sister. How old is each now ?
14. A woman is five times as old as her daughter. In 9 years, the woman will be three times as old as her daughter. How old is each now ?
15. A boy is 4 years older than his sister. Ten years ago, he was twice as old as his sister. How old is each now ?
16. A father is 25 years older than his son. In 10 years, the father will be twice as old as his son. How old is each now ?

Lever Problems

A rigid bar rotating about a fixed point is called a **lever**. The fixed point is called a **fulcrum** of the lever. If a force F_1 is applied to one side of a lever at a distance d_1 from the fulcrum, and a force F_2 is applied on the other side of the lever at a distance d_2 from the fulcrum, then the lever will be in equilibrium if and only if

$$F_1 d_1 = F_2 d_2 .$$

This relationship is called the **Law of the Lever** and has many practical applications.

EXAMPLE

What force must be applied to the end of a 10-foot beam to balance a weight of 6 pounds on the other end, if the beam rests on a fulcrum located 1 1/2 feet from the weight ?

SOLUTION

(a) Let $x = F_2$; force to be applied. In the law of the lever, let $F_1 = 6$, $d_1 = 3/2$, $d_2 = 10 - 3/2 = 17/2$, and write the resulting equation

$$6 \cdot \frac{3}{2} = \frac{17}{2} \cdot x.$$

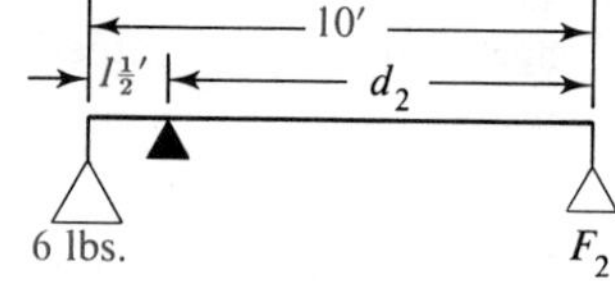

(b) Solve the equation

$$9 = \frac{17}{2} x$$

$$x = \frac{18}{17}$$

Therefore, the applied force must be **1 1/17** lb.

Check: Does $6 \cdot 3/2 = 17/2 \cdot 18/17$? Yes.

17. How far from the fulcrum of a see-saw must a man weighing 180 lbs be seated to balance the see-saw with his 82-lb son who is seated 6 feet from the fulcrum on the other side of the see-saw ?
18. One weight is 10 lbs greater than another. If the lighter weight is placed 10 ft from the fulcrum of a lever and just balances the heavier weight placed 6 ft from the fulcrum, find the two weights.
19. A 24-lb weight is placed 2 feet further from the fulcrum on one side of a lever than a 32-lb weight on the other side. How far is each weight from the fulcrum if the lever is in balance ?
20. A 48-lb weight and a 36-lb weight are placed on opposite ends of a beam, each weight at a distance of 8 ft from the fulcrum. Where should a 24-lb weight be placed to balance the beam ? (*Hint*: $F_1d_1 = F_2d_2 + F_3d_3$.)

Coin Problems

The basic idea of problems involving coins (or bills) is that the value of a number of coins or bills of the same denomination is equal to the product of the value of a single coin (or bill) and the total number of coins (or bills). That is,

$$\begin{bmatrix}\text{Value of}\\ n\\ \text{coins}\end{bmatrix} = \begin{bmatrix}\text{Value of}\\ 1\\ \text{coin}\end{bmatrix} \times \begin{bmatrix}\text{Number}\\ \text{of}\\ \text{coins}\end{bmatrix}.$$

It is helpful in most coin problems to think in terms of cents rather than dollars. Tables can also be helpful in solving coin problems.

EXAMPLE

A collection of coins consisting of dimes and quarters has a value of \$11.60. How many dimes and quarters are in the collection if there are 32 more dimes than quarters?

SOLUTION

(a) Represent the unknown quantities symbolically.
Let x represent a number of quarters; then $x + 32$ represents the number of dimes. Make a table.

Denomination	Value of 1 coin	Number of coins	Value of coins
Dimes	10	$x + 32$	$10(x + 32)$
Quarters	25	x	$25x$

Solution continued overleaf.

Write an equation relating the value of the quarters and value of the dimes to the value of the entire collection.

$$\begin{bmatrix}\text{value of}\\ \text{quarters}\\ \text{in cents}\end{bmatrix} + \begin{bmatrix}\text{value of}\\ \text{dimes}\\ \text{in cents}\end{bmatrix} = \begin{bmatrix}\text{value of}\\ \text{collection}\\ \text{in cents}\end{bmatrix}$$

$$\mathbf{25x \quad + 10(x + 32) \quad = \quad 1160.}$$

(b) Solve for x.

$$\begin{aligned} 25x + 10x + 320 &= 1160 \\ 35x &= 840 \\ x &= 24 \\ x + 32 &= 56 \end{aligned}$$

Therefore there are **24 quarters** and **56 dimes** in the collection.

Check: Do 24 quarters and 56 dimes have a value of \$11.60? Yes.

21. A man has \$1.80 in dimes and nickels with three more dimes than nickels. How many of each coin does he have?
22. A man has \$2.80 in dimes and quarters. If he has 7 more dimes than quarters, how many of each kind of coin does he have?
23. A collection of 16 coins consisting entirely of nickels and quarters has a value of \$2.20. How many of each coin are in the collection?
24. A man has \$446 in ten-dollar, five-dollar, and one-dollar bills. There are 94 bills in all and 10 more five-dollar bills than ten-dollar bills. How many of each kind does he have?
25. A vendor bought a supply of ice cream bars at three for 20 cents. He ate one and sold the remainder at 10 cents each. If he made \$2.00 profit, how many bars did he buy?
26 The admission at a baseball game was \$1.50 for adults and \$.85 for children. The receipts were \$93.10 for 82 paid admissions. How many adults and children attended the game?

Mixture Problems

Another type of problem similar to coin problems is the mixture problems. The key to solving mixture problems is to recognize that the amount of a given substance in a mixture is obtained by multiplying the amount of the mixture by the percent (or rate) of the given substance in the mixture. Thus,

$$\begin{bmatrix}\text{Amount}\\ \text{of}\\ \text{substance}\end{bmatrix} = \begin{bmatrix}\text{Amount}\\ \text{of}\\ \text{mixture}\end{bmatrix} \times \begin{bmatrix}\text{Percent (rate)}\\ \text{of substance}\\ \text{in mixture}\end{bmatrix}.$$

Tables are especially helpful in mixture problems.

EXAMPLE

How many gallons of a 10% solution of acid should be added to 20 gallons of a 60% solution of acid to obtain a 50% solution?

SOLUTION

(a) Represent the unknown quantity symbolically.
Let n represent a number of gallons of 10% solution. Make a table.

Mixture	Part of acid in mixture	Amount of mixture	Amount of acid
10%	0.10	n	$0.10n$
60%	0.60	20	$0.60(20)$
50%	0.50	$n+20$	$0.50(n+20)$

Write an equation relating the amount of *pure* acid before and after combining the solutions.

$$\begin{bmatrix}\text{pure acid in}\\ 10\%\text{ solution}\end{bmatrix} + \begin{bmatrix}\text{pure acid in}\\ 60\%\text{ solution}\end{bmatrix} = \begin{bmatrix}\text{pure acid in}\\ 50\%\text{ solution}\end{bmatrix}$$

$$\mathbf{0.10}\boldsymbol{n} \quad + \quad \mathbf{0.60(20)} \quad = \quad \mathbf{0.50\ (}\boldsymbol{n}\mathbf{+20)}$$

(b) Solve for n. First multiply each member by 100.

$$\begin{aligned} 10n + 60(20) &= 50(n+20) \\ 10n + 1200 &= 50n + 1000 \\ -40n &= -200 \\ n &= 5 \end{aligned}$$

Therefore **5 gallons** of the 10% solution are needed.

Check: Does 5 gallons of a 10% solution when added to 20 gallons of a 60% solution form a 50% solution? Does $0.10(5) + 0.60(20) = 0.50(25)$? Does $0.5 + 12.0 = 12.5$? Yes.

27. How many gallons of a 30% salt solution must be added to 40 gallons of 12% salt solution to obtain a 20% solution?
28. How many pounds of an alloy containing 45% silver must be melted with an alloy containing 60% silver to obtain 40 pounds of an alloy containing 48% silver?
29. How much pure alcohol should be added to 12 ounces of a 45% solution to obtain a 60% solution?
30. How much water should be added to one gallon of pure acid to obtain a 15% solution?

31. A druggist wishes to make 30 pounds of a mixed candy to sell at 63 cents per pound. How many pounds of candy worth 85 cents per pound should be mixed with candy worth 55 cents per pound to obtain the desired mixture?
32. A grocer wishes to stock 12 pounds of nuts to sell at 72 cents per pound. How many pounds each of peanuts, worth 60 cents per pound, and pecans, worth 96 cents per pound, must be mixed to obtain the desired stock?

Interest Problems

Simple interest problems involve the fact that the amount of interest earned during a single year is equal to the amount invested times the rate of interest. Thus, $I = P \times r$.

EXAMPLE

A man has an annual income of $12,000 from two investments. He has $10,000 more invested at 8% than he has invested at 6%. How much does he have invested at each rate?

SOLUTION

Represent the amount invested at each rate symbolically.
Let A represent an amount invested at 6%; then $A + 10{,}000$ represents an amount invested at 8%. Make a table.

Investment	Rate	Amount	Interest
6%	0.06	A	$0.06A$
8%	0.08	$A + 10{,}000$	$0.08(A + 10{,}000)$

(a) Write an equation relating the interest from each investment and the total interest received.

$$\begin{bmatrix}\text{interest from}\\ 6\%\text{ investment}\end{bmatrix} + \begin{bmatrix}\text{interest from}\\ 8\%\text{ investment}\end{bmatrix} = [\text{total interest}].$$

$$\mathbf{0.06A \quad + 0.08(A + 10{,}000) \ = \ 12{,}000}$$

(b) Solve for A. First multiply each member by 100.

$$\begin{aligned} 6A + 8(A + 10{,}000) &= 1{,}200{,}000 \\ 6A + 8A + 80{,}000 &= 1{,}200{,}000 \\ 14A &= 1{,}120{,}000 \\ A &= 80{,}000 \\ A + 10{,}000 &= 90{,}000 \end{aligned}$$

Therefore **$80,000** is invested at 6% and **$90,000** is invested at 8%.

Check. Does $0.06(80{,}000) + 0.08(90{,}000) = 12{,}000$? Yes.

33. A sum of \$2000 is invested, part at 6% and the remainder at 8%. Find the amount invested at each rate if the yearly income from the two investments is \$132.
34. A sum of \$2200 is invested, part at 5% and the remainder at 6%. Find the yearly interest on both investments if the interest on each investment is the same.
35. A man invests equal amounts of money at 5% and 7%. How much has he invested at each rate if his annual return is \$144?
36. A man has three times as much money invested in 8% bonds as he has in stocks paying 5%. How much does he have invested in each if his yearly income from the investments is \$1740?
37. A man has \$1000 more invested at 5% than he has invested at 8%. If his annual income from the two investments is \$739, how much does he have invested at each rate?
38. A man has \$500 more invested at 6% than he does at 8%. If his annual return on his investment is \$114, how much does he have invested at each rate?

Uniform-Motion Problems

Solving uniform-motion problems requires applying the fact that the distance traveled at a uniform rate is equal to the product of the rate and the time traveled. This relationship may be expressed by any one of the three equations

$$d = rt, \quad r = \frac{d}{t}, \quad \text{and} \quad t = \frac{d}{r}.$$

It is frequently helpful to use a table such as that shown in the following example.

EXAMPLE

An express train travels 150 miles in the same time that a freight train travels 100 miles. If the express goes 20 miles per hour faster than the freight, find the rate of each.

SOLUTION

(a) Represent the unknown rates in symbols.
Let r represent a rate for the freight train; then $r + 20$ represents the rate for the express train. Make a table.

	d	r	t
Freight	100	r	$\frac{100}{r}$
Express	150	$r + 20$	$\frac{150}{r + 20}$

Solution continued overleaf.

The fact that the times are equal is the significant equality in the problem.

$$[t \text{ freight}] = [t \text{ express}]$$

$$\frac{\mathbf{100}}{r} = \frac{\mathbf{150}}{r + \mathbf{20}}$$

(b) Solve the equation.

$$\begin{aligned} r(r+20)\frac{100}{r} &= r(r+20)\frac{150}{r+20} \\ (r+20)100 &= (r)150 \\ 100r + 2000 &= 150r \\ -50r &= -2000 \\ r &= 40 \\ r + 20 &= 60 \end{aligned}$$

Thus, the freight train's rate is **40 mph** and the express train's rate is **60 mph.**

Check: The time of the freight ($100/40$) equals the time of the express train ($150/60$).

39. An airplane travels 1260 miles in the same time that an automobile travels 420 miles. If the rate of the airplane is 120 miles per hour greater than the rate of the automobile, find the rate of each.
40. Two cars start together and travel in the same direction, one going twice as fast as the other. At the end of three hours they are 96 miles apart. How fast is each traveling?
41. A freight train leaves town A for town B, traveling at an average rate of 40 miles per hour. Three hours later a passenger train also leaves town A for town B, traveling at an average rate of 80 miles per hour. How far from town A does the passenger train pass the freight train?
42. Two planes leave an airport at the same time and travel in opposite directions. If one plane averages 440 miles per hour over the ground and the other 520 miles per hour, in how long will they be 2880 miles apart?
43. A car leaves a town at the rate of 42 miles per hour. Two hours later a second car follows at the rate of 54 miles per hour. How far from the town will they pass each other?
44. A boat sails due west from a harbor at 36 knots. An hour later, another boat leaves the harbor on the same course at 45 knots. How far out at sea will the second boat overtake the first?

Work Problems

Some problems involve the accomplishment of a task in some fixed time when a steady rate of work is assumed. For example, if it takes four hours for a pipe to empty a tank, then in one hour the pipe empties $1/4$ of the tank; or if it takes a man three days to paint a house, then in one day he paints $1/3$ of the house.

Problems involving tasks of this kind can be represented by equations in which the right-hand member is 1, which denotes completion of the total task.

EXAMPLE

One pipe can empty a tank in six hours and another can empty the same tank in nine hours. How long will it take both pipes to empty the tank?

SOLUTION

(a) Represent the unknown time in symbols.
Let x represent time for both pipes to empty the tank. Write an equation expressing the conditions of the problem.

$$\begin{bmatrix}\text{portion of tank}\\ \text{emptied by pipe 1}\end{bmatrix} + \begin{bmatrix}\text{portion of tank}\\ \text{emptied by pipe 2}\end{bmatrix} = \begin{bmatrix}\text{portion of tank}\\ \text{emptied by both}\end{bmatrix}$$

$$\left(\frac{1}{6}\right)x \quad + \quad \left(\frac{1}{9}\right)x \quad = \quad 1.$$

(b) Solve the equation.

$$(18)\left(\frac{1}{6}\right)x + (18)\left(\frac{1}{9}\right)x = (18)1$$

$$3x + 2x = 18$$

$$5x = 18$$

$$x = \frac{18}{5}$$

The tank will be emptied in **$3^3/_5$ hours** when both pipes are open together.

Check: Does $(1/6)(18/5) + (1/9)(18/5) = 1$? Does $3/5 + 2/5 = 1$? Yes.

45. A man can paint a house in five days working alone, and his son can do the same job in 15 days working alone. How long will it take them to paint the house if they work together?

46. It takes one pipe 30 hours to fill a tank, while a second pipe can fill the same tank in 45 hours. How long will it take both pipes running together to fill the tank?

47. One typist can complete a job in four hours, while two other typists, each working alone, require six hours and 12 hours, respectively, to do the same job. How long will it take all three typists to complete the job if they work together?

48. One pipe can fill a tank in four hours and another can empty it in six hours. If both pipes are opened, how long will it take to fill the empty tank?

49. A printing press can complete a run on a job in 15 hours. When the press and a second press work together, the same job can be run in six hours. How long would it take the second press to make the run alone?
50. A farmer can plow a field in eight hours, but when he and his son work together on the job, they can complete it in three hours. How long would it take the son to plow the field alone?

Inequalities

Sentences describing inequalities are set up in the same way as those describing equalities, except that symbols such as $\leq$ or $>$ replace the symbol $=$.

EXAMPLE

A student must have an average of 80% to 90% on five tests in a course to receive a B. His grades on the first four tests were 98%, 76%, 86%, and 92%. What grade on the fifth test would give him a B in the course?

SOLUTION

(a) Represent the unknown quantity symbolically.
Let x represent a grade (in percent) on the fifth test. Write an inequality expressing the word sentence.

$$80 \leq \frac{98+76+86+92+x}{5} < 90$$

(b) Solve.

$$400 \leq 352 + x < 450$$
$$48 \leq x < 98$$
$$\{x \mid 48 \leq x < 98\}$$

Therefore, **any grade equal to or greater than 48 and less than 98** would give the student a B.

51. In the preceding example, what grade on the fifth test would give the student a B if his grades on the first four tests were 78%, 64%, 88%, and 76%?
52. The Fahrenheit and centigrade temperatures are related by the formula $C = \frac{5}{9}(F - 32)$. Within what range must the temperature be in Fahrenheit degrees for the temperature in centigrade degrees to lie between $-9°$ and $27°$?
53. A man wishes to invest $10,000, part at 5% and part at 7%. What is the least amount he can invest at 7% if he wishes a minimum annual return of $616?
54. A man can sail upstream in a river at an average rate of four miles per hour, and downstream at an average rate of six miles per hour. What is

the greatest length of time he can sail upstream if he starts at 8:00 a.m. and must be back no later than 6:00 p.m.?

55. A chemist wishes to produce 20 ounces of a solution containing between $7\frac{1}{2}\%$ and 8% of a certain chemical. What are the greatest and least amounts of a 10% solution of the chemical he can mix with a 5% solution of the chemical to produce the desired solution?

56. A man has \$5000 invested at 4%. What is the least amount of money he can invest at 6% in order to receive at least 5% return on his total investment?

Chapter review

[4.1] 1. Is -3 a solution of $\frac{2x+1}{5} - x = 4$?

2. Is $\frac{1}{2}$ a solution of $\frac{1}{R} + \frac{1}{2R} = 3$?

[4.2] Solve each equation.

3. $2(x-3) = 12$

4. $3[x - 2(x+1)] = 4 + x$

5. $\frac{x}{4} - 2 = \frac{x+1}{3}$

6. $\frac{2}{x-1} + \frac{1}{3(x-1)} = \frac{7}{2}$

7. $\frac{2T+6}{3} = \frac{1}{2} + T$

8. $\frac{1}{r} + \frac{1}{2r} = \frac{3}{2}$

[4.3] Solve each equation for the specified variable.

9. $I = prt$, for r

10. $r = k + gt$, for g

11. $S = 3\pi d + 5\pi D$, for D

12. $\frac{1}{r} = \frac{1}{r_1} + \frac{1}{r_2}$, for r

[4.4] Solve each inequality and represent the solution set on a line graph.

13. $\frac{3x-2}{2} \leq 5$

14. $\frac{2}{5} + T > \frac{T}{2}$

[4.5] 15. A man has \$20,000 to invest. He invests \$5000 at 4% and \$10,000 at 3%. At what rate should he invest the remainder in order to have a yearly income of \$800?

16. A man has \$1.60 in dimes and nickels, with two more nickels than dimes. How many of each coin does he have?

17. One weight is 16 lbs greater than another. When the heavier weight is placed 4 ft from the fulcrum of a lever and the other weight is placed 12 ft from the fulcrum, the lever is balanced. Find each weight.

18. A balloon, using sand bags for ballast, needs 25 bags of a certain weight. If each bag is increased in weight by five pounds, five fewer bags will be needed to produce the same ballast. How heavy is each of the original 25 bags?

19. One freight train travels six miles an hour faster than a second freight train. The first freight travels 280 miles in the same time the second freight travels 210 miles. Find the rate of each.

20. Clerk A can process 50 applications in four hours, and clerk B can process 50 applications in eight hours. How long will it take both clerks working together to process 50 applications?

5

Exponents

5.1 Laws for positive-integral exponents

In Chapter 2 the expression a^n, where a is any real number and n is a positive integer, was defined by

$$a^n = a \cdot a \cdot a \cdot a \cdot \cdots \cdot a \text{ (}n\text{ factors).}$$

The following laws were developed from this definition:

I. $a^m \cdot a^n = a^{m+n}$.

II. $\dfrac{a^m}{a^n} = a^{m-n} \quad (m > n, \quad a \neq 0)$,

The following four laws are also very useful in simplifying expressions involving positive-integral exponents.

IIa. $\dfrac{a^m}{a^n} = \dfrac{1}{a^{n-m}} \quad (m < n, \quad a \neq 0)$.

This follows from the fact that

$$\frac{a^m}{a^n} = \frac{a \cdot a \cdot \cdots \cdot a \quad (m \text{ factors})}{a \cdot a \cdot a \cdot \cdots \cdot a \quad (n \text{ factors})}$$
$$= \frac{a \cdot a \cdot \cdots \cdot a \quad (m \text{ factors})}{a \cdot a \cdot \cdots \cdot a \quad (m \text{ factors})} \cdot \frac{1}{[a \cdot a \cdot \cdots \cdot a \quad (n-m) \text{ factors}]}$$
$$= \frac{1}{a^{n-m}}.$$

For example,

$$\frac{x^5}{x^8} = \frac{1}{x^3} \quad \text{and} \quad \frac{x^2y}{x^3y^2} = \frac{1}{xy} \quad (x, y \neq 0).$$

III. $\boldsymbol{(a^m)^n = a^{mn}}$.

We have

$$(a^m)^n = a^m \cdot a^m \cdot a^m \cdot \cdots \cdot a^m \quad (n \text{ factors})$$
$$= a^{m+m+m+\cdots+m \ (n \text{ terms})}$$
$$= a^{mn}.$$

For example,

$$(x^2)^3 = x^6 \quad \text{and} \quad (x^5)^2 = x^{10}.$$

IV. $\boldsymbol{(ab)^n = a^nb^n}$.

This follows from the fact that

$$(ab)^n = (ab)(ab)(ab) \cdot \cdots \cdot (ab) \quad (n \text{ factors})$$
$$= [a \cdot a \cdot a \cdot \cdots \cdot a \quad (n \text{ factors})][b \cdot b \cdot b \cdot \cdots \cdot b \quad (n \text{ factors})]$$
$$= a^nb^n.$$

For example,

$$(xy)^3 = x^3y^3 \quad \text{and} \quad (2x^2y^3)^3 = 8x^6y^9.$$

V. $\boldsymbol{\left(\frac{a}{b}\right)^n = \frac{a^n}{b^n} \quad (b \neq 0)}$.

This is so because

$$\left(\frac{a}{b}\right)^n = \left(\frac{a}{b}\right) \cdot \left(\frac{a}{b}\right) \cdot \left(\frac{a}{b}\right) \cdot \cdots \cdot \left(\frac{a}{b}\right) \quad (n \text{ factors})$$
$$= \frac{a \cdot a \cdot a \cdot \cdots \cdot a \quad (n \text{ factors})}{b \cdot b \cdot b \cdot \cdots \cdot b \quad (n \text{ factors})}$$
$$= \frac{a^n}{b^n}.$$

For example,

$$\left(\frac{x}{y}\right)^3 = \frac{x^3}{y^3} \quad \text{and} \quad \left(\frac{2x^2}{y^3}\right)^4 = \frac{(2x^2)^4}{(y^3)^4} \quad (y \neq 0).$$

In order to avoid the necessity of constantly noting exceptions, we shall assume that in the exercises in this chapter the variables are restricted so that no denominator vanishes.

EXERCISE 5.1

Simplify.

EXAMPLES

a. $\dfrac{x^3y^2}{xy^5}$ b. $\left(\dfrac{2x^3}{y}\right)^2$ c. $\dfrac{(-xy^2)^3}{(-x^2y)^2}$

SOLUTIONS

a. $\mathbf{\dfrac{x^2}{y^3}}$

b. $\dfrac{2^2 \cdot x^6}{y^2}$

$\mathbf{\dfrac{4x^6}{y^2}}$

c. $\dfrac{-x^3y^6}{x^4y^2}$

$\mathbf{\dfrac{-y^4}{x}}$

1. $x^2 \cdot x^3$
2. $a^5 \cdot a$
3. $y^3 \cdot y^4$
4. $b^5 \cdot b^5$
5. $\dfrac{x^5}{x^2}$
6. $\dfrac{y^3}{y^5}$
7. $\dfrac{x^3y^5}{xy^2}$
8. $\dfrac{a^2y^2}{ay}$
9. $(x^2)^4$
10. $(y^3)^2$
11. $(-x^2)^4$
12. $(-x^4)^3$
13. $(x^2y^3)^3$
14. $(rst^2)^4$
15. $(x^3y^2z)^2$
16. $(a^2b^2c^2)^{10}$
17. $\left(\dfrac{x^2}{y^3}\right)^3$
18. $\left(\dfrac{a^3}{b^3}\right)^2$
19. $\left(\dfrac{3y}{x^2}\right)^4$
20. $\left(\dfrac{4x}{y}\right)^3$
21. $-\left(\dfrac{2x}{5y^2}\right)^3$
22. $-\left(\dfrac{3a}{7b^2}\right)^2$
23. $(3e^2)^2(2e)^3$
24. $(4mv^2)^3(2v^4)^2$
25. $(\pi R^2)^2(2\pi R)^2$
26. $(n^2c^2)^3(-nc)^3$
27. $\dfrac{(2H^2)^3}{(2H)^2}$
28. $\dfrac{(6PV^2)^2}{(3P^2V)^3}$
29. $\dfrac{(KT^2)^3}{(K^2T)^2}$
30. $\dfrac{(KP^2L)^2}{(PL^2)^3}$

5.2 Positive rational exponents

If the laws developed in the preceding section are to hold for rational exponents, meanings consistent with these laws must be assigned to powers with rational exponents. Let us examine exponents that are the reciprocals of natural numbers, that is, exponents of the form $1/n$, where n is a natural number. We first assume that there exists a number $a^{1/n}$, where a is positive. If the third law of exponents is to hold for these new numbers, we have

$$(a^{1/n})^n = a^{n/n} = a,$$

and $a^{1/n}$ *must be defined as one of n equal factors of a*. The number $a^{1/n}$ is called an **nth root of** a. When n is 2 or 3, $a^{1/n}$ is called a **square root** or a **cube root** of a, respectively. Any positive number has both a positive and a negative nth root when n is an even number; however, we shall refer only to the positive nth root when using $a^{1/n}$ and $a > 0$. If we wish to refer to the negative nth root, we shall write $-a^{1/n}$. Thus, the positive square root of 81 can be written as $81^{1/2}$, or 9, and the negative square root can be written as $-81^{1/2}$, or -9.

If a is negative and n is a natural number, there are two possibilities to be considered for $a^{1/n}$.

1. If n is even, we can have no real root, for an even power of either a positive or a negative real number is always positive.
2. If n is odd, there will be one negative root, since an odd power of a negative real number is negative.

Thus,

$$(-8)^{1/3} = -2, \quad \text{since} \quad (-2)(-2)(-2) = -8.$$

However,

$$(-4)^{1/2} \neq 2, \quad \text{since} \quad (2)(2) \neq -4;$$

and

$$(-4)^{1/2} \neq -2, \quad \text{since} \quad (-2)(-2) \neq -4.$$

There is no square root of -4 in the real numbers. Note the difference in meaning between such symbols as $(-4)^{1/2}$ and $-4^{1/2}$. The symbol $(-4)^{1/2}$ is undefined in the set of real numbers. On the other hand, $-4^{1/2}$ is a real number, the negative square root of 4.

We define the general rational exponent to obey the third law of exponents. Thus, for non-negative real numbers a, the number $a^{m/n}$ is defined by

$$a^{m/n} = (a^{1/n})^m.$$

It can be shown that for $a \geq 0$, $(a^m)^{1/n}$ is also equal to $a^{m/n}$, so that

$$a^{m/n} = (a^{1/n})^m = (a^m)^{1/n},$$

and we can look at $a^{m/n}$ in two ways, either as the mth power of the nth root of a, or as the nth root of the mth power of a. For example,

$$8^{2/3} = (8^{1/3})^2 = (2)^2 = 4$$

or

$$8^{2/3} = (8^2)^{1/3} = (64)^{1/3} = 4.$$

Hereafter, we shall use whichever viewpoint is most convenient for the purpose at hand.

Because we defined $a^{1/n}$ to be the positive nth root of a for a positive and n an even natural number, and a^m to be negative for a negative and m an odd natural number, we must also agree that, for m and n even natural numbers and a any real number,

$$(a^m)^{1/n} = |a|^{m/n}.$$

For the special case of m and n *even* and $m = n$,

$$(a^n)^{1/n} = |a|$$

for any real number a. For example,

$$(a^2)^{1/2} = |a|.$$

If the absolute-value notation were not used, we would, for example, have the inconsistent result

$$[(-4)^2]^{1/2} = [(-4)^2]^{(1/2)},$$

$$(16)^{1/2} = (-4)^1,$$

and

$$4 = -4.$$

We have defined powers with rational exponents so that they obey the third law for positive integral exponents. However, they can also be shown to obey the other laws pertaining to positive integral exponents, so that Laws I–V, on pages 105 and 106, are also valid for all positive rational exponents.

Recall that any number that can be expressed as the quotient of two integers is called a rational number. Any real number that cannot be so expressed is called an **irrational number**; any expression such as $a^{1/n}$ represents a rational number if and only if a is the nth power of a rational number. Thus $4^{1/2}$, $(-27)^{1/3}$, and $(81)^{1/4}$ are rational, while $2^{1/2}$, $5^{1/3}$, and $7^{1/4}$ are irrational.

EXERCISE 5.2

Throughout these problems, assume that all bases are positive unless otherwise specified.

Simplify.

EXAMPLES

a. $64^{1/2}$ b. $8^{2/3}$ c. $(-27)^{4/3}$

SOLUTIONS

a. **8**

b. $(8^{1/3})^2$
2^2
4

c. $[(-27)^{1/3}]^4$
$(-3)^4$
81

1. $9^{1/2}$
2. $(25)^{1/2}$
3. $(32)^{1/5}$
4. $(27)^{1/3}$
5. $(-27)^{1/3}$
6. $(64)^{1/3}$
7. $(27)^{2/3}$
8. $(32)^{3/5}$
9. $(81)^{3/4}$
10. $(125)^{2/3}$
11. $(-8)^{4/3}$
12. $(-64)^{2/3}$
13. $\left(\frac{1}{8}\right)^{5/3}$
14. $\left(\frac{1}{16}\right)^{3/4}$
15. $\left(\frac{4}{9}\right)^{3/2}$
16. $\left(\frac{25}{16}\right)^{3/2}$

Write each of the following expressions as a product or quotient of powers in which each variable occurs but once and all exponents are positive.

EXAMPLES

a. $x^{1/4} \cdot x^{1/2}$ b. $\dfrac{x^{5/6}}{x^{2/3}}$ c. $\dfrac{(x^{1/2}y^2)^2}{(x^{2/3} \cdot y)^3}$

SOLUTIONS

a. $\mathbf{x^{3/4}}$

b. $\dfrac{x^{5/6}}{x^{4/6}}$
$\mathbf{x^{1/6}}$

c. $\dfrac{xy^4}{x^2y^3}$
$\mathbf{\dfrac{y}{x}}$

17. $x^{1/3} \cdot x^{1/3}$
18. $y^{1/2} \cdot y^{3/2}$
19. $\dfrac{x^{2/3}}{x^{1/3}}$
20. $\dfrac{x^{3/4}}{x^{1/4}}$
21. $(n^{1/2})^3$
22. $(x^6)^{2/3}$
23. $(m^4)^{1/8}$
24. $(x^{1/2}y^{1/3})^6$
25. $\left(\dfrac{a^4}{b}\right)^{1/4}$
26. $\left(\dfrac{y^4}{x^2}\right)^{1/2}$
27. $\left(\dfrac{a^6}{c^3}\right)^{2/3}$
28. $\left(\dfrac{x^{1/2}}{y^2}\right)^2 \left(\dfrac{y^4}{x^2}\right)^{1/4}$

29. $(xy^{1/2})(x^{3/4}y^{1/4})$ 30. $(x^{1/3}y^{1/4})(x^{5/6}y)$ 31. $\dfrac{x^{2/3}y}{x^{1/6}y^{1/3}}$

32. $\dfrac{xy^{1/2}}{x^{1/2}y^{1/4}}$ 33. $\dfrac{(x^{1/3}y^{1/4})^2}{x^{1/4}y^{1/3}}$ 34. $\left(\dfrac{x^{3/4}y^{1/3}}{xy^{1/4}}\right)^2$

35. $(PV^2)^2(PV^{1/2})$ 36. $(R^{1/2}t^2)^4(Rt^2)$ 37. $\left(\dfrac{N^2Q}{t}\right)^{1/2}\left(\dfrac{Nt}{Q}\right)^{1/2}$

38. $\left(\dfrac{L^2R^3}{8t}\right)^{1/3}\left(\dfrac{t^{1/3}}{L^{1/3}R}\right)$ 39. $\left[\left(\dfrac{9V}{4r^{1/2}}\right)^{1/2}\right]^4$ 40. $\left[\left(\dfrac{kP^2}{m^3}\right)^{1/6}\right]^2$

5.3 Zero and negative rational exponents

In order for the first law of exponents to hold for zero as an exponent—that is, for

$$a^0a^n = a^{0+n} = a^n,$$

it is clear that a^0 must be defined as 1. This also follows if the second law of exponents is to hold for the case $m = n$. Thus, because

$$\frac{a^n}{a^n} = a^{n-n} = a^0 \quad (a \neq 0)$$

and

$$\frac{a^n}{a^n} = 1 \quad (a \neq 0),$$

it follows that a logical definition of a^0 is

$$\mathbf{a^0 = 1} \quad (a \neq 0).$$

In accord with this definition, we have, for example,

$$3^0 = 1, \quad \left(\frac{1}{2}\right)^0 = 1, \quad (-4)^0 = 1, \quad \text{and} \quad x^0 = 1 \quad (x \neq 0).$$

If the first law of exponents is to hold for negative exponents, we have

$$a^n \cdot a^{-n} = a^{n-n} = a^0 = 1.$$

Since

$$a^n \cdot \frac{1}{a^n} = 1 \quad (a \neq 0),$$

it is evident that a^{-n} must be defined by

$$\boldsymbol{a^{-n} = \frac{1}{a^n}} \quad (a \neq 0).$$

It follows from this definition that

$$\frac{1}{a^{-n}} = \frac{1}{\frac{1}{a^n}} = a^n \quad (a \neq 0).$$

For example,

$$3^{-2} = \frac{1}{3^2} = \frac{1}{9},$$

$$\frac{1}{2^{-3}} = 2^3,$$

$$\frac{x^{-2}}{y^{-1}} = \frac{\frac{1}{x^2}}{\frac{1}{y}} = \frac{y}{x^2} \quad (x, y \neq 0).$$

It can be shown that zero and negative exponents have now been defined so that their properties are consistent with all of the basic properties of positive rational exponents, and therefore these laws can be applied in the case of *all* rational exponents.

EXERCISE 5.3

Simplify.

EXAMPLES

a. $\dfrac{3^{-2}}{2^{-3}}$ b. $\dfrac{16^{1/2}}{8^{-2/3}}$ c. $\dfrac{3^{-3}(4)^2}{2^{-1}}$

SOLUTIONS

a. $\dfrac{\frac{1}{3^2}}{\frac{1}{2^3}}$ b. $\dfrac{16^{1/2}}{\frac{1}{8^{2/3}}}$ c. $\dfrac{\frac{1}{3^3}(4)^2}{\frac{1}{2^1}}$

$\frac{2^3}{3^2}$ $\qquad$ $(16)^{1/2}(8)^{2/3}$ $\qquad$ $\frac{\frac{16}{27}}{\frac{1}{2}}$

$\mathbf{\frac{8}{9}}$ $\qquad$ $4(4)$

$\mathbf{16}$ $\qquad$ $\mathbf{\frac{32}{27}}$

1. 2^{-1}
2. $\frac{1}{3^{-1}}$
3. $\left(\frac{3}{5}\right)^{-1}$
4. $\left(\frac{1}{3}\right)^{-2}$
5. $\frac{2^0}{3^{-2}}$
6. $\frac{5^{-1}}{3^{-2}}$
7. $(-8)^{-1/3}$
8. $16^{-1/4}$
9. $\frac{3^{-2}(5)^2}{2^{-3}}$
10. $\frac{5^{-1}(2)^{-2}}{3^{-2}}$
11. $\frac{16^{-1/2}(16)^{1/2}}{16^{1/4}}$
12. $\frac{8^{-2/3}(4)^{-1/2}}{16^{-1/4}}$

Perform the operations indicated and express each power with a positive exponent. Assume that all variables represent positive real numbers only.

EXAMPLES

a. $x^{-3} \cdot x^5$ $\qquad$ b. $(x^2y^{-3})^{-1}$ $\qquad$ c. $\left(\frac{x^{-1}y^2z^0}{x^3y^{-4}z^2}\right)^{-1}$

SOLUTIONS

a. x^{-3+5} $\qquad$ b. $x^{-2}y^3$ $\qquad$ c. $\frac{xy^{-2}z^0}{x^{-3}y^4z^{-2}}$

$\mathbf{x^2}$ $\qquad$ $\mathbf{\frac{y^3}{x^2}}$ $\qquad$ $\mathbf{\frac{x^4z^2}{y^6}}$

13. $x^{-3} \cdot x^4$
14. $x^2 \cdot x^{-3}$
15. $\frac{x^5}{x^{-2}}$
16. $\frac{x^{-2}}{x^3}$
17. $(x^{-3})^2$
18. $(x^2)^{-4}$
19. $(x^{1/2})^{-3}$
20. $(x^3)^{-1/3}$
21. $(m^2p^{-1})^{-1/2}$
22. $(k^2V^{-2}t)^{-1/6}$
23. $\frac{R^{-2}}{r^{-3}}$
24. $(n^0t^2)^0$

25. $\dfrac{x^{-1}}{y^{-1}}$ 26. $\dfrac{x^{-2}}{y^{-3}}$ 27. $\dfrac{a^2x^{-3}}{b^2y^{-2}}$

28. $\dfrac{a^{-2}b^{-2}c}{ab^{-3}c^0}$ 29. $\dfrac{8^{-1}x^0y^{-3}}{(2xy)^{-5}}$ 30. $\dfrac{16^{-1/4}r^2s^{-3}}{(2)^{-2}rs^{-2}}$

31. $\left(\dfrac{x^{-1}y^3}{2x^0y^{-5}}\right)^{-2}$ 32. $\left(\dfrac{a^{-1}b^{-2}}{3^0ab}\right)^{-1}$

5.4 Expressions containing more than one term

In applying the laws of exponents to simplify expressions containing more than one term, one must be careful to distinguish between sums and products. While $(x^2y^3)^{-1} = x^{-2}y^{-3}$, it is not true that $(x^2 + y^3)^{-1}$ is equal to $x^{-2} + y^{-3}$.

EXERCISE 5.4

Represent each expression as a single fraction involving positive exponents only.

EXAMPLES

a. $(a + b)^{-3}$ b. $(x + y)^3(r - s)^{-1}$

SOLUTIONS

a. $\mathbf{\dfrac{1}{(a + b)^3}}$

b. $(x + y)^3\left(\dfrac{1}{r - s}\right)$

$\mathbf{\dfrac{(x + y)^3}{r - s}}$

1. $(x - 2)^{-2}$
2. $(x - 3y)^{-1}$
3. $(2x + y)^{-1}(x - y)^2$
4. $(x + y)^{-2}(x + 2y)^{-3}$
5. $y(y - 2)^{-1/2}(y + 1)^2$
6. $y^2(y + 3)^{-2/3}(y + 1)^{-2}$

EXAMPLES

a. $4^{1/2} + 4^{-1/2}$ b. $x^{-1} + y^{-2}$ c. $(x^{-1} + x^{-2})^{-1}$

SOLUTIONS

a. $4^{1/2} + \frac{1}{4^{1/2}}$

$2 + \frac{1}{2}$

$\mathbf{\frac{5}{2}}$

b. $\frac{1}{x} + \frac{1}{y^2}$

$\frac{(y^2)1}{(y^2)x} + \frac{1(x)}{y^2(x)}$

$\mathbf{\frac{y^2 + x}{xy^2}}$

c. $\left(\frac{1}{x} + \frac{1}{x^2}\right)^{-1}$

$\left(\frac{(x)1}{(x)x} + \frac{1}{x^2}\right)^{-1}$

$\left(\frac{x+1}{x^2}\right)^{-1}$

$\mathbf{\frac{x^2}{x+1}}$

7. $2^{-1} + 2^{-2}$
8. $2^{-2} + 4^{-1}$
9. $2^3 + 8^{-3}$
10. $16^{-1/2} + 8^{-1/3}$
11. $2(2)^{-2} - 4^{-1/2}$
12. $8^{-2/3} - 2^2(3^{-1})$
13. $x^{-1} - y^{-1}$
14. $a^{-2} + b^{-2}$
15. $\frac{r^{-1}}{s^{-1}} + \frac{s}{r}$
16. $\frac{r}{s^{-1}} + \frac{r^{-1}}{s}$
17. $R^{-1}S^{-1} - \frac{R^{-1}}{S^{-1}}$
18. $\frac{R^2}{S^{-2}} - \frac{S^{-1}}{R^{-2}}$
19. $ER^{-1} + E^{-1}R$
20. $E^{-1}R - ER^{-1}$
21. $\frac{E^{-1} + R^{-1}}{(ER)^{-1}}$
22. $\frac{E}{R^{-1}} + \frac{E^{-1}}{R}$
23. $(P^{-1} - Q^{-1})^{-1}$
24. $\frac{P^{-1} + Q^{-1}}{P^{-1} - Q^{-1}}$

5.5 Scientific notation

In certain scientific applications of mathematics, very small and very large numbers have to be considered. Such numbers may be represented very conveniently by means of an exponential form called **scientific notation**. For example,

$$\begin{aligned} 3796 &= 3.796 \times 10^3, \\ 30.21 &= 3.021 \times 10^1, \\ 0.0214 &= 2.14 \times 10^{-2}, \\ 0.000301 &= 3.01 \times 10^{-4}. \end{aligned}$$

In each case we have represented a number as the product of a number between 1 and 10, and a power of 10; that is, we have factored a power of 10 from each number. Furthermore, the exponent of the power of 10 is identical to the number of places we have moved the decimal point in going from the first digital form to the second.

A number written in scientific notation may be written in standard form by moving the decimal point in the first factor the number of places indicated by the exponent on 10—to the left if the exponent is negative and to the right if it is positive. We simply perform the indicated multiplication. For example,

$$3.75 \times 10^{4} = 37{,}500,$$
$$7.34 \times 10^{-4} = 0.000734,$$
$$2.98 \times 10^{-1} = 0.298.$$

Scientific notation can frequently be used to simplify numerical calculations. For example, if we write the quotient

$$\frac{248{,}000}{0.0124}$$

in the form

$$\frac{2.48 \times 10^{5}}{1.24 \times 10^{-2}},$$

we may perform the computation as follows:

$$\frac{\overset{2}{\cancel{2.48}}}{\underset{1}{\cancel{1.24}}} \times \frac{10^{5}}{10^{-2}} = 2 \times 10^{7} = 20{,}000{,}000.$$

Sometimes it is more convenient to express a number in an exponential form involving a power of ten in which the other factor is not a number between 1 and 10. For example, under certain circumstances, any of the following forms may be a more useful representation for 6.28×10^{5}:

$$62.8 \times 10^{4},$$
$$628 \times 10^{3},$$

or

$$0.628 \times 10^{6}.$$

EXERCISE 5.5

Express in scientific notation.

EXAMPLES

a. 680,000 b. 0.0000431 c. 0.2

SOLUTIONS

a. $\mathbf{6.8 \times 10^{5}}$ b. $\mathbf{4.31 \times 10^{-5}}$ c. $\mathbf{2 \times 10^{-1}}$

1. 34,000 2. 253 3. 21 4. 3190
5. 8,372,000 6. 25,300,000 7. 0.0014 8. 0.3
9. 0.0000006 10. 0.0234 11. 0.0000230 12. 0.5020

Write in standard form.

EXAMPLES

a. 1.01×10^3 b. 6.3×10^{-4} c. 8×10^{-2}

SOLUTIONS

a. **1010** b. **0.00063** c. **0.08**

13. 1.6×10^3 14. 2.1×10^3 15. 6×10^5
16. 3.8×10^1 17. 1.95×10^4 18. 2.02×10^{-6}
19. 2.3×10^{-7} 20. 4.8×10^{-1} 21. 1.234×10^4
22. 2.68×10^{-1} 23. 4×10^2 24. 3×10^{-4}

Simplify.

EXAMPLES

a. $\dfrac{4 \times 10^3 \times 6 \times 10^{-5}}{8 \times 10^{-3}}$ b. $\dfrac{0.006 \times 3 \times 0.0008}{0.009}$

SOLUTIONS

a. $\dfrac{24 \times 10^{-2}}{8 \times 10^{-3}}$

3×10, or **30**

b. $\dfrac{6 \times 10^{-3} \times 3 \times 8 \times 10^{-4}}{9 \times 10^{-3}}$

1.6×10^{-3}, or **0.0016**

25. $\dfrac{10^3 \times 10^5}{10^2}$ 26. $\dfrac{10^3 \times 10^{-6}}{10^2}$

27. $\dfrac{10^3 \times 10^{-7} \times 10^2}{10^{-2} \times 10^4}$ 28. $\dfrac{10^2 \times 10^5 \times 10^{-3}}{10^2 \times 10^2}$

29. $\dfrac{(4 \times 10^3) \times (6 \times 10^{-2})}{3 \times 10^{-7}}$ 30. $\dfrac{(2 \times 10^2)^2 \times (3 \times 10^{-3})}{2 \times 10^4}$

31. $\dfrac{0.6 \times 0.00084 \times 0.093}{0.00021 \times 0.00031}$ 32. $\dfrac{0.065 \times 2.2 \times 50}{1.30 \times 0.011 \times 0.05}$

33. $\dfrac{28 \times 0.0006 \times 450}{1.5 \times 700 \times 0.018}$ 34. $\dfrac{0.0054 \times 0.05 \times 300}{0.0015 \times 0.27 \times 80}$

Evaluate the given formula for the given values of the variables.

EXAMPLE

$R = \frac{1.08l}{d^2}$; $l = 2 \times 10^3$, $d = 1.2 \times 10^{-3}$

SOLUTION

Replace l with 2×10^3 and d with 1.2×10^{-3}.

$$R = \frac{1.08 \times (2 \times 10^3)}{(1.2 \times 10^{-3})}$$

$$= \frac{2.16 \times 10^3}{1.44 \times 10^{-6}} = \frac{\overset{3}{\cancel{2.16}} \times 10^{-2} \times 10^3}{\underset{2}{\cancel{1.44}} \times 10^{-2} \times 10^{-6}}$$

$$= \frac{3}{2} \times 10^9 = \mathbf{1.5 \times 10^9}$$

35. $e = \frac{Pl}{AE}$; $P = 5 \times 10^5$, $l = 1.2 \times 10^2$, $A = 25$, $E = 3 \times 10^6$

36.. $F = \frac{P}{EI}$; $P = 26 \times 10^5$, $E = 1.3 \times 10^2$, $I = 4 \times 10^{-5}$

37. $T = 2\pi(LC)^{1/2}$; $\pi = 3.14$, $L = 25 \times 10^{-4}$, $C = 1 \times 10^{-6}$

38. $P = V^2g$; $V = 3 \times 10^3$; $g = 6 \times 10^{-4}$

39. $k = \frac{AE}{l}$; $A = 2.4$, $E = 15 \times 10^6$, $l = 7.2 \times 10^2$

40. $f = \frac{10^6}{2\pi(LC)^{1/2}}$; $\pi = 3.14$, $L = 2.5 \times 10^2$, $C = 4 \times 10^{-5}$

5.6 Products and factors of expressions containing rational exponents

In Chapter 2 we used the distributive law to multiply expressions in which the exponents on factors were limited to positive integers. In this section we shall extend the applications of the distributive law to include expressions containing rational exponents. For example, by an application of the distributive law, the product

$$x^{-1/3}(x^{2/3} - x^{1/3})$$

can first be written as

$$x^{-1/3} \cdot x^{2/3} - x^{-1/3} \cdot x^{1/3},$$

and then, by using the laws of exponents, it can be simplified as

$$\begin{aligned} x^{-1/3+2/3} - x^{-1/3+1/3} &= x^{1/3} - x^0 \\ &= x^{-1/3} - 1. \end{aligned}$$

The distributive law can also be applied in the form

$$ax + ay = a(x + y)$$

to rewrite sums of terms involving rational exponents as products. For example, we can rewrite the expression

$$x^{3/5} + x^{1/5}$$

as a product in which one factor is $x^{1/5}$ by first writing

$$x^{1/5}(? + ?)$$

and then inserting within the parentheses the appropriate expression which contains two terms. This expression can be determined by inspection. We ask ourselves for the terms that multiply $x^{1/5}$ to yield $x^{3/5}$ and $x^{1/5}$ in turn. The final result appears as

$$x^{3/5} + x^{1/5} = x^{1/5}(x^{2/5} + 1).$$

EXERCISE 5.6

Find each product.

EXAMPLES

a. $x^2(x^3)$ b. $y(y^{3/4})$ c. $y^{-1/2}(y^{3/2} + y^{1/2})$

SOLUTIONS

a. x^{2+3}
$\mathbf{x^5}$

b. $y^{1+3/4}$
$\mathbf{y^{7/4}}$

c. $y^{-1/2+3/2} + y^{-1/2+1/2}$
$y^1 + y^0$
$\mathbf{y + 1}$

1. $y^3(y^4)$
2. $x^2(x^5)$
3. $x(x^{1/2})$
4. $y(y^{2/3})$
5. $y^{1/3}(y^{1/3})$
6. $x^{1/4}(x^{3/4})$
7. $x^{1/2}(x^{3/4})$
8. $y^{1/3}(y^{1/6})$
9. $y^{1/2}(y^{1/3})$
10. $x^{1/3}(x^{1/4})$
11. $x^{2/3}(x^{1/4})$
12. $y^{3/4}(y^{2/3})$

13. $x^{-1/2}(x)$
14. $y(y^{-3/2})$
15. $y^{-1/2}(y^{3/2})$
16. $x^{-1/4}(x^{7/4})$
17. $y^{-1/2}(y^{-1/3})$
18. $x^{-2/3}(x^{-1/4})$
19. $x^{1/2}(x + x^{1/2})$
20. $x^{1/5}(x + x^{4/5})$
21. $x^{1/3}(x^{1/3} + x^{2/3})$
22. $x^{1/4}(x^{1/4} + x^{3/4})$
23. $x^{3/4}(x^{1/4} + x^{3/4})$
24. $x^{2/3}(x^{2/3} - x^{1/3})$
25. $R^{1/2}(R - R^{3/4})$
26. $R^{2/3}(R^{1/6} - R)$
27. $S^{-3/4}(S^{-1/4} + S^{3/4})$
28. $S^{-1/4}(S^{3/4} + S^{-2})$
29. $T^{-1/2}(T^{-3} - T^{-2/3})$
30. $T^{-2/3}(T^{-1/2} - T^{-1})$

Factor as indicated.

EXAMPLES

a. $y^5 = y^2(?)$ b. $y^{3/4} = y^{1/4}(?)$ c. $y^{-1/2} + y^{1/2} = y^{-1/2}(\ ?\ )$

SOLUTIONS

a. $\mathbf{y^2(y^3)}$ b. $y^{1/4}(y^{2/4})$ $\mathbf{y^{1/4}y^{1/2}}$ c. $\mathbf{y^{-1/2}(1 + y)}$

31. $y^7 = y^3(?)$
32. $x^4 = x^3(?)$
33. $x^{-5} = x^{-2}(?)$
34. $y^{-6} = y^{-5}(?)$
35. $y^{-2} = y(?)$
36. $x^3 = x^{-2}(?)$
37. $x^{3/5} = x^{1/5}(?)$
38. $x^{7/8} = x^{3/8}(?)$
39. $x^{-1/3} = x^{-2/3}(?)$
40. $x^{-1/4} = x(?)$
41. $y^{1/3} = y(?)$
42. $y^{3/5} = y(?)$
43. $y^{3/2} + y = y(\ ?\ )$
44. $y^{1/2} + y = y(\ ?\ )$
45. $R - R^{2/3} = R^{2/3}(\ ?\ )$
46. $T^{2/3} + T^{1/3} = T(\ ?\ )$
47. $R^{1/2} + R^{3/2} = R^{3/2}(\ ?\ )$
48. $T^{3/5} + T^{6/5} = T^{6/5}(\ ?\ )$
49. $M^{-3/2} + M^{-1/2} = M^{-1/2}(\ ?\ )$
50. $M^{1/2} + M^{-1/2} = M^{-1/2}(\ ?\ )$

Chapter review

[5.1] Simplify.

1. $(ab^2)^2$ 2. $(-a^2b)^3$ 3. $\left(\frac{2x}{y^2}\right)^3$ 4. $\left(\frac{r^2}{3s^2}\right)$

[5.2] Simplify.

5. $8^{1/3}$ 6. $(-8)^{1/3}$ 7. $\left(\frac{1}{16}\right)^{1/2}$ 8. $\left(\frac{9}{4}\right)^{3/2}$

9. $y^{1/4}y^{3/4}$ 10. $\frac{y^{3/2}}{y^{1/2}}$ 11. $(a^{3/4}b^{1/4})^4$ 12. $\left(\frac{a^2}{b^{1/3}}\right)^{3/2}$

[5.3] Simplify.

13. 3^{-2} 14. $8^{-2/3}$ 15. $3^0 4^{-1/2}$ 16. $\frac{1}{16^{-3/2}}$

Perform the operations indicated and express each power with a positive exponent.

17. $y^{-2}y^3$ 18. $(y^{-2})^3$ 19. $(T^{-1}n^2)^{1/2}$ 20. $\frac{R^2L^{-2}}{R^{-3}L^{-3}}$

[5.4] Represent as a single fraction involving positive exponents only.

21. $3^{-1} + 3^{-2}$ 22. $\frac{x^{-1} + y^{-1}}{x + y}$

[5.5] Express in scientific notation.

23. 0.00028 24. 205000

Simplify.

25. $\frac{10^3 \times 10^{-2} \times 10^5}{10^7 \times 10^{-4} \times 10}$ 26. $\frac{0.003 \times 4000}{0.000012}$

[5.6] Find each product.

27. $x^{1/2}(x^2 - x^{-1/2})$ 28. $R^{-2/3}(R + R^{-1/2})$

Factor as indicated.

29. $y^{4/5} = y^{1/5}(?)$ 30. $S^{-1/2} - S^{-3/2} = S^{-1/2}(\ ?\ )$

Roots and radicals

6.1 Radical notation

In Section 5.2 we referred to $a^{1/n}$ as the nth root of a; that is, $a^{1/n}$ is one of n equal factors of a ($a \geq 0$). Another representation is given by

$$\boldsymbol{a^{1/n} = \sqrt[n]{a},}$$

and in many cases the latter form is more convenient to use. In such a representation, the symbol $\sqrt{}$ is called a **radical**, a is called the **radicand,** n is called the **index**, and the expression is said to be a **radical of order n**. We require that the index be a natural number. If no index is written, the index is understood to be 2, and the expression is called the **square root** of the radicand.

Since, from Section 5.2, for $a \geq 0$,

$$a^{m/n} = (a^m)^{1/n} = (a^{1/n})^m,$$

we may write such a power

$$\boldsymbol{a^{m/n} = \sqrt[n]{a^m} = (\sqrt[n]{a})^m,}$$

where the denominator of the exponent is the index of the radical and the numerator of the exponent is either the exponent of the radicand or the exponent of the root. For example,

$$x^{2/3} = \sqrt[3]{x^2} = (\sqrt[3]{x})^2,$$
$$8^{2/3} = \sqrt[3]{8^2} = (\sqrt[3]{8})^2.$$

Since we have restricted the index of a radical to be a natural number, we must always express a fractional exponent in standard form (m/n or $-m/n$) before writing the power in radical form. Thus

$$x^{-(3/4)} = x^{(-3/4)} = \sqrt[4]{x^{-3}}.$$

Removing the restriction that $a \geq 0$ for n *even*, we define

$$\sqrt[n]{a^n} = |a|,$$

because we wish to use the symbol $\sqrt[n]{a^n}$ only for the positive nth root of a^n. Considering odd indices, there is no ambiguous interpretation possible. That is, *for n odd*

$$\sqrt[n]{a^n} = a$$

for *all values of a*. For example,

$$\sqrt{2^2} = |2| = 2, \qquad \sqrt{(-2)^2} = |-2| = 2,$$
$$\sqrt[3]{8} = 2, \quad \text{and} \quad \sqrt[3]{-8} = -2.$$

We shall continue to agree that *in the exercises a radicand containing one or more variables represents a non-negative number, and that the replacement set for any such variable contains non-negative numbers only.*

EXERCISE 6.1

Write in radical form.

EXAMPLES

a. $5^{1/2}$ b. $yx^{2/3}$ c. $(x - y^2)^{-1/2}$

SOLUTIONS

a. $\sqrt{5}$ b. $y\sqrt[3]{x^2}$ c. $\dfrac{1}{\sqrt{x - y^2}}$

1. $3^{1/2}$	2. $5^{1/3}$	3. $x^{3/2}$
4. $y^{2/3}$	5. $3x^{1/3}$	6. $5x^{1/5}$
7. $xy^{1/2}$	8. $(xy)^{1/2}$	9. $2x^{2/3}$
10. $(2x)^{2/3}$	11. $3xy^{2/3}$	12. $3(xy)^{2/3}$
13. $(x + 2y)^{1/2}$	14. $(x - y)^{1/3}$	15. $(x - y)^{2/3}$
16. $(2r - 3s)^{3/5}$	17. $x^{1/2} - y^{1/2}$	18. $a^{1/3} - b^{2/3}$
19. $4^{-1/3}$	20. $6^{-2/5}$	21. $3x^{-2/3}$
22. $(3x)^{-2/3}$	23. $(x^2 - y^2)^{-1/2}$	24. $(a^3 - b^3)^{-1/3}$

Represent with positive fractional exponents.

EXAMPLES

a. $\sqrt{2^3}$ b. $\sqrt[3]{7a^2}$ c. $\dfrac{1}{\sqrt{x-1}}$

SOLUTIONS

a. $\mathbf{2^{3/2}}$ b. $\mathbf{(7a^2)^{1/3}}$, or $\mathbf{7^{1/3}a^{2/3}}$ c. $\mathbf{\dfrac{1}{(x-1)^{1/2}}}$

25. $\sqrt{3}$	26. $\sqrt[3]{7}$	27. $\sqrt[3]{x^2}$
28. $\sqrt[4]{x^3}$	29. $\sqrt[3]{xy}$	30. $\sqrt[5]{x^2y^3}$
31. $3\sqrt{x}$	32. $\sqrt{3x}$	33. $\sqrt[3]{2xy^2}$
34. $2(\sqrt[3]{xy^2})$	35. $2x\sqrt[3]{y^2}$	36. $2y^2(\sqrt[3]{x})$
37. $\sqrt{x - y}$	38. $\sqrt[4]{x + 2y}$	39. $3\sqrt[3]{x^2 - y}$
40. $2(\sqrt[5]{x^4 - y^4})$	41. $\sqrt{x} - 2\sqrt{y}$	42. $\sqrt[3]{x} + 2(\sqrt[3]{y})$
43. $\dfrac{1}{\sqrt{x}}$	44. $\dfrac{2}{\sqrt[3]{y}}$	45. $\dfrac{x}{\sqrt[3]{y}}$
46. $\sqrt[3]{\dfrac{x}{y}}$	47. $\dfrac{2}{\sqrt{x + y}}$	48. $\dfrac{3}{\sqrt{x} + \sqrt{y}}$

Find each root.

EXAMPLES

a. $\sqrt{49}$ b. $\sqrt[3]{-8}$ c. $\sqrt[3]{x^6y^3}$ d. $-\sqrt[4]{81x^4}$

SOLUTIONS

a. **7** b. **−2** c. $\mathbf{x^2y}$ d. $\mathbf{-3x}$

49. $\sqrt{16}$ 50. $\sqrt{144}$ 51. $-\sqrt{25}$ 52. $-\sqrt{169}$

53. $\sqrt[3]{27}$ 54. $\sqrt[3]{125}$ 55. $\sqrt[3]{-64}$ 56. $\sqrt[5]{32}$

57. $-\sqrt[4]{16}$ 58. $-\sqrt[6]{64}$ 59. $\sqrt{x^4}$ 60. $\sqrt{y^6}$

61. $\sqrt[3]{8T^6}$ 62. $\sqrt[3]{27T^9}$ 63. $\sqrt{S^4T^6}$ 64. $\sqrt{S^8T^{10}}$

65. $\sqrt{\frac{4}{9}M^2R^8}$ 66. $\sqrt{\frac{9}{16}M^2R^4}$ 67. $\sqrt[3]{\frac{-8}{125}M^3}$ 68. $\sqrt[3]{\frac{8}{27}M^3R^6}$

Evaluate each expression for the given values of the variables.

EXAMPLES

a. $\sqrt[3]{\frac{V}{F}}$; $V=32$ and $F=4$

b. $\sqrt{2R+T^3}$; $R=40$ and $T=4$

SOLUTIONS

a. $\sqrt[3]{\frac{32}{4}}$

$\sqrt[3]{8}$

2

b. $\sqrt{2(40)+(4)^3}$

$\sqrt{80+64}$

12

69. $\sqrt{M^3N}$; $M=2$ and $N=8$

70. $\sqrt[3]{R^2+2T}$; $R=5$ and $T=1$

71. $2\sqrt{P^3}+\sqrt[3]{T}$; $P=4$ and $T=27$

72. $R\sqrt{2Q}-\sqrt{R}$; $Q=18$ and $R=4$

73. $\sqrt{\frac{P^2}{Q^3}}$; $P=5$ and $Q=4$

74. $\sqrt[3]{\frac{4-L}{R}}$; $L=3$ and $R=64$

6.2 Changing forms of radicals

From the definition of a radical and from the laws of exponents, we can derive two important relationships. (In all cases, $a, b > 0$ and n is a natural number.) We have, first,

$$\sqrt[n]{ab} = \sqrt[n]{a}\,\sqrt[n]{b} \tag{1}$$

or

$$\sqrt[n]{a}\,\sqrt[n]{b} = \sqrt[n]{ab}. \tag{1a}$$

This relationship follows from

$$\sqrt[n]{ab} = (ab)^{1/n} = a^{1/n}b^{1/n} = \sqrt[n]{a}\,\sqrt[n]{b}.$$

Relationship (1) can be used to write a radical in a form in which the radicand contains no prime factor or polynomial factor raised to a power greater than or equal to the index of the radical. Thus we can write

$$\begin{aligned}\sqrt{18} &= \sqrt{3^2}\,\sqrt{2}\\ &= 3\sqrt{2},\end{aligned}$$

or

$$\begin{aligned}\sqrt[3]{16x^3y^5} &= \sqrt[3]{2^3x^3y^3}\,\sqrt[3]{2y^2}\\ &= 2xy\,\sqrt[3]{2y^2},\end{aligned}$$

where in each case the radicand is factored into two factors; one factor consists of bases raised to the same power as the index of the radical. This factor is then "removed" from the radicand.

The second important relationship is

$$\sqrt[n]{\frac{a}{b}} = \frac{\sqrt[n]{a}}{\sqrt[n]{b}} \qquad (2)$$

or

$$\frac{\sqrt[n]{a}}{\sqrt[n]{b}} = \sqrt[n]{\frac{a}{b}}. \qquad (2a)$$

This relationship follows from the fact that

$$\sqrt[n]{\frac{a}{b}} = \left(\frac{a}{b}\right)^{1/n} = \frac{a^{1/n}}{b^{1/n}} = \frac{\sqrt[n]{a}}{\sqrt[n]{b}}.$$

We use relationship (2) to write a radical in a form in which the radicand contains no fraction. For example,

$$\begin{aligned}\sqrt{\frac{3}{4}} &= \frac{\sqrt{3}}{\sqrt{4}}\\ &= \frac{\sqrt{3}}{2}.\end{aligned}$$

If the denominator of the radicand is not the square of a monomial, we can use the fundamental principle of fractions to obtain an equal fraction that has such a denominator. Thus,

$$\sqrt{\frac{2}{5x}} = \frac{\sqrt{2}}{\sqrt{5x}}$$

$$= \frac{\sqrt{2}\sqrt{5x}}{\sqrt{5x}\sqrt{5x}}$$

$$= \frac{\sqrt{10x}}{5x}.$$

The foregoing process is called "rationalizing the denominator" of the fraction, because the result is a fraction with a denominator free of radicals. In the event the radical is of order n, we must build to a denominator that is the nth power of a monomial. For example,

$$\sqrt[5]{\frac{6}{16x^3}} = \frac{\sqrt[5]{6}}{\sqrt[5]{16x^3}}$$

$$= \frac{\sqrt[5]{6}\sqrt[5]{2x^2}}{\sqrt[5]{16x^3}\sqrt[5]{2x^2}}$$

$$= \frac{\sqrt[5]{12x^2}}{\sqrt[5]{32x^5}}$$

$$= \frac{\sqrt[5]{12x^2}}{2x}.$$

Application of (1) or (2) above can be used to rewrite radical expressions in various ways, and, in particular, to write them in what is called "simplest" form. A radical expression is said to be in simplest form if the following conditions exist:

a. The radicand contains no polynomial factor raised to a power equal to or greater than the index of the radical.
b. The radicand contains no fractions.
c. No radical expressions are contained in denominators of fractions.
d. The index of the radical and exponents on factors in the radicand have no common factors.

Although we generally change the form of radicals containing fractional radicands, there are times when such forms are preferred. For example, in certain situations, $\sqrt{1/2}$ or $1/\sqrt{2}$ may be more useful than the equivalent form $\sqrt{2}/2$. We rationalize denominators in this and the next section primarily to develop the ability to recognize equal radical expressions.

EXERCISE 6.2

Simplify.

EXAMPLES

a. $\sqrt{300}$ b. $\sqrt[3]{2x^7y^3}$ c. $\sqrt{2xy}\sqrt{8x}$ d. $\sqrt{0.0049}$

SOLUTIONS

a. $\sqrt{100}\sqrt{3}$

$\mathbf{10\sqrt{3}}$

b. $\sqrt[3]{x^6y^3}\sqrt[3]{2x}$

$\mathbf{x^2y(\sqrt[3]{2x})}$

c. $\sqrt{16x^2y}$

$\sqrt{16x^2}\sqrt{y}$

$\mathbf{4x\sqrt{y}}$

d. $\sqrt{49 \times 10^{-4}}$

$\sqrt{49}\sqrt{10^{-4}}$

$\mathbf{7 \times 10^{-2}}$

1. $\sqrt{9}$
2. $\sqrt{36}$
3. $\sqrt{200}$
4. $\sqrt{160}$
5. $\sqrt{x^4}$
6. $\sqrt{y^6}$
7. $\sqrt{x^3}$
8. $\sqrt{y^{11}}$
9. $\sqrt{4x^5}$
10. $\sqrt{16y^3}$
11. $\sqrt{8x^6}$
12. $\sqrt{18z^8}$
13. $\sqrt[3]{64}$
14. $\sqrt[3]{125}$
15. $\sqrt[3]{-8x^6}$
16. $\sqrt[5]{-32x^5}$
17. $\sqrt[4]{x^5}$
18. $\sqrt[4]{3x^5y^5}$
19. $\sqrt[5]{x^{11}y}$
20. $\sqrt[4]{32xy^7}$
21. $\sqrt{18}\sqrt{2}$
22. $\sqrt{3}\sqrt{27}$
23. $\sqrt{xy}\sqrt{x^5y}$
24. $\sqrt{a}\sqrt{ab^2}$
25. $\sqrt[3]{2}\sqrt[3]{4}$
26. $\sqrt[4]{3}\sqrt[4]{27}$
27. $\sqrt[4]{x^3}\sqrt[4]{x}$
28. $\sqrt[5]{a^3}\sqrt[5]{a^4}$
29. $\sqrt{9 \times 10^2}$
30. $\sqrt{25 \times 10^4}$
31. $\sqrt{40{,}000}$
32. $\sqrt{64{,}000{,}000}$
33. $\sqrt{36 \times 10^{-4}}$
34. $\sqrt{9 \times 10^{-6}}$
35. $\sqrt{0.000012}$
36. $\sqrt{0.0072}$

EXAMPLES

a. $\sqrt{\frac{1}{3}}$ b. $\frac{3x}{\sqrt{2x}}$ c. $\sqrt[3]{\frac{2}{y^2}}$ d. $\frac{\sqrt{6x}\sqrt{5x}}{\sqrt{15}}$

SOLUTIONS

a. $\frac{\sqrt{1}}{\sqrt{3}}\left(\frac{\sqrt{3}}{\sqrt{3}}\right)$

$\mathbf{\frac{\sqrt{3}}{3}}$

b. $\frac{3x}{\sqrt{2x}}\left(\frac{\sqrt{2x}}{\sqrt{2x}}\right)$

$\mathbf{\frac{3\sqrt{2x}}{2}}$

c. $\frac{\sqrt[3]{2}}{\sqrt[3]{y^2}}\left(\frac{\sqrt[3]{y}}{\sqrt[3]{y}}\right)$

$\mathbf{\frac{\sqrt[3]{2y}}{y}}$

d. $\frac{\sqrt{15x^2}\sqrt{2}}{\sqrt{15}}$

$\mathbf{x\sqrt{2}}$

37. $\sqrt{\frac{1}{5}}$ 38. $\sqrt{\frac{2}{3}}$ 39. $\frac{1}{\sqrt{2}}$ 40. $\frac{2}{\sqrt{7}}$

41. $\sqrt{\frac{x}{2}}$ 42. $\sqrt{\frac{y}{3}}$ 43. $\sqrt{\frac{L}{C}}$ 44. $\sqrt{\frac{2R}{L}}$

45. $\frac{x}{\sqrt{x}}$ 46. $\frac{y}{\sqrt{y}}$ 47. $\frac{-ab}{\sqrt{b}}$ 48. $\frac{-n}{\sqrt{nt}}$

49. $\sqrt[3]{\frac{y}{2x}}$ 50. $\sqrt[3]{\frac{1}{6x^2}}$ 51. $\frac{1}{\sqrt[4]{8}}$ 52. $\frac{3}{\sqrt[4]{9}}$

53. $\frac{\sqrt{x^5y^3}}{\sqrt{xy}}$ 54. $\frac{\sqrt{xy^3}}{\sqrt{y}}$ 55. $\frac{\sqrt{xy}\sqrt{xy^4}}{\sqrt{y}}$ 56. $\frac{\sqrt[3]{xy}\sqrt[3]{y^2}}{\sqrt[3]{x}}$

Reduce the order of each radical.

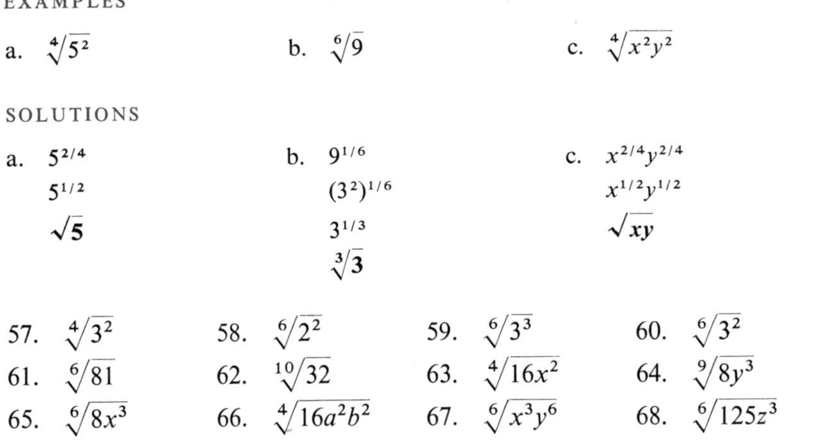

EXAMPLES

a. $\sqrt[4]{5^2}$ b. $\sqrt[6]{9}$ c. $\sqrt[4]{x^2y^2}$

SOLUTIONS

a. $5^{2/4}$
$5^{1/2}$
$\sqrt{5}$

b. $9^{1/6}$
$(3^2)^{1/6}$
$3^{1/3}$
$\sqrt[3]{3}$

c. $x^{2/4}y^{2/4}$
$x^{1/2}y^{1/2}$
$\sqrt{xy}$

57. $\sqrt[4]{3^2}$ 58. $\sqrt[6]{2^2}$ 59. $\sqrt[6]{3^3}$ 60. $\sqrt[6]{3^2}$

61. $\sqrt[6]{81}$ 62. $\sqrt[10]{32}$ 63. $\sqrt[4]{16x^2}$ 64. $\sqrt[9]{8y^3}$

65. $\sqrt[6]{8x^3}$ 66. $\sqrt[4]{16a^2b^2}$ 67. $\sqrt[6]{x^3y^6}$ 68. $\sqrt[6]{125z^3}$

6.3 Expressions containing radicals

The distributive law,

$$a(b + c) = ab + ac, \tag{1}$$

is assumed to hold for all real numbers. By the symmetric law of equality and the commutative law of multiplication, (1) can be written as

$$ba + ca = (b + c)a.$$

Since at this time all radical expressions have been defined so that they represent real numbers, the distributive law holds for radical expressions, and we can invoke it to write sums containing radicals of the same order as a single term. For example,

$$3\sqrt{3} + 4\sqrt{3} = (3 + 4)\sqrt{3} = 7\sqrt{3}$$

and

$$\sqrt{x} + 7\sqrt{x} = (1 + 7)\sqrt{x} = 8\sqrt{x}.$$

Sometimes it may be necessary to simplify radical expressions before they can be combined. Such examples are shown in the Exercise set below.

EXERCISE 6.3

Add.

EXAMPLES

a. $\sqrt{20} + \sqrt{45}$

b. $\sqrt{32x} + \sqrt{2x} - \sqrt{18x}$

SOLUTIONS

a. $2\sqrt{5} + 3\sqrt{5}$

$\mathbf{5\sqrt{5}}$

b. $4\sqrt{2x} + \sqrt{2x} - 3\sqrt{2x}$

$\mathbf{2\sqrt{2x}}$

1. $\sqrt{3} + 2\sqrt{3}$
2. $3\sqrt{5} - 6\sqrt{5}$
3. $2\sqrt{3} + \sqrt{27}$
4. $\sqrt{75} - 2\sqrt{27}$
5. $\sqrt{50x} + 2\sqrt{32x} - \sqrt{2x}$
6. $\sqrt{4y} - \sqrt{25y} + \sqrt{y}$
7. $\sqrt{4x^2y} + \sqrt{9x^2y} - \sqrt{x^2y}$
8. $\sqrt{8y^3} - 2\sqrt{18y^3} + 3\sqrt{50y^3}$
9. $2\sqrt{x^3y} - \sqrt{4x^3y} + 3\sqrt{x^3y}$
10. $\sqrt{9xy^3} + 2\sqrt{xy^3} - \sqrt{4xy^3}$
11. $2(\sqrt[3]{5}) + 4(\sqrt[3]{5}) - \sqrt[3]{5}$
12. $\sqrt[4]{2} - 3(\sqrt[4]{2}) + 6(\sqrt[4]{2})$
13. $3(\sqrt[3]{16}) - \sqrt[3]{2}$
14. $\sqrt[3]{54} + 2(\sqrt[3]{128})$
15. $5(\sqrt[3]{2x}) + 2(\sqrt[3]{16x})$
16. $\sqrt[3]{250x} + 2(\sqrt[3]{16x})$

Write as a single fraction.

EXAMPLES

a. $\frac{\sqrt{2}}{4} - \frac{\sqrt{2}}{2}$ b. $\sqrt{3} - \frac{\sqrt{3}}{3}$ c. $\frac{\sqrt{5}}{2} + \sqrt{5}$

SOLUTIONS

a. $\frac{\sqrt{2}}{4} - \frac{\sqrt{2}}{2}\left(\frac{2}{2}\right)$

$\frac{\sqrt{2} - 2\sqrt{2}}{4}$

$\mathbf{\frac{-\sqrt{2}}{4}}$

b. $\left(\frac{3}{3}\right)\frac{\sqrt{3}}{1} - \frac{\sqrt{3}}{3}$

$\frac{3\sqrt{3} - \sqrt{3}}{3}$

$\mathbf{\frac{2\sqrt{3}}{3}}$

c. $\frac{\sqrt{5}}{2} + \frac{\sqrt{5}}{1}\left(\frac{2}{2}\right)$

$\frac{\sqrt{5} + 2\sqrt{5}}{2}$

$\mathbf{\frac{3\sqrt{5}}{2}}$

17. $\frac{\sqrt{2}}{3} + \frac{\sqrt{2}}{3}$
18. $\frac{\sqrt{3}}{7} + \frac{\sqrt{3}}{7}$
19. $\frac{3\sqrt{5}}{6} - \frac{\sqrt{5}}{6}$
20. $\frac{3\sqrt{2}}{5} + \frac{2\sqrt{2}}{5}$
21. $\frac{\sqrt{11}}{2} + \frac{\sqrt{11}}{3}$
22. $\frac{\sqrt{10}}{3} - \frac{\sqrt{10}}{4}$
23. $\frac{3\sqrt{3}}{2} + \frac{\sqrt{3}}{6}$
24. $\frac{5\sqrt{2}}{6} - \frac{\sqrt{2}}{3}$
25. $\frac{2\sqrt{5}}{5} - \sqrt{5}$
26. $\frac{\sqrt{7}}{3} + \sqrt{7}$
27. $\sqrt{3} + \frac{3\sqrt{3}}{8}$
28. $\sqrt{3} - \frac{2\sqrt{3}}{5}$
29. $\frac{3\sqrt{R}}{4} - \frac{\sqrt{R}}{3}$
30. $\frac{\sqrt{L}}{3} + \frac{2\sqrt{L}}{5}$
31. $\sqrt{4L} - \frac{\sqrt{9L}}{2}$
32. $\frac{\sqrt{4T}}{3} + \sqrt{T}$
33. $\frac{\sqrt{a^2b}}{2} - \sqrt{4a^2b}$
34. $\sqrt{ab^2} + \frac{\sqrt{25ab^2}}{4}$

6.4 Products and factors of radical expressions

Since all radical expressions have thus far been defined to represent real numbers, the distributive law holds for radical expressions. For example,

$$x(\sqrt{2} + \sqrt{3}) = x\sqrt{2} + x\sqrt{3},$$
$$\sqrt{3x}(2 - \sqrt{2}) = 2\sqrt{3x} - \sqrt{6x},$$
$$\sqrt{3}(\sqrt{2x} + \sqrt{6}) = \sqrt{6x} + \sqrt{18}$$
$$= \sqrt{6x} + 3\sqrt{2}.$$

In Chapter 2 we agreed to factor from each term of an expression only those common factors that were integers or positive integral powers of variables. However, we can (if we wish) consider other real numbers for factors. Thus radicals common to each term in an expression may be factored from the expression. For example,

$$\sqrt{a} + \sqrt{ab} = \sqrt{a} + \sqrt{a}\sqrt{b}$$
$$= \sqrt{a}(1 + \sqrt{b}).$$

EXERCISE 6.4

Multiply factors and simplify.

EXAMPLES

a. $4(\sqrt{3} + 1)$ b. $\sqrt{x}(\sqrt{2x} - \sqrt{x})$ c. $(1 + \sqrt{3})(2 - \sqrt{3})$

SOLUTIONS

a. $\mathbf{4\sqrt{3} + 4}$

b. $\mathbf{x\sqrt{2} - x}$

c. $1(2 - \sqrt{3}) + \sqrt{3}(2 - \sqrt{3})$
$2 - \sqrt{3} + 2\sqrt{3} - 3$
$\mathbf{-1 + \sqrt{3}}$

1. $2(3 - \sqrt{5})$
2. $5(2 - \sqrt{7})$
3. $\sqrt{2}(3 + \sqrt{3})$
4. $\sqrt{3}(5 - \sqrt{2})$
5. $\sqrt{2}(\sqrt{6} + 3)$
6. $\sqrt{6}(3 + \sqrt{6})$
7. $(3 + \sqrt{5})(2 - \sqrt{5})$
8. $(1 - \sqrt{2})(2 + \sqrt{2})$
9. $(\sqrt{x} + 1)(\sqrt{x} - 3)$
10. $(\sqrt{y} + 3)(\sqrt{y} - 4)$
11. $(\sqrt{x} - 3)(\sqrt{x} + 3)$
12. $(2 + \sqrt{x})(2 - \sqrt{x})$
13. $(\sqrt{x} - 3)^2$
14. $(\sqrt{x} + y)^2$
15. $(\sqrt{2} - \sqrt{3})(\sqrt{2} + 2\sqrt{3})$
16. $(\sqrt{3} - \sqrt{5})(2\sqrt{3} + \sqrt{5})$
17. $(\sqrt{5} - \sqrt{2})(\sqrt{5} + \sqrt{2})$
18. $(\sqrt{7} - \sqrt{3})(\sqrt{7} + \sqrt{3})$
19. $(\sqrt{5} - \sqrt{2})^2$
20. $(\sqrt{2} - 2\sqrt{3})^2$
21. $(\sqrt{x} - 2\sqrt{3})(\sqrt{x} + \sqrt{3})$
22. $(\sqrt{2} - 2\sqrt{x})(2\sqrt{2} + \sqrt{x})$
23. $(\sqrt{x} - 2\sqrt{y})(\sqrt{x} + 3\sqrt{y})$
24. $(\sqrt{y} + 3\sqrt{x})(\sqrt{y} - 3\sqrt{x})$

Change each expression to the form indicated.

EXAMPLES

a. $3+\sqrt{18}=3(\ \ ?\ \ )$ b. $\sqrt{x}+\sqrt{xy}=\sqrt{x}(\ \ ?\ \ )$

SOLUTIONS

a. $3+3\sqrt{2}$

$\mathbf{3(1+\sqrt{2})}$

b. $\sqrt{x}+\sqrt{x}\sqrt{y}$

$\mathbf{\sqrt{x}(1+\sqrt{y})}$

25. $2+2\sqrt{3}=2(\ \ ?\ \ )$
26. $5+10\sqrt{2}=5(\ \ ?\ \ )$
27. $3\sqrt{7}-3=3(\ \ ?\ \ )$
28. $6\sqrt{3}+\sqrt{9}=3(\ \ ?\ \ )$
29. $4+2\sqrt{3}=2(\ \ ?\ \ )$
30. $12-6\sqrt{5}=6(\ \ ?\ \ )$
31. $2\sqrt{27}+6=6(\ \ ?\ \ )$
32. $5\sqrt{5}-\sqrt{25}=5(\ \ ?\ \ )$
33. $4+\sqrt{16y}=4(\ \ ?\ \ )$
34. $3+\sqrt{18x}=3(\ \ ?\ \ )$
35. $y\sqrt{3}-x\sqrt{3}=\sqrt{3}(\ \ ?\ \ )$
36. $x\sqrt{2}-\sqrt{8}=\sqrt{2}(\ \ ?\ \ )$
37. $\sqrt{2}-\sqrt{6}=\sqrt{2}(\ \ ?\ \ )$
38. $\sqrt{12}-2\sqrt{6}=2\sqrt{3}(\ \ ?\ \ )$
39. $\sqrt{3x}+\sqrt{5x}=\sqrt{x}(\ \ ?\ \ )$
40. $\sqrt{2x}-\sqrt{x}=\sqrt{x}(\ \ ?\ \ )$

Simplify fractions.

EXAMPLES

a. $\dfrac{4+6\sqrt{3}}{2}$ b. $\dfrac{2x-\sqrt{8x^2}}{4x}$ c. $\dfrac{\sqrt{6}-\sqrt{8}}{\sqrt{2}}$

SOLUTIONS

a. $\dfrac{2(2+3\sqrt{3})}{2}$

$\mathbf{2+3\sqrt{3}}$

b. $\dfrac{2x-2x\sqrt{2}}{4x}$

$\dfrac{2x(1-\sqrt{2})}{4x}$

$\mathbf{\dfrac{1-\sqrt{2}}{2}}$

c. $\dfrac{\sqrt{2}\sqrt{3}-2\sqrt{2}}{\sqrt{2}}$

$\dfrac{\sqrt{2}(\sqrt{3}-2)}{\sqrt{2}}$

$\mathbf{\sqrt{3}-2}$

41. $\dfrac{2+2\sqrt{3}}{2}$ 42. $\dfrac{6+2\sqrt{5}}{2}$ 43. $\dfrac{2-\sqrt{12}}{2}$

44. $\dfrac{3-\sqrt{18}}{3}$ 45. $\dfrac{6+\sqrt{12}}{2}$ 46. $\dfrac{5+2\sqrt{125}}{5}$

47. $\dfrac{6+2\sqrt{18}}{6}$ 48. $\dfrac{8-2\sqrt{12}}{4}$ 49. $\dfrac{x-\sqrt{x^3}}{x}$

50. $\dfrac{xy-x\sqrt{xy^2}}{xy}$ 51. $\dfrac{x\sqrt{y}-\sqrt{y^3}}{\sqrt{y}}$ 52. $\dfrac{\sqrt{x}-y\sqrt{x^3}}{\sqrt{x}}$

6.5 Rationalizing denominators

In Section 6.2 you rationalized denominators of fractions in which the denominators were monomials. For example,

$$\frac{a}{\sqrt{b}} = \frac{a}{\sqrt{b}}\frac{\sqrt{b}}{\sqrt{b}} = \frac{a\sqrt{b}}{b}.$$

The distributive law also provides us with a means of rationalizing denominators of fractions in which radicals occur in one or both of two terms. To accomplish this, we first recall that

$$(a-b)(a+b) = a^2 - b^2,$$

where the product contains no linear term. Each of the two factors of a product exhibiting this property is said to be the **conjugate** of the other. Now consider a fraction of the form

$$\frac{a}{b+\sqrt{c}}.$$

If we multiply the numerator and denominator of this fraction by the conjugate of the denominator, the denominator of the resulting fraction will contain no term linear in $\sqrt{c}$ and, hence, will be free of radicals. That is,

$$\frac{a(b-\sqrt{c})}{(b+\sqrt{c})(b-\sqrt{c})} = \frac{ab-a\sqrt{c}}{b^2-c},$$

where the denominator has been rationalized. This process is equally applicable to radical fractions of the form

$$\frac{a}{\sqrt{b}+\sqrt{c}},$$

since

$$\frac{a(\sqrt{b}-\sqrt{c})}{(\sqrt{b}+\sqrt{c}\,(\sqrt{b}-\sqrt{c})} = \frac{a\sqrt{b}-a\sqrt{c}}{b-c}.$$

It is sometimes helpful to rewrite a fraction in a form in which the numerator has been rationalized. Procedures similar to the ones described above can be used.

EXERCISE 6.5

Rationalize denominators.

EXAMPLES

a. $\frac{3}{\sqrt{5}}$ b. $\frac{3}{\sqrt{2}-1}$ c. $\frac{1}{\sqrt{x}-\sqrt{y}}$

SOLUTIONS

a. $\frac{3}{\sqrt{5}}\frac{\sqrt{5}}{\sqrt{5}}$

$\mathbf{\frac{3\sqrt{5}}{5}}$

b. $\frac{3(\sqrt{2}+1)}{(\sqrt{2}-1)(\sqrt{2}+1)}$

$\frac{3\sqrt{2}+3}{2-1}$

$\mathbf{3\sqrt{2}+3}$

c. $\frac{1(\sqrt{x}+\sqrt{y})}{(\sqrt{x}-\sqrt{y})(\sqrt{x}+\sqrt{y})}$

$\mathbf{\frac{\sqrt{x}+\sqrt{y}}{x-y}}$

1. $\frac{1}{\sqrt{2}}$
2. $\frac{3}{\sqrt{3}}$
3. $\frac{x}{\sqrt{5}}$
4. $\frac{y}{\sqrt{2}}$
5. $\frac{4}{\sqrt{2y}}$
6. $\frac{6}{\sqrt{3x}}$
7. $\frac{1}{1+\sqrt{3}}$
8. $\frac{1}{2-\sqrt{2}}$
9. $\frac{2}{\sqrt{7}-2}$
10. $\frac{2}{4-\sqrt{5}}$
11. $\frac{4}{1+\sqrt{x}}$
12. $\frac{1}{2-\sqrt{y}}$
13. $\frac{x}{\sqrt{x}-3}$
14. $\frac{y}{\sqrt{3}-y}$
15. $\frac{\sqrt{x}}{\sqrt{x}-\sqrt{y}}$
16. $\frac{\sqrt{y}}{\sqrt{x}+\sqrt{y}}$
17. $\frac{\sqrt{y-1}}{1+\sqrt{y-1}}$
18. $\frac{\sqrt{x+2}}{1-\sqrt{x+2}}$

Rationalize numerators.

19. $\dfrac{\sqrt{3}}{3}$ 20. $\dfrac{\sqrt{2}}{2}$ 21. $\dfrac{\sqrt{a}}{3}$

22. $\dfrac{\sqrt{b}}{2}$ 23. $\dfrac{\sqrt{2T}}{T}$ 24. $\dfrac{\sqrt{5C}}{10C}$

25. $\dfrac{1-\sqrt{2}}{2}$ 26. $\dfrac{\sqrt{3}+\sqrt{2}}{\sqrt{3}}$ 27. $\dfrac{\sqrt{x}-1}{3}$

28. $\dfrac{4-\sqrt{2y}}{2}$ 29. $\dfrac{\sqrt{x}-\sqrt{y}}{x}$ 30. $\dfrac{2\sqrt{x}+\sqrt{y}}{\sqrt{xy}}$

6.6 Irrational numbers

Recall that any number that can be expressed as the quotient of two integers is a rational number. Any real number that cannot be so expressed is called an **irrational number**. Any radical expression $\sqrt[n]{a}$ represents a rational number if and only if a is the nth power of a rational number. Thus $\sqrt{4}$, $-\sqrt[3]{27}$, $\sqrt[4]{81}$, and $\sqrt[5]{-32}$ are rational numbers equal to 2, -3, 3, and -2, respectively, while $\sqrt{5}$, $\sqrt[3]{9}$, $\sqrt[4]{15}$, and $\sqrt[5]{61}$ are irrational numbers. Although an irrational number does not have an exact decimal representation, we can obtain a decimal approximation correct to any desired degree of accuracy. For example,

$$\sqrt{2} \approx 1.4,$$

$$\sqrt{2} \approx 1.41,$$

$$\sqrt{2} \approx 1.414,$$

where the symbol $\approx$ is read "is approximately equal to." A table of square roots appears on page 316, and a brief excerpt from it is shown below.

No.	Sq.	Sq. Root	Prime Factors
23	529	4.796	23
24	576	4.899	$2^3 \cdot 3$
25	625	5.000	5^2
26	676	5.099	$2 \cdot 13$

Notice that the table can be used in two ways to obtain square roots. For example, we can read an approximation for $\sqrt{24}$ at once by looking under the "Sq. Root" column directly opposite 24. Thus $\sqrt{24} \approx 4.899$. However, since $24^2 = 576$, it is also true that $\sqrt{576} = 24$. In other words, we can obtain the square root of any number in the "Sq." column by reading to the left under the "No." column. Moreover, a shift in the position of the decimal point in the numeral for a given number by *one position* results in a shift in the decimal point in the numeral for its square by *two positions*. This means that the table can also be used to find, for example, $\sqrt{0.23}$ or $\sqrt{5.29}$ by an appropriate shift in the decimal point. Since $(4.796)^2 \approx 23$, we have $(0.4796)^2 \approx 0.23$, from which $\sqrt{0.23} \approx 0.4796$. Similarly, $\sqrt{5.29} \approx 2.3$.

By estimating, we can exploit the table still further. Because 535 is about 1/8 of the way between 529 and 576, we can estimate that $\sqrt{535}$ is about 1/8 of the way between $\sqrt{529}$ and $\sqrt{576}$, to obtain $\sqrt{535} \approx 23.1$ or 23.2. Actually, to 5 decimal places, $\sqrt{535} \approx 23.13007$, but either 23.1 or 23.2 can be a useful approximation of this square root.

A more precise method of approximation called linear interpolation will be discussed in Chapter 11.

EXERCISE 6.6

Graph each set of real numbers on a separate line graph. Use the table on page 313 to obtain approximations for irrational numbers.

EXAMPLE

$\{\sqrt{4}, -\sqrt{3}, \sqrt{17}, -\sqrt{9}\}$

SOLUTION

$-\sqrt{9}$ $-\sqrt{3}$ $\sqrt{4}$ $\sqrt{17}$

−5 0 5

1. $\{-\sqrt{7}, -\sqrt{1}, \sqrt{5}, \sqrt{9}\}$
2. $\left\{-\frac{2}{3}, 0, \sqrt{3}, -\sqrt{1}, -\sqrt{11}\right\}$
3. $\{-\sqrt{20}, -\sqrt{6}, \sqrt{1}, 6\}$
4. $\left\{\sqrt{41}, \sqrt{7}, -\sqrt{7}, \frac{3}{4}\right\}$
5. $\left\{\frac{27}{4}, 6, -\sqrt{6}, -1\right\}$
6. $\left\{\frac{7}{4}, \sqrt{12}, -\sqrt{12}, 12\right\}$
7. $\left\{-8, \sqrt{7}, -\frac{3}{4}, 8\right\}$
8. $\left\{\sqrt{16}, -4, 0, \frac{3}{4}\right\}$
9. $\{-10, \sqrt{5}, \sqrt{37}, -\sqrt{2}\}$
10. $\{-\sqrt{74}, \sqrt{32}, \sqrt{49}, -6\}$
11. $\{\sqrt{24}, -\sqrt{24}, 24, -24\}$
12. $\{13, -\sqrt{13}, \sqrt{13}, -13\}$

Use the table on page 415 to obtain values or approximations for each of the following irrational numbers.

13. $\sqrt{729}$ 14. $\sqrt{2809}$ 15. $\sqrt{0.23}$ 16. $\sqrt{0.87}$
17. $\sqrt{46.24}$ 18. $\sqrt{88.36}$ 19. $\sqrt{0.0784}$ 20. $\sqrt{0.2916}$
21. $\sqrt{1253}$ 22. $\sqrt{213}$ 23. $\sqrt{47.73}$ 24. $\sqrt{0.6807}$

Find a decimal approximation for the variable in the left-hand member of each formula.

EXAMPLE

$l = \sqrt{\dfrac{2K}{P}}$; $K = 3.02$ and $P = 4$

SOLUTION

$$l = \sqrt{\frac{2(3.02)}{4}}$$

$$= \sqrt{\frac{6.04}{4}} = \sqrt{1.51}$$

From the table on page 415, we have $\sqrt{1.44} \approx 1.2$ and $\sqrt{1.69} \approx 1.3$. Since 1.51 is a little more than ¼ of the way between 1.44 and 1.69, we estimate $\sqrt{1.51} \approx$ **1.23.**

25. $v = \sqrt{2gh}$; $g = 32.2$ and $h = 8$

26. $I = \sqrt{\dfrac{P}{r}}$; $P = 21{,}290$ and $r = 10$

27. $d = \sqrt{\dfrac{K}{I}}$; $K = 5.73$ and $I = 0.3$

28. $p = 2\pi\sqrt{\dfrac{W}{2g}}$; $\pi = 3.14$, $W = 24{,}075$, and $g = 32.2$

29. $R = \sqrt{P^2 + Q^2}$; $P = 100$ and $Q = 50$

30. $r = 2\sqrt{\dfrac{L}{C}}$; $L = 3.3$ and $C = 11$

31. $i = \sqrt{\dfrac{lg}{W}}$; $l = 0.005$, $g = 32.2$, and $W = 0.04$

32. $Q = \sqrt{\dfrac{3g}{l}}$; $g = 32.2$ and $l = 9$

6.7 Sums and differences of complex numbers

To this point, our discussion has been concerned entirely with the set of real numbers. In this and the next section we introduce and briefly examine a different kind of number, one that we shall encounter again in later chapters.

First note that there is no real number denoted by $\sqrt{-2}$, since the square of any real number is non-negative. Indeed, for $b > 0$, there is no real number corresponding to $\sqrt{-b}$. We can, however, consider a new set of numbers which contain elements that are square roots of negative numbers.

Let us define $\sqrt{-b}$, where b is a positive real number, to be the number such that

$$\boldsymbol{\sqrt{-b}\sqrt{-b} = -b.}$$

In particular then, the number $\sqrt{-1}$ satisfies

$$\sqrt{-1}\sqrt{-1} = -1.$$

Sometimes the number $\sqrt{-1}$ is designated by the symbol i, so that

$$\boldsymbol{i^2 = -1.}$$

Since for $b > 0$,

$$\sqrt{-b}\sqrt{-b} = -b,$$

and, since we assume that

$$(i\sqrt{b})(i\sqrt{b}) = i^2 b = -b,$$

it follows that

$$\boldsymbol{\sqrt{-b} = \sqrt{-1}\sqrt{b} = i\sqrt{b}.}$$

Hence, the square root of any negative real number can be represented as the product of a real number and the number $\sqrt{-1}$ or i. For example,

$$\sqrt{-4} = \sqrt{-1} \cdot \sqrt{4} = i\sqrt{4} = 2i$$

and

$$\sqrt{-3} = \sqrt{-1} \cdot \sqrt{3} = i\sqrt{3} = \sqrt{3}\, i.$$

The number represented by the symbol $\sqrt{-b}$, where $b \in R$ and $b > 0$, is called a **pure imaginary number**. Now, consider all possible expressions of the form $a + bi$, where $a, b \in R$ and $i = \sqrt{-1}$. We assume that each such expression names a number called a **complex number**. The set of all such numbers is denoted by C. If $b = 0$, then $a + bi = a$, and it is evident that the set R of real numbers is a subset of the set C of complex numbers. If $b \neq 0$, then each such number in the set C is called an **imaginary number**.

The sum and difference of two complex numbers are defined by

$$\begin{aligned}(a + bi) + (c + di) &= (a + c) + (b + d)i,\\(a + bi) - (c + di) &= (a - c) + (b - d)i.\end{aligned}$$

For example,

$$(3 + 4i) + (6 - 2i) = (3 + 6) + [4 + (-2)]i = 9 + 2i$$

and

$$(3 + 4i) - (6 - 2i) = (3 - 6) + [4 - (-2)]i = -3 + 6i.$$

Sometimes, to avoid ambiguity in reading symbols, we use the forms $a + ib$ and $a - ib$ for $a + bi$ and $a - bi$, respectively. Also, if $b = 0$, we write a for $a + 0i$, and if $a = 0$, we write bi for $0 + bi$.

EXERCISE 6.7

Write each expression in the form $a + bi$ or $a + ib$.

EXAMPLES

a. $\sqrt{-16}$ b. $2 - 3\sqrt{-16}$ c. $\dfrac{2 + \sqrt{-8}}{2}$

SOLUTIONS

a. $\sqrt{-1 \cdot 16}$

$\sqrt{-1}\sqrt{16}$

$4i$

b. $2 - 3\sqrt{-1 \cdot 16}$

$2 - 3\sqrt{-1}\sqrt{16}$

$2 - 12i$

c. $\dfrac{2 + 2i\sqrt{2}}{2}$

$\dfrac{2(1 + i\sqrt{2})}{2}$

$1 + i\sqrt{2}$

1. $\sqrt{-9}$
2. $\sqrt{-25}$
3. $3\sqrt{-4}$
4. $2\sqrt{-16}$
5. $4\sqrt{-32}$
6. $3 - \sqrt{-48}$
7. $\dfrac{\sqrt{-16}}{2}$
8. $\dfrac{4\sqrt{-9}}{3}$
9. $\dfrac{2 - \sqrt{-4}}{2}$
10. $\dfrac{4 - \sqrt{-32}}{2}$
11. $\dfrac{-2 - \sqrt{-8}}{2}$
12. $\dfrac{-3 + \sqrt{-18}}{-3}$

EXAMPLES

a. $(3 - i) - (6 + 2i)$

b. $(4 - 3\sqrt{-1}) + (2 + \sqrt{-1})$

SOLUTIONS

a. $(3 - 6) + (-1 - 2)i$
$\mathbf{-3 - 3i}$

b. $(4 - 3i) + (2 + i)$
$\mathbf{6 - 2i}$

13. $(2 + i) + (3 - 2i)$
14. $(1 - 5i) + (3 + i)$
15. $(1 - 3i) - (6 + 2i)$
16. $(5 - 2i) - (6 + 3i)$
17. $(2 - i) - (-3 + 4i)$
18. $(1 + 6i) - (-2 + i)$
19. $i + (3 + i)$
20. $2i + (2 + i)$
21. $4 - (2 - 3i)$
22. $6 - (3 + 2i)$
23. $(1 - 2i) + 2$
24. $(3 + i) - 6$
25. $(5 - 2\sqrt{-9}) + (3 + 2\sqrt{-9})$
26. $(3 - \sqrt{-1}) + (3 + \sqrt{-1})$
27. $(\sqrt{-4} - 3) - (\sqrt{-4} + 3)$
28. $(4 - \sqrt{-9}) - (2 - \sqrt{-16})$
29. $(3 - \sqrt{-8}) + (4 + 2\sqrt{-2})$
30. $(2 + \sqrt{-12}) + (3 - \sqrt{-3})$
31. $2\sqrt{-32} + (4 - \sqrt{-18})$
32. $3\sqrt{-12} + (2 - \sqrt{-27})$
33. $(1 - \sqrt{-45}) - 2\sqrt{-20}$
34. $(4 + 2\sqrt{-50}) - 3\sqrt{-8}$
35. $\left(\frac{1}{2} + \frac{\sqrt{-12}}{3}\right) + \left(\frac{1}{4} + \frac{\sqrt{-3}}{2}\right)$
36. $\left(\frac{2}{3} + \frac{\sqrt{-18}}{6}\right) + \left(\frac{1}{2} - \frac{\sqrt{-32}}{2}\right)$

6.8 Products and quotients of complex numbers

The product of two complex numbers is defined in such fashion that it conforms with the product of two binomials. Thus,

$$\begin{aligned}(a + bi)(c + di) &= a(c + di) + bi(c + di)\\ &= ac + adi + bci + bdi^2\\ &= ac + adi + bci - bd;\end{aligned}$$

$$\mathbf{(a + bi)(c + di) = (ac - bd) + (ad + bc)i.}$$

Ordinarily, we simplify such products mentally by duplicating the foregoing sequence. For example,

$$\begin{aligned}(2 + 3i)(4 - 2i) &= 8 - 4i + 12i - 6i^2\\ &= 8 - 4i + 12i + 6\\ &= 14 + 8i.\end{aligned}$$

Direct application of the definition of the product of two complex numbers to the same factors yields

$$\begin{aligned}(2 + 3i)(4 - 2i) &= [(2)(4) - (3)(-2)] + [(2)(-2) + (3)(4)]i \\ &= (8 + 6) + (-4 + 12)i \\ &= 14 + 8i.\end{aligned}$$

To simplify $7(6 + 3i)$, we have

$$\begin{aligned}(7 + 0i)(6 + 3i) &= 42 + 21i + 0i + 0i^2 \\ &= 42 + 21i.\end{aligned}$$

Notice that in the definition of a product we tacitly assume that multiplication distributes over addition in C. Notice also that the result obtained in the second example above can be obtained simply by applying the distributive law to $7(6 + 3i)$ directly. That is,

$$7(6 + 3i) = 7 \cdot 6 + 7 \cdot 3i = 42 + 21i,$$

and it is not necessary to write 7 in the form $7 + 0i$.

Assuming that the fundamental principle of fractions is valid in C and that

$$\frac{a + bi}{c} = \frac{a}{c} + \frac{b}{c}i \quad (c \neq 0),$$

we can express the quotient of two complex numbers in the form $a + bi$ by using the rationalizing process developed in Section 6.5.
For example, since the conjugate of $4 - i$ is $4 + i$,

$$\begin{aligned}\frac{3 + 2i}{4 - i} &= \frac{(3 + 2i)(4 + i)}{(4 - i)(4 + i)} \\ &= \frac{10 + 11i}{17} \\ &= \frac{10}{17} + \frac{11}{17}i.\end{aligned}$$

The powers of i exhibit an interesting periodic property. Observe that

$$\begin{aligned}i &= i \\ i^2 &= -1, \\ i^3 &= i^2 \cdot i = -i, \\ i^4 &= (i^2)^2 = 1, \\ i^5 &= i^4 \cdot i = i,\end{aligned}$$

and so forth. Thus, any integral power of i is equal to one of the four numbers i, -1, $-i$, or 1.

Certain relationships valid for real numbers are not valid for complex numbers. For instance, if $\sqrt{a}$ and $\sqrt{b}$ are both real,

$$\sqrt{a}\sqrt{b} = \sqrt{ab},$$

but if $\sqrt{a}$ and $\sqrt{b}$ are both imaginary,

$$\sqrt{a}\sqrt{b} = -\sqrt{ab}.$$

Specifically,

$$\sqrt{2}\sqrt{3} = \sqrt{6},$$

but

$$\sqrt{-2}\sqrt{-3} = (i\sqrt{2})(i\sqrt{3}) = i^2\sqrt{6} = -\sqrt{6}.$$

To avoid difficulty with this, all factors of the form $\sqrt{-a} \quad (a > 0)$ should be changed to the form $i\sqrt{a}$ before simplifying expressions for products or quotients involving such symbols.

EXERCISE 6.8

Write each expression in the form $a + bi$ or $a - bi$.

EXAMPLES

a. $3i \cdot i$ b. $i(3 + 2i)$ c. $(2 - i)(3 + 2i)$

SOLUTIONS

a. $3i^2$
$3(-1)$
$\mathbf{-3}$

b. $3i + 2i^2$
$\mathbf{-2 + 3i}$

c. $6 + 4i - 3i - 2i^2$
$6 + i + 2$
$\mathbf{8 + i}$

1. $2i \cdot i$
2. $i \cdot 4i$
3. $3i \cdot 2i$
4. $4i \cdot 5i$
5. $3i(-i)$
6. $-2i \cdot i$
7. $3(2 + i)$
8. $i(2 - i)$
9. $-2i(3 + 6i)$
10. $-3i(4 - 2i)$
11. $2(3 - i) - 4 \cdot 2i$
12. $5(3 - 2i) + 2 \cdot 3i$
13. $2i(3 - i) + 2(4 - 2i)$
14. $i(2 - i) + 2i(3 - 2i)$
15. $(2 - i)(3 + i)$
16. $(3 + 2i)(2 - i)$
17. $(1 + i)(2 + 3i)$
18. $(4 - 2i)(3 - i)$
19. $(1 + i)^2 - (1 - i)^2$
20. $(3 + 2i)^2 + (3 - 2i)^2$

EXAMPLES

a. $\sqrt{-2}\sqrt{-18}$

b. $\sqrt{-3}(3-\sqrt{-3})$

SOLUTIONS

a. $i\sqrt{2}\cdot 3i\sqrt{2}$

$3i^2\cdot 2$

$3(-1)\cdot 2$

$\mathbf{-6}$

b. $i\sqrt{3}(3-i\sqrt{3})$

$3i\sqrt{3}-3i^2$

$3i\sqrt{3}-3(-1)$

$\mathbf{3+3i\sqrt{3}}$

21. $\sqrt{-4}\sqrt{-9}$
22. $\sqrt{-16}\sqrt{-1}$
23. $\sqrt{-8}\sqrt{-6}$
24. $\sqrt{-5}\sqrt{-12}$
25. $\sqrt{-2}(1+\sqrt{-2})$
26. $\sqrt{-5}(3-\sqrt{-5})$
27. $(4+2\sqrt{-3})(4-\sqrt{-3})$
28. $(2-\sqrt{-2})(3-\sqrt{-2})$
29. $(3-\sqrt{-5})(3+\sqrt{-5})$
30. $(1+2\sqrt{-3})(1-2\sqrt{-3})$

EXAMPLES

a. $\dfrac{-2}{i}$

b. $\dfrac{i}{1+2i}$

c. $\dfrac{1}{2-\sqrt{-9}}$

SOLUTIONS

a. $\dfrac{-2(i)}{i(i)}$

$\dfrac{-2i}{i^2}$

$\dfrac{-2i}{-1}$

$\mathbf{2i}$

b. $\dfrac{i(1-2i)}{(1+2i)(1-2i)}$

$\dfrac{i-2i^2}{1-4i^2}$

$\dfrac{i+2}{5}$

$\mathbf{\dfrac{2}{5}+\dfrac{1}{5}i}$

c. $\dfrac{1(2+3i)}{(2-3i)(2+3i)}$

$\dfrac{2+3i}{4-9i^2}$

$\dfrac{2+3i}{13}$

$\mathbf{\dfrac{2}{13}+\dfrac{3}{13}i}$

31. $\dfrac{1}{i}$
32. $\dfrac{-3}{i}$
33. $\dfrac{-3}{\sqrt{-4}}$
34. $\dfrac{7}{\sqrt{-9}}$
35. $\dfrac{-2}{1-i}$
36. $\dfrac{3}{2+i}$
37. $\dfrac{3}{3+2i}$
38. $\dfrac{-5}{2-i}$

39. $\dfrac{\sqrt{-1}}{2+\sqrt{-9}}$ 40. $\dfrac{\sqrt{-9}}{1-\sqrt{-1}}$ 41. $\dfrac{\sqrt{-1}-1}{\sqrt{-1}+1}$ 42. $\dfrac{1+\sqrt{-4}}{1-\sqrt{-4}}$

43. Simplify.
a. i^6 b. i^{12} c. i^{15}

44. Express each of the following with positive exponents and simplify.
a. i^{-1} b. i^{-2} c. i^{-3} d. i^{-4} e. i^{-5}

Chapter review

[6.1] Write in radical form.

1. $x^{3/2}$ 2. $x^{1/3} - y^{1/3}$ 3. $(x-y)^{1/3}$ 4. $5x^{-1/3}$

Find each root.

5. $\sqrt{64}$ 6. $\sqrt[3]{-8}$ 7. $\sqrt{S^2R^4}$ 8. $\sqrt[3]{\dfrac{8}{27}T^6}$

[6.2] Simplify.

9. $\sqrt{12}$ 10. $\sqrt[3]{x^4y^5}$ 11. $\sqrt{a}\sqrt{ab}$ 12. $\sqrt{0.0032}$

13. $\sqrt{\dfrac{3}{5}}$ 14. $\dfrac{6}{\sqrt{2x}}$ 15. $\dfrac{6}{\sqrt[3]{2x}}$ 16. $\sqrt[4]{9y^2}$

[6.3] Add.

17. $2\sqrt{50}+\sqrt{18}$ 18. $\sqrt{75a^3} - a\sqrt{3a} + \sqrt{12a}$

Write as a single fraction.

19. $\sqrt{\dfrac{1}{2}} - \dfrac{\sqrt{2}}{2}$ 20. $\sqrt{9x} - \dfrac{\sqrt{16x}}{3}$

[6.4] Multiply factors and simplify.

21. $(\sqrt{2}-3)(\sqrt{2}+3)$ 22. $(\sqrt{x}+3)(2\sqrt{x}-1)$

Change each expression to the form indicated.

23. $2-\sqrt{8} = 2(\quad ? \quad)$ 24. $\sqrt{y}+\sqrt{5y} = \sqrt{y}(\quad ? \quad)$

[6.5] Rationalize denominators.

25. $\dfrac{1}{2-\sqrt{3}}$ 26. $\dfrac{a-b}{\sqrt{a}+\sqrt{b}}$

Rationalize numerators.

27. $\dfrac{\sqrt{3}}{2}$ 28. $\dfrac{1+\sqrt{3}}{2}$

[6.6] 29. Graph $\{\sqrt{9}, -\sqrt{2}, \sqrt{29}, -\sqrt{25}\}$ on a line graph.

30. If $b = \sqrt{\dfrac{2R}{a}}$, $R = 17.2$, and $a = 4$, find b.

[6.7] Write each expression in the form $a + bi$ or $a + ib$.

31. $\sqrt{-36}$

32. $\dfrac{3+\sqrt{-18}}{3}$

33. $(5-3i)-(3-2i)$

34. $(2+\sqrt{-9})+(4-\sqrt{-1})$

[6.8] Write each expression in the form $a + bi$ or $a + ib$.

35. $2i \cdot 5i$

36. $(5-3i)(2+i)$

37. $\sqrt{-3}\sqrt{-6}$

38. $(2+\sqrt{-3})(3-2\sqrt{-3})$

39. $\dfrac{4}{1+i}$

40. $\dfrac{2}{1+\sqrt{-3}}$

7

Second-degree equations: one variable

An open sentence in one variable that, in simplest form, contains the second, but no higher, power of the variable is commonly called a second-degree or **quadratic equation** in that variable. We shall designate as standard form for such equations

$$ax^2 + bx + c = 0,$$

where a, b, and c are constants representing real numbers and $a \neq 0$.

In Chapter 4 we solved linear equations by performing certain elementary transformations. The transformations were based on the assumption that we always obtain an equivalent equation when we add (or subtract) the same polynomial to (or from) each member of an equation or when we multiply or divide each member by the same polynomial (not representing zero). These transformations are equally applicable to equations of higher degree and, in particular, to quadratic equations. For example,

$$2x = 6 - 8x^2$$

is equivalent to

$$8x^2 + 2x - 6 = 0,$$

which is equivalent to

$$4x^2 + x - 3 = 0.$$

7.1 Solution of quadratic equations by factoring

If the left member of a quadratic equation in standard form is factorable, we may solve the equation by making use of the principle:

The product of two factors equals zero if and only if one or both of the factors equals zero.

That is,

$$\boldsymbol{ab = 0 \text{ if and only if } a = 0 \text{ or } b = 0.}$$

We shall use the word "or" in an inclusive sense to mean either one or the other or both. Thus, if

$$x^2 + 2x - 15 = 0, \tag{1}$$

then

$$(x + 5)(x - 3) = 0, \tag{2}$$

which will be true if and only if

$$x + 5 = 0 \quad \text{or} \quad x - 3 = 0.$$

We observe that $x + 5$ equals zero if $x = -5$, and $x - 3$ equals zero if $x = 3$; and either -5 or 3, when substituted for x in (2) or (1), will make the left member zero. The solution set of (1) is $\{-5, 3\}$.

In general, the solution set of a quadratic equation can be expected to contain two members. However, if the left member of a quadratic equation in standard form is the square of a binomial, the solution set contains but one member. Consider the equation

$$x^2 - 2x + 1 = 0.$$

If we factor the left member, we have

$$(x - 1)(x - 1) = 0,$$

and the solution set is $\{1\}$, which contains but one member. Because, for reasons of convenience and consistency in more advanced work, we wish to consider each quadratic equation to have two roots, a solution of this sort is said to be of **multiplicity two**; that is, it is considered twice as a solution.

Since the solution set of the quadratic equation

$$(x - r_1)(x - r_2) = 0 \tag{3}$$

is $\{r_2, r_2\}$, if r_1 and r_2 are given as solutions of a quadratic equation, the equation can be written directly as (3). Completion of the indicated multiplication transforms the equation to standard form. For example, if 2 and -3 are the solutions of a quadratic equation, then

$$[x-(2)][x-(-3)] = 0,$$

or

$$(x - 2)(x + 3) = 0,$$

from which

$$x^2 + x - 6 = 0.$$

If $\frac{1}{4}$ and $\frac{3}{2}$ are the solutions of a quadratic equation, then we have

$$\left(x - \frac{1}{4}\right)\left(x - \frac{3}{2}\right) = 0,$$

from which

$$x^2 - \frac{7}{4}x + \frac{3}{8} = 0.$$

This equation can be transformed to one with integral coefficients by multiplying the left member and the right member by 8 to obtain the equivalent equation

$$8x^2 - 14x + 3 = 0.$$

EXERCISE 7.1

What is the set of values of x for which each of the following products yields zero? Determine your answer by inspection or follow the procedure in the examples.

EXAMPLE

$x(x - 3)(2x + 1)$

SOLUTION

Set each factor equal to zero.

$$x = 0; \quad x - 3 = 0; \quad 2x + 1 = 0$$

Solution continued overleaf.

Solve each linear equation.

$$x = 0; \quad x = 3; \quad x = -\frac{1}{2}$$

The set is $\left\{\mathbf{0, 3, -\frac{1}{2}}\right\}$.

1. $(x - 2)(x - 3)$
2. $2x(x + 1)$
3. $x(2x - 3)(x + 2)$
4. $3x(3x - 4)(2x - 1)$
5. $4(x - 1)(2x + 5)$
6. $3x(x - 3)(2x - 5)$
7. $(x - 1)(x + 2)(x - 3)$
8. $(2x + 1)(2x - 1)(x - 2)$
9. $(5x + 2)(3x - 1)(4x - 3)$
10. $(6x - 1)(5x + 2)(3x - 2)$

Solve for x.

EXAMPLE

$x(x + 3) = 0$

SOLUTION

Set each factor equal to zero.

$$x = 0; \quad x + 3 = 0$$

Solve each linear equation.

$$x = 0; \qquad x = -3$$

The solution set is $\{\mathbf{0, -3}\}$.

11. $(x - 3)(x + 2) = 0$
12. $x(x - 5) = 0$
13. $2(2x - 3)(x + 1) = 0$
14. $3(3x - 1)(2x - 1) = 0$
15. $4x(2x + 1)(x - 3) = 0$
16. $2x(x + 1) = 0$
17. $(2x - 5)(x + 3) = 0$
18. $(4x - 3)(3x - 4) = 0$

EXAMPLE

$x^2 + x = 30$

SOLUTION

Write in standard form.

$$x^2 + x - 30 = 0$$

Factor left member.

$$(x + 6)(x - 5) = 0$$

Set each factor equal to zero.

$$x+6=0; \qquad x-5=0$$

Solve each linear equation.

$$x=-6; \qquad x=5$$

The solution set is **{−6, 5}**.

19. $x^2+2x=0$
20. $x^2-4=0$
21. $2x^2-18=0$
22. $3x^2-3=0$
23. $x^2-\frac{4}{9}=0$
24. $x^2-\frac{1}{4}=0$
25. $R^2-3R-4=0$
26. $R^2+3R+2=0$
27. $S^2+5S-14=0$
28. $S^2+S-42=0$
29. $3T^2-6T=-3$
30. $12T^2=8T+15$

EXAMPLE

$3P(P+1)=2P+2$

SOLUTION

Write in standard form.

$$3P^2+P-2=0$$

Factor left member.

$$(3P-2)(P+1)=0$$

Set each factor equal to zero.

$$3P-2=0; \quad P+1=0$$

Solve each linear equation.

$$P=\frac{2}{3}; \qquad P=-1$$

The solution set is $\left\{\mathbf{\frac{2}{3}, -1}\right\}$.

31. $L(2L-3)=-1$
32. $2L(L-2)=L+3$
33. $(M-2)(M+1)=4$
34. $M(3M+2)=(M+2)^2$
35. $\frac{2S^2}{3}+\frac{S}{3}-2=0$
36. $S-1=\frac{S^2}{4}$
37. $\frac{S^2}{6}+\frac{S}{3}=\frac{1}{2}$
38. $\frac{S^2}{4}-\frac{3S}{4}=1$
39. $3=\frac{10}{T^2}-\frac{7}{T}$
40. $\frac{4}{3T}+\frac{3}{3T+1}+2=0$
41. $\frac{2}{T-3}-\frac{6}{T-8}=-1$
42. $\frac{T}{T-1}-\frac{T}{T+1}=\frac{4}{3}$

Given the solutions of a quadratic equation, write the equation in standard form with integral coefficients. Use x as the variable.

EXAMPLE

$\frac{3}{4}$ and -2

SOLUTION

Write in the form $(x - r_1)(x - r_2) = 0$.

$$\left(x - \frac{3}{4}\right)[x - (-2)] = 0$$

Simplify left member.

$$\left(x - \frac{3}{4}\right)(x + 2) = 0$$

$$x^2 + \frac{5}{4}x - \frac{3}{2} = 0$$

Multiply each member by 4.

$$\mathbf{4x^2 + 5x - 6 = 0}$$

43. 3 and 2

44. 5 and 1

45. $-\frac{1}{2}$ and 3

46. $\frac{2}{3}$ and 2

47. $\frac{1}{2}$ and $\frac{3}{4}$

48. $-\frac{2}{3}$ and $\frac{1}{2}$

49. $\frac{1}{3}$ and $\frac{1}{3}$

50. $-\frac{2}{5}$ and $-\frac{2}{5}$

51. i and $-i$

52. $3i$ and $-3i$

53. $1 + i$ and $1 - i$

54. $2 + i$ and $2 - i$

7.2 Solution of equations of the form $x^2 = a$

Quadratic equations of the form

$$x^2 = a,$$

where a is any real number, may be solved by a method often termed the **extraction of roots.** If the equation has a solution, then from the definition of a square root, x must be a square root of a. Since each nonzero real number a has two square roots (either real or imaginary), we have two solutions. These are $\sqrt{a}$ and $-\sqrt{a}$, and the solution set is $\{\sqrt{a}, -\sqrt{a}\}$, whose elements are real if $a \geq 0$ and imaginary if $a < 0$. If $a = 0$, we have one number, 0, satisfying the equation, and the solution set is $\{0\}$. The same conclusion can be reached by noting that, if we factor

$$x^2 - a = 0$$

over the complex numbers, we have

$$(x - \sqrt{a})(x + \sqrt{a}) = 0,$$

which, when solved by the methods of the preceding section, also leads to the solution set $\{\sqrt{a}, -\sqrt{a}\}$.

Equations of the form

$$(x - a)^2 = b$$

can also be solved by the same method. For example,

$$(x - 2)^2 = 16$$

implies that

$$x - 2 = 4 \quad \text{or} \quad x - 2 = -4,$$

from which we have

$$x = 6 \quad \text{or} \quad x = -2$$

and the solution set is $\{6, -2\}$.

EXERCISE 7.2

Solve for x by the extraction of roots.

EXAMPLE

$7x^2 - 63 = 0$

SOLUTION

Obtain an equivalent equation with x^2 as the only term in the left member.

$$7x^2 = 63$$

$$x^2 = 9$$

Set x equal to each square root of 9.

$$x = +3; \quad x = -3$$

The solution set is $\mathbf{\{3, -3\}}$.

1. $x^2 = 4$
2. $x^2 = 16$
3. $3x^2 = 12$
4. $5x^2 = 125$
5. $9x^2 - 100 = 0$
6. $9x^2 - 4 = 0$
7. $x^2 = 5$
8. $3x^2 = 21$
9. $\dfrac{x^2}{4} = 3$
10. $\dfrac{2x^2}{7} = 8$
11. $x^2 = P$
12. $x^2 - R = 0$
13. $Px^2 - Q = 0$
14. $Lx^2 - C = 0$
15. $2Mx^2 - N = 0$
16. $Rx^2 - 3LC = 0$
17. $\dfrac{x^2}{R} = T$
18. $\dfrac{Rx^2}{P} = V$

EXAMPLE

$(x + 3)^2 = 7$

SOLUTION

Set $x + 3$ equal to each square root of 7.

$$x + 3 = \sqrt{7}; \qquad x + 3 = -\sqrt{7}$$
$$x = -3 + \sqrt{7}; \qquad x = -3 - \sqrt{7}$$

The solution set is $\{-3 + \sqrt{7}, -3 - \sqrt{7}\}$.

19. $(x - 1)^2 = 4$
20. $(x - 3)^2 = 16$
21. $(2x + 5)^2 = 9$
22. $(2x + 1)^2 = 25$
23. $(x - 6)^2 = 5$
24. $(x + 2)^2 = 7$
25. $(x - 2)^2 = 3$
26. $(x - 3)^2 = 18$
27. $(x - R)^2 = 4$
28. $(x + S)^2 = 5$
29. $(Px + Q)^2 = 16$
30. $(Px - Q)^2 = 9$
31. $(Sx - T)^2 = 5$
32. $(Px - L)^2 = 3$
33. $(x + B)^2 = C$
34. $(Ax + 1)^2 = C$
35. $(Rx - C)^2 = L$
36. $P(Rx - S)^2 = T$

7.3 Solution by completing the square

The technique of the preceding section can be used to find the solution set of any quadratic equation. Let us first consider the general quadratic equation in standard form

$$ax^2 + bx + c = 0,$$

for the special case where $a = 1$; that is,

$$x^2 + bx + c = 0. \tag{1}$$

If we can factor the left member of Equation (1), we can solve the equation by factoring; if not, we can write it in the form

$$(x - p)^2 = q,$$

which we can solve by the extraction of roots. We begin the latter process by adding $-c$ to each member of (1), which yields

$$x^2 + bx \qquad = -c. \tag{2}$$

If we then add $(b/2)^2$ to each member of Equation (2),

$$x^2 + bx + \left(\frac{b}{2}\right)^2 = -c + \left(\frac{b}{2}\right)^2, \tag{3}$$

the left member is equivalent to $\left(x + \frac{b}{2}\right)^2$ and we have

$$\left(x + \frac{b}{2}\right)^2 = -c + \frac{b^2}{4}. \tag{4}$$

Since we have performed only elementary transformations, Equation (4) is equivalent to Equation (2), and we can solve (4) by the method of the preceding section.

The technique used to obtain Equations (3) and (4) is called **completing the square.**

To determine the term necessary to complete the square in $x^2 + bx$, divide the coefficient b of the linear term by 2 and square the result.

The expression obtained,

$$x^2 + bx + \left(\frac{b}{2}\right)^2,$$

is called a **perfect square** and may be written in the form

$$\left(x + \frac{b}{2}\right)^2.$$

We began with the special case

$$x^2 + bx + c = 0, \tag{1}$$

rather than the general form

$$ax^2 + bx + c = 0,$$

because the term necessary to complete the square is more obvious when $a = 1$. However, a quadratic equation in standard form can always be written in the form (1) by dividing each member of

$$ax^2 + bx + c = 0$$

by a and obtaining

$$x^2 + \frac{b}{a}x + \frac{c}{a} = 0.$$

EXERCISE 7.3

(a) State what term must be added to each expression to make a perfect square.

(b) Write the perfect square obtained in the form $\left(x + \frac{b}{2}\right)^2$.

EXAMPLE

$x^2 - 9x$

SOLUTION

Square one-half of the coefficient of the first-degree term (-9).

$$\left(\frac{-9}{2}\right)^2 = \mathbf{\frac{81}{4}}$$

Rewrite $x^2 - 9x + \frac{81}{4}$ as the square of an expression.

$$\mathbf{\left(x - \frac{9}{2}\right)^2}$$

1. $x^2 + 2x$
2. $x^2 - 4x$
3. $x^2 - 6x$
4. $x^2 + 10x$
5. $x^2 + 3x$
6. $x^2 - 5x$
7. $x^2 - 7x$
8. $x^2 + 11x$
9. $x^2 - x$
10. $x^2 + 15x$
11. $x^2 + \frac{1}{2}x$
12. $x^2 - \frac{3}{4}x$

Solve by completing the square.

EXAMPLE

$2x^2 + x - 1 = 0$

SOLUTION

Rewrite equation with the constant term as the right member and the coefficient of x^2 equal to 1.

$$x^2 + \frac{1}{2}x \qquad = \frac{1}{2}$$

Add the square of one-half of the coefficient of the first-degree term to each member.

$$x^2 + \frac{1}{2}x + \frac{1}{16} = \frac{1}{2} + \frac{1}{16}$$

Rewrite left member as the square of an expression.

$$\left(x + \frac{1}{4}\right)^2 = \frac{9}{16}$$

Set $x + \frac{1}{4}$ equal to each square root of $\frac{9}{16}$.

$$x + \frac{1}{4} = \frac{3}{4}; \quad x + \frac{1}{4} = -\frac{3}{4}$$

$$x = \frac{1}{2} \qquad x = -1$$

The solution set is $\left\{\mathbf{\frac{1}{2}}, \mathbf{-1}\right\}$.

13. $x^2 + 4x - 12 = 0$
14. $x^2 - x - 6 = 0$
15. $x^2 - 2x + 1 = 0$
16. $x^2 + 4x + 4 = 0$
17. $x^2 + 9x + 20 = 0$
18. $x^2 - x - 20 = 0$
19. $x^2 - 2x - 1 = 0$
20. $x^2 + 3x - 1 = 0$
21. $2x^2 = 2 - 3x$
22. $2x^2 = 3 - 5x$
23. $2x^2 + 4x = -1$
24. $3x^2 + x = 1$

Reduce each equation in Exercises 25–32 to the form $y = (x - h)^2 + k$ by completing the square in x.

EXAMPLE

$y = x^2 - 3x + 5$

SOLUTION

Write equation in the following form.

$$y = (x^2 - 3x \qquad) + 5$$

Complete the square in x by adding and subtracting $\frac{9}{4}$ in the right member.

$$y = \left(x^2 - 3x + \frac{9}{4}\right) + 5 - \frac{9}{4}$$

$$\mathbf{y = \left(x - \frac{3}{2}\right)^2 + \frac{11}{4}}$$

25. $y = x^2 + 2x - 3$
26. $y = x^2 - 6x + 7$
27. $y = x^2 + x + 1$
28. $y = x^2 - 3x + 7$
29. $y - x^2 = 5 - 2x$
30. $y - x^2 = 18 + 10x$
31. $3y - 3x^2 = 6x + 5$
32. $2y = 2x^2 + 8x - 3$

Reduce each of the following equations to the form $(x - h)^2 + (y - k)^2 = r^2$.

EXAMPLE

$x^2 + y^2 - 4x + 6y = 5$

SOLUTION

Write equation in the following form.

$$[x^2 - 4x + (\quad)] + [y^2 + 6y + (\quad)] = 5$$

Complete the squares in x and y by adding 4 and 9 to each member.

$$[x^2 - 4x + 4] + [y^2 + 6y + 9] = 5 + 4 + 9$$

$$\mathbf{(x - 2)^2 + (y + 3)^2 = 18} \quad \text{or} \quad \mathbf{(x - 2)^2 + (y + 3)^2 = (\sqrt{18})^2}$$

33. $x^2 + y^2 - 4x - 4y - 17 = 0$
34. $x^2 + y^2 + 6x - 6y + 18 = 0$
35. $x^2 + y^2 + 6x - 2y + 6 = 0$
36. $x^2 + y^2 - 2x + 4y + 2 = 0$
37. $x^2 + y^2 - 2x + 8y + 15 = 0$
38. $x^2 + y^2 + 2x - 2y + 1 = 0$
39. $4x^2 + 4y^2 - 4x + 8y = 11$
40. $16x^2 + 16y^2 - 8x + 16y - 59 = 0$
41. Solve $ax^2 + bx + c = 0$ for x in terms of a, b, and c.

7.4 The quadratic formula

We can complete the square on the general quadratic equation

$$ax^2 + bx + c = 0 \quad (a \neq 0)$$

as follows:

$$ax^2 + bx + c = 0,$$

$$x^2 + \frac{b}{a}x + \frac{c}{a} = 0,$$

$$x^2 + \frac{b}{a}x + \left(\frac{b}{2a}\right)^2 = -\frac{c}{a} + \left(\frac{b}{2a}\right)^2,$$

$$\left(x + \frac{b}{2a}\right)^2 = \frac{b^2}{4a^2} - \frac{c}{a},$$

$$\left(x + \frac{b}{2a}\right)^2 = \frac{b^2 - 4ac}{4a^2},$$

$$x + \frac{b}{2a} = \pm\sqrt{\frac{b^2 - 4ac}{4a^2}},$$

$$x = -\frac{b}{2a} \pm \frac{\sqrt{b^2 - 4ac}}{2a}$$

The right-hand member of the last equation can now be written as a single fraction to obtain

$$\boldsymbol{x = \frac{-b \pm \sqrt{b^2 - 4ac}}{2a}}.$$

This result, called the **quadratic formula**, expresses the solutions of a quadratic equation in terms of the coefficients. The symbol "$\pm$" is used to condense the writing of the two equations

$$x = \frac{-b + \sqrt{b^2 - 4ac}}{2a} \quad \text{and} \quad x = \frac{-b - \sqrt{b^2 - 4ac}}{2a}$$

into a single equation. We need only substitute the coefficients a, b, and c of a given quadratic equation in the formula to find the solution set for the equation.

In the quadratic formula, the number represented by $b^2 - 4ac$ is called the **discriminant** of the equation, and it affects the solution set in the following ways:

1. If $b^2 - 4ac = 0$, there is one real solution.
2. If $b^2 - 4ac < 0$, there are two imaginary solutions.
3. If $b^2 - 4ac > 0$, there are two unequal real solutions.

We note from the form

$$x = \frac{-b}{2a} \pm \frac{\sqrt{b^2 - 4ac}}{2a}$$

that if the solutions of a quadratic equation are complex, they are conjugates of each other.

EXERCISE 7.4

Solve for x, y, or z using the quadratic formula.

EXAMPLE

$$\frac{x^2}{4} + \frac{x}{4} = 3$$

SOLUTION

Write in standard form.

$$x^2 + x = 12$$
$$x^2 + x - 12 = 0$$

Substitute 1 for a, 1 for b, and -12 for c in the quadratic formula.

$$x = \frac{-1 \pm \sqrt{1 - 4(1)(-12)}}{2}$$

Simplify.

$$x = \frac{-1 \pm \sqrt{1 + 48}}{2} = \frac{-1 \pm 7}{2}$$

The solution set is **{3, −4}**.

1. $x^2 - 3x + 2 = 0$
2. $x^2 + 4x + 4 = 0$
3. $x^2 - 2x = -1$
4. $z^2 - 5z = 6$
5. $y^2 + y = 1$
6. $x^2 - 3x = -1$
7. $2x^2 = 7x - 6$
8. $3y^2 = 5y - 1$
9. $2x^2 - x + 1 = 0$

10. $6z^2 = 5z + 6$ 11. $x^2 = \frac{15}{4} - x$ 12. $\frac{y^2 - y}{2} + 1 = 0$

13. $\frac{x^2}{3} = \frac{x}{2} + \frac{3}{2}$ 14. $\frac{x^2 - 3}{2} + \frac{x}{4} = 1$ 15. $x^2 - kx - 2k^2 = 0$

16. $2x^2 - kx + 3 = 0$ 17. $ax^2 - x + c = 0$ 18. $x^2 + 2x + c + 3 = 0$

In Problems 19–28, find only the discriminant and determine whether the solution set contains

(a) one real solution,
(b) two unequal real solutions, or
(c) two imaginary solutions.

EXAMPLE

$x^2 - x - 3 = 0$

SOLUTION

Substitute 1 for a, -1 for b, and -3 for c.

$$\begin{aligned} b^2 - 4ac &= (-1)^2 - 4(1)(-3) = 1 + 12 \\ &= \mathbf{13} \end{aligned}$$

Since $b^2 - 4ac > 0$, there are **two unequal real solutions.**

19. $x^2 - 7x + 12 = 0$ 20. $y^2 - 2y - 3 = 0$ 21. $5x^2 + 2x - 1 = 0$
22. $2y^2 + 3y + 7 = 0$ 23. $x^2 - 2x = -1$ 24. $2z^2 - z = 12$
25. $8x^2 + 17 = x$ 26. $y^2 + 4 = -4y$ 27. $3x^2 = -2x - 1$
28. $3y^2 = 2 - y$
29. Determine k so that $kx^2 + 4x + 1 = 0$ has one solution.
30. Determine k so that $x^2 - kx + 9 = 0$ has one solution.
31. Determine k so that the solutions of $x^2 + 2x + k + 3 = 0$ are real numbers.
32. Determine k so that the solutions of $x^2 - x + k - 2 = 0$ are imaginary.

7.5 Equations involving radicals

In order to solve equations containing radicals, let us make the following assumption.

If each member of an equation is raised to the same power, the solution set of the resulting equation contains all of the solutions of the original equation.

For example, if $x = 3$, then the solution set of $x^2 = 9$ contains 3 as a member. In general, if

$$x = \sqrt[n]{a}, \tag{1}$$

then the solution set of the equation

$$x^n = a, \tag{2}$$

resulting from raising each member of Equation (1) to the nth power, contains $\sqrt[n]{a}$ as a member.

The application of the foregoing assumption does not result in an equivalent equation, and is not an elementary transformation. Equation (2) actually has $(n - 1)$ other solutions that are not solutions of (1). With respect to equation (1), these are called **extraneous solutions.** This situation derives from our restricting $\sqrt[n]{a}$ to a unique number. Thus the solution set of the equation $x^2 = 9$, obtained from $x = 3$ by squaring each member, contains -3 as an extraneous solution, since -3 does not satisfy the original equation. Because the result of applying the foregoing assumption is not an equivalent equation, each solution obtained through its use *must* be checked in the original equation to verify its validity.

EXERCISE 7.5

Solve. If there is no solution, so state.

EXAMPLE

$\sqrt{x+2}+4=x$

SOLUTION

Obtain $\sqrt{x+2}$ as the only term in one member.

$$\sqrt{x+2}=x-4$$

Square each member.

$$x+2=x^2-8x+16$$

Solve quadratic equation.

$$x^2-9x+14=0$$

$$(x-7)(x-2)=0$$

$$x=7; \quad x=2$$

Substitute each number in the original equation.

Does $\sqrt{7+2}+4=7$?	Does $\sqrt{2+2}+4=2$?
Does $7=7$?	Does $6=2$? 2 is not a solution.

The solution set is $\{7\}$.

1. $\sqrt{x}=8$
2. $\sqrt{y}-4=1$
3. $2\sqrt{y}-3=1.$
4. $3+5\sqrt{y}=-13$
5. $\sqrt{y+8}+4=3$
6. $\sqrt{x-3}=5$
7. $\sqrt{2x+1}-4=0$
8. $3\sqrt{3y-2}-6=0$
9. $2x-3=\sqrt{7x-3}$
10. $\sqrt{3x+10}=x+4$
11. $\sqrt{x+3}\sqrt{x-9}=8$
12. $\sqrt{x-4}\sqrt{x+4}=3$

EXAMPLE

$\sqrt{y-5}-\sqrt{y}=1$

SOLUTION

Write with $\sqrt{y-5}$ as left member.

$$\sqrt{y-5}=1+\sqrt{y}$$

Square each member.

$$y-5=1+2\sqrt{y}+y$$

Write with $\sqrt{y}$ as right member.

$$-3=\sqrt{y}$$

(*Note*: At this point it is obvious that the equation has no solution, since $\sqrt{y}$ cannot be negative. We continue the solution, however, for illustrative purposes.) Square each member.

$$9=y$$

Substitute 9 in the original equation.

Does $\sqrt{4}-\sqrt{9}=1$?
Does $2-3=1$?
Does $-1=1$? 9 is not a solution of the original equation.

The solution set is $\emptyset$.

13. $\sqrt{y+4}=\sqrt{y+20}-2$
14. $4\sqrt{y}+\sqrt{1+16y}=5$
15. $\sqrt{x}+\sqrt{2}=\sqrt{x+2}$
16. $\sqrt{5x+15}=4-\sqrt{x-1}$
17. $\sqrt{5+R}+\sqrt{R}=5$
18. $\sqrt{T+7}+\sqrt{T+4}=3$

Solve. Leave the results in the form of an equation.

19. $r = \sqrt{\dfrac{A}{\pi}}$, for A

20. $t = \sqrt{\dfrac{2v}{g}}$, for g

21. $R\sqrt{RL} = 1$, for L

22. $P = \pi\sqrt{\dfrac{e}{g}}$, for g

23. $P = \sqrt{Q^2 - R^2}$, for R

24. $L = \dfrac{1}{\sqrt{1 - C}}$, for C

7.6 Word problems

Some procedures for writing equations for word problems that were examined in Chapter 4 should be reviewed at this time.

In some cases, the mathematical model we obtain for a physical situation is a quadratic equation and consequently has two solutions. It may be that one solution to the equation, but not both, fits the physical situation. For example, if we were asked to find two consecutive *natural numbers* whose product is 72, we would write the equation

$$x(x + 1) = 72$$

as our model. Solving this equation, we have

$$x^2 + x - 72 = 0,$$
$$(x + 9)(x - 8) = 0,$$

where the solution set is $\{8, -9\}$. Since -9 is not a natural number, we must reject it as a possible answer to our original question; however, the solution 8 leads to the consecutive natural numbers 8 and 9. As additional examples, we observe that we would not accept -6 feet as the height of a man or $27/4$ for the number of people in a room.

A quadratic equation used as a model for a physical situation may have two, one, or no meaningful solutions—meaningful, that is, in a physical sense. Answers to word problems should always be checked against the original problem.

EXERCISE 7.6

Solve.

1. Find two numbers whose sum is 13 and whose product is 36.
2. Find two numbers whose sum is -16 and whose product is 48.
3. Find two consecutive integers whose product is 56.
4. Find two consecutive odd integers whose product is 63.

5. Find two consecutive positive integers the sum of whose squares is 85.
6. Find two consecutive positive integers the difference of whose squares is 15.
7. The sum of a number and its reciprocal is $17/4$. What is the number?
8. The difference of a number subtracted from twice its reciprocal is $31/4$. What is the number?
9. The sum of the squares of two negative integers is 193. One of the integers is -7. What is the other ?
10. The difference of the squares of two positive integers is 161. What is the larger integer if the smaller integer is 8 ?
11. The sum of two positive integers is 120 and their product is 3200. What are the integers ?
12. A positive integer is increased by 12. If the square of the resulting, larger integer equals nine times the square of the smaller integer, what is the smaller integer ?
13. An object, dropped from a height, falls a distance of s feet given by the equation $s = 16t^2$, where t is the time the object is falling. Approximately how long will it take the object to reach the ground if it is dropped from a height of 800 feet ?
14. A ball thrown vertically upward reaches a height h in feet given by the equation $h = 32t - 8t^2$, where t is the time in seconds after the throw. How long will it take the ball to reach a height of 24 feet on its way up? How long after the throw will the ball return to the ground?
15. The distance s a body falls in a vacuum is given by $s = v_0 t + \frac{1}{2}gt^2$, where s is in feet, t is in seconds, v_0 is the initial velocity in feet per second, and g is the constant of acceleration due to gravity (approximately 32 ft/sec/sec). How long will it take a body to fall 150 feet if v_0 is 20 feet per second? If the body starts from rest?
16. A man sailed a boat across a lake and back in $2\frac{1}{2}$ hours. If his rate returning was 2 miles per hour less than his rate going, and if the distance each way was 6 miles, find his rate each way.
17. A man rode a bicycle for six miles and then walked an additional four miles. The total time for his trip was six hours. If his rate walking was two miles per hour less than his rate on the bicycle, what was each rate?
18. A man and his son working together can paint their house in 4 days. The man can do the job alone in six days less than the son can do it. How long would it take each of them to paint the house alone?
19. A theatre that is rectangular in shape seats 720 people. The number of rows needed to seat the people would be four fewer if each row held six more people. How many people would then be in each row?

20. Two tanks, each cylindrical in shape and ten feet in length, are to be replaced by a single tank of the same length. If the two original tanks have radii that measure 6 feet and 8 feet, respectively, what must be the length of the radius of the single tank replacing them if it is to hold the same volume of liquid?

Chapter review

[7.1] Solve by factoring.

1. $x^2 + 3x - 10 = 0$ 2. $R^2 - 3R = 0$ 3. $T^2 - 2T = 15$

4. Write the quadratic equation in standard form that has solutions $\frac{2}{3}$ and $-\frac{1}{2}$.

[7.2] Solve for x by the extraction of roots.

5. $4x^2 - 9 = 0$ 6. $(x - c)^2 - b = 0$

[7.3] Solve by completing the square.

7. $2x^2 - 3x - 2 = 0$ 8. $x^2 - 4x - 2 = 0$

9. Write $y = x^2 - 4x + 9$ in the form $y = (x - h)^2 + k$.

10. Write $x^2 + y^2 + 4x - 6y = 3$ in the form $(x - h)^2 + (y - k)^2 = r^2$.

[7.4] Solve for x by using the quadratic formula.

11. $2x^2 = 3 - 5x$ 12. $x^2 = \frac{7}{3} - x$ 13. $2x^2 - kx - k^2 = 0$

14. Determine k so that $x^2 + kx + 9 = 0$ has one solution.

[7.5] Solve.

15. $\sqrt{x} - 5 = 3$ 16. $\sqrt{R + 1} + \sqrt{R + 8} = 7$

[7.6] Solve.

17. The sum of a number and two times its reciprocal is $\frac{73}{6}$. What is the number?

18. The product of two positive numbers is 136, and one of the numbers is one greater than twice the other. What are the numbers?

19. A man drives his car for 20 miles at a certain speed. He then increases his speed by 20 miles per hour and continues for an additional 30 miles. If the total trip takes one hour, how fast was the man traveling originally?

20. One pipe will fill a tank in three hours less than another. If, running together, both pipes fill the tank in two hours, how long will it take each pipe alone to fill the tank?

Functions and graphs I

8.1 Functions

The notion of a **function** is one of the most basic and useful in all of mathematics. Essentially, a function is a relationship between the elements of two sets. For example, consider a circle. For each possible radius, there corresponds an area. A correspondence can be established between the elements in the set of all possible radii and the elements in the set of all possible areas. Figure 8.1 pictures this idea schematically. The area of a circle is said to be a function of the radius of the circle, or the radius and the area are said to be functionally related.

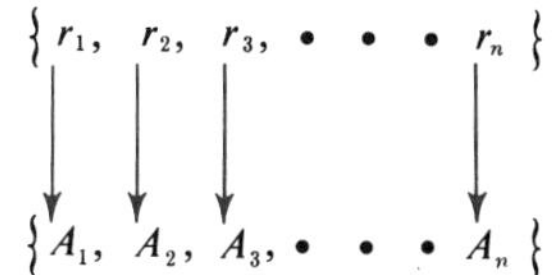

FIGURE **8.1**

To prevent misunderstanding, let us agree that when we speak of such things as radius, diameter, base, altitude, and so forth, we are referring to the numbers that measure these things. If we intend to refer to the thing itself rather than the measure, we shall say so, because, except in rare instances, it is the numerical measure with which we shall be concerned. While it is perfectly correct to speak of a function relating sets of non-numerical elements, we shall work only with functions relating the elements of two sets of numbers.

Given this background, a function may be defined as follows:

A function is an association or correspondence between the elements of two sets of numbers such that each element of the first set is associated with one and only one element of the second.

The first set of numbers is called the **domain** of the function and the second set is called the **range** of the function. A function is defined whenever we have a set of numbers (the domain) and a rule or scheme for associating each element of this set with a number in the range.

The rule for associating the elements of the domain with those of the range is usually given in the form of an equation in two variables. Thus the assignment of an area A of a circle to a radius r is made by the equation

$$A = \pi r^2.$$

The domain of the function defined by this equation is the set of positive real numbers. The variable representing an element in the domain of a function is often referred to as the **independent variable**, while the variable representing an element in the range is called the **dependent variable**. In the foregoing example, r is the independent variable and A is the dependent variable.

In functions relating sets whose elements are represented by x and y—for example, the function defined by $3x + y = 2$—we shall agree that x *will be the independent variable and* y *will be the dependent variable.*

In this book we shall be concerned only with functions whose domain and range consist of sets of real numbers. For example, in the function defined by

$$A = \pi r^2,$$

the replacement set of r must be restricted to the set of positive real numbers if we are to have a physically meaningful domain.

In any equation involving two variables in which the left or right member consists of the dependent variable alone, the dependent variable is said to be expressed **explicitly** in terms of the independent variable. For example, in the equations

$$y = 3x - 2, \quad y = x + 4, \quad \text{and} \quad y = x^2 + 2x - 1,$$

y is explicitly related to x. In equations such as

$$x + 2y = 3, \quad xy = 4, \quad \text{and} \quad x^2 + 2y - 3 = 0,$$

x and y are said to be **implicitly** related.

Functions are usually designated by means of a single symbol. The symbol f is generally used for this purpose. Thus we may speak of the function f defined by the equation

$$y = x + 2.$$

If the discussion includes a consideration of more than one function, g or h or F or some other symbol may be used to designate an individual function.

The symbol for the function can be used in conjunction with the variable representing an element in the domain to represent the associated element in the range. Thus $f(x)$ (read "f of x" or "the value of f at x") is the element in the range of f associated with the element x in the domain. [Note that $f(x)$ does not designate the product of f and x!] Suppose f is the function defined by the equation

$$y = x + 3;$$

then we can write

$$f(x) = x + 3,$$

where $f(x)$ is playing the same role as y; that is, $f(x)$ is the dependent variable. This notation is particularly useful because, when x is replaced by a specific number a in the domain, $f(a)$ denotes the related element in the range. For example, if

$$f(x) = x^2 + 3,$$

then

$$\begin{aligned}
f(3) &= (3)^2 + 3 = 12, \\
f(0) &= (0)^2 + 3 = 3, \\
f(h) &= h^2 + 3, \\
f(h + 2) &= (h + 2)^2 + 3 = h^2 + 4h + 4 + 3 \\
&= h^2 + 4h + 7.
\end{aligned}$$

Furthermore, it is frequently convenient to use such symbols as $f(x)$ $g(x)$, $Q(x)$, and so forth, to represent expressions in x. For example, we may write $f(x) = 0$, $g(x) = 0$, or $Q(x) = 0$ whenever we wish to discuss equations obtained by setting an expression in x equal to 0.

EXERCISE 8.1

If $f(x) = x + 2$, find the given element in the range.

EXAMPLE

$f(3)$

SOLUTION

Substitute 3 for x

$$f(3) = (3) + 2$$
$$= \mathbf{5}$$

1. $f(0)$
2. $f(1)$
3. $f(-3)$
4. $f(-4)$
5. $f\left(\frac{1}{2}\right)$
6. $f\left(\frac{2}{3}\right)$
7. $f(a)$
8. $f(a + 1)$

If $g(x) = x^2 - 2x + 1$, find the given element in the range.

9. $g(-2)$
10. $g(-1)$
11. $g(0)$
12. $g(1)$
13. $g(2)$
14. $g(3)$
15. $g(a)$
16. $g(a + 1)$

If $f(x) = x + 4$ defines a function, find the number in the domain of f associated with the given element in the range.

EXAMPLE

$f(x) = 5$

SOLUTION

Replace $f(x)$ with $x + 4$

$$x + 4 = 5$$
$$x = 1; \quad \text{hence the number is } \mathbf{1}.$$

17. $f(x) = 3$
18. $f(x) = 2$
19. $f(x) = 0$
20. $f(x) = -1$
21. $(x) = -2$
22. $f(x) = -3$

If $g(x) = x^2 - 1$, find all elements in the domain of g associated with the given element in the range.

23. $g(x) = 0$ 24. $g(x) = 3$ 25. $g(x) = 8$ 26. $g(x) = 5$

27. If $F(x) = x^2 + 1$, find
    a. $F(a) - F(b)$. *b.* $F(x + 1) - F(x)$.

28. If $f(x) = x^2 - x + 1$, find
    a. $f(x + h) - f(x)$. *b.* $\dfrac{f(x + h) - f(x)}{h}$.

State a rule in the form of an equation and specify a meaningful domain for a function relating the quantities in each of the following. See pages 288, 292, 296, and 299 for a list of formulas.

EXAMPLE

The area (A) of a triangle and the base (b), given a height of 6 inches.

SOLUTION

The area of a triangle is given by the formula $A = \frac{1}{2}bh$. Substituting 6 for the height yields

$$A = \frac{1}{2}b(6)$$

$$A = 3b, \quad \{b \mid b > 0\}.$$

29. The circumference (C) and radius (r) of a circle.
30. The area (A) and the side (s) of a square.
31. The perimeter (P) and the side (s) of a square.
32. The perimeter (P) of a rectangle of length 5 and the width (w) of the rectangle.
33. The distance (d) traveled by a car moving at a constant rate of 40 mph and the time (t) it has traveled.
34. The area (A) and the height (h) of a triangle whose base is 8 inches.
35. The hypotenuse (h) and the leg (b) of a right triangle whose other leg is 6 inches.
36. The leg (a) of a right triangle whose hypotenuse is 10 inches and the other leg (b) inches.

EXAMPLE

The area (A) and circumference (C) of a circle.

SOLUTION

Since

$$A = \pi r^2, \tag{1}$$

the problem is to obtain an expression for r in terms of C. If each member of

$$C = 2\pi r, \tag{2}$$

is multiplied by $\frac{1}{2\pi}$, it follows that

$$r = \frac{C}{2\pi}. \tag{3}$$

Substituting the value of r in Equation (3) for r in Equation (1) gives

$$A = \pi\left(\frac{C}{2\pi}\right)^2 = \frac{\pi \cdot C^2}{4\pi^2} = \frac{C^2}{4\pi}.$$

$$A = \frac{C^2}{4\pi}; \quad \{C \mid C > 0\}$$

37. The area (A) and side (s) of an equilateral triangle. *Hint:* Use Pythagorean theorem.
38. The area (A) and perimeter (P) of a square.
39. The perimeter (P) and the area (A) of a rectangle whose length is 10 inches.
40. The area (A) of a rectangle and its width (w) if the perimeter of the rectangle is 20 inches.

8.2 Solutions of equations—ordered pairs

An equation in two variables, such as $y = x + 2$ [or $f(x) = x + 2$], is said to be satisfied if the variables are replaced with a pair of numbers, one from the replacement set of x and one from the replacement set of y [or $f(x)$], and the resulting statement is true. The pair of numbers, usually written in the form (x, y) [or $(x, f(x))$], is called a **solution** of the equation. The pair (x, y) is called an **ordered pair**, since it is understood that the numbers will be written in a particular order, the number from the domain first, and the number from the range second. These numbers are then called the first and second **components** of the ordered pair, respectively.

We can find ordered pairs that satisfy a given equation in two variables by assigning numbers to one variable and determining the associated numbers for the other. Thus, for

$$y = x + 2, \tag{1}$$

we can obtain any number of solutions by assigning values to x and determining in each case the related value for y. For example, if x is assigned the value 3, we have

$$\begin{aligned} y &= 3 + 2 \\ &= 5, \end{aligned}$$

and the ordered pair (3, 5) is a solution of Equation (1).

In the same way, we can find a value for y for any arbitrary value of x. Restricting the replacement set of x to

$$\{3, -1, 0\},$$

we obtain the ordered pairs

$$(3, 5), \quad (-1, 1), \quad \text{and} \quad (0, 2)$$

as solutions of Equation (1). In fact, when the replacement set is as specified,

$$\{(3, 5), (-1, 1), (0, 2)\}$$

is the complete solution set.

If an equation in two variables defines a function whose domain is the set of real numbers, it is clearly impossible to list every ordered pair in the function. The set notation used earlier with inequalities is useful in such cases. For example, the function defined by

$$y = 2x + 1$$

can be written

$$\{(x, y) \mid y = 2x + 1\}$$

[read "the set of ordered pairs (x, y) such that $y = 2x + 1$"]. There are infinitely many ordered pairs in this set.

Since a solution set of this kind is a complete description of the association between the elements of the domain and those of the range, we have the alternative definition:

A function is a set of ordered pairs in which each first component is paired with only one second component.

In taking this ordered-pair viewpoint of a function, we see that the solution set of the defining equation and the function are one and the same thing.

EXERCISE 8.2

Find the missing component so that the ordered pair will satisfy the equation.

EXAMPLE

$y - 2x = 4$

a. (0,) b. (, 0) c. (3,)

SOLUTION

a.
$$y - 2x = 4$$
$$y - 2(0) = 4$$
$$y = 4$$
(0, 4)

b.
$$y - 2x = 4$$
$$0 - 2x = 4$$
$$x = -2$$
(−2, 0)

c.
$$y - 2x = 4$$
$$y - 2(3) = 4$$
$$y = 10$$
(3, 10)

1. $y = x + 7$
   a. (0,) b. (2,) c. (−2,)
2. $y = x - 3$
   a. (0,) b. (−1,) c. (1,)
3. $y = 2x + 1$
   a. (2,) b. (, 2) c. (0,)
4. $y = 3x - 1$
   a. (0,) b. (, 0) c. ($^2/_3$,)
5. $2x + 2y = 3$
   a. (0,) b. (, 0) c. ($^1/_2$,)
6. $2x + 3y = 4$
   a. (0,) b. (, 0) c. (3,)

In each of the remaining problems a rule and a domain are stated to define a function.

In each Exercise 7–14, express each function as a set of ordered pairs in the form (x, y).

EXAMPLE

$y = \frac{12}{x + 3}$, {3, 6, 9}

SOLUTION

If $x = 3$, $y = \frac{12}{3+3} = 2$.

If $x = 6$, $y = \frac{12}{6+3} = \frac{4}{3}$.

If $x = 9$, $y = \frac{12}{9+3} = 1$.

$\{(3, 2), (6, 4/3), (9, 1)\}$

7. $y = -x$, $\{-1, 0, 1\}$
8. $y = 2x + 1$, $\{1, 2, 3, 4\}$
9. $y = x^2$, $\{-2, -1, 0, 1, 2\}$
10. $y = x^2 + 2x + 1$, $\{1, 2, 3\}$
11. $y = \frac{1}{x}$, $\{1, 3, 5\}$
12. $y = \frac{1}{x-2}$, $\{-2, -1, 0\}$
13. $y = x + 1$, {real numbers}. *Hint:* Use $\{(x, y) \mid \qquad \}$.
14. $y = x + 1$, {positive real numbers}

In each Exercise 15–24, express each function as a set of ordered pairs in the form specified.

15. $C = 6.28r$, $\{2, 4, 6\}$; (r, C)
16. $E = 0.02R$, $\{5, 10, 15\}$; (R, E)
17. $A = 3.14r^2$, $\{1, 3, 5\}$; (r, A)
18. $s = 16.1t^2$, $\{1, 2, 3\}$; (t, s)
19. $P = 1000 + 40t$, $\{5, 10, 20\}$; (t, P)
20. $l = 4 + (n - 1)6$, $\{4, 8, 12\}$; (n, l)
21. $T = \sqrt{\frac{l}{16}}$, $\{1, 4, 9\}$; (l, T)
22. $t = \sqrt{\frac{s}{16}}$, $\{1, 2, 3\}$; (s, t)
23. $F = \frac{9}{5}C + 32$, $\{0, 100, 150\}$; (C, F)
24. $F = \frac{9}{5}C + 32$, $\{-40, -20, -10\}$; (C, F)

8.3 Graphical representation of a function

A **Cartesian**, or **rectangular coordinate, system** is formed by establishing a pair of perpendicular line graphs (called **axes**) at some point in a geometric plane. An arbitrary scaling of the line graphs from their point of intersection (the **origin**) permits the establishment of a one-to-one correspondence between the points in the plane and ordered pairs of real numbers (x, y). That is, for each point in the plane, there corresponds a unique ordered pair and vice versa. This correspondence is established by using x, the first component of the ordered pair, to denote the directed (perpendicular) distance of the point from the vertical axis (the **abscissa** of the point), to the right if the distance is positive, to the left if it is negative. The second component, y, of the ordered pair is used to denote the directed distance of the point from the horizontal axis (the **ordinate** of the point), above the axis if the distance is positive and below if negative. Each axis is labeled with the variable it represents. Most commonly, the horizontal axis is called the ***x*-axis** and the vertical axis is called the ***y*-axis**. The components of the ordered pair that corresponds to a given point are called the **coordinates** of the point. The four regions into which the axes divide the plane are called **quadrants**, and they are referred to by number as illustrated in Figure 8.2.

In Section 8.2 we observed that a function can be defined as a set of ordered pairs. Consequently, a function—or at least a part of a function—can now be displayed as a set of points in the plane. The restriction "part" is used, because the domain or range or both may be the set of real numbers, which cannot be represented on a finite straight line. For example, consider

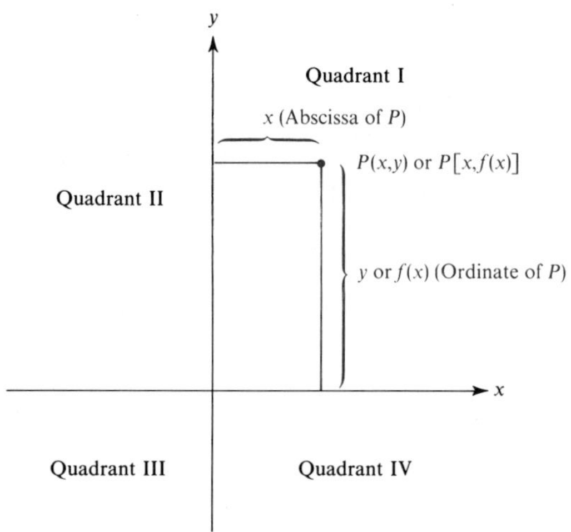

FIGURE **8.2**

the function defined by

$$(x) = x - 3. \tag{1}$$

Solutions to this equation can be obtained by assigning values to x and computing corresponding values for $f(x)$. For instance, for values of x equal to -1, 0, 1, 2, 3, and 4, six solutions to Equation (1) are

$$(-1, -4), (0, -3), (1, -2), (2, -1), (3, 0), \text{ and } (4, 1).$$

Sometimes, the components of ordered pairs are displayed in tabular form. In this example a table might appear as

x	-1	0	1	2	3	4
y	-4	-3	-2	-1	0	1

Locating the points corresponding to the above ordered pairs on a coordinate system, we have Figure 8.3a. The points appear to lie on a straight line, as shown in Figure 8.3b. It can be shown that the coordinates of each point on this line constitute a solution of Equation (1); and, conversely, every solution of (1) corresponds to a point on the line. The line is referred to as the **graph of the function** defined by (1), or, alternatively, as the **graph of the equation**.

Generally it is easier to find ordered pairs that are solutions of equations in two variables if the dependent variable is expressed explicitly in terms of

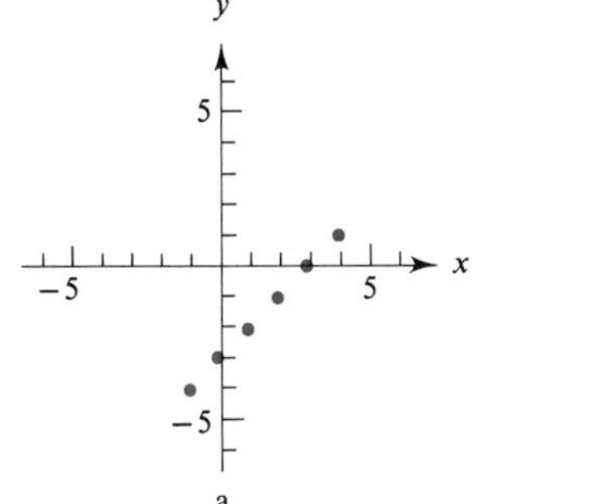

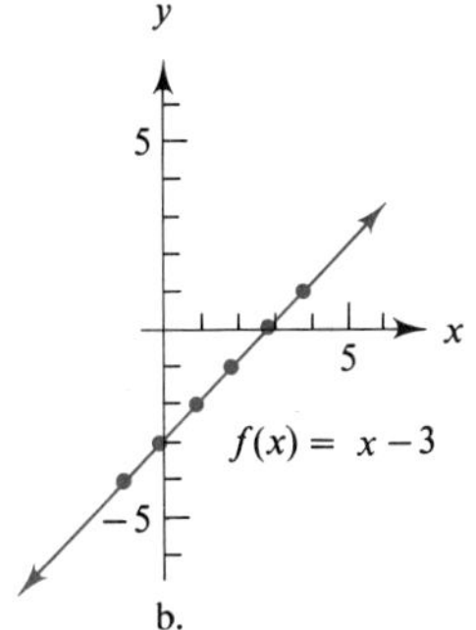

FIGURE **8.3**

the independent variable. For example, it is easier to find some ordered pairs, say,

$$(-2, ?), (-1, ?), (0, ?), (1, ?), \text{ and } (2, ?),$$

that are solutions of

$$\frac{x + 2y}{3} = 4$$

if y is first expressed explicitly in terms of x as follows:

$$(3)\frac{x + 2y}{3} = (3)4,$$

$$x + 2y = 12,$$

$$x + 2y + (-x) = 12 + (-x),$$

$$2y = 12 - x,$$

$$\frac{1}{2}(2y) = \frac{1}{2}(12 - x),$$

$$y = \frac{12 - x}{2}.$$

The second components of the ordered pairs above can now be obtained by substituting -2, -1, 0, 1, and 2 for x in the right-hand member of the last equation and obtaining the corresponding values of y.

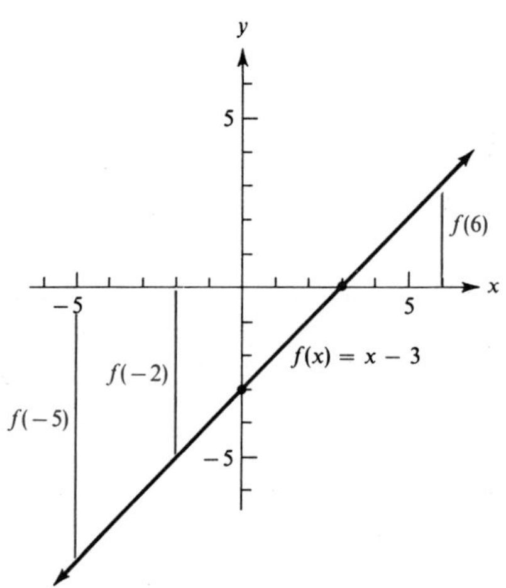

FIGURE **8.4**

Function notation can be used to denote the ordinate associated with a given abscissa on a graph. For example, the graph of

$$f(x) = x - 3$$

that we obtained above and several line segments marked with their lengths in function notation are shown in Figure 8.4. It is often useful to think of the ordinate of a point in terms of the length of the perpendicular line segment from the point to the x-axis.

EXERCISE 8.3

State the quadrant in which each of the following points is located.

1. $(3, 4)$
2. $(-2, 4)$
3. $(5, -3)$
4. $(-6, -2)$
5. $(-3, 1)$
6. $(2, 8)$
7. $(-3, -5)$
8. $(4, -1)$

Find ordered pairs in the solution set of each equation for x equal to -3, -1, 0, 1, and 3. Graph these ordered pairs and draw a straight line through the points.

EXAMPLE

$y - x = 3$

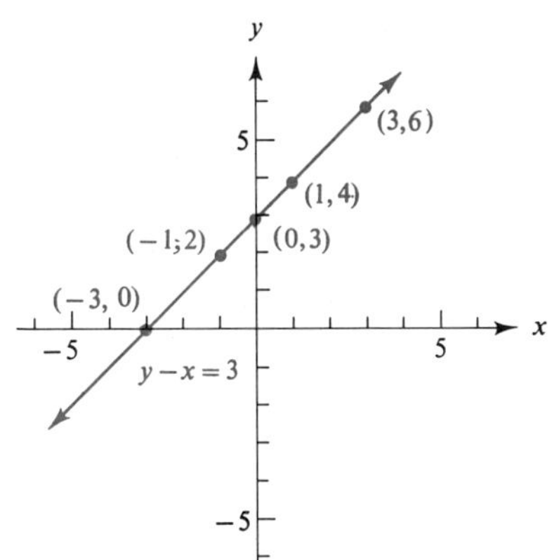

SOLUTION

Solve explicitly for y in terms of x.

$$y = x + 3$$

Substitute $-3, -1, 0, 1$, and 3 for x.

If $x = -3$, $y = (-3) + 3 = 0$;
If $x = -1$, $y = (-1) + 3 = 2$;
If $x = 0$, $y = (0) + 3 = 3$;
If $x = 1$, $y = (1) + 3 = 4$;
If $x = 3$, $y = (3) + 3 = 6$.

9. $y = x + 1$
10. $y = 2x - 4$
11. $y - x = 5$
12. $y - 2x = 3$
13. $2x + y = 4$
14. $3x + y = 6$
15. $x + 2y = 4$
16. $x - 3y = 9$
17. $2x + 3y = 6$

18. $3x + 2y = 6$
19. $4x - y = 6$
20. $3x - y = 5$
21. $3x - 4y = 12$
22. $4x - 3y = 12$
23. $\frac{1}{2}x + y = 4$
24. $\frac{2}{3}x + 2y = 1$
25. $\frac{3}{4}x + \frac{1}{2}y = 1$
26. $\frac{1}{3}x + \frac{1}{2}y = \frac{1}{4}$

EXAMPLE

Graph $f(x) = x - 1$. Represent $f(5)$ and $f(3)$ as line segments on the graph.

SOLUTION

$f(5)$ is the ordinate at $x = 5$.
$f(3)$ is the ordinate at $x = 3$.

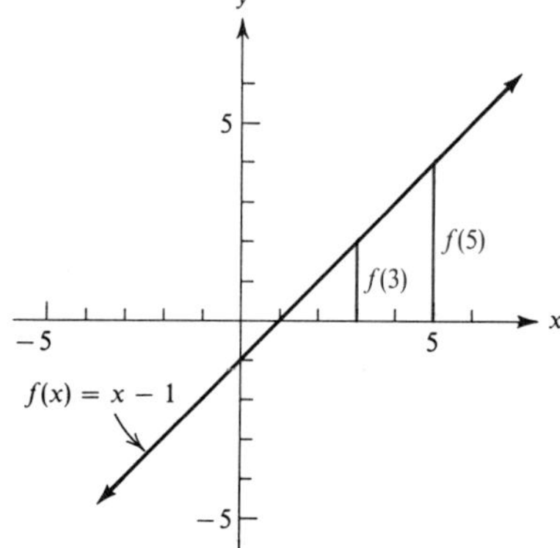

27. Graph $f(x) = 2x + 1$. Represent $f(3)$ and $f(-2)$ as line segments on the graph.
28. Graph $f(x) = 2x - 5$. Represent $f(4)$ and $f(0)$ as line segments on the graph.
29. Graph $f(x) = 4 + x$. Represent $f(5)$ and $f(3)$ as line segments on the graph.
30. Graph $f(x) = 3x - 6$. Represent $f(6)$ and $f(-3)$ as line segments on the graph.
31. Suppose the function f defined by $y = x$ has as domain $\{1, 2, 3, 4\}$. Graph the ordered pairs in the function on a rectangular coordinate system. What is the range of f?
32. Suppose the function defined by $y = 2x + 1$ has as domain the set of real numbers between and including -1 and 1. Graph the function on a rectangular coordinate system. What is the range of the function?

8.4 Linear functions

Any first-degree equation in two variables—that is, any equation that can be written equivalently in the form

$$Ax + By + C = 0 \quad (B \neq 0) \tag{1}$$

—defines a function and, although we do not prove it here, its graph is a straight line. For this reason, such equations are often called **linear equations**, and functions defined by such equations are called **linear functions**. Since any two distinct points determine a straight line, it is evident that we need find only two solutions to such an equation to determine its graph. In practice, the two solutions easiest to find are usually those with first and second components, respectively, zero—that is, the solutions $(0, y)$ and $(x, 0)$. Since these two points are the points where the graph crosses the y- and x-axes, respectively, they are easy to locate. The x- and y-coordinates, respectively, of these points are called the ***x*- and *y*-intercepts** of the graph, and are usually denoted by a and b. As an example, consider the function defined by

$$3x + 4y = 12. \tag{2}$$

If $y = 0$,

$$3x = 12,$$
$$x = 4,$$

and the x-intercept is 4. If $x = 0$,

$$4y = 12,$$
$$y = 3,$$

and the y-intercept is 3. Thus the graph of the function defined by Equation (2) appears as in Figure 8.5. Observe that the intercepts in this example were readily obtained from Equation (2). It would not have been particularly helpful to have solved explicitly for y first.

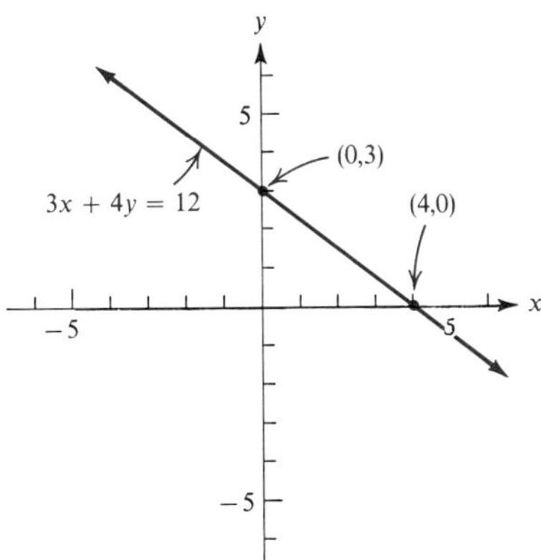

FIGURE **8.5**

If the graph intersects the axes at or near the origin, the intercepts either do not represent two separate points or the points are too close together to be of much use in drawing the graph. It is then necessary to determine at least one other point at a distance far enough removed from the origin to establish the line with accuracy.

There are two special cases of linear equations worth noting. First, an equation such as

$$y = 4$$

may be considered as an equation in two variables,

$$0x + y = 4.$$

For each x, this equation assigns $y = 4$. That is, any ordered pair of the form $(x, 4)$ is a solution of the equation. For instance,

$$(-1, 4), (2, 4), (4, 4), \quad \text{etc.},$$

are all solutions of the equation. The graphs of these ordered pairs determine the straight line shown in Figure 8.6. Because the equation

$$y = 4$$

assigns to each x the same value for y, the function defined by this equation is called a **constant function**.

The other special case of the linear equation is of the type

$$x = 3,$$

which may also be considered as an equation in two variables,

$$x + 0y = 3.$$

Here, only one value is permissible for x, namely 3, while any value may be assigned to y. That is, any ordered pair of the form $(3, y)$ is a solution of this equation. If we obtain two solutions, say $(3, 1)$ and $(3, 3)$, their graphs determine the line shown in Figure 8.7. Since infinitely many values of y are associated with the number 3 by this equation, the equation does not define a function.

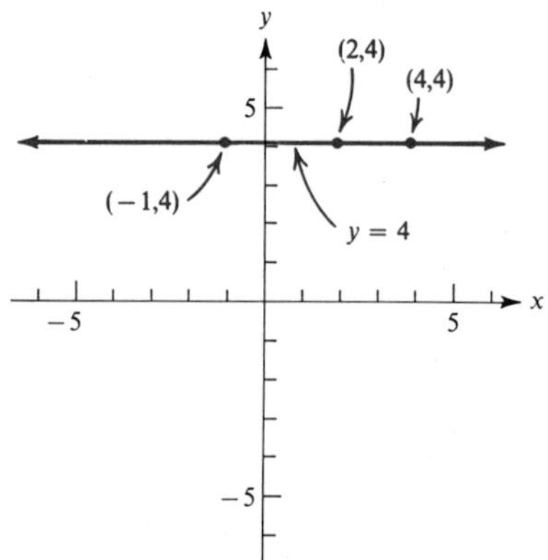

FIGURE **8.6**

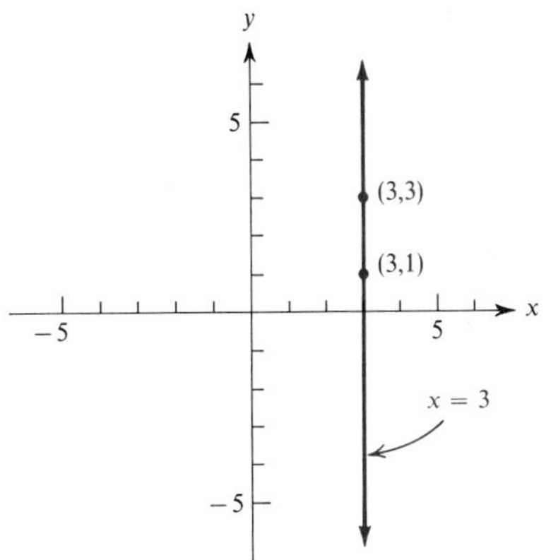

FIGURE **8.7**

EXERCISE 8.4

Graph each equation.

EXAMPLE

$3x + 4y = 24$

SOLUTION

Determine the intercepts.

If $x = 0$, $y = 6$;
if $y = 0$, $x = 8$.

6 is the y-intercept;
8 is the x-intercept.

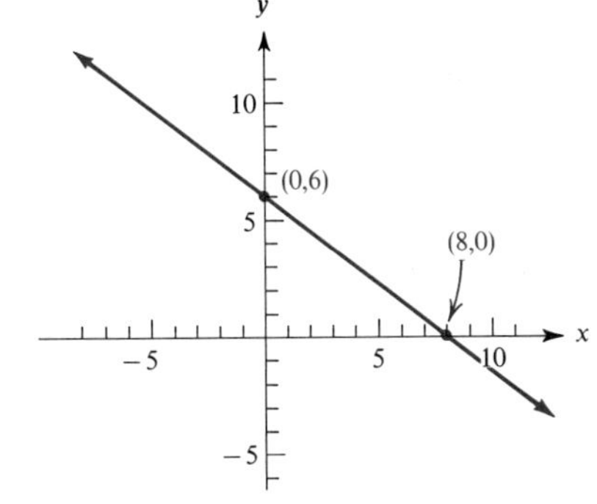

1. $y = 3x + 1$	2. $y = x - 5$	3. $y = 2x + 3$
4. $y = 2x - 3$	5. $3x + y = -2$	6. $3x - y = 4$
7. $2x + 3y = 6$	8. $3x - 2y = 8$	9. $2x + 5y = 10$
10. $2x - 5y = 10$	11. $y = x$	12. $y = 2x$
13. $y = -x$	14. $y = -2x$	15. $x = -2$
16. $y = -3$	17. $y - 10 = 0$	18. $x - 50 = 0$

Graph each equation. Designate the horizontal axis by the independent variable and the vertical axis by the dependent variable. Scale each axis in convenient units.

EXAMPLE

$P = 400 - 0.02Q$

SOLUTION

Determine the intercepts.

If $Q = 0$,

$$P = 400 - 0.02(0),$$
$$P = 400.$$

If $P = 0$,

$$0 = 400 - 0.02Q,$$
$$0.02Q = 400,$$
$$Q = 20{,}000.$$

Convenient scalings for the P-axis and the Q-axis are shown in the figure. These units were chosen in order to include the intercepts.

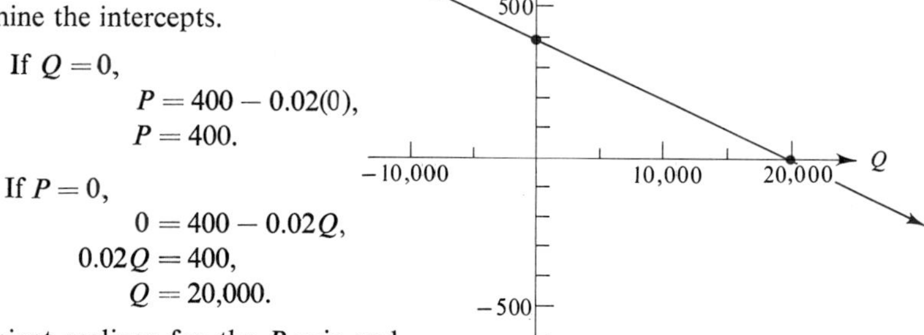

19. $P = 200 + 0.04Q$
20. $P = 500 - 0.01Q$
21. $E = 0.02I + .20$
22. $E = 0.03I - 0.60$
23. $H = 0.02 + 100L$
24. $H = 0.04 - 800L$
25. $C = 6.28r$
26. $F = \frac{9}{5}C + 32$

8.5 Distance and slope formulas

Any two distinct points in a plane can be looked upon as the end points of a line segment. Two line segments are shown in Figure 8.8. Letters with subscripts, such as x_1, y_1, x_2, and y_2, are used to indicate specific numbers—numbers that are to be considered fixed during a discussion.

We shall discuss two fundamental properties of a line segment—its **length** and its **inclination** with respect to the x-axis. We first observe that given any

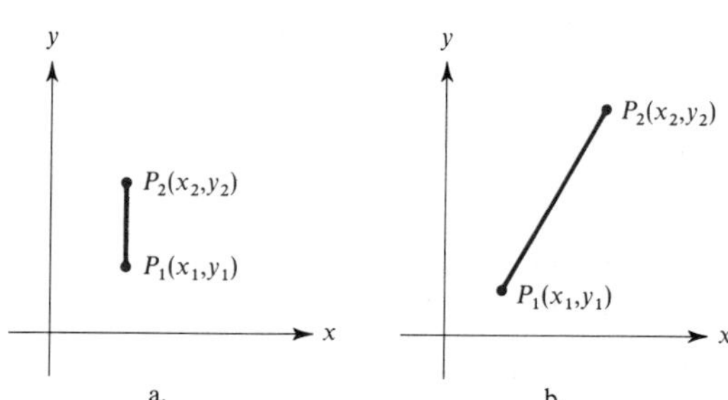

FIGURE **8.8**

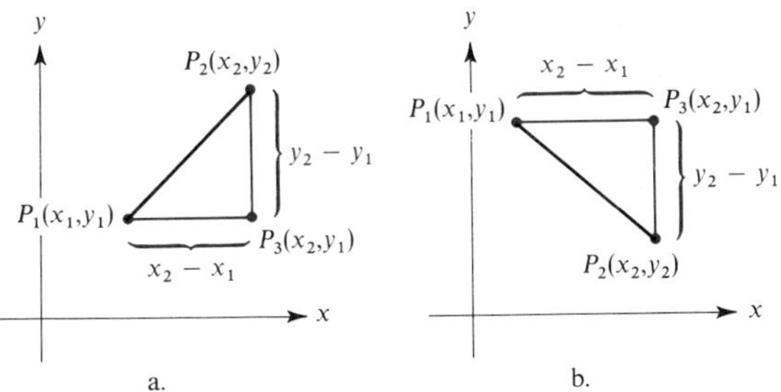

FIGURE **8.9**

two distinct points they either lie on the same vertical line or one is to the right of the other, the one on the right having a greater x-coordinate. Let us first consider the latter case. If we construct through P_2 a line parallel to the y-axis and through P_1 a line parallel to the x-axis, the lines will meet at a point P_3 as shown in either a or b of Figure 8.9. The x-coordinate of P_3 is evidently the same as the x-coordinate of P_2, while the y-coordinate of P_3 is the same as that of P_1; hence the coordinates of P_3 are (x_2, y_1). By inspection, we observe that the distance between P_2 and P_3 is $(y_2 - y_1)$ and the distance between P_1 and P_3 is $(x_2 - x_1)$.

In general, since $y_2 - y_1$ is positive or negative as $y_2 > y_1$ or $y_2 < y_1$, respectively, and $x_2 - x_1$ is positive or negative as $x_2 > x_1$ or $x_2 < x_1$, respectively, it is also convenient to designate the distances represented by $x_2 - x_1$ and $y_2 - y_1$ as positive or negative.

The Pythagorean theorem can be used to find the length of the line segment from P_1 to P_2. As you know, this theorem asserts that the square of the length of the hypotenuse of any right triangle is equal to the sum of the squares of the lengths of the legs. Thus,

$$d^2 = (x_2 - x_1)^2 + (y_2 - y_1)^2,$$

and if we consider only the positive square root of the right-hand member,

$$\boldsymbol{d = \sqrt{(x_2 - x_1)^2 + (y_2 - y_1)^2}}. \tag{1}$$

Since the distances $(x_2 - x_1)$ and $(y_2 - y_1)$ are squared, it makes no difference whether they are positive or negative—the result is the same. Equation (1) is a formula for the distance between any two points in the plane in terms of the coordinates of the points. The distance is always taken as positive. If the points P_1 and P_2 lie on the same horizontal line $(y = y_1)$ and if we are concerned only with distance and not direction, then

$$d = \sqrt{(x_2 - x_1)^2 + 0^2} = |x_2 - x_1|,$$

and, if they lie on the same vertical line $(x_2 = x_1)$, then

$$d = \sqrt{0^2 + (y_2 - y_1)^2} = |y_2 - y_1|.$$

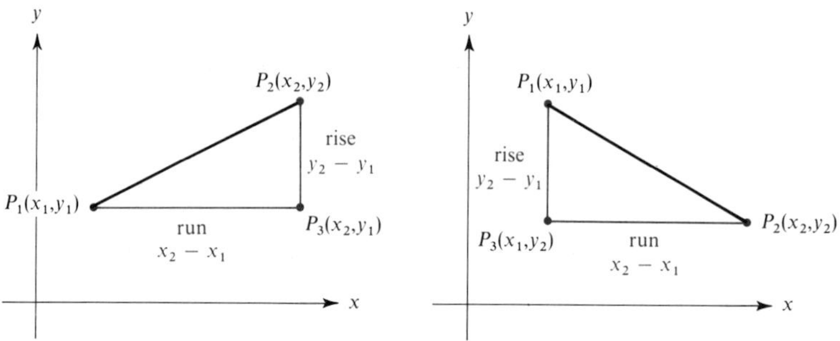

FIGURE **8.10**

The second useful property of the line segment joining two points is that of its inclination. The inclination of a line segment can be measured by comparing the *rise* of the segment with a given *run*, as shown in Figure 8.10.

The ratio of *rise to run* is called the **slope** of the line segment and is designated by the letter m. Thus

$$m = \frac{\text{rise}}{\text{run}}.$$

Since the rise is simply $y_2 - y_1$ and the run is $x_2 - x_1$, the slope of the line segment joining P_1 and P_2 is given by

$$\boldsymbol{m = \frac{y_2 - y_1}{x_2 - x_1} \quad (x_2 \neq x_1).} \tag{2}$$

Taking P_2 to the right of P_1, the run, $x_2 - x_1$, is positive and the slope is positive or negative as the rise, $y_2 - y_1$, is positive or negative—that is, according to whether the line slopes upward or downward from P_1 to P_2. A positive slope indicates that a line is rising to the right; a negative slope indicates that it is falling to the right. Since

$$\frac{y_2 - y_1}{x_2 - x_1} = \frac{-(y_1 - y_2)}{-(x_1 - x_2)} = \frac{y_1 - y_2}{x_1 - x_2},$$

the restriction that P_2 be to the right of P_1 is not necessary, and the order in which the points are considered is immaterial.

If a segment is parallel to the x-axis, as shown in Figure 8.11a, then $y_2 - y_1 = 0$ and it has slope 0. If a segment is parallel to the y-axis, as shown in b, then $x_2 - x_1 = 0$ and its slope is not defined.

Two line segments with slopes m_1 and m_2 are parallel if $m_1 = m_2$. It can be shown that two line segments with slopes m_1 and m_3 are perpendicular

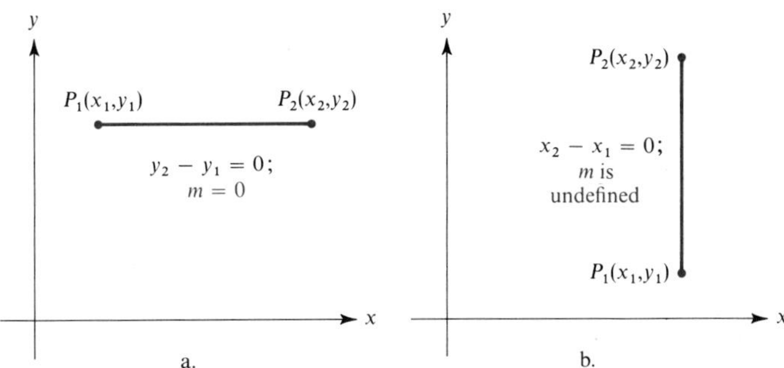

FIGURE **8.11**

if $m_1 = -1/m_3$ or equivalently $m_1 m_3 = -1$. For example, in Figure 8.12, line segment AB with slope

$$m_1 = \frac{5-3}{5-2} = \frac{2}{3},$$

is parallel to line segment CD with slope

$$m_2 = \frac{1-(-1)}{4-1} = \frac{2}{3}.$$

Line segment DE with slope

$$m_3 = \frac{1-7}{4-0} = \frac{-6}{4} = \frac{-3}{2}$$

is perpendicular to AB and CD, because

$$\begin{aligned} m_1 m_3 &= m_2 m_3 \\ &= \frac{2}{3}\left(\frac{-3}{2}\right) = -1. \end{aligned}$$

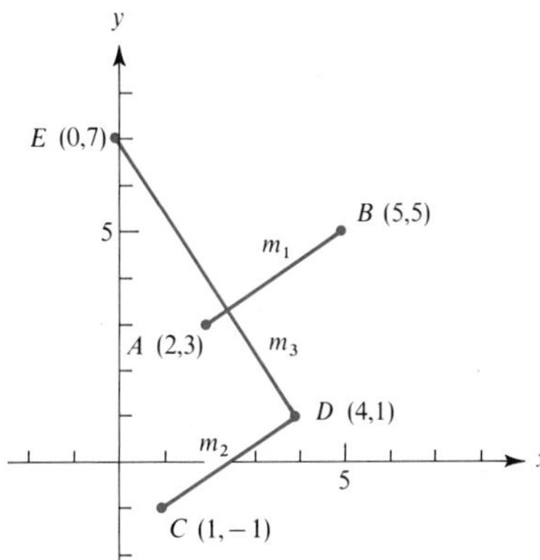

FIGURE **8.12**

EXERCISE 8.5

Find the distance between each of the given pairs of points, and find the slope of the line segment joining them. Sketch each line segment on the coordinate plane.

EXAMPLE

(3, −5), (2, 4)

SOLUTION

Consider (3, −5) as P_1 and (2, 4) as P_2.

$$d = \sqrt{(x_2 - x_1)^2 + (y_2 - y_1)^2} \qquad m = \frac{y_2 - y_1}{x_2 - x_1}$$

$$= \sqrt{[2-3]^2 + [4-(-5)]^2} \qquad = \frac{4-(-5)}{2-3}$$

$$= \sqrt{1+81} \qquad = \frac{9}{-1}$$

$$\mathbf{d = \sqrt{82}} \qquad \mathbf{m = -9}$$

1. (1, 1), (4, 5)
2. (−1, 1), (5, 9)
3. (−3, 2), (2, 14)
4. (−4, −3), (1, 9)
5. (2, 1), (1, 0)
6. (−3, 2), (0, 0)
7. (5, 4), (−1, 1)
8. (2, −3), (−2, −1)
9. (3, 5), (−2, 5)
10. (2, 0), (−2, 0)
11. (0, 5), (0, −5)
12. (−2, −5), (−2, 3)

Find the perimeter of the triangle whose vertices are given. Sketch each triangle on the coordinate plane.

13. (0, 6), (9, −6), (−3, 0)
14. (10, 1), (3, 1), (5, 9)
15. (5, 6), (11, −2), (−10, −2)
16. (−1, 5), (8, −7), (4, 1)
17. Show that the triangle described in Problem 13 is a right triangle. *Hint:* Use the converse of the Pythagorean theorem; that is, if $c^2 = a^2 + b^2$, the triangle is a right triangle.
18. Show that the triangle with vertices at (0, 0), (6, 0), and (3, 3) is a right isosceles triangle—that is, a right triangle with two sides that have the same length.

19. Use the fact that two line segments are perpendicular if and only if the product of their slopes equals -1 to show that the two line segments whose end points are $(0, -7)$, $(8, -5)$, and $(5,7)$, $(8, -5)$ are perpendicular.
20. Show that the two line segments whose end points are $(8, 0)$, $(6, 6)$ and $(-3, 3)$, $(6, 6)$ are perpendicular.
21. Use the fact that two line segments are parallel if and only if their slopes are equal to show that the points $(2, 4)$, $(3, 8)$, $(5, 1)$, and $(4, -3)$ are the vertices of a parallelogram.
22. Show that the point $(-5, 4)$, $(7, -11)$, $(12, 25)$, and $(0, 40)$ are the vertices of a parallelogram.

8.6 Forms of linear equations

We have previously used the equation

$$\boldsymbol{Ax + By + C = 0 \quad (B \neq 0)} \tag{1}$$

to define a linear function. Let us designate Equation (1) as **standard form** for a linear equation in two variables and then consider two alternative forms that display useful aspects.

Assuming that the slope of the line segment joining any two points on a line does not depend upon the points, consider a line on the plane with given slope m and passing through a given point (x_1, y_1), shown in Figure 8.13. If we choose any other point on the line and assign to it the coordinates (x, y), the slope of the segment is given by

$$\frac{y - y_1}{x - x_1} = m \quad (x \neq x_1), \tag{2}$$

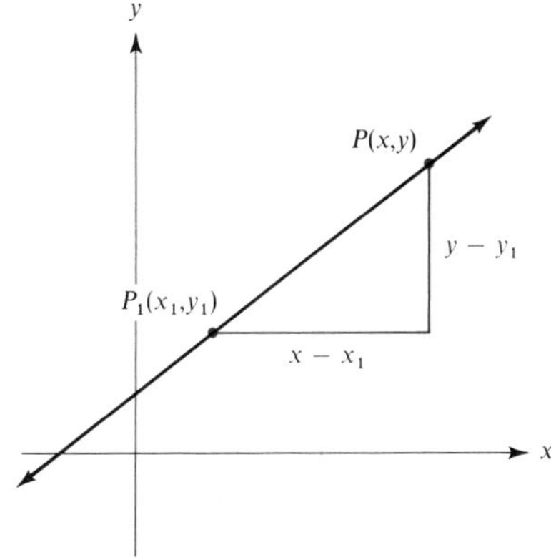

FIGURE **8.13**

from which

$$y - y_1 = m(x - x_1) \quad (x \neq x_1). \tag{3}$$

Notice that Equations (2) and (3) are equivalent for all values of x except x_1, in which case the left-hand member of (2) is not defined. The ordered pair (x_1, y_1) corresponds to a point on the line, and the coordinates of (x_1, y_1) also satisfy (3). Hence the graph of (3) contains all those points in the graph of (2) and, in addition, the point (x_1, y_1). Since x and y are the coordinates of *any* point on the line, Equation (3), with the restriction that $x \neq x_1$ removed, is the equation of the line passing through (x_1, y_1) with slope m. The equation

$$\mathbf{y - y_1 = m(x - x_1)}$$

is called the **point-slope form** for a linear equation.

Now consider the equation of the line passing through a given point on the y-axis whose coordinates are $(0, b)$ and having slope m shown in Figure 8.14.

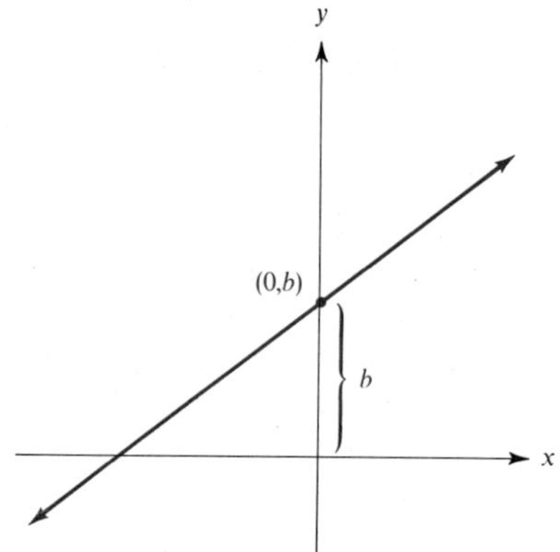

FIGURE **8.14**

Substituting $(0, b)$ in the point-slope form of a linear equation

$$y - y_1 = m(x - x_1),$$

we obtain

$$y - b = m(x - 0),$$

from which

$$\mathbf{y = mx + b.} \tag{4}$$

Equation (4) is called the **slope-intercept form** for the equation of a straight line. Any equation in standard form, such as

$$2x + 3y - 6 = 0,$$

can be written in the slope-intercept form by solving for y explicitly in terms of x. Thus,

$$3y = -2x + 6,$$

from which

$$y = -\frac{2}{3}x + 2.$$

The slope of the line, $-2/3$, and the y-intercept, 2, can now be read directly from the last form of the equation by comparing it with Equation (4).

EXERCISE 8.6

Sketch the line through each of the given points and having the given slope. Find the equation and write it in standard form.

EXAMPLE

$(3, -5)$, $m = -2$

SOLUTION

Substitute given values in the point-slope form of the linear equation.

$$y - y_1 = m(x - x_1)$$
$$y - (-5) = -2(x - 3)$$
$$y + 5 = -2x + 6$$
$$\mathbf{2x + y - 1 = 0}$$

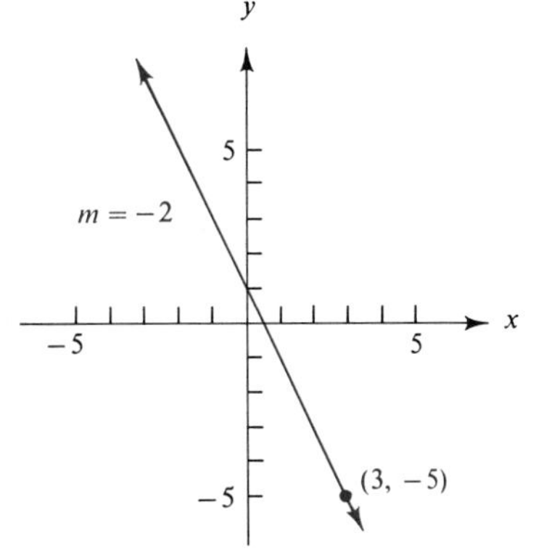

1. $(2, 1)$, $m = 4$
2. $(-2, 3)$, $m = 5$
3. $(5, 5)$, $m = -1$
4. $(-3, -2)$, $m = \frac{1}{2}$
5. $(0, 0)$, $m = 3$
6. $(-1, 0)$, $m = 1$
7. $(0, -1)$, $m = -\frac{1}{2}$
8. $(2, -1)$, $m = \frac{3}{4}$
9. $(-2, -2)$, $m = -\frac{3}{4}$
10. $(2, -3)$, $m = 0$
11. $(-4, 2)$, $m = 0$
12. $(-1, -2)$, parallel to y-axis

Write each of the equations in slope-intercept form; specify the slope of the line and the y-intercept.

EXAMPLE

$2x - 3y = 5$

SOLUTION

Solve explicitly for y.

$$-3y = 5 - 2x$$
$$3y = 2x - 5$$
$$y = \frac{2}{3}x - \frac{5}{3}$$

Compare with the general slope-intercept form $y = mx + b$.

$$\textbf{slope, } \frac{2}{3}\textbf{; } y\textbf{-intercept, } -\frac{5}{3}$$

13. $x + y = 3$
14. $2x + y = -1$
15. $3x + 2y = 1$
16. $3x - y = 7$
17. $x - 3y = 2$
18. $2x - 3y = 0$
19. $8x - 3y = 0$
20. $-x = 2y - 5$
21. $y + 2 = 0$
22. $y - 3 = 0$
23. $x + 5 = 0$
24. $x - 4 = 0$
25. Write the equation of the line that has the same slope as the graph of
$$x - 2y = 5$$
and passes through the origin. Draw the graph of both equations.
26. Write the equation of the line that has the same slope as the graph of
$$2y - 3x = 5$$
and passes through (0, 5). Draw the graph of both equations.
27. Write the equation of the line that is perpendicular to the graph of
$$x - 2y = 5$$
and that passes through the origin. Draw the graph of both equations.
28. Write the equation of the line that is perpendicular to the graph of
$$2y - 3x = 5$$
and passes through (0, 5). Draw the graph of both equations.
29. Show that
$$y - y_1 = \left(\frac{y_2 - y_1}{x_2 - x_1}\right)(x - x_1)$$
is the equation of the line joining the points (x_1, y_1) and (x_2, y_2). *Hint:* Use the point-slope form and the slope formula from Section 8.5.

30. Using the form

$$y - y_1 = \left(\frac{y_2 - y_1}{x_2 - x_1}\right)(x - x_1),$$

find the equation of the lines through the given points.

a. (2, 1) and (−1, 3).
b. (3, 0) and (5, 0).
c. (−2, 1) and (3, −2).
d. (−1, −1) and (1, 1).

8.7 Graphs of first-degree relations

An open sentence of the form

$$\boldsymbol{Ax + By + C \leq 0},$$

where A, B, and C are real numbers, is an inequality of the first degree. The graph of such an inequality is a region of the plane rather than a straight line. For example, consider the inequality

$$2x + y - 3 < 0. \tag{1}$$

If it is rewritten in the form

$$y < -2x + 3, \tag{2}$$

we have that y is less than $-2x + 3$ for each x. The graph of the equation

$$y = -2x + 3 \tag{3}$$

is simply a straight line, as illustrated in Figure 8.15a. Therefore, to graph inequality (2), we need only observe that any point below this line has x- and y-coordinates that satisfy (2), and consequently the solution set of (2)

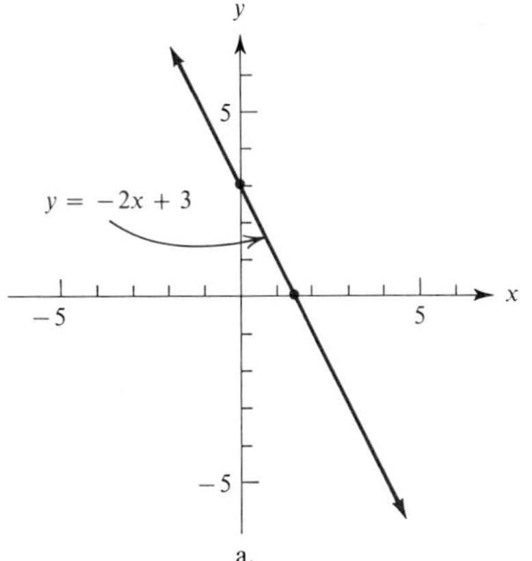

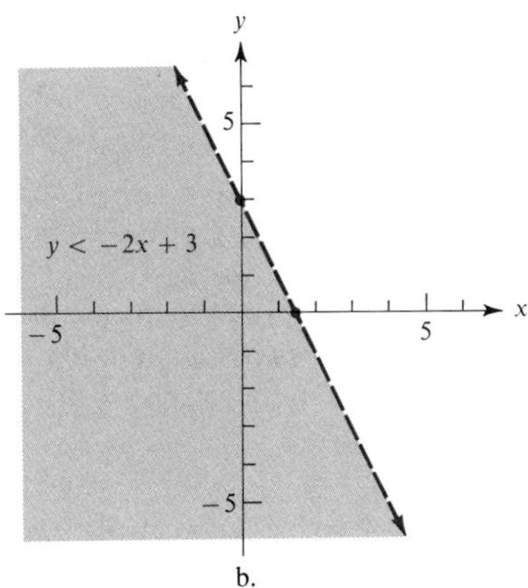

FIGURE **8.15**

corresponds to the entire region below the line. The region is indicated on the graph with shading. That the points on the line do not correspond to elements in the solution set of the inequality is shown by means of a broken line, as in Figure 8.15b. Had the inequality been

$$2x + y - 3 \leq 0,$$

the line would be a part of the graph of the solution set and would be shown as a solid line.

In general, if $B \neq 0$,

$$Ax + By + C < 0 \quad \text{or} \quad Ax + By + C > 0$$

will have as a solution set all ordered pairs associated with all points in a half-plane either above or below the line with equation

$$Ax + By + C = 0,$$

depending upon the inequality symbols involved. We can determine which of the half-planes should be shaded by substituting the coordinates of any point not on the line and noting whether or not they satisfy the inequality. If they do, then the half-plane containing the point is shaded; if the coordinates do not satisfy the inequality, then the other half-plane is shaded. A good point to use in this process is the origin with the coordinates (0, 0).

EXERCISE 8.7

Graph the solution set of the inequality.

EXAMPLE

$2x + y \geq 4$

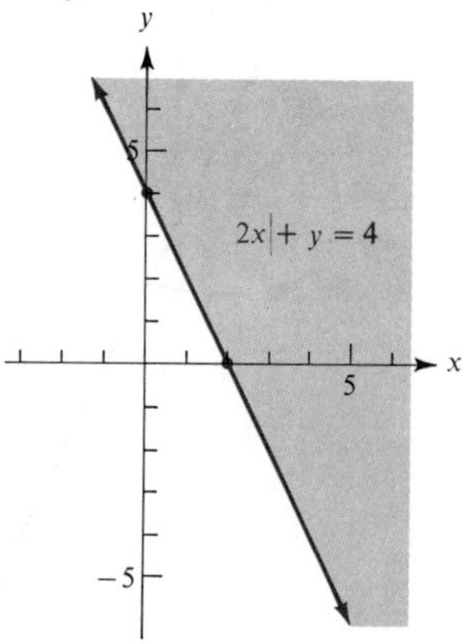

SOLUTION

Graph the equality $2x + y = 4$.
Substitute 0 for x and y in the inequality.

$$2(0) + 0 \geq 4$$

Since this is not a true statement, shade the half-plane that does not contain the origin.
Line is included in the graph.

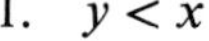

1. $y < x$
2. $y > x$
3. $y \leq x + 2$
4. $y \geq x - 2$
5. $x + y < 5$
6. $x - y < 3$
7. $2x + y < 2$
8. $x - 2y < 5$

9. $x \leq 2y + 4$
10. $2x \leq y + 1$
11. $0 \geq x - y$
12. $0 \geq x + 3y$

Graph each set of ordered pairs.

EXAMPLE

$\{(x, y) \mid x > 2\}$

SOLUTION

Graph the equality $x = 2$.
Shade region to the right of the line representing $x = 2$; this region includes all the points for which $x > 2$.
Line is excluded from the graph.

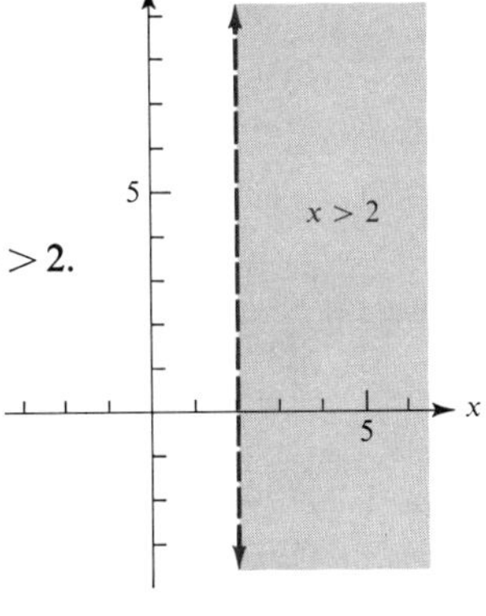

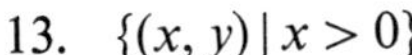

13. $\{(x, y) \mid x > 0\}$
14. $\{(x, y) \mid y < 0\}$
15. $\{(x, y) \mid y \geq 3\}$
16. $\{(x, y) \mid x < -2\}$
17. $\{(x, y) \mid -1 < x < 5\}$
18. $\{(x, y\} \mid 0 \leq y \leq 1\}$

Show the region that contains elements common to both solution sets.

EXAMPLE

$\{(x, y) \mid y > x\}$ and $\{(x, y) \mid y > 2\}$

SOLUTION

Graph each set. Region containing the elements common to both sets is doubly shaded.

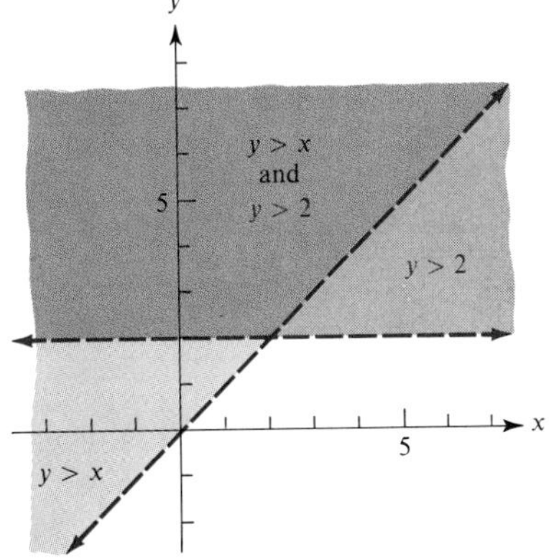

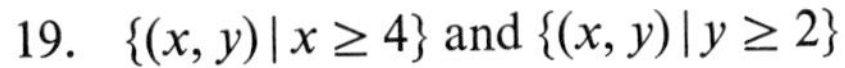

19. $\{(x, y) \mid x \geq 4\}$ and $\{(x, y) \mid y \geq 2\}$
20. $\{(x, y) \mid x \leq 2\}$ and $\{(x, y) \mid y \leq 2\}$
21. $\{(x, y) \mid y \geq x\}$ and $\{(x, y) \mid y \leq 2\}$
22. $\{(x, y) \mid x \leq 2\}$ and $\{(x, y) \mid y < x\}$
23. $\{(R, S\} \mid S > 3\}$ and $\{(R, S) \mid R < S\}$
24. $\{(P, Q) \mid Q \leq P\}$ and $\{(P, Q) \mid Q < 2\}$
25. $\{(P, V) \mid P + V \leq 6$ and $\{(P, V) \mid P + V \geq 4\}$
26. $\{(M, N) \mid N \leq 2 - M\}$ and $\{(M, N) \mid N > M - 2\}$

Chapter review

[8.1] 1. If $f(x) = x^2 + 3x + 1$, find $f(2)$.

2. If $g(x) = x^2 - 2x + 3$, find $g(-3)$.

3. If $F(x) = x^2 + 2x + 3$, find $F(x+1) - F(x)$.

4. State a rule in the form of an equation that relates the perimeter (P) of a rectangle of width 7 to the length (l).

[8.2] 5. Find the missing components in each solution of $2x - 6y = 1$.
a. (0, ?) b. (?, 0) c. (3, ?)

6. Specify the function defined by $F = \frac{9}{5}C + 32$ with domain $\{-10, 20, 80\}$ as a set of ordered pairs (C, F).

[8.3] Graph each equation.

7. $f(x) = 2x - 5$

8. $f(x) = 3x - 6$

9. Represent $f(2)$ as a line segment on the graph of the equation in Problem 7.

10. Represent $f(-3)$ as a line segment on the graph of the equation in Problem 8.

[8.4] Graph each equation using the intercept method.

11. $x - 6y = 6$

12. $E = 0.02I + 0.40$

[8.5] 13. Find the distance between the points $(3, -5)$ and $(6, 8)$.

14. Find the slope of the line segment between the points $(3, -5)$ and $(6, 8)$.

[8.6] 15. Find the equation in standard form of the line through (3, 5) with slope 2.

16. Write $2x + 3y = 6$ in slope-intercept form and specify the slope and the y-intercept of its graph.

17. Write the equation in standard form of the line through $(-1, 1)$ parallel to the graph of $2x = 5 - 3y$.

18. Find the equation in standard form of the line through $(-2, 7)$ and $(3, 5)$.

[8.7] 19. Graph $y > x + 2$.

20. Graph $\{(P, V) \mid P < 3\}$.

Functions and graphs II

9.1 The graph of a quadratic function

Consider the quadratic equation in two variables

$$y = x^2 - 4. \qquad (1)$$

As with linear equations in two variables, solutions of this equation are ordered pairs. We need replacements for both x and y in order to obtain a statement we can adjudge to be true or false. Such ordered pairs can be found by arbitrarily assigning values to x and computing related values for y. For instance, assigning the value -3 to x in Equation (1), we have

$$\begin{aligned} y &= (-3)^2 - 4 \\ &= 5, \end{aligned}$$

and $(-3, 5)$ is a solution. Similarly, we find that

$$(-2, 0), (-1, -3), (0, -4), (1, -3), (2, 0), \text{ and } (3, 5)$$

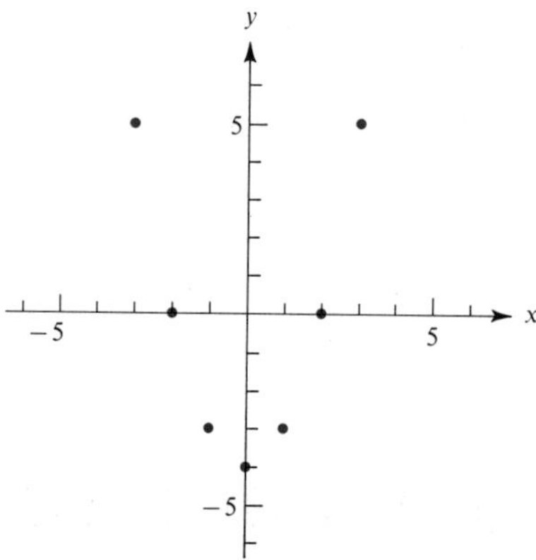

FIGURE **9.1**

are also solutions of Equation (1). Plotting the corresponding points on the plane, we have the graph in Figure 9.1. Clearly, these points do not lie on a straight line, and we might reasonably inquire whether the graph of the solution set of Equation (1),

$$\{(x, y) \mid y = x^2 - 4\},$$

forms any kind of meaningful pattern on the plane. By plotting additional solutions—solutions with x-components between those already found—we may be able to obtain a clearer picture. Accordingly, we find the solutions

$$\left(\frac{-5}{2}, \frac{9}{4}\right), \left(\frac{-3}{2}, \frac{-7}{4}\right), \left(\frac{-1}{2}, \frac{-15}{4}\right), \left(\frac{1}{2}, \frac{-15}{4}\right), \left(\frac{3}{2}, \frac{-7}{4}\right), \text{ and } \left(\frac{5}{2}, \frac{9}{4}\right),$$

and by plotting these points in addition to those found earlier, we have the graph in Figure 9.2a. It now appears reasonable to connect these points in

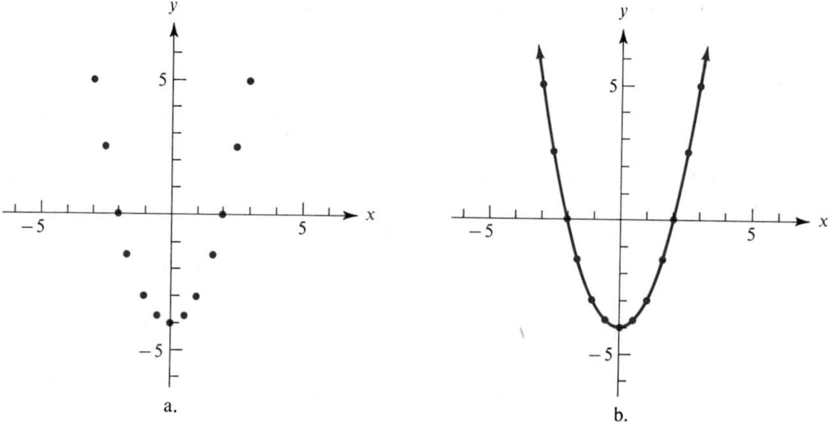

FIGURE **9.2**

sequence from left to right, by a smooth curve as in Figure 9.2b, and to assume that the curve is a good approximation to the graph of Equation (1). This curve is an example of a **parabola**. Note that, regardless of the number of individual points we plot, we have no absolute assurance that connecting these points with a smooth curve will produce the correct graph, although the more points we plot and locate, the more reasonable it should seem that connecting them will do this. Proving that the graph is indeed a parabola requires the use of advanced concepts and is not attempted herein. More generally, the graph of any quadratic equation of the form

$$y = ax^2 + bx + c, \tag{2}$$

where a, b, and c are real and $a \neq 0$, is a parabola.

In graphing parabolas, it is helpful to note that if $a > 0$, the parabola "opens" upward, as illustrated in Figure 9.2. If $a < 0$, the parabola "opens" downward. For example, graphing the equation

$$f(x) = -x^2 + 16$$

by first obtaining the ordered pairs $(-5, -9)$, $(-4, 0)$, $(-3, 7)$, $(-2, 12)$, $(-1, 15)$, $(0, 16)$, $(1, 15)$, $(2, 12)$, $(3, 7)$, $(4, 0)$, and $(5, -9)$, we obtain the parabola in Figure 9.3.

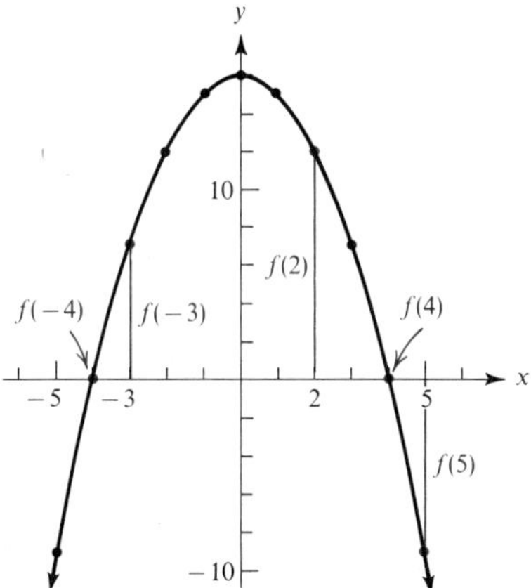

FIGURE **9.3**

As in the case of the graph of a linear equation in Section 8.4, the ordinate of a point on the graph of a quadratic equation is the directed length of the line segment drawn from the point on the curve perpendicular to the x-axis. Several representative ordinates using function notation are shown in Figure 9.3.

EXERCISE 9.1

In each Problem 1–14,

(a) find solutions of the equation using integral values for x where $-4 \leq x \leq 4$;
(b) use the solutions to graph the equation.

EXAMPLE

$y = -x^2 + 3x$

SOLUTION

a. $(-4, -28)$
$(-3, -18)$
$(-2, -10)$
$(-1, -4)$
$(0, 0)$
$(1, 2)$
$(2, 2)$
$(3, 0)$
$(4, -4)$

b.

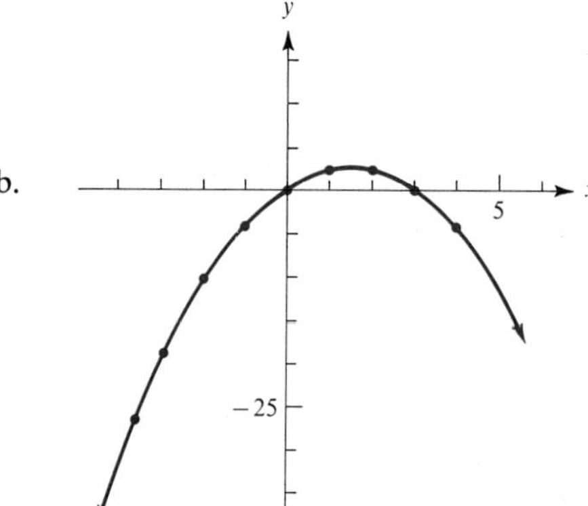

1. $y = x^2 + 1$	2. $y = x^2 + 4$
3. $f(x) = x^2 - 3$	4. $f(x) = x^2 - 5$
5. $y = -x^2 + 4$	6. $y = -x^2 + 5$
7. $y = 3x^2 + x$	8. $y = 5x^2 - x$
9. $y = x^2 + 2x + 1$	10. $y = x^2 - 2x + 1$
11. $f(x) = -2x^2 + x - 3$	12. $f(x) = -x^2 + 2x - 1$
13. $f(x) = -x^2 + x + 1$	14. $f(x) = 3x^2 + x - 2$

15. Graph the equation $f(x) = x^2 + 2$. Represent $f(0)$ and $f(4)$ by drawing appropriate line segments.
16. Graph the equation $f(x) = x^2 - 1$. Represent $f(-3)$ and $f(2)$ by drawing appropriate line segments.

9.2 Conic sections

In addition to $y = ax^2 + bx + c$, there are three other types of quadratic equations in two variables whose graphs are of particular interest. We shall discuss each of them separately. First, consider the equation

$$x^2 + y^2 = 25. \tag{1}$$

Solving this equation explicitly for y, we have

$$y = \pm\sqrt{25 - x^2}.$$

Assigning values to x, we find the following ordered pairs in the solution set:

$$(-5, 0),\quad (-4, 3),\quad (-3, 4),\quad (0, 5),\quad (3, 4),\quad (4, 3),\quad (5, 0),$$
$$(-4, -3), (-3, -4), (0, -5), (3, -4), (4, -3).$$

Plotting these points on the plane, we have the graph in Figure 9.4a. Connecting these points with a smooth curve, we have the graph in Figure 9.4b, a circle with radius 5 and center at the origin.

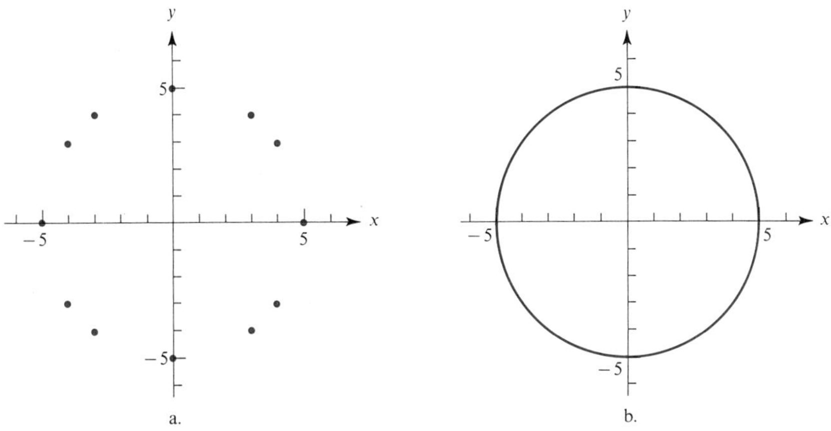

FIGURE **9.4**

Since the number 25 in the right-hand member of Equation (1) clearly determines the length of the radius of the circle, we can generalize and observe that any equation in the form

$$x^2 + y^2 = r^2$$

graphs into a circle with radius r and with center at the origin.

Note that in the preceding example it is not necessary to assign any values of x greater than 5 or less than -5, because y is imaginary for these values of x.

The second quadratic equation in two variables of interest is typified by

$$4x^2 + 9y^2 = 36. \tag{2}$$

We obtain solutions of Equation (2) by first solving explicitly for y,

$$y = \pm\frac{2}{3}\sqrt{9 - x^2},$$

and assigning to x some values $-3 \leq x \leq 3$. We obtain, for example,

$$(-3, 0), \left(-2, \frac{2}{3}\sqrt{5}\right), \left(-1, \frac{4}{3}\sqrt{2}\right), (0, 2), \left(1, \frac{4}{3}\sqrt{2}\right), \left(2, \frac{2}{3}\sqrt{5}\right), (3, 0),$$

$$\left(-2, -\frac{2}{3}\sqrt{5}\right), \left(-1, -\frac{4}{3}\sqrt{2}\right), (0, -2), \left(1, -\frac{4}{3}\sqrt{2}\right), \left(2, -\frac{2}{3}\sqrt{5}\right).$$

Locating the corresponding points on the plane and connecting them with a smooth curve, we have the graph in Figure 9.5. This curve is called an

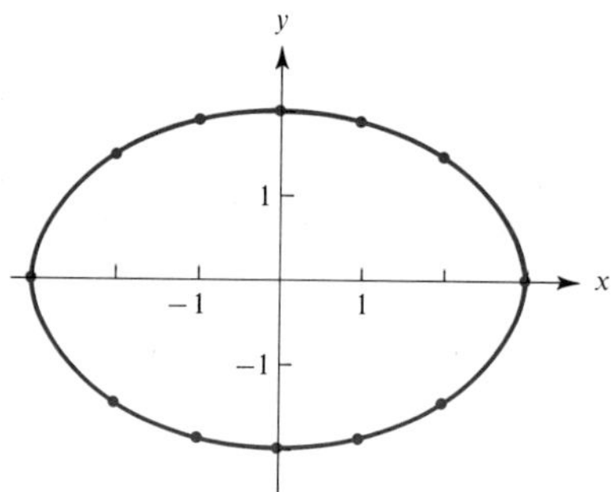

FIGURE **9.5**

ellipse. The graph of any equation of the form

$$Ax^2 + By^2 = C \quad (A, B, C > 0, A \neq B)$$

is an ellipse with center at the origin.

The third quadratic equation with which we are presently concerned is typified by

$$x^2 - y^2 = 9.$$

Solving this equation for y, we obtain

$$y = \pm\sqrt{x^2 - 9},$$

which has as a part of its solution set the ordered pairs

$$(-5, 4), \quad (-4, \sqrt{7}), \quad (-3, 0), \quad (3, 0), \quad (4, \sqrt{7}), \quad (5, 4),$$
$$(-5, -4), \quad (-4, -\sqrt{7}), \quad (4, -\sqrt{7}), \quad (5, -4).$$

Plotting the corresponding points and connecting them with a smooth curve, we obtain Figure 9.6. This curve is called a **hyperbola**. Any equation of the

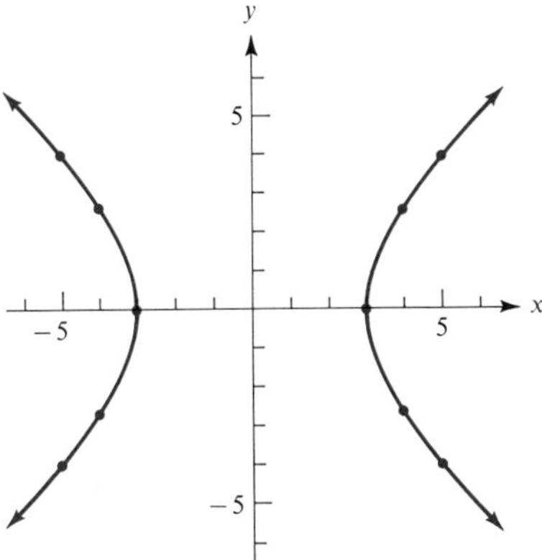

FIGURE **9.6**

form

$$Ax^2 - By^2 = C \quad (A, B, C > 0)$$

or

$$By^2 - Ax^2 = C \quad (A, B, C > 0)$$

graphs into a hyperbola with center at the origin.

The graphs of the equations dealt with in this section, together with the parabola of the preceding section, are called **conic sections** or **conics** because such curves are intersections of a plane and a cone, as shown in Figure 9.7.

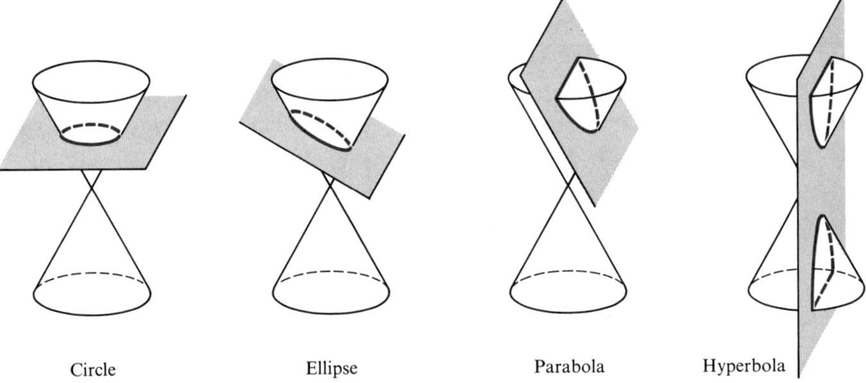

FIGURE **9.7**

EXERCISE 9.2

Graph each equation.

EXAMPLE

$4x^2 + y^2 = 36$

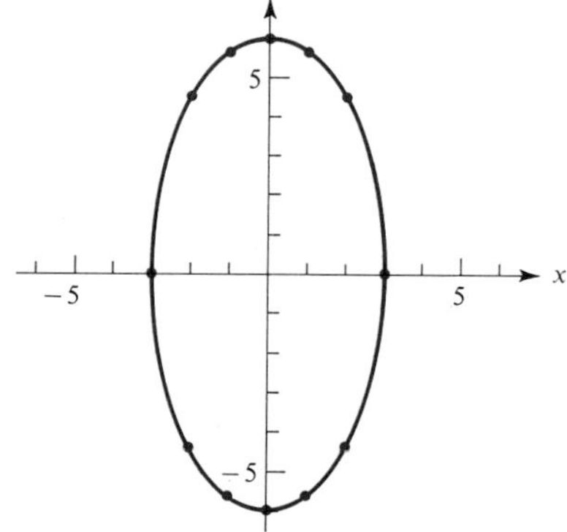

SOLUTION

$$y^2 = 36 - 4x^2$$
$$y = \pm 2\sqrt{9 - x^2}$$

Solutions are:

$(3, 0)$ $(-3, 0)$
$(2, \pm 2\sqrt{5})$ $(-2, \pm 2\sqrt{5})$
$(1, \pm 4\sqrt{2})$ $(-1, \pm 4\sqrt{2})$
$(0, \pm 6)$

1. $x^2 + y^2 = 4$
2. $x^2 + y^2 = 9$
3. $9x^2 + y^2 = 36$
4. $4x^2 + y^2 = 4$
5. $x^2 + 4y^2 = 16$
6. $x^2 + 9y^2 = 4$
7. $x^2 - y^2 = 1$
8. $4x^2 - y^2 = 1$
9. $y^2 - x^2 = 9$
10. $4y^2 - 9x^2 = 36$
11. $2x^2 + 3y^2 = 24$
12. $4x^2 + 3y^2 = 12$

13. Graph $xy = 4$ and $xy = 12$ on the same set of axes. (These curves are also hyperbolas.)
14. Graph $xy = -4$ and $xy = -12$ on the same set of axes.
15. Graph $x^2 + y^2 = 25$ and $4x^2 + y^2 = 36$ on the same set of axes. What is the significance of the coordinates of the points of intersection?
16. Graph $y^2 - x^2 = 0$. Discuss the graph in relation to Figure 9.7.
17. Graph $4x^2 - y^2 = 0$. Generalize from the result and discuss the graph of any equation of the form $Ax^2 - By^2 = C$, where $A, B > 0$ and $C = 0$.
18. Graph $4x^2 + y^2 = 0$. Generalize from the result and discuss the graph of any equation of the form $Ax^2 + By^2 = C$, where $A, B > 0$ and $C = 0$.

9.3 Sketching graphs of conic sections

We should make use of the form of the equation to aid in graphing quadratic equations in two variables. With this in mind, the ideas developed in Sections 9.1 and 9.2 may be summarized as follows:

1. A quadratic equation of the form

$$y = ax^2 + bx + c \tag{1}$$

has a graph that is a parabola, opening upward if $a > 0$ and downward if $a < 0$.

2. A quadratic equation of the form

$$Ax^2 + By^2 = C \quad (A^2 + B^2 \neq 0) \tag{2}$$

has a graph that is

(a) a circle if $A = B$ and A, B, and C have like signs;
(b) an ellipse if $A \neq B$ and A, B, and C have like signs;
(c) a hyperbola if A and B are opposite in sign and $C \neq 0$;
(d) two distinct lines through the origin if A and B are opposite in sign and $C = 0$ (see Problem 17, Exercise 9.2);
(e) a point if A and B are both greater than 0 or both less than 0 and $C = 0$ (see Problem 18, Exercise 9.2);
(f) imaginary if A and B are both greater than 0 and C less than 0 or if A and B are both greater than or equal to 0 and C greater than 0.

After the general form of the curve is recognized, the graph of a few points should suffice to sketch the graph. The intercepts, for instance, are always easy to locate.

In obtaining the x-intercepts of the parabola with Equation (1),

$$y = ax^2 + bx + c,$$

we identify those values of x for which $y = 0$, that is, values of x for which

$$ax^2 + bx + c = 0.$$

For example, the x-intercepts of

$$y = -2x^2 - 5x + 3 \tag{3}$$

are the solutions of

$$-2x^2 - 5x + 3 = 0,$$

which can be easily found by first writing the equation equivalently as

$$-1(x + 3)(2x - 1) = 0.$$

By inspection, we observe that these solutions, and therefore the x-intercepts of the graph of Equation (3), are -3 and $\frac{1}{2}$. The y-intercept, 3, can be found readily by inspection by assigning the value 0 to x in Equation (3).

Note that we have two different names for a single idea:

1. The elements of the solution set of the equation $ax^2 + bx + c = 0$.
2. The x-intercepts of the graph of the function $\{(x, y) \mid y = ax^2 + bx + c\}$.

Regarding this idea, you should recall that a quadratic equation in one variable may have no real solution, one real solution, or two real solutions. If the equation has no real solution, we find that the graph of the related quadratic equation in two variables does not touch the x-axis; if there is one solution, the graph is tangent to the x-axis; if there are two real solutions, the graph crosses the x-axis at two distinct points, as shown in Figure 9.8.

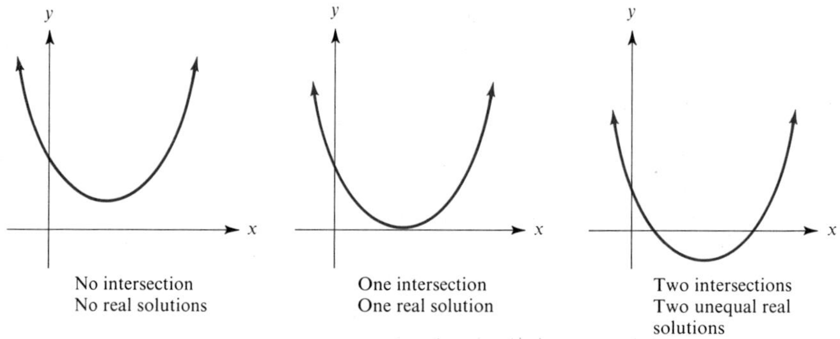

FIGURE **9.8**

Now consider the equation

$$x^2 + 4y^2 = 8. \tag{4}$$

By comparing this equation with 2(b), we note immediately that its graph is an ellipse. If $y = 0$, then $x = \pm\sqrt{8}$, and if $x = 0$, then $y = \pm\sqrt{2}$. We can then sketch the graph of Equation (4) as in Figure 9.9.

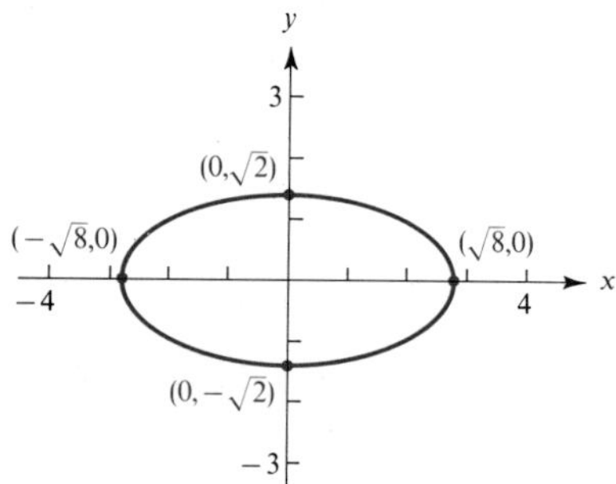

FIGURE **9.9**

As another example, consider the equation

$$x^2 - y^2 = 3. \tag{5}$$

By comparing this equation with 2(c), we see that its graph is a hyperbola. If $y = 0$, then $x = \pm\sqrt{3}$, and if $x = 0$, then y is imaginary (the graph will not cross the y-axis). By assigning a few other arbitrary values to one of the variables—say x, for example (4,) and (−4,)—we can find the additional ordered pairs

$$(4, \sqrt{13}), \quad (4, -\sqrt{13}), \quad (-4, \sqrt{13}), \quad \text{and} \quad (-4, -\sqrt{13})$$

which satisfy Equation (5). The graph can then be sketched as shown in Figure 9.10.

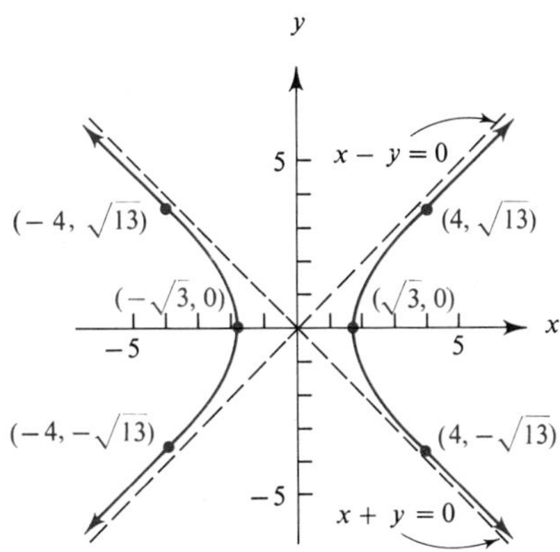

FIGURE **9.10**

The dashed lines shown in the figure are called **asymptotes** of the graph and they themselves constitute the graph of $x^2 - y^2 = 0$. While we shall not discuss the notion in detail here, it is true that, in general, the graph of $Ax^2 - By^2 = C$ will "approach" the two straight lines in the graph of $Ax^2 - By^2 = 0$ for each $A, B, C > 0$.

EXERCISE 9.3

Name and sketch the graph of each of the following equations.

EXAMPLE

$4x^2 = 36 - 9y^2$

SOLUTION

Rewrite in standard form.

$$4x^2 + 9y^2 = 36$$

By inspection, the graph is an ellipse.

The x-intercepts are 3 and -3.
The y-intercepts are 2 and -2.

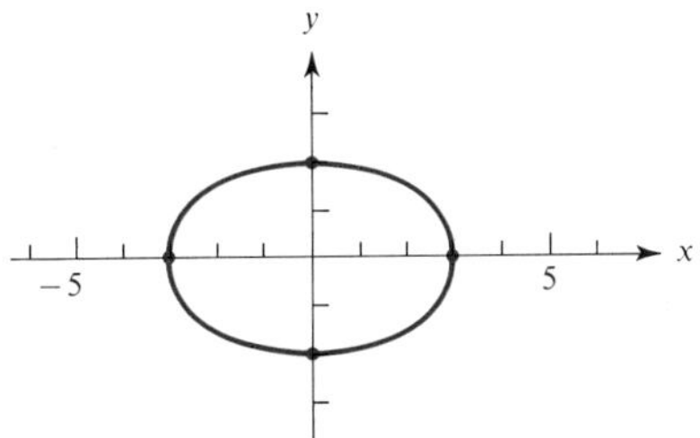

1. $x^2 + y^2 = 49$	2. $x^2 + y^2 = 64$	3. $x^2 + 9y = 0$
4. $2x^2 - y = 0$	5. $4x^2 + 25y^2 = 100$	6. $x^2 + 2y^2 = 8$
7. $x^2 = 9 + y^2$	8. $-2y^2 = 8 - x^2$	9. $4x^2 - 4y^2 = 0$
10. $x^2 - 9y^2 = 0$	11. $y = x^2 - 4$	12. $y = 4 - 2x^2$
13. $4x^2 + 4y^2 = 1$	14. $9x^2 + 9y^2 = 2$	15. $3x^2 = 12 - 4y^2$
16. $4x^2 = 12 - 3y^2$	17. $y = x^2 - 3x + 2$	18. $y = 4 + 3x - x^2$

19. Graph the set of points satisfying *both* $x^2 + y^2 = 25$ *and* $y = x^2$.
20. Graph the set of points satisfying *both* $x^2 + y^2 = 16$ *and* $y = x^2 - 4$.

9.4 Variation as a functional relationship

There are two types of functional relationships, widely used in the sciences, to which custom has assigned special names. First, any function defined by the equation

$$y = kx \quad (k \text{ a positive constant}) \tag{1}$$

is an example of **direct variation**. The variable y is said to **vary directly** as the variable x. Another example of direct variation is

$$y = kx^2 \quad (k \text{ a positive constant}), \tag{1a}$$

where we say that y varies directly as the square of x, and, in general,

$$y = kx^n \quad (k \text{ a positive constant and } n > 0) \tag{1b}$$

asserts that y varies directly as the nth power of x. We find examples of such variation in the relationships existing between the length of the radius of a circle and the circumference and area. Thus

$$C = 2\pi r$$

asserts that the circumference of a circle varies directly as the length of the radius, while

$$A = \pi r^2$$

expresses the fact that the area of a circle varies directly as the square of the length of the radius.

The second important type of variation arises from the equation

$$xy = k \quad (k \text{ a positive constant}),$$

where x and y are said to **vary inversely**. When this equation is written in the form

$$y = \frac{k}{x}, \tag{2}$$

y is said to vary inversely as x. Similarly, if

$$y = \frac{k}{x^2},$$

y is said to vary inversely as the square of x, etc. As an example of inverse variation, consider the set of rectangles with area 24 square units. Since the area of a rectangle is given by

$$lw = A,$$

we have

$$lw = 24,$$

and the length and width of the rectangle can be seen to vary inversely.

The constant involved in equations describing direct or inverse variation is called the **constant of variation**. If we know that one variable varies directly or inversely as another, and if we have one set of associated values for the variables, we can find the constant of variation involved. For example, suppose we know that y varies directly as x^2, and that $y = 4$ when $x = 7$. The fact that y varies directly as x^2 tells us that

$$y = kx^2.$$

The fact that $y = 4$ when $x = 7$ tells us that (7, 4) is a solution of the equation and hence that

$$4 = k(7)^2 = k(49),$$

from which

$$k = \frac{4}{49}.$$

The equation specifically expressing the direct variation is

$$y = \frac{4}{49}x^2.$$

In the event that one variable varies as the product of two or more other variables, we refer to the relationship as **joint variation**. Thus, if y varies jointly as u, v, and w, we have

$$y = kuvw. \tag{3}$$

Also, **joint** (direct and inverse) variation may take place concurrently. That is, y may vary directly as x and inversely as z, giving rise to the equation

$$y = k\frac{x}{z}.$$

It should be pointed out that the way in which the word "variation" is used herein is a technical one, and when the ideas of direct, inverse, or joint variation are encountered, we should always think of equations of the form (1), (2), or (3). For instance, the equations

$$y = 2x + 1,$$

$$y = \frac{1}{x} - 2,$$

and

$$y = xz + 2$$

do not describe examples of variation within our meaning of the word.

EXERCISE 9.4

Write an equation expressing the relationship between the variables, using k as the constant of variation.

EXAMPLE

At constant temperature, the volume (V) of a gas varies inversely as the pressure (P).

SOLUTION

$$V = \frac{k}{P}$$

1. The distance (d) traveled by a car moving at a constant rate varies directly as the time (t).
2. The tension (T) on a spring varies directly as the distance (s) it is stretched.
3. The current (I) in an electrical circuit with constant voltage varies inversely as the resistance (R) of the circuit.
4. The time (t) required by a car to travel a fixed distance varies inversely as the rate (r) at which it travels.
5. The volume (V) of a rectangular box of fixed depth varies jointly as its length (l) and width (w).

6. The power (P) in an electric circuit varies jointly as the resistance (R) and the square of the current (I).

Find the constant of variation for each of the following stated conditions.

EXAMPLE

If V varies inversely as P and $V = 100$ when $P = 30$.

SOLUTION

Write an equation expressing the relationship between the variables.

$$V = \frac{k}{P}$$

Substitute the known values for the variables and solve for k.

$$100 = \frac{k}{30}$$

$$k = \mathbf{3000}$$

7. If y varies directly as x and $y = 6$ when $x = 2$.
8. If y varies directly as x and $y = 2$ when $x = 5$.
9. If u varies inversely as the square of v and $u = 2$ when $v = 10$.
10. If r varies inversely as the cube of t and $r = 8$ when $t = 10$.
11. If z varies jointly as x and y and $z = 8$ when $x = 2$ and $y = 2$.
12. If p varies jointly as q and r and $p = 5$ when $q = 2$ and $r = 7$.
13. If z varies directly as the square of x and inversely as the cube of y and $z = 4$ when $x = 3$ and $y = 2$.
14. If z varies directly as the cube of x and inversely as the square of y and $z = 2$ when $x = 2$ and $y = 4$.
15. If z varies inversely as the sum of x^2 and y and $z = 12$ when $x = 4$ and $y = 6$.
16. If z varies directly as the sum of x and y and inversely as their product and $z = 8$ when $x = 3$ and $y = 4$.

EXAMPLE

If V varies directly as T and inversely as P and $V = 40$ when $T = 300$ and $P = 30$, find V when $T = 324$ and $P = 24$.

SOLUTION

Write an equation expressing the relationship between the variables.

$$V = \frac{kT}{P} \qquad (1)$$

Solution continued overleaf.

Substitute the initially known values for V, T, and P. Solve for k.

$$40 = \frac{k300}{30}$$

$$4 = k$$

Rewrite Equation (1) with k replaced by 4.

$$V = \frac{4T}{P}$$

Substitute second set of values for T and P and solve for V.

$$V = \frac{4(324)}{24} = \mathbf{54}$$

17. If y varies directly as x^2 and $y = 9$ when $x = 3$, find y when $x = 4$.
18. If r varies directly as s and inversely as t and $r = 12$ when $s = 8$ and $t = 2$, find r when $s = 3$ and $t = 6$.
19. The distance a particle falls in a certain medium is directly proportional to the square of the length of time it falls. If the particle falls 16 feet in 2 seconds, how far will it fall in 10 seconds?
20. In Problem 19, how far will the particle fall in 20 seconds?
21. The pressure exerted by a liquid at a given point varies directly as the depth of the point beneath the surface of the liquid. If a certain liquid exerts a pressure of 40 pounds per square foot at a depth of 10 feet, what would be the pressure at 40 feet?
22. The volume (V) of a gas varies directly as its temperature (T) and inversely as its pressure (P). A gas occupies 20 cubic feet at a temperature of 300° A (absolute) and a pressure of 30 pounds per square inch. What will the volume be if the temperature is raised to 360° A and the pressure decreased to 20 pounds per square inch?
23. The maximum-safe uniformly distributed load (L) for a horizontal beam varies jointly as its breadth (b) and square of its depth (d) and inversely as its length (l). An 8-foot beam with $b = 2$ and $d = 4$ will safely support a uniformly distributed load of up to 740 pounds. How many uniformly distributed pounds will an 8-foot beam support if $b = 2$ and $d = 6$?
24. The resistance (R) of a wire varies directly as its length (l) and inversely as the square of its diameter (d). 50 feet of wire of diameter 0.012 inches has a resistance of 10 ohms. What is the resistance of 50 feet of the same type of wire if the diameter is increased to 0.015 inches?

9.5 Proportions

There is an alternative term used to describe the variation relationships discussed in Section 9.4. The word "proportional" is frequently used in this sense. To say that "y is directly proportional to x" or "y is inversely proportional to x" is another way of describing direct and inverse variation. The use of the word "proportion" arises from the fact that any two solutions (a, b) and (c, d) of an equation expressing a variation satisfy an equation of the form

$$\frac{a}{b} = \frac{c}{d},$$

which is commonly called a **proportion**. For example, consider the situation in which the volume of a gas varies directly with the absolute temperature under a constant pressure and which can be represented by the relationship

$$V = kT.$$

For any set of values T_1 and V_1,

$$k = \frac{V_1}{T_1}, \tag{1}$$

and for any other set of values T_2 and V_2,

$$k = \frac{V_2}{T_2}. \tag{2}$$

Equating the right-hand members of (1) and (2),

$$\frac{V_1}{T_1} = \frac{V_2}{T_2},$$

from which the value of any variable can be determined if the values of the other variables are known.

One particularly useful application of proportion is that of converting from one unit of measure to another. For example, given that 16 fluid ounces are equivalent to one pint in liquid measure, we can determine the number of ounces in $3\frac{1}{4}$ pints by solving the proportion

$$\frac{16}{1} = \frac{x}{3\frac{1}{4}}.$$

Similarly, to convert 45 cubic feet to cubic yards, we use the fact that 27 cubic feet are equivalent to one cubic yard and solve the proportion

$$\frac{1}{27}=\frac{x}{45}.$$

A conversion table with some common units of measurement that are needed in the exercise set is given on page 414.

EXERCISE 9.5

Solve Problems 17–24 of Exercise 9.4 by first expressing each relationship as a proportion.

EXAMPLE

If V varies directly as T and inversely as P, and $V=40$ when $T=300$ and $P=30$, find V when $T=324$ and $P=24$.

SOLUTION

Write an equation expressing the relationship between the variables.

$$V=\frac{kT}{P}$$

Solve for k.

$$k=\frac{VP}{T}$$

Write a proportion relating the variables for two different sets of conditions.

$$\frac{V_1P_1}{T_1}=\frac{V_2P_2}{T_2}$$

Substitute the known values of the variables and solve for V_2.

$$\frac{(40)(30)}{300}=\frac{V_2(24)}{324},$$

$$V_2=\frac{(324)(40)(30)}{300(24)}=\mathbf{54}$$

1. Problem 17.
2. Problem 18.
3. Problem 19.
4. Problem 20.
5. Problem 21.
6. Problem 22.
7. Problem 23.
8. Problem 24.

Use the table of equivalents on page 414 and make the given conversion. Give answers to nearest tenth of a unit.

EXAMPLE

375 square inches to square feet.

SOLUTION

Since 144 sq in. is equivalent to 1 sq ft., we have

$$\frac{1}{144} = \frac{x}{375}$$

from which

$$x = \frac{375}{144} \approx 2.6.$$

Thus, 375 sq in. is approximately equal to **2.6 sq ft.**

9. $2\frac{1}{5}$ miles to yards.
10. 8800 feet to miles.
11. 54 ounces to pounds.
12. 4450 pounds to tons.
13. 4 cubic feet to cubic inches.
14. $8\frac{1}{3}$ cubic yards to cubic feet.
15. 243 fluid ounces to quarts.
16. $3\frac{1}{2}$ gallons to pints.
17. 800 inches to yards.
18. 12,000 feet to nautical miles.
19. 28.2 centimeters to inches.
20. 3.25 feet to centimeters.
21. 217 inches to meters.
22. 1243 centimeters to yards.
23. 27 square meters to square feet.
24. 20 square yards to square centimeters.
25. 15 pints to liters.
26. 1000 liters to gallons.

Chapter review

[9.1 -9.3] Name the graph of each equation.

1. $x^2 - 3y^2 = 8$
2. $x^2 = 4 - y^2$
3. $x^2 - y + 4 = 0$
4. $2y^2 = 4 - x^2$

Graph each equation.

5. $x^2 + y^2 = 36$
6. $4x^2 + y^2 = 36$
7. $9x^2 - y^2 = 0$
8. $4x^2 - y^2 = 16$
9. $y^2 - 4x^2 = 16$
10. $y = x^2 - 4$
11. $y = x^2 + 6x + 5$
12. $2x^2 + 2y^2 = 12$

[9.4] 13. If y varies inversely as t^2, and $y = 16$ when $t = 3$, find y when $t = 4$.

14. The weight of a body above the surface of the earth varies inversely as the square of its distance from the center of the earth. If we assume the radius of the earth to be 4000 miles, how much would a man weigh 500 miles above the earth's surface if he weighed 200 pounds on the surface?

15. The speed of a gear varies directly as the number of teeth it contains. If a gear with 10 teeth rotates at 240 revolutions per minute (rpm), with what rpm would a gear with 18 teeth revolve under the same conditions?

16. The number of posts required to string a telephone line over a given distance varies inversely as the distance between the posts. If it takes 80 posts separated by 120 feet to string a wire between two points, how many posts would be required if the posts were 150 feet apart?

[9.5] 17. Solve Problem 15 above by first eliminating the constant of variation.

18. Solve Problem 16 above by first eliminating the constant of variation.

19. Convert 3740 yards to miles.

20. Convert 16.5 gallons to fluid ounces.

Systems of equations

In Chapter 8, we observed that the solution set of an equation in two variables, such as

$$ax + by + c = 0$$

or

$$Ax^2 + By^2 = C,$$

might contain infinitely many ordered pairs of numbers. It is often necessary to consider pairs of such sentences and to inquire whether or not the solution sets of the sentences contain ordered pairs in common.

10.1 Systems of linear equations in two variables

Let us begin by considering the system

$$a_1 x + b_1 y + c_1 = 0 \quad (a_1, b_1 \text{ not both } 0)$$
$$a_2 x + b_2 y + c_2 = 0 \quad (a_2, b_2 \text{ not both } 0).$$

In a geometric sense, because the graphs of both equations are straight lines, there are three possibilities, as illustrated in Figure 10.1:

(a) The graphs are the same line.
(b) The graphs are parallel but distinct lines.
(c) The graphs intersect in one and only one point.

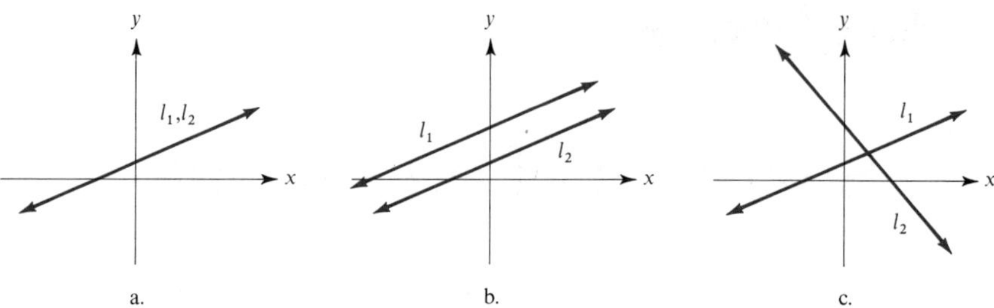

FIGURE **10.1**

These possibilities lead, correspondingly, to the conclusion that one and only one of the following is true for any given system of two such linear equations in x and y:

(a) The solution sets of the equations are equal.
(b) The two solution sets contain no common ordered pairs.
(c) The two solution sets contain one and only one common ordered pair.

In case (a), the two linear equations in x and y are said to be **dependent** and **consistent**; in case (b), the equations are **independent** and **inconsistent**; and in case (c), the equations are **independent** and **consistent**.

To identify common solutions in a system of independent and consistent equations such as

$$\begin{aligned} x + y &= 5 \\ x - y &= 1, \end{aligned} \tag{1}$$

we can use the fact that for any solution common to both equations, the same variables in the equation represent the same numbers. Therefore, the sum of the left-hand members of the equations must be equal to the sum of the right-hand members, and the result of adding the corresponding members is the equation

$$2x + 0y = 6, \quad \text{or} \quad x = 3.$$

Hence any solution of the System (1) must be of the form $(3, y)$. We can obtain the corresponding value of y by substituting 3 for x in either equation in (1). Let us use $x + y = 5$, to obtain

$$3 + y = 5, \quad \text{or} \quad y = 2.$$

Therefore, the solution set of the system contains only the ordered pair (3, 2). The graphs of the equations in (1) are shown in Figure 10.2, and in some cases

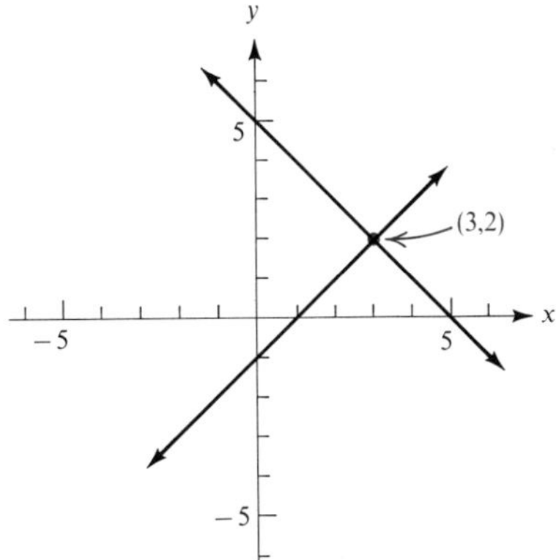

FIGURE **10.2**

the graphs alone are sufficient to determine the common solution. That is, in cases where both of the components of the ordered pair in the solution are integral, we can usually determine the solution by inspecting the graphs of the equations.

However, because graphing equations is a time-consuming process, and, more importantly, because graphical results are not always precise, solutions to systems of linear equations are usually sought by other methods, such as the **addition method** illustrated above.

In using the addition method to solve a system of equations, it is sometimes necessary first to multiply one or both equations by a constant in order to produce equations in which the coefficients of one of the variables are of opposite sign. For example, consider the system

$$2x - y = 5 \tag{2}$$

$$3x + 2y = 4. \tag{3}$$

We can multiply each member of Equation (2) by 2 and add the resulting equations, member by member, to get

$$2(2x) + 3x - 2(y) + 2y = 2(5) + 4 \quad \text{or} \quad 7x = 14,$$

from which we obtain 2 as the x-component of a solution of Equations (2) and (3). Thus $(2, y)$ is the form of the required solution, and, upon substituting 2 for x in (2), we have

$$2(2) - y = 5,$$

$$y = -1.$$

The solution set of the system is, therefore, $\{(2, -1)\}$. In this case, we chose the multiplier 2, so that the coefficient of y in the sum of the equations was zero.

If the coefficients of the variables in one equation in a system are proportional to the coefficients in the other equation, the equations are either dependent or inconsistent. For example, consider the system

$$2x + 3y = 2 \tag{4}$$

$$4x + 6y = 7. \tag{5}$$

Since the coefficients of x and y in Equations (4) and (5) are such that

$$\frac{2}{4} = \frac{3}{6},$$

it follows that any attempt to add the left-hand members of these equations to eliminate one variable will result in an equation free of both variables. Thus, if we multiply -2 times Equation (4) we have the system

$$-4x - 6y = -4$$

$$4x + 6y = 7,$$

which, upon addition of left-hand members and right-hand members, yields

$$0 = 3.$$

Since this equation is not true for any values of x and y, the equations do not have any common solutions and they are inconsistent. Now, consider the system

$$2x + 3y = 2 \tag{6}$$

$$4x + 6y = 4, \tag{7}$$

where the coefficients of the variables and the constant terms are proportional. The result of adding -2 times Equation (6) to Equation (7) is

$$0 = 0.$$

Because this equation is true for all values of x and y, the solution sets of the given equations must be identical and the equations are dependent.

Systems of equations are quite useful in expressing relationships in practical applications. By the assignment of separate variables to represent separate physical quantities, the difficulty encountered in symbolically representing these relationships can usually be decreased. In writing systems of equations, we must be careful that the conditions giving rise to one equation are independent of the conditions giving rise to any other equation.

EXERCISE 10.1

Find the solution set for each of the following systems by analytic methods. Check your solutions of Problems 1–10 by graphical methods. If the equations are inconsistent or dependent, so state.

EXAMPLE

$$\frac{2}{3}x - y = 2 \qquad (1)$$

$$x + \frac{1}{2}y = 7 \qquad (2)$$

SOLUTION

Multiply each member of Equation (1) by -3 and each member of Equation (2) by 2.

$$-2x + 3y = -6 \qquad (1')$$

$$2x + y = 14 \qquad (2')$$

Add (1′) to (2′) and solve for y.

$$4y = 8$$

$$y = 2$$

Substitute 2 for y in either (1), (2), (1′), or (2′) and solve for x. In this example, Equation (2) is used.

$$x + \frac{1}{2}(2) = 7$$

$$x = 6$$

The solution set is **{(6, 2)}**.

1. $x - y = 1$
   $x + y = 5$
2. $2x - 3y = 6$
   $x + 3y = 3$
3. $3x + y = 7$
   $2x - 5y = -1$
4. $2x - y = 7$
   $3x + 2y = 14$
5. $5x - y = -29$
   $2x + 3y = 2$
6. $x + 4y = -14$
   $3x + 2y = -2$
7. $5x + 2y = 3$
   $x = 0$
8. $2x - y = 0$
   $x = -3$
9. $3x - 2y = 4$
   $y = -1$
10. $x + 2y = 6$
    $y = 2$
11. $\frac{1}{4}x - \frac{1}{3}y = -\frac{5}{12}$
    $\frac{1}{10}x + \frac{1}{5}y = \frac{1}{2}$
12. $\frac{2}{3}x - y = 4$
    $x - \frac{3}{4}y = 6$

13. $\frac{1}{7}x - \frac{3}{7}y = -1$
$2x - y = -4$

14. $\frac{1}{3}x - \frac{2}{3}y = 2$
$x - 2y = 6$

15. $x + 3y = 6$
$2x + 6y = 12$

16. $3x - 2y = 6$
$6x - 4y = 8$

17. $2x - y = 1$
$8x - 4y = 3$

18. $6x + 2y = 1$
$12x + 4y = 2$

Solve.

EXAMPLE

The sum of two numbers is 17 and one of the numbers is four less than twice the other. Find the numbers.

SOLUTION

Represent each number by a separate variable.

Let x and y represent numbers.

Represent two independent conditions stated in the problem by two equations.

$$x + y = 17$$
$$x = 2y - 4$$

Rewrite the equations in standard form.

$$x + y = 17$$
$$x - 2y = -4$$

Solve the resulting system.
The numbers are **10** and **7**.

Check: Does $(10) + (7) = 17$? Yes.
Does $(10) = 2(7) - 4$? Yes.

19. The difference of two numbers is 6, and the greater number is ten less than twice the lesser number. Find the numbers.

20. One number is one less than three times another, and their sum is 15. Find the numbers.

21. The sum of two weights is 28 lbs. When the lighter weight is placed 18 ft from the fulcrum of a lever, it just balances the heavier weight placed 6 ft from the fulcrum. Find the weights.

22. How far from the fulcrum of a lever would a weight of 4 lbs have to be placed to balance a weight of 5 lbs located 8 ft from the fulcrum?

23. How much of a total amount of \$1800 must be invested at 6% and how much at 5% if the annual income is to be \$100?

24. A man has \$2000 more invested at 7% than he does at 8%. How much does he have invested at each rate if his annual income from the investments is \$740?

25. A boat can go 4 miles upstream in 2 hours and can return in 1 hour. Find the speed of the current and the rate of the boat in still water.
26. If a boat can go 10 miles downstream in 50 minutes and 12 miles upstream in an hour and a half, find the speed of the current and the rate of the boat in still water.
27. How many ounces of a 75% alcohol must be added to how many ounces of a 90% alcohol to produce 50 ounces of 80% alcohol?
28. How many pounds of candy worth 70 cents per pound must be mixed with candy worth 90 cents per pound to produce 120 pounds of a mixture worth 85 cents per pound?
29. Find a and b so that the graph of $ax + by + 3 = 0$ passes through the points $(-1, 2)$ and $(-3, 0)$.
30. Find a and b so that the graph of $ax + by = 12$ passes through $(2,1)$ and $(6, -1)$.

10.2 Linear systems in three variables—analytic solution

A solution of an equation in three variables, such as

$$x + 2y - 3z + 4 = 0, \tag{1}$$

is an ordered triple of numbers (x, y, z), because all three of the variables in Equation (1) must be replaced before we can decide whether the result is an equality. Thus $(0, -2, 0)$ and $(-1, 0, 1)$ are solutions of (1), while $(1, 1, 1)$ is not. There are, of course, infinitely many members in the solution set.

The solution set of a system of three linear equations in three variables, such as

$$x + 2y - 3z + 4 = 0 \tag{1}$$

$$2x - y + z - 3 = 0 \tag{2}$$

$$3x + 2y + z - 10 = 0, \tag{3}$$

contains the common members of the solution sets of all three of the equations in the system.

We seek solution sets of systems such as this by methods analogous to those used in solving linear systems in two variables. Since graphical treatments would be three-dimensional (linear equations in three variables can be represented by planes, and their common intersection, if any, would represent the solution set), we shall consider analytic solutions only. In the foregoing system, we might begin by multiplying Equation (1) by -2 and adding the result to Equation (2) to produce

$$-5y + 7z - 11 = 0, \tag{4}$$

which is satisfied by any ordered triple (x, y, z) that satisfies Equations (1) and (2). Similarly, we can add -3 times Equation (1) to Equation (3) to obtain

$$-4y + 10z - 22 = 0, \tag{5}$$

which is satisfied by any ordered triple (x, y, z) that satisfies both Equations (1) and (3). We can now argue that any ordered triple satisfying the original system

$$-5y + 7z - 11 = 0 \tag{4}$$

$$-4y + 10z - 22 = 0 \tag{5}$$

will also satisfy Equations (4) and (5). Since the system (4) and (5) does not depend on x, the problem has been reduced to one of finding the y- and z-components of the solution only. This system can be solved by the method of Section 10.1, which leads to the values $y = 2$ and $z = 3$. Since any solution of (1), (2), and (3) must be of the form $(x, 2, 3)$, 2 can be substituted for y and 3 for z in (1) to obtain $x = 1$, so that the desired solution set is

$$\{(1, 2, 3)\}.$$

If at any step in the procedure just outlined either an identity or a contradiction results, the system contains dependent equations or else two or three inconsistent equations; and it either has no members in its solution set or else it has an infinite number of members.

The process of solving a system of consistent and independent equations can be reduced to a series of mechanical procedures, as illustrated by the first example in Exercise 10.2.

EXERCISE 10.2

Solve. If the equations in the system are not consistent or independent, so state.

EXAMPLE

$$x + 2y - z + 1 = 0 \tag{1}$$

$$x - 3y + z - 2 = 0 \tag{2}$$

$$2x + y + 2z - 6 = 0 \tag{3}$$

SOLUTION

Multiply (1) by -1 and add the result to (2); multiply (1) by -2 and add the result to (3).

$$-5y + 2z - 3 = 0 \tag{4}$$

$$-3y + 4z - 8 = 0 \tag{5}$$

Multiply (4) by -2 and add the result to (5).

$$7y - 2 = 0 \tag{6}$$

$$y = \frac{2}{7}$$

Substitute $\frac{2}{7}$ for y in either (4) or (5), say (4).

$$-5\left(\frac{2}{7}\right) + 2z - 3 = 0$$

$$z = \frac{31}{14}$$

Substitute $\frac{2}{7}$ for y and $\frac{31}{14}$ for z in either (1), (2), or (3), say (1).

$$x + 2\left(\frac{2}{7}\right) - \frac{31}{14} + 1 = 0$$

$$x = \frac{9}{14}$$

The solution set is $\left\{\left(\mathbf{\frac{9}{14}, \frac{2}{7}, \frac{31}{14}}\right)\right\}$.

1. $x + y + z = 2$
   $2x - y + z = -1$
   $x - y - z = 0$

2. $x + y + z = 1$
   $2x - y + 3z = 2$
   $2x - y - z = 2$

3. $x + y + 2z = 0$
   $2x - 2y + z = 8$
   $3x + 2y + z = 2$

4. $5y - 8z = -19$
   $5x - 8z = 6$
   $3x - 2y = 12$

5. $x - 2y + z = -1$
   $2x + y - 3z = 3$
   $3x + 3y - 2z = 10$

6. $x - 2y + 4z = -3$
   $3x + y - 2z = 12$
   $2x + y - 3z = 11$

7. $x - 2y + 3z = 4$
   $2x - y + z = 1$
   $3x - 3y + 4z = 5$

8. $x + 5y - z = 2$
   $3x - 9y + 3z = 6$
   $x - 3y + z = 4$

9. $2x + z = 7$
$y - z = -2$
$x + y = 2$

10. $2x - 3y + z = 3$
$x - y - 2z = -1$
$-x + 2y - 3z = -4$

11. $x - \frac{1}{2}y - \frac{1}{2}z = 4$
$x - \frac{3}{2}y - 2z = 3$
$\frac{1}{4}x + \frac{1}{4}y - \frac{1}{4}z = 0$

12. $x + 2y + \frac{1}{2}z = 0$
$x + \frac{3}{5}y - \frac{2}{5}z = \frac{1}{5}$
$4x - 7y - 7z = 6$

EXAMPLE

The sum of three numbers is 12. Twice the first number is equal to the second, and the third number is equal to the sum of the other two. Find the numbers.

SOLUTION

Represent each number by a separate variable.

Let x, y, and z represent numbers.

Write the three independent conditions stated in the problem as three equations.

$$x + y + z = 12$$
$$2x = y$$
$$x + y = z$$

Rewrite equations in standard form.

$$x + y + z = 12$$
$$2x - y = 0$$
$$x + y - z = 0$$

Find the solution set of the system by any convenient means.

The numbers are **2**, **4**, and **6**.

13. The sum of three numbers is 15. The second equals two times the first and the third equals the second. Find the numbers.
14. The sum of three numbers is 2. The first number is equal to the sum of the other two, and the third number is the result of subtracting the first from the second. Find the numbers.
15. A box contains \$6.25 in nickels, dimes, and quarters. There are 85 coins in all with three times as many nickels as dimes. How many coins of each kind are there?

16. The sum of three distances, d_1, d_2, and d_3, is 184 miles. The distance d_1 is 32 miles less than d_2 and one-half of d_3. Find the three distances.
17. The perimeter of a triangle is 155 inches. The side x is 20 inches shorter than the side y, and the side y is 5 inches longer than the side z. Find the lengths of the sides of the triangle.
18. The equation for a circle can be written $x^2 + y^2 + ax + by + c = 0$. Find the equation of the circle whose graph contains the points (2, 3), (3, 2), and (−4, −5).
19. Find values for a, b, and c so that the graph of $x^2 + y^2 + ax + by + c = 0$ will contain the points (−2, 3), (1, 6), and (2, 4).
20. Find values for a, b, and c so that the graph of $y = ax^2 + bx + c$ will contain the points (−1, 2), (1, 6), and (2, 11).

10.3 Linear systems in two variables—solution by determinants

An expression of the form

$$\begin{vmatrix} a_1 & b_1 \\ a_2 & b_2 \end{vmatrix}$$

is called a **determinant**. The numbers a_1, b_1, a_2, and b_2 are called **elements** of the determinant. Because this determinant has two rows and two columns of elements, it is called a two-by-two (2 × 2) determinant, or a determinant of order two. We define the value of the determinant to be

$$a_1b_2 - a_2b_1.$$

This value is obtained by multiplying the elements on the diagonals and adding the negative of the second product to the first product. This process can be shown schematically as

$$\begin{vmatrix} a_1 & b_1 \\ a_2 & b_2 \end{vmatrix} = a_1b_2 - a_2b_1.$$

For example,

$$\begin{vmatrix} 1 & 2 \\ -1 & 3 \end{vmatrix} = 3 - (-2) = 5$$

and

$$\begin{vmatrix} 0 & -1 \\ -1 & 7 \end{vmatrix} = 0 - 1 = -1.$$

A determinant, therefore, is simply another way to represent a single number.

Determinants can be used to solve linear systems. In this section, we shall confine our attention to linear systems of two equations in two variables of the form

$$a_1 x + b_1 y = c_1, \tag{1}$$

$$a_2 x + b_2 y = c_2. \tag{2}$$

If this system is solved by means of addition, we have, upon multiplication of Equation (1) by $-a_2$ and Equation (2) by a_1, the equations

$$-a_2 a_1 x - a_2 b_1 y = -a_2 c_1, \tag{1'}$$

$$a_1 a_2 x + a_1 b_2 y = a_1 c_2, \tag{2'}$$

the sum of whose members is

$$a_1 b_2 y - a_2 b_1 y = a_1 c_2 - a_2 c_1.$$

Now, factoring y from each term in the left-hand member, we have

$$(a_1 b_2 - a_2 b_1) y = a_1 c_2 - a_2 c_1,$$

from which

$$y = \frac{a_1 c_2 - a_2 c_1}{a_1 b_2 - a_2 b_1}. \tag{3}$$

But the numerator of the right-hand member of Equation (3) is just the value of the determinant

$$\begin{vmatrix} a_1 & c_1 \\ a_2 & c_2 \end{vmatrix},$$

which we designate as D_y, and the denominator is the value of the determinant

$$\begin{vmatrix} a_1 & b_1 \\ a_2 & b_2 \end{vmatrix},$$

which we designate as D, so that (3) can be written

$$y = \frac{D_y}{D} = \frac{\begin{vmatrix} a_1 & c_1 \\ a_2 & c_2 \end{vmatrix}}{\begin{vmatrix} a_1 & b_1 \\ a_2 & b_2 \end{vmatrix}}. \tag{4}$$

The elements of the determinant in the denominator are the coefficients of the variables in Equations (1) and (2). The elements of the determinant in the numerator of (4) are identical to those in the denominator, except that the elements in the column containing the coefficients of y have been replaced by the constant terms of (1) and (2).

By a similar procedure, we can show that

$$\boldsymbol{x} = \frac{\boldsymbol{D_x}}{\boldsymbol{D}} = \frac{\begin{vmatrix} c_1 & b_1 \\ c_2 & b_2 \end{vmatrix}}{\begin{vmatrix} a_1 & b_1 \\ a_2 & b_2 \end{vmatrix}},$$

and (4) and (5) together yield the components of the ordered pair in the solution set of the system. The use of determinants in this way is known as *Cramer's rule* for the solution of a system of linear equations. As an example of the use of Cramer's rule, consider the system

$$2x + y = 4$$
$$x - 3y = -5.$$

We find

$$D = \begin{vmatrix} 2 & 1 \\ 1 & -3 \end{vmatrix} = -6 - 1 = -7,$$

$$D_x = \begin{vmatrix} 4 & 1 \\ -5 & -3 \end{vmatrix} = -12 + 5 = -7,$$

and

$$D_y = \begin{vmatrix} 2 & 4 \\ 1 & -5 \end{vmatrix} = -10 - 4 = -14.$$

Therefore,

$$x = \frac{D_x}{D} = \frac{-7}{-7} = 1,$$

$$y = \frac{D_y}{D} = \frac{-14}{-7} = 2,$$

and the solution set is $\{(1, 2)\}$.

If $D = 0$ in this procedure, the equations in the system are either dependent or inconsistent as D_y and D_x are zero or not zero.

EXERCISE 10.3

Evaluate.

EXAMPLE

$$\begin{vmatrix} 2 & -3 \\ 1 & 4 \end{vmatrix}$$

Solution overleaf.

SOLUTION

$$\begin{vmatrix} 2 & -3 \\ 1 & 4 \end{vmatrix} = (2)(4) - (1)(-3) = \mathbf{11}$$

1. $\begin{vmatrix} 1 & 0 \\ 2 & 1 \end{vmatrix}$
2. $\begin{vmatrix} 3 & -2 \\ 4 & 1 \end{vmatrix}$
3. $\begin{vmatrix} -5 & -1 \\ 3 & 3 \end{vmatrix}$
4. $\begin{vmatrix} 1 & -2 \\ -1 & 2 \end{vmatrix}$
5. $\begin{vmatrix} -1 & 6 \\ 0 & -2 \end{vmatrix}$
6. $\begin{vmatrix} 20 & 3 \\ -20 & -2 \end{vmatrix}$
7. $\begin{vmatrix} \frac{2}{3} & \frac{3}{5} \\ \frac{2}{5} & -\frac{4}{3} \end{vmatrix}$
8. $\begin{vmatrix} \frac{3}{4} & -\frac{1}{3} \\ -\frac{2}{3} & -\frac{1}{4} \end{vmatrix}$

Find the solution set of each of the following systems by Cramer's rule.

EXAMPLE

$$2x - 3y = 6$$
$$2x + y = 14$$

SOLUTION

$$D = \begin{vmatrix} 2 & -3 \\ 2 & 1 \end{vmatrix} = (2)(1) - (2)(-3) = 8$$

The elements in D_x are obtained from the elements in D by replacing the elements in the column containing the coefficients of x with the constant terms 6 and 14.

$$D_x = \begin{vmatrix} 6 & -3 \\ 14 & 1 \end{vmatrix} = (6)(1) - (14)(-3) = 48$$

The elements in D_y are obtained from the elements in D by replacing the elements in the column containing the coefficients of y with the constant terms 6 and 14.

$$D_y = \begin{vmatrix} 2 & 6 \\ 2 & 14 \end{vmatrix} = (2)(14) - (2)(6) = 16$$

Values for x and y can now be determined by Cramer's rule.

$$x = \frac{D_x}{D} = \frac{48}{8} = 6$$

$$y = \frac{D_y}{D} = \frac{16}{8} = 2$$

The solution set is $\{(\mathbf{6, 2})\}$.

9. $2x - 3y = -1$
   $x + 4y = 5$
10. $3x - 4y = -2$
    $x - 2y = 0$
11. $3x - 4y = -2$
    $x + 2y = 6$

12. $2x - 4y = 7$
$x - 2y = 1$

13. $\frac{1}{3}x - \frac{1}{2}y = 0$
$\frac{1}{2}x + \frac{1}{4}y = 4$

14. $\frac{2}{3}x + y = 1$
$x - \frac{4}{3}y = 0$

15. $x - 2y = 5$
$\frac{2}{3}x - \frac{4}{3}y = 6$

16. $-\frac{1}{2}x + y = 3$
$-\frac{1}{4}x - y = -3$

17. $x - 3y = 1$
$y = 1$

18. $2x - 3y = 12$
$x = 4$

19. $ax + by = 1$
$bx + ay = 1$

20. $x + y = a$
$x - y = b$

Show that for every real value of the variables:

21. $\begin{vmatrix} a & a \\ b & b \end{vmatrix} = 0.$

22. $\begin{vmatrix} a_1 & b_1 \\ a_2 & b_2 \end{vmatrix} = -\begin{vmatrix} a_2 & b_2 \\ a_1 & b_1 \end{vmatrix}.$

23. $\begin{vmatrix} a_1 & b_1 \\ a_2 & b_2 \end{vmatrix} = -\begin{vmatrix} b_1 & a_1 \\ b_2 & a_2 \end{vmatrix}.$

24. $\begin{vmatrix} ka_1 & b_1 \\ ka_2 & b_2 \end{vmatrix} = k\begin{vmatrix} a_1 & b_1 \\ a_2 & b_2 \end{vmatrix}.$

25. $\begin{vmatrix} ka & a \\ kb & b \end{vmatrix} = 0.$

26. $\begin{vmatrix} a_1 + ka_2 & b_1 + kb_2 \\ a_2 & b_2 \end{vmatrix} = \begin{vmatrix} a_1 & b_1 \\ a_2 & b_2 \end{vmatrix}.$

10.4 Third-order determinants

A determinant of the form

$$\begin{vmatrix} a_1 & b_1 & c_1 \\ a_2 & b_2 & c_2 \\ a_3 & b_3 & c_3 \end{vmatrix}$$

is said to be a three-by-three (3×3), or third-order, determinant. The value of a third-order determinant is defined to be

$$\begin{vmatrix} a_1 & b_1 & c_1 \\ a_2 & b_2 & c_2 \\ a_3 & b_3 & c_3 \end{vmatrix} = a_1 b_2 c_3 - a_1 b_3 c_2 + a_3 b_1 c_2 - a_2 b_1 c_3 + a_2 b_3 c_1 - a_3 b_2 c_1. \quad (1)$$

Again, we note that a 3×3 determinant is simply another way of writing a single number, namely, that number represented by the expression in the right-hand member of (1).

The **minor** of an element in a determinant is defined to be the determinant that remains after deletion of the row and column in which the element appears. In the determinant in Equation (1),

the minor of the element a_1 is $\begin{vmatrix} b_2 & c_2 \\ b_3 & c_3 \end{vmatrix}$;

the minor of the element b_1 is $\begin{vmatrix} a_2 & c_2 \\ a_3 & c_3 \end{vmatrix}$;

the minor of the element c_1 is $\begin{vmatrix} a_2 & b_2 \\ a_3 & b_3 \end{vmatrix}$; etc.

If, by suitably factoring pairs of terms in the right-hand member, (1) is rewritten in the form

$$\begin{vmatrix} a_1 & b_1 & c_1 \\ a_2 & b_2 & c_2 \\ a_3 & b_3 & c_3 \end{vmatrix} = a_1(b_2c_3 - b_3c_2) - b_1(a_2c_3 - a_3c_2) + c_1(a_2b_3 - a_3b_2), \quad (2)$$

we observe that the sums enclosed in parentheses in the right-hand member of Equation (2) arc the respective minors (second-order determinants) of the elements a_1, b_1, and c_1. Therefore, (2) can be written

$$\begin{vmatrix} a_1 & b_1 & c_1 \\ a_2 & b_2 & c_2 \\ a_3 & b_3 & c_3 \end{vmatrix} = a_1\begin{vmatrix} b_2 & c_2 \\ b_3 & c_3 \end{vmatrix} - b_1\begin{vmatrix} a_2 & c_2 \\ a_3 & c_3 \end{vmatrix} + c_1\begin{vmatrix} a_2 & b_2 \\ a_3 & b_3 \end{vmatrix}.$$

The right-hand member of (3) is called the **expansion** of the determinant by minors about the first row.

Suppose, instead of factoring the right-hand member of (1) into the right-hand member of (2), we factor it as

$$\begin{vmatrix} a_1 & b_1 & c_1 \\ a_2 & b_2 & c_2 \\ a_3 & b_3 & c_3 \end{vmatrix} = a_1(b_2c_3 - b_3c_2) - a_2(b_1c_3 - b_3c_1) + a_3(b_1c_2 - b_2c_1). \quad (4)$$

Then we have the expansion of the determinant by minors about the first column,

$$\begin{vmatrix} a_1 & b_1 & c_1 \\ a_2 & b_2 & c_2 \\ a_3 & b_3 & c_3 \end{vmatrix} = a_1\begin{vmatrix} b_2 & c_2 \\ b_3 & c_3 \end{vmatrix} - a_2\begin{vmatrix} b_1 & c_1 \\ b_3 & c_3 \end{vmatrix} + a_3\begin{vmatrix} b_1 & c_1 \\ b_2 & c_2 \end{vmatrix}. \quad (5)$$

With the proper use of signs it is possible to expand a determinant by minors about *any* row or *any* column and obtain an expression equivalent to a factored form of the right-hand member of (1). A helpful device for determining the signs of the terms in an expansion of a third-order determinant by minors is the array of alternating signs

$$\begin{matrix} + & - & + \\ - & + & - \\ + & - & + \end{matrix}$$

which we will call the **sign array** for the determinant. To obtain an expansion of a determinant about a given row or column, the appropriate sign from the sign array is prefixed to each term in the expansion.

As an example, consider the expansion of a determinant about the second row:

$$\begin{vmatrix} 1 & 2 & -3 \\ 0 & 2 & -1 \\ 1 & 1 & 0 \end{vmatrix} = -0\begin{vmatrix} 2 & -3 \\ 1 & 0 \end{vmatrix} + 2\begin{vmatrix} 1 & -3 \\ 1 & 0 \end{vmatrix} - (-1)\begin{vmatrix} 1 & 2 \\ 1 & 1 \end{vmatrix}$$

$$= 0 + 2(0 + 3) + 1(1 - 2)$$

$$= 6 - 1$$

$$= 5.$$

You should expand this determinant about each row and about each column to verify that the result is the same in each expansion.

The expansion of a higher-order determinant by minors can be accomplished in the same way. By continuing the pattern of alternating signs used for third-order determinants, we can extend the sign array to higher-order determinants. The determinants in each term in the expansion will be of order one less than the order of the original determinant.

EXERCISE 10.4

Evaluate.

EXAMPLE

$$\begin{vmatrix} 1 & 2 & 0 \\ 3 & -1 & 4 \\ -2 & 1 & 3 \end{vmatrix}$$

SOLUTION

Expand about any row or column; the first row is used here.

$$\begin{vmatrix} 1 & 2 & 0 \\ 3 & -1 & 4 \\ -2 & 1 & 3 \end{vmatrix} = 1\begin{vmatrix} -1 & 4 \\ 1 & 3 \end{vmatrix} - 2\begin{vmatrix} 3 & 4 \\ -2 & 3 \end{vmatrix} + 0\begin{vmatrix} 3 & -1 \\ -2 & 1 \end{vmatrix}$$

$$= 1[(-1)(3) - (1)(4)] - 2[(3)(3) - (-2)(4)] + 0$$

$$= (-3 - 4) - 2(9 + 8)$$

$$= -7 - 34$$

$$= \mathbf{-41}$$

1. $\begin{vmatrix} 2 & 0 & 1 \\ 1 & 1 & 2 \\ -1 & 0 & 1 \end{vmatrix}$

2. $\begin{vmatrix} 1 & 3 & 1 \\ -1 & 2 & 1 \\ 0 & 2 & 0 \end{vmatrix}$

3. $\begin{vmatrix} 2 & -1 & 0 \\ -3 & 1 & 2 \\ 1 & -3 & 1 \end{vmatrix}$

4. $\begin{vmatrix} 2 & 4 & -1 \\ -1 & 3 & 2 \\ 4 & 0 & 2 \end{vmatrix}$

5. $\begin{vmatrix} 1 & 2 & 3 \\ 4 & -1 & 2 \\ 2 & 0 & 2 \end{vmatrix}$

6. $\begin{vmatrix} 1 & 0 & 0 \\ 0 & 1 & 2 \\ 0 & 3 & 4 \end{vmatrix}$

7. $\begin{vmatrix} -1 & 0 & 2 \\ -2 & 1 & 0 \\ 0 & 1 & -3 \end{vmatrix}$

8. $\begin{vmatrix} 2 & 1 & 4 \\ 3 & 2 & 6 \\ 5 & -3 & 10 \end{vmatrix}$

9. $\begin{vmatrix} 2 & 5 & -1 \\ 1 & 0 & 2 \\ 0 & 0 & 1 \end{vmatrix}$

10. $\begin{vmatrix} 2 & 3 & 1 \\ 0 & 1 & 0 \\ -4 & 2 & 1 \end{vmatrix}$

11. $\begin{vmatrix} 2 & -1 & 3 \\ 0 & 2 & 1 \\ -1 & 3 & 4 \end{vmatrix}$

12. $\begin{vmatrix} 2 & 0 & 1 \\ 3 & 4 & 0 \\ 0 & -2 & 3 \end{vmatrix}$

13. $\begin{vmatrix} 2 & 2 & 2 \\ 0 & 1 & 0 \\ -2 & -2 & -2 \end{vmatrix}$

14. $\begin{vmatrix} 1 & 3 & -5 \\ 2 & 0 & 1 \\ 0 & 2 & -2 \end{vmatrix}$

15. $\begin{vmatrix} 4 & 2 & 1 \\ 8 & 4 & 2 \\ -1 & 2 & 1 \end{vmatrix}$

16. $\begin{vmatrix} 3 & -2 & 0 \\ 1 & 0 & -2 \\ 5 & 3 & 1 \end{vmatrix}$

17. $\begin{vmatrix} 0 & 6 & 2 \\ 8 & 1 & 2 \\ -3 & 1 & 4 \end{vmatrix}$

18. $\begin{vmatrix} 8 & 0 & 2 \\ 6 & -1 & 0 \\ 3 & 2 & 1 \end{vmatrix}$

19. $\begin{vmatrix} x & 0 & 0 \\ 0 & x & 0 \\ 0 & 0 & x \end{vmatrix}$

20. $\begin{vmatrix} 0 & 0 & x \\ 0 & x & 0 \\ x & 0 & 0 \end{vmatrix}$

21. $\begin{vmatrix} x & y & 0 \\ x & y & 0 \\ 0 & 0 & 1 \end{vmatrix}$

22. $\begin{vmatrix} 0 & a & b \\ a & 0 & a \\ b & a & 0 \end{vmatrix}$

23. $\begin{vmatrix} a & b & 0 \\ b & 0 & b \\ 0 & b & a \end{vmatrix}$

24. $\begin{vmatrix} 0 & b & 0 \\ b & a & b \\ 0 & b & 0 \end{vmatrix}$

Solve for x.

25. $\begin{vmatrix} x & 0 & 0 \\ 2 & 1 & 3 \\ 0 & 1 & 4 \end{vmatrix} = 3$

26. $\begin{vmatrix} x^2 & 0 & 1 \\ 2 & -1 & 3 \\ 3 & 2 & 0 \end{vmatrix} = 1$

27. $\begin{vmatrix} x^2 & x & 1 \\ 0 & 2 & 1 \\ 3 & 1 & 4 \end{vmatrix} = 28$

28. $\begin{vmatrix} x & 1 & 1 \\ 0 & x & 1 \\ 0 & x & 0 \end{vmatrix} = -4$

Extend the procedure in the text and evaluate.

29. $\begin{vmatrix} 0 & 1 & 0 & 0 \\ 1 & 0 & 3 & 2 \\ 5 & -1 & 2 & 1 \\ 1 & 0 & 1 & 1 \end{vmatrix}$

30. $\begin{vmatrix} 1 & 2 & 0 & -1 \\ 1 & 0 & -1 & 2 \\ 0 & 1 & 1 & 1 \\ 2 & -1 & 0 & 1 \end{vmatrix}$

31. $\begin{vmatrix} 2 & -1 & 3 & 1 \\ 1 & 1 & 3 & 1 \\ 0 & 0 & 2 & 0 \\ 2 & -1 & 5 & 2 \end{vmatrix}$

32. $\begin{vmatrix} 1 & 0 & -3 & 5 \\ -1 & 1 & 0 & -1 \\ -2 & 2 & 1 & 0 \\ 0 & 3 & 2 & 1 \end{vmatrix}$

10.5 Properties of determinants

Determinants have several useful properties that permit us to form equal but simpler determinants. These properties hold for determinants of any order. Although we shall not prove the properties here, let us state them for the general $n \times n$ determinant and take a more detailed look at several examples of them as they apply to 3×3 determinants.

Property 1. *If each entry in any row (or column) of a determinant is* 0, *then the determinant is equal to* 0.

For example,

$$\begin{vmatrix} 1 & 1 & 0 \\ 3 & 5 & 0 \\ 2 & 7 & 0 \end{vmatrix} = 0 \quad \text{and} \quad \begin{vmatrix} 3 & -2 & 4 \\ 0 & 0 & 0 \\ 1 & 2 & 0 \end{vmatrix} = 0.$$

Property 2. *If any two rows (or columns) of a determinant are interchanged, the resulting determinant is the negative of the original determinant.*

For example,

$$\begin{vmatrix} 1 & 3 & 4 \\ 2 & 5 & 6 \\ 7 & 8 & 9 \end{vmatrix} = -\begin{vmatrix} 3 & 1 & 4 \\ 5 & 2 & 6 \\ 8 & 7 & 9 \end{vmatrix} \quad \text{and} \quad \begin{vmatrix} 1 & 3 & 4 \\ 2 & 5 & 6 \\ 7 & 8 & 9 \end{vmatrix} = -\begin{vmatrix} 7 & 8 & 9 \\ 2 & 5 & 6 \\ 1 & 3 & 4 \end{vmatrix}.$$

In the first example the first and second columns were interchanged, and in the second example the first and third rows were interchanged.

Property 3. *If two rows (or two columns) of a determinant have corresponding entries that are equal, the determinant is* 0.

For example,

$$\begin{vmatrix} 1 & 2 & 1 \\ 4 & 5 & 6 \\ 1 & 2 & 1 \end{vmatrix} = 0 \quad \text{and} \quad \begin{vmatrix} 7 & 4 & 4 \\ 2 & -3 & -3 \\ -1 & 5 & 5 \end{vmatrix} = 0.$$

Property 4. *If each of the entries of one row (or column) of a determinant is multiplied by k, the determinant is multiplied by k.*

For example,

$$\begin{vmatrix} 1 & 0 & 0 \\ 2 & 1 & 3 \\ 1\times 2 & 3\times 2 & 4\times 2 \end{vmatrix} = 2\begin{vmatrix} 1 & 0 & 0 \\ 2 & 1 & 3 \\ 1 & 3 & 4 \end{vmatrix} \quad\text{and}\quad \begin{vmatrix} 4 & 5 & 8 \\ 1 & 1 & 2 \\ 3 & 1 & 6 \end{vmatrix} = 2\begin{vmatrix} 4 & 5 & 4 \\ 1 & 1 & 1 \\ 3 & 1 & 3 \end{vmatrix}.$$

Property 5. *If each entry in a row (or column) of a determinant is written as the sum of two terms, the determinant can be written as the sum of two determinants as illustrated in the following examples:*

$$\begin{vmatrix} 1 & 3 & 6 \\ a_2+d_2 & b_2+e_2 & c_2+f_2 \\ 2 & 5 & 4 \end{vmatrix} = \begin{vmatrix} 1 & 3 & 6 \\ a_2 & b_2 & c_2 \\ 2 & 5 & 4 \end{vmatrix} + \begin{vmatrix} 1 & 3 & 6 \\ d_2 & e_2 & f_2 \\ 2 & 5 & 4 \end{vmatrix}$$

and

$$\begin{vmatrix} 2 & 4 & c_1+d_1 \\ 3 & 1 & c_2+d_2 \\ 6 & 5 & c_3+d_3 \end{vmatrix} = \begin{vmatrix} 2 & 4 & c_1 \\ 3 & 1 & c_2 \\ 6 & 5 & c_3 \end{vmatrix} + \begin{vmatrix} 2 & 4 & d_1 \\ 3 & 1 & d_2 \\ 6 & 5 & d_3 \end{vmatrix}.$$

Property 6. *If to each entry of one row (or column) of a determinant is added k times the corresponding entry of any other row (or column, respectively) in the determinant, the resulting determinant is equal to the original determinant.*

For example,

$$\begin{vmatrix} 1 & 2 & 3 \\ 4 & 5 & 6 \\ 7 & 8 & 9 \end{vmatrix} = \begin{vmatrix} 1+2(3) & 2 & 3 \\ 4+2(6) & 5 & 6 \\ 7+2(9) & 8 & 9 \end{vmatrix} = \begin{vmatrix} 7 & 2 & 3 \\ 16 & 5 & 6 \\ 25 & 8 & 9 \end{vmatrix}$$

and

$$\begin{vmatrix} 1 & 2 & 3 \\ 4 & 5 & 6 \\ 7 & 8 & 9 \end{vmatrix} = \begin{vmatrix} 1+4(4) & 2+4(5) & 3+4(6) \\ 4 & 5 & 6 \\ 7 & 8 & 9 \end{vmatrix} = \begin{vmatrix} 17 & 22 & 27 \\ 4 & 5 & 6 \\ 7 & 8 & 9 \end{vmatrix}.$$

The preceding statements can be used to write sequences of equal determinants, leading from one form to another, more useful form. For example, we may wish to expand

$$D = \begin{vmatrix} 2 & -1 & 1 \\ 1 & 3 & -4 \\ 3 & -1 & 5 \end{vmatrix}.$$

As a step toward expanding the determinant, we shall use Property 6 to produce an equal determinant with a row or a column containing zero entries in all but one place. Let us arbitrarily select the second column for this role.

Multiplying elements of the first row by 3 and adding the result to elements of the second row, we obtain

$$D = \begin{vmatrix} 2 & -1 & 1 \\ 1+3(2) & 3+3(-1) & -4+3(1) \\ 3 & -1 & 5 \end{vmatrix} = \begin{vmatrix} 2 & -1 & 1 \\ 7 & 0 & -1 \\ 3 & -1 & 5 \end{vmatrix}.$$

Next, multiplying elements of the first row by -1 and adding the result to elements of the third row, we find that

$$D = \begin{vmatrix} 2 & -1 & 1 \\ 7 & 0 & -1 \\ 3-1(2) & -1-1(-1) & 5-1(1) \end{vmatrix} = \begin{vmatrix} 2 & -1 & 1 \\ 7 & 0 & -1 \\ 1 & 0 & 4 \end{vmatrix}.$$

If we now expand the determinant about the second column, we have

$$D = \begin{vmatrix} 2 & -1 & 1 \\ 7 & 0 & -1 \\ 1 & 0 & 4 \end{vmatrix} = -(-1)\begin{vmatrix} 7 & -1 \\ 1 & 4 \end{vmatrix} + 0 - 0.$$

We can now expand directly and obtain

$$D = 28 - (-1) = 29.$$

EXERCISE 10.5

Without evaluating, state (using Properties 1–6 why each statement is true.

1. $\begin{vmatrix} 2 & 3 & 1 \\ 0 & 0 & 0 \\ -1 & 2 & 0 \end{vmatrix} = 0$

2. $\begin{vmatrix} 3 & 1 & 3 \\ 0 & 1 & 0 \\ 1 & 2 & 1 \end{vmatrix} = 0$

3. $\begin{vmatrix} 3 & 1 & -1 \\ 0 & 1 & 2 \\ 3 & 1 & -1 \end{vmatrix} = 0$

4. $\begin{vmatrix} 3 & 2 & 0 \\ -1 & 2 & 0 \\ 1 & 1 & 0 \end{vmatrix} = 0$

5. $\begin{vmatrix} -2 & 1 & 0 \\ 3 & 4 & 1 \\ -4 & 2 & 0 \end{vmatrix} = 0$

6. $\begin{vmatrix} 0 & 1 & 4 \\ 6 & 1 & 2 \\ 0 & 2 & 8 \end{vmatrix} = 0$

7. $\begin{vmatrix} 2 & 3 \\ 1 & -1 \end{vmatrix} = -\begin{vmatrix} 3 & 2 \\ -1 & 1 \end{vmatrix}$

8. $\begin{vmatrix} -2 & 3 & 1 \\ -1 & 0 & 1 \\ -2 & 1 & 0 \end{vmatrix} = -\begin{vmatrix} 2 & 3 & 1 \\ 1 & 0 & 1 \\ 2 & 1 & 0 \end{vmatrix}$

9. $\begin{vmatrix} 4 & 2 & 1 \\ 0 & -1 & -2 \\ 1 & 0 & 2 \end{vmatrix} = -\begin{vmatrix} 4 & 2 & 1 \\ 0 & 1 & 2 \\ 1 & 0 & 2 \end{vmatrix}$

10. $\begin{vmatrix} 3 & 1 & 0 \\ -2 & 1 & 1 \\ 0 & 2 & -1 \end{vmatrix} = -\begin{vmatrix} 0 & 1 & 3 \\ 1 & 1 & -2 \\ -1 & 2 & 0 \end{vmatrix}$

11. $2\begin{vmatrix} 1 & 0 & 2 \\ -1 & 2 & 0 \\ 1 & 1 & 1 \end{vmatrix} = \begin{vmatrix} 1 & 0 & 2 \\ -1 & 2 & 0 \\ 2 & 2 & 2 \end{vmatrix}$

12. $\begin{vmatrix} 3 & -4 & 2 \\ 1 & -2 & 0 \\ 0 & 8 & 1 \end{vmatrix} = -2\begin{vmatrix} 3 & 2 & 2 \\ 1 & 1 & 0 \\ 0 & -4 & 1 \end{vmatrix}$

13. $\begin{vmatrix} 3 & 0 & 6 \\ 2 & 1 & 2 \\ 0 & 1 & -2 \end{vmatrix} = 6\begin{vmatrix} 1 & 0 & 1 \\ 2 & 1 & 1 \\ 0 & 1 & -1 \end{vmatrix}$

14. $\begin{vmatrix} 1 & 2 & 1 \\ -1 & 0 & -2 \\ 2 & 4 & 1 \end{vmatrix} = -2\begin{vmatrix} 1 & 1 & 1 \\ 1 & 0 & 2 \\ 2 & 2 & 1 \end{vmatrix}$

15. $\begin{vmatrix} 3 & 5 \\ 1 & 4 \end{vmatrix} = \begin{vmatrix} 3 & 3 \\ 1 & 2 \end{vmatrix} + \begin{vmatrix} 3 & 2 \\ 1 & 2 \end{vmatrix}$

16. $\begin{vmatrix} 3 & 1 & 4 \\ 2 & 2 & 2 \\ 1 & 0 & 1 \end{vmatrix} = \begin{vmatrix} 3 & 1 & 4 \\ 1 & 1 & 1 \\ 1 & 0 & 1 \end{vmatrix} + \begin{vmatrix} 3 & 1 & 4 \\ 1 & 1 & 1 \\ 1 & 0 & 1 \end{vmatrix}$

17. $\begin{vmatrix} 1 & 2 \\ 3 & 4 \end{vmatrix} = \begin{vmatrix} 1+2 & 2 \\ 3+4 & 4 \end{vmatrix}$

18. $\begin{vmatrix} 1 & 2 \\ 3 & 4 \end{vmatrix} = \begin{vmatrix} 1+4 & 2 \\ 3+8 & 4 \end{vmatrix}$

19. $\begin{vmatrix} 1 & 2 \\ 3 & 4 \end{vmatrix} = \begin{vmatrix} 1 & 2 \\ 3-3 & 4-6 \end{vmatrix}$

20. $\begin{vmatrix} 1 & 2 \\ 3 & 4 \end{vmatrix} = \begin{vmatrix} 1 & 2-2 \\ 3 & 4-6 \end{vmatrix}$

21. $\begin{vmatrix} 1 & 2 & 1 \\ 0 & 2 & 3 \\ 2 & -1 & 2 \end{vmatrix} = \begin{vmatrix} 1 & 2 & 1 \\ 0 & 2 & 3 \\ 0 & -5 & 0 \end{vmatrix}$

22. $\begin{vmatrix} -1 & 1 & 0 \\ 2 & 3 & -1 \\ 2 & 1 & 2 \end{vmatrix} = \begin{vmatrix} 0 & 1 & 0 \\ 5 & 3 & -1 \\ 3 & 1 & 2 \end{vmatrix}$

One of the properties of determinants was used on the left-hand member of each of the following equalities to produce the elements in the right-hand member. Complete the entries.

23. $\begin{vmatrix} 1 & 3 \\ 2 & 2 \end{vmatrix} = \begin{vmatrix} 1 & 3 \\ 0 & \quad \end{vmatrix}$

24. $\begin{vmatrix} 2 & -1 \\ 3 & 1 \end{vmatrix} = \begin{vmatrix} \quad & 0 \\ 3 & 1 \end{vmatrix}$

25. $\begin{vmatrix} 1 & -2 & 1 \\ 3 & 1 & 4 \\ 0 & 2 & 1 \end{vmatrix} = \begin{vmatrix} 1 & -2 & 1 \\ 0 & \quad & 1 \\ 0 & 2 & 1 \end{vmatrix}$

26. $\begin{vmatrix} 3 & -1 & 0 \\ 1 & 2 & 1 \\ 2 & 3 & 1 \end{vmatrix} = \begin{vmatrix} 3 & -1 & 0 \\ 1 & 2 & 1 \\ \quad & 1 & 0 \end{vmatrix}$

27. $\begin{vmatrix} 2 & 3 & 4 \\ 0 & 2 & 2 \\ 1 & 1 & 3 \end{vmatrix} = \begin{vmatrix} 0 & 1 & \quad \\ 0 & 2 & 2 \\ 1 & 1 & 3 \end{vmatrix}$

28. $\begin{vmatrix} 2 & 1 & 1 \\ 1 & 2 & 0 \\ 2 & -1 & 4 \end{vmatrix} = \begin{vmatrix} 2 & 1 & 1 \\ 1 & 2 & 0 \\ \quad & -5 & 0 \end{vmatrix}$

Replace each determinant with an equivalent 2×2 determinant and then evaluate.

29. $\begin{vmatrix} 2 & 1 & 0 \\ 3 & 2 & 1 \\ -1 & 2 & 0 \end{vmatrix}$

30. $\begin{vmatrix} 1 & 2 & 1 \\ 2 & -1 & 3 \\ 0 & 1 & 0 \end{vmatrix}$

31. $\begin{vmatrix} 1 & 0 & 3 \\ 2 & -1 & 1 \\ 1 & 2 & 1 \end{vmatrix}$

32. $\begin{vmatrix} 1 & 2 & -1 \\ 2 & 1 & 3 \\ 0 & 1 & 2 \end{vmatrix}$

33. $\begin{vmatrix} 1 & 2 & 1 \\ -1 & 2 & 3 \\ 2 & -1 & 1 \end{vmatrix}$

34. $\begin{vmatrix} 3 & -1 & 2 \\ 1 & 2 & 1 \\ -2 & 1 & 3 \end{vmatrix}$

35. $\begin{vmatrix} 2 & 3 & -1 \\ 1 & -2 & 1 \\ 2 & 3 & 4 \end{vmatrix}$

36. $\begin{vmatrix} 22 & 23 & 19 \\ 3 & -1 & 2 \\ 2 & 1 & 3 \end{vmatrix}$

37. $\begin{vmatrix} 45 & 44 & 43 \\ 4 & 5 & 6 \\ 1 & 2 & 1 \end{vmatrix}$

38. $\begin{vmatrix} 0 & 0 & 1 & 2 \\ 6 & 0 & 0 & 1 \\ 6 & 1 & 0 & -1 \\ 6 & 1 & 0 & 2 \end{vmatrix}$ 39. $\begin{vmatrix} 4 & 2 & 0 & 2 \\ -1 & 0 & 2 & 1 \\ 3 & 0 & -1 & 1 \\ 0 & 0 & 2 & 1 \end{vmatrix}$ 40. $\begin{vmatrix} -3 & 1 & 2 & 2 \\ 1 & 4 & 0 & 2 \\ 2 & -1 & 0 & 3 \\ 1 & 0 & 0 & -1 \end{vmatrix}$

10.6 Linear systems in three variables—solution by determinants

Consider the linear system in three variables

$$a_1 x + b_1 y + c_1 z = d_1 \tag{1}$$

$$a_2 x + b_2 y + c_2 z = d_2 \tag{2}$$

$$a_3 x + b_3 y + c_3 z = d_3 . \tag{3}$$

By solving this system by the methods of Section 10.2, it can be shown that Cramer's rule is applicable to such systems and, in fact, to all similar systems as well as to linear systems in two variables. That is,

$$x = \frac{D_x}{D}, \qquad y = \frac{D_y}{D}, \qquad z = \frac{D_z}{D}, \quad D \neq 0$$

where

$$D = \begin{vmatrix} a_1 & b_1 & c_1 \\ a_2 & b_2 & c_2 \\ a_3 & b_3 & c_3 \end{vmatrix}, \quad D_x = \begin{vmatrix} d_1 & b_1 & c_1 \\ d_2 & b_2 & c_2 \\ d_3 & b_3 & c_3 \end{vmatrix},$$

$$D_y = \begin{vmatrix} a_1 & d_1 & c_1 \\ a_2 & d_2 & c_2 \\ a_3 & d_3 & c_3 \end{vmatrix}, \quad D_z = \begin{vmatrix} a_1 & b_1 & d_1 \\ a_2 & b_2 & d_2 \\ a_3 & b_3 & d_3 \end{vmatrix}.$$

Note that the elements of the determinant D in each denominator are the coefficients of the variables in Equations (1), (2), and (3) and that the numerators are formed from D by replacing the elements in the x, y, or z column, respectively, by d_1, d_2, and d_3. We illustrate the application of Cramer's rule by example. Consider the system

$$x + 2y - 3z = -4$$
$$2x - y + z = 3$$
$$3x + 2y + z = 10$$

The determinant D, whose elements are the coefficients of the variables, is given by

$$D = \begin{vmatrix} 1 & 2 & -3 \\ 2 & -1 & 1 \\ 3 & 2 & 1 \end{vmatrix}.$$

We arbitrarily expand the determinant about the first column, and get

$$D = \begin{vmatrix} 1 & 2 & -3 \\ 2 & -1 & 1 \\ 3 & 2 & 1 \end{vmatrix} = 1\begin{vmatrix} -1 & 1 \\ 2 & 1 \end{vmatrix} - 2\begin{vmatrix} 2 & -3 \\ 2 & 1 \end{vmatrix} + 3\begin{vmatrix} 2 & -3 \\ -1 & 1 \end{vmatrix}$$

$$= -3 - 16 - 3$$

$$= -22.$$

Replacing the first column in D with -4, 3, and 10, we obtain

$$D_x = \begin{vmatrix} -4 & 2 & -3 \\ 3 & -1 & 1 \\ 10 & 2 & 1 \end{vmatrix}.$$

Expanding D_x about the third column, we have

$$D_x = \begin{vmatrix} -4 & 2 & -3 \\ 3 & -1 & 1 \\ 10 & 2 & 1 \end{vmatrix} = -3\begin{vmatrix} 3 & -1 \\ 10 & 2 \end{vmatrix} - 1\begin{vmatrix} -4 & 2 \\ 10 & 2 \end{vmatrix} + 1\begin{vmatrix} -4 & 2 \\ 3 & -1 \end{vmatrix}$$

$$= -48 + 28 - 2$$

$$= -22.$$

D_y and D_z can be computed in similar fashion.

$$D_y = \begin{vmatrix} 1 & -4 & -3 \\ 2 & 3 & 1 \\ 3 & 10 & 1 \end{vmatrix} = -44, \quad D_z = \begin{vmatrix} 1 & 2 & -4 \\ 2 & -1 & 3 \\ 3 & 2 & 10 \end{vmatrix} = -66.$$

We then have

$$x = \frac{D_x}{D} = \frac{-22}{-22} = 1,$$

$$y = \frac{D_y}{D} = \frac{-44}{-22} = 2,$$

$$z = \frac{D_z}{D} = \frac{-66}{-22} = 3,$$

and the solution set of the system is $\{(1, 2, 3)\}$.

If $D = 0$ for a linear system in three variables, the system either has an empty solution set or a solution set with infinitely many members.

EXERCISE 10.6

Solve by Cramer's rule. If the solution set has no members or an infinite number of members, so state.

EXAMPLE

$4x + 10y - z = 2$
$2x + 8y + z = 4$
$x - 3y - 2z = 3$

SOLUTION

Determine values for D, D_x, D_y, and D_z. The elements of D are the coefficients of the variables in the order they occur. For D_x, D_y, and D_z, the respective column of elements in D is replaced by the constants 2, 4, and 3.

$$D = \begin{vmatrix} 4 & 10 & -1 \\ 2 & 8 & 1 \\ 1 & -3 & -2 \end{vmatrix} = 12; \qquad D_x = \begin{vmatrix} 2 & 10 & -1 \\ 4 & 8 & 1 \\ 3 & -3 & -2 \end{vmatrix} = 120;$$

$$D_y = \begin{vmatrix} 4 & 2 & -1 \\ 2 & 4 & 1 \\ 1 & 3 & -2 \end{vmatrix} = -36; \qquad D_z = \begin{vmatrix} 4 & 10 & 2 \\ 2 & 8 & 4 \\ 1 & -3 & 3 \end{vmatrix} = 96.$$

Use Cramer's rule to determine x, y, and z.

$$x = \frac{D_x}{D} = \frac{120}{12} = 10, \qquad y = \frac{D_y}{D} = \frac{-36}{12} = -3, \qquad z = \frac{D_z}{D} = \frac{96}{12} = 8$$

The solution set is $\{(\mathbf{10}, \mathbf{-3}, \mathbf{8})\}$.

1. $x + y = 2$
   $2x - z = 1$
   $2y - 3z = -1$

2. $2x - 6y + 3z = -12$
   $3x - 2y + 5z = -4$
   $4x + 5y - 2z = 10$

3. $x - 2y + z = -1$
   $3x + y - 2z = 4$
   $y - z = 1$

4. $2x + 5z = 9$
   $4x + 3y = -1$
   $3y - 4z = -13$

5. $2x + 2y + z = 1$
   $x - y + 6z = 21$
   $3x + 2y - z = -4$

6. $4x + 8y + z = -6$
   $2x - 3y + 2z = 0$
   $x + 7y - 3z = -8$

7. $x + y + z = 0$
   $2x - y - 4z = 15$
   $x - 2y - z = 7$

8. $x + y - 2z = 3$
   $3x - y + z = 5$
   $3x + 3y - 6z = 9$

9. $x - 2y + 2z = 3$
$2x - 4y + 4z = 1$
$3x - 3y - 3z = 4$

10. $3x - 2y + 5z = 6$
$4x - 4y + 3z = 0$
$5x - 4y + z = -5$

11. $x - 4z = -1$
$3x + 3y = 2$
$3x + 4z = 5$

12. $x - 3y = -1$
$3y - z = -9$
$x - 4y = 1$

13. $x + 4z = 3$
$y + 3z = 9$
$2x + 5y - 5z = -5$

14. $2x + y = 18$
$y + z = -1$
$3x - 2y - 5z = 38$

10.7 Second-degree systems in two variables—solution by substitution

Real solutions of systems of equations in two variables, where one or both of the equations are quadratic, can often be found by graphing both equations and estimating the coordinates of any points they have in common. For example, to find the solution set of the system

$$x^2 + y^2 = 26 \tag{1}$$

$$x + y = 6, \tag{2}$$

we graph the equations on the same set of axes, as shown in Figure 10.3, and

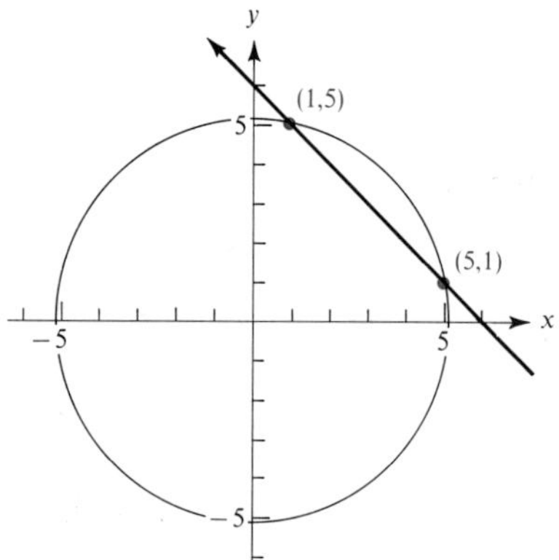

FIGURE **10.3**

observe that the graphs intersect at the points (1, 5) and (5, 1). Thus the solution set of the system (1) and (2) is

$$\{(1, 5), (5, 1)\}.$$

However, solving second-degree systems graphically on the real plane may produce only approximations to such real solutions as exist, and furthermore, we cannot expect to locate solutions among imaginary numbers. It is therefore more practical to concentrate on analytic methods of solution, since the results are exact and we can obtain all complex solutions. It is suggested, however, that you sketch the graphs of the equations as a rough check on an analytic solution. We can expect to find at most four and as few as no points of intersection in graphing second-degree systems, depending upon the types of equations involved and upon the coefficients and constants in the given equations.

One of the most useful techniques available for finding solution sets for systems of equations is that of **substitution**. Consider the system above,

$$x^2 + y^2 = 26 \tag{1}$$

$$x + y = 6. \tag{2}$$

Equation (2) can be written in the form

$$y = 6 - x \tag{3}$$

and we can argue that for any ordered pair (x, y) in the solution set of both Equations (1) and (2), x and y in (1) represent the same numbers as x and y in (3), and hence the substitution axiom can be invoked to replace y in (1) by its equal $(6 - x)$ from (3). This will produce

$$x^2 + (6 - x)^2 = 26, \tag{4}$$

which will have as a solution set those values of x for which the ordered pair (x, y) is a common solution of Equations (1) and (2). Rewriting (4) equivalently, we have

$$x^2 + 36 - 12x + x^2 = 26,$$

$$2x^2 - 12x + 10 = 0,$$

$$x^2 - 6x + 5 = 0,$$

$$(x - 5)(x - 1) = 0,$$

from which x is either 1 or 5. Now by replacing x in Equation (3) by each of these numbers in turn, we have

$$y = 6 - (1) = 5$$

and

$$y = 6 - (5) = 1,$$

so that the solution set of the system (1) and (2) is $\{(1, 5), (5, 1)\}$.

Check this solution and notice that these ordered pairs are also solutions of (1). Using Equation (1) rather than Equation (2) or (3) to obtain values for the y-component, we would have

$$(1)^2 + y^2 = 26 \qquad (5)^2 + y^2 = 26$$
$$y = \pm 5 \qquad y = \pm 1$$

and the solutions obtained are (1, 5), (1, −5), (5, 1), and (5, −1). However, (1, −5) and (5, −1) are not solutions of Equation (2). Therefore the solution set is again {(1, 5), (5, 1)}. This example suggests that if the degrees of equations differ, one component of a solution should be substituted in the equation of lower degree in order to find *only* those ordered pairs that are solutions of *both* equations.

The technique of solution by substitution can be used very easily with systems containing two first-degree equations, or one first-degree and one higher-degree equation, but is less satisfactory for systems where both of the equations are of degree greater than one in both variables.

EXERCISE 10.7

Solve by the method of substitution. In Problems 1–10, check the solutions by sketching the graphs of the equations and estimating the coordinates of any points of intersection.

EXAMPLE

$$y = x^2 + 2x + 1 \qquad (1)$$
$$y - x = 3 \qquad (2)$$

SOLUTION

Solve Equation (2) explicitly for y.

$$y = x + 3 \qquad (2')$$

Substitute $(x + 3)$ for y in (1).

$$x + 3 = x^2 + 2x + 1 \qquad (3)$$

Solve for x.

Check:

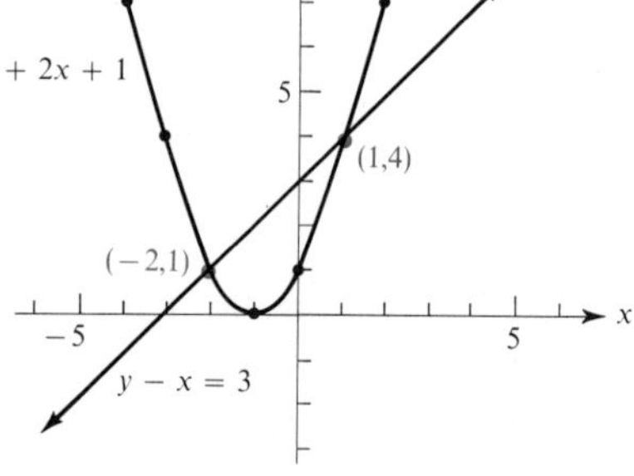

$$x^2 + x - 2 = 0$$
$$(x + 2)(x - 1) = 0$$
$$x = -2,\ x = 1$$

Substitute each of these values in turn in (2′) to determine values for y.

If $x = -2$, then $y = 1$;
if $x = 1$, then $y = 4$.

The solution set is **{(−2, 1), (1, 4)}**.

1. $y = x^2 - 5$
   $y = 4x$
2. $y = x^2 - 2x + 1$
   $y + x = 3$
3. $x^2 + y^2 = 13$
   $x + y = 5$
4. $x^2 + 2y^2 = 12$
   $2x - y = 2$
5. $x + y = 1$
   $xy = -12$
6. $2x - y = 9$
   $xy = -4$
7. $xy = 4$
   $x^2 + y^2 = 8$
8. $x^2 - y^2 = 35$
   $xy = 6$
9. $x^2 + y^2 = 9$
   $y = 4$
10. $x^2 - 2y^2 = 6$
    $x = 4$
11. $x^2 - xy - 2y^2 = 4$
    $x - y = 2$
12. $x^2 - 2x + y^2 = 3$
    $2x + y = 4$
13. $2x^2 - 5xy + 2y^2 = 5$
    $2x - y = 1$
14. $2x^2 + xy + y^2 = 9$
    $-x + 3y = 9$
15. $x^2 - 2x + y^2 - 1 = 0$
    $2x - 3y = 3$
16. $2x^2 - xy - y^2 = 20$
    $2x + y = 4$
17. The sum of the squares of two positive numbers is 13. If twice the first number is added to the second, the sum is 7. Find the numbers.
18. The sum of two numbers is 6 and their product is $35/4$. Find the numbers.
19. The sum of two numbers is 13, and their product is 12. Find the numbers.
20. The product of two numbers is 216. Twice the sum of the two numbers is 60. Find the numbers.
21. Twice the sum of two numbers is 18. If one of the numbers is decreased by 5 and the other is increased by 12, the product of the resultant two numbers is twice the product of the original two numbers. Find the numbers.
22. At a constant temperature, the pressure (P) and volume (V) of a gas are related by the equation $PV = K$, where K is a constant. The product of the pressure (in pounds per square inch) and the volume (in cubic inches) of a certain gas is 30 inch-pounds. If the temperature remains constant as the pressure is increased 4 pounds per square inch, the volume is decreased by 2 cubic inches. Find the original pressure and volume.

10.8 Second-degree systems in two variables—solution by other methods

If both of the equations in a system are of second degree in both variables, the method of addition often provides a simpler means of solution than does substitution. For example, consider the system

$$4x^2 + y^2 = 25 \tag{1}$$

$$x^2 - y^2 = -5. \tag{2}$$

By adding members of (1) to corresponding members of (2), we have

$$5x^2 = 20,$$

from which

$$x = 2 \quad \text{or} \quad x = -2,$$

and we have the x-components of the members of the solution set of the system of Equations (1) and (2). Substituting 2 for x in either (1) or (2), say (1),

$$\begin{aligned} 4(2)^2 + y^2 &= 25, \\ y^2 &= 25 - 16, \\ y^2 &= 9, \end{aligned}$$

from which

$$y = 3 \quad \text{or} \quad y = -3.$$

Thus, the ordered pairs (2, 3) and (2, -3) are in the solution set of the system. Substituting -2 for x in Equation (1) or (2) [this time we shall use (2)] gives us

$$\begin{aligned} (-2)^2 - y^2 &= -5, \\ -y^2 &= -5 - 4, \\ y^2 &= 9, \end{aligned}$$

so that

$$y = 3 \quad \text{or} \quad y = -3.$$

Thus, the ordered pairs (-2, 3) and (-2, -3) are also solutions of the system, and the complete solution set is

$$\{(2, 3), (2, -3), (-2, 3), (-2, -3)\}.$$

For an example of a slightly different procedure, consider the system

$$x^2 + y^2 = 5 \tag{3}$$

$$x^2 - 2xy + y^2 = 1. \tag{4}$$

By forming the sum of Equation (3) and -1 times Equation (4), we have

$$2xy = 4$$

$$xy = 2, \tag{5}$$

whose solution set contains all of the ordered pairs that satisfy both (3) and (4). Therefore, forming the new system,

$$x^2 + y^2 = 5 \tag{3}$$

$$xy = 2, \tag{5}$$

we can be sure that the solution set of this system will include the solution set of the system (3) and (4). This latter system can be solved by substitution. We have from Equation (5)

$$y = \frac{2}{x}.$$

Replacing y in (3) by $\frac{2}{x}$, we find

$$x^2 + \left(\frac{2}{x}\right)^2 = 5,$$

$$x^2 + \frac{4}{x^2} = 5. \qquad (6)$$

Multiplying each member by x^2, we have

$$x^4 + 4 = 5x^2, \qquad (7)$$

$$x^4 - 5x^2 + 4 = 0, \qquad (7')$$

which is quadratic in x^2. Factoring the left-hand member of Equation (7′), we obtain

$$(x^2 - 1)(x^2 - 4) = 0,$$

from which

$$x^2 - 1 = 0 \quad \text{or} \quad x^2 - 4 = 0,$$

and

$$x = 1, \qquad x = -1, \qquad x = 2, \qquad x = -2.$$

Since the step from (6) to (7) was a nonelementary transformation, we are careful to note that these all satisfy (6). Now substituting 1, −1, 2, and −2 in turn for x in Equation (5), we have

$$\begin{aligned} &\text{for } x = 1, \quad y = 2; \\ &\text{for } x = -1, \quad y = -2; \\ &\text{for } x = 2, \quad y = 1; \\ &\text{for } x = -2, \quad y = -1; \end{aligned}$$

and the solution set of the system (3) and (4) is

$$\{(1, 2), (-1, -2), (2, 1), (-2, -1)\}.$$

There are other techniques involving substitution in conjunction with addition that are useful in handling systems of higher-degree equations, but they all bear similarity to those illustrated. Each system should be scrutinized for some means of finding an equivalent system that will lend itself to solution by addition or substitution.

EXERCISE 10.8

Solve analytically.

EXAMPLE

$4x^2 + y^2 = 25$ (1)

$x^2 + 4y^2 = 40$ (2)

SOLUTION

Add the members of the equation term by term using 1 times Equation (1) and -4 times Equation (2) and solve for y.

$$-15y^2 = -135$$
$$y^2 = 9$$
$$y = 3 \quad \text{or} \quad y = -3$$

Substitute values for y in (1) or (2), say (1), to obtain associated values for x.

$$4x^2 + (3)^2 = 25 \qquad 4x^2 + (-3)^2 = 25$$
$$4x^2 + 9 = 25 \qquad 4x^2 + 9 = 25$$
$$x^2 = 4 \qquad x^2 = 4$$
$$x = \pm 2 \qquad x = \pm 2$$

The solution set is **{(2, 3), (2, −3), (−2, 3), (−2, −3)}**.

1. $x^2 + y^2 = 10$
   $9x^2 + y^2 = 18$
2. $x^2 + 4y^2 = 52$
   $x^2 + y^2 = 25$
3. $x^2 + 4y^2 = 17$
   $3x^2 - y^2 = -1$
4. $9x^2 + 16y^2 = 100$
   $x^2 + y^2 = 8$
5. $x^2 - y^2 = 7$
   $2x^2 + 3y^2 = 24$
6. $x^2 + 4y^2 = 25$
   $4x^2 + y^2 = 25$
7. $3x^2 + 4y^2 = 16$
   $x^2 - y^2 = 3$
8. $4x^2 + 3y^2 = 12$
   $x^2 + 3y^2 = 12$
9. $4x^2 - 9y^2 + 132 = 0$
   $x^2 + 4y^2 - 67 = 0$
10. $16y^2 + 5x^2 - 26 = 0$
    $25y^2 - 4x^2 - 17 = 0$
11. $x^2 - xy + y^2 = 7$
    $x^2 + y^2 = 5$
12. $3x^2 - 2xy + 3y^2 = 43$
    $x^2 + y^2 = 17$
13. $2x^2 + xy - 4y^2 = -12$
    $x^2 - 2y^2 = -4$
14. $x^2 + 2xy - y^2 = 14$
    $x^2 - y^2 = 8$

Chapter review

[10.1] Solve each system by addition.

1. $x + 5y = 18$
   $x - y = -3$

2. $x + 5y = 11$
   $2x + 3y = 8$

State whether the equations in each system are dependent, inconsistent, or consistent and independent.

3. a. $2x - 3y = 4$
   $x + 2y = 7$

   b. $2x - 3y = 4$
   $6x - 9y = 12$

4. a. $2x - 3y = 4$
   $6x - 9y = 4$

   b. $x - y = 6$
   $x + y = 6$

[10.2] Solve each system by addition.

5. $x + 3y - z = 3$
   $2x - y + 3z = 1$
   $3x + 2y + z = 5$

6. $x + y + z = 2$
   $3x - y + z = 4$
   $2x + y + 2z = 3$

[10.3] 7. Evaluate $\begin{vmatrix} 3 & -2 \\ 1 & -5 \end{vmatrix}$.

8. Solve the system by Cramer's rule.
   $2x + 3y = -2$
   $x - 8y = -39$

[10.4] Evaluate each determinant.

9. $\begin{vmatrix} 2 & 1 & 3 \\ 0 & 4 & -1 \\ 2 & 0 & 3 \end{vmatrix}$.

10. $\begin{vmatrix} 2 & -1 & 3 \\ 4 & 2 & 1 \\ 0 & -2 & 4 \end{vmatrix}$

[10.5] Each of the following statements, 11–14, is a result of one or more of the properties in Section 10.5. State the property (or properties) applicable in each case.

11. $\begin{vmatrix} 3 & -1 & 2 \\ 0 & 0 & 0 \\ 2 & 1 & -3 \end{vmatrix} = 0$

12. $\begin{vmatrix} 2 & 4 & 0 \\ 1 & -2 & 3 \\ 0 & 1 & 2 \end{vmatrix} = \begin{vmatrix} 0 & 8 & -6 \\ 1 & -2 & 3 \\ 0 & 1 & 2 \end{vmatrix}$

13. $\begin{vmatrix} 0 & 4 & 0 \\ -2 & 0 & -6 \\ 1 & 0 & 3 \end{vmatrix} = 0$

14. $\begin{vmatrix} 2 & 4 & 1 \\ 3 & 2 & 2 \\ -2 & 0 & 10 \end{vmatrix} = -2 \begin{vmatrix} 2 & 4 & 1 \\ 3 & 2 & 2 \\ 1 & 0 & -5 \end{vmatrix}$

15. Reduce to an equal 2×2 determinant and evaluate.

$$\begin{vmatrix} 0 & 2 & 1 \\ 3 & -1 & 1 \\ 1 & 0 & -2 \end{vmatrix}.$$

[10.6] Solve each system by Cramer's rule.

16. $2x + 3y - z = -2$
$x - y + z = 6$
$3x - y + z = 10$

17. $2x - y + 3z = 4$
$x + y - 2z = 0$
$4x - 2y + 6z = 3$

[10.7 –10.8] Solve each of the following by substitution or addition.

18. $x^2 + y = 3$
$5x + y = 7$

19. $x^2 + 3xy + x = -12$
$2x - y = 7$

20. $2x^2 + 5y^2 - 53 = 0$
$4x^2 + 3y^2 - 43 = 0$

11

Exponential and logarithmic functions

11.1 The exponential function

In Chapter 2, powers b^x were defined for any real b, and x a natural number. In Chapter 5, the definition was extended to include x negative or zero, and then x a rational number p/q. For rational exponents, the base b was restricted to positive values to ensure that $b^{p/q}$ be real.

We now inquire whether or not we can interpret powers with irrational exponents, such as

$$b^{\pi}, b^{\sqrt{2}}, \quad \text{and} \quad b^{-\sqrt{3}},$$

to be real numbers. In Section 6.6, we observed that irrational numbers can be approximated by rational numbers to as great a degree of accuracy as desired. That is, $\sqrt{2} \approx 1.4$ or $\sqrt{2} \approx 1.414$, and so forth.

Now, because 2^x is defined for rational x, such powers as $2^{1.4} = 2^{7/5}$, $2^{1.41} = 2^{141/100}$, etc., are clearly real numbers. Such powers can be used to give meaning to powers with irrational exponents. Thus, we can write the sequence of inequalities

$$2^1 < 2^{\sqrt{2}} < 2^2,$$
$$2^{1.4} < 2^{\sqrt{2}} < 2^{1.5},$$
$$2^{1.41} < 2^{\sqrt{2}} < 2^{1.42},$$
$$2^{1.414} < 2^{\sqrt{2}} < 2^{1.415},$$

and so on, where $2^{\sqrt{2}}$ is a number lying between the number on the left and that on the right. It is clear that this process can be continued indefinitely and that the difference between the number on the left and the one on the right can be made as small as we please. This being the case, we assume that there is just one number, $2^{\sqrt{2}}$, that will satisfy this inequality no matter how long this process is carried on. Since we can produce the same type of argument for any irrational exponent x, we shall assume that b^x $(b > 0)$ is defined for all real values of x.

Since for each real x there is one and only one number b^x, the equation

$$f(x) = b^x \quad (b > 0) \tag{1}$$

defines a function. Because $1^x = 1$ for all real values of x, Equation (1) defines a constant function if $b = 1$. If $b \neq 1$, we say that (1) defines an **exponential function.**

Exponential functions can perhaps be visualized more clearly by considering their graphs. We illustrate two typical examples, in which $0 < b < 1$ and $b > 1$, respectively. Assigning values to x in the equations

$$f(x) = \left(\frac{1}{2}\right)^x \quad \text{and} \quad f(x) = (2)^x,$$

we find some ordered pairs in each function and sketch the graphs in Figure 11.1. Notice that the graph of the function determined by $f(x) = (\frac{1}{2})^x$ goes

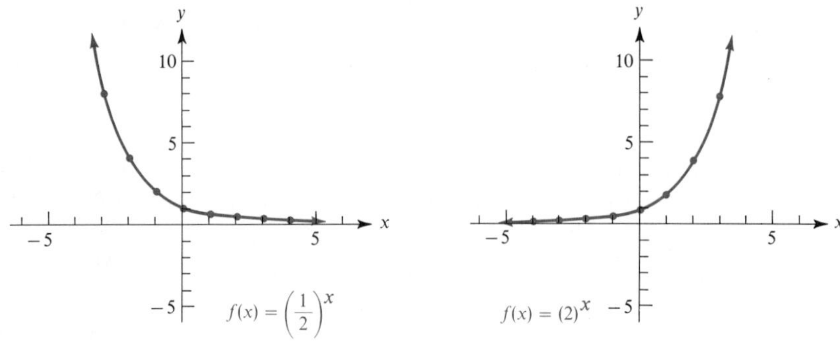

FIGURE **11.1**

down to the right, and the graph of the function determined by $f(x) = (2)^x$ goes *up* to the right. For this reason, we say that the former function is a **decreasing function** and that the latter is an **increasing function.**

EXERCISE 11.1

Find the second component of each of the ordered pairs that makes the pair a solution of the equation.

EXAMPLE

$y = 2^x$; $(-3,\quad)$, $(0,\quad)$, $(3,\quad)$

SOLUTION

For $x = -3$, $y = 2^{(-3)} = \frac{1}{8}$.

For $x = 0$, $y = 2^{(0)} = 1$.

For $x = 3$, $y = 2^{(3)} = 8$.

The ordered pairs are $\left(-3, \frac{1}{8}\right)$, **(0, 1)**, and **(3, 8)**.

1. $y = 3^x$; $(0,\quad)$, $(1,\quad)$, $(2,\quad)$
2. $y = 4^x$; $(0,\quad)$, $(1,\quad)$, $(2,\quad)$
3. $y = 2^x$; $(-2,\quad)$, $(0,\quad)$, $(2,\quad)$
4. $y = 5^x$; $(-2,\quad)$, $(0,\quad)$, $(2,\quad)$
5. $f(x) = \left(\frac{1}{2}\right)^x$; $(-3,\quad)$, $(0,\quad)$, $(3,\quad)$
6. $f(x) = \left(\frac{1}{3}\right)^x$; $(-3,\quad)$, $(0,\quad)$, $(3,\quad)$
7. $g(x) = 10^x$; $(-2,\quad)$, $(-1,\quad)$, $(0,\quad)$
8. $g(x) = 10^x$; $(0,\quad)$, $(1,\quad)$, $(2,\quad)$

Graph the function defined by each of the following exponential equations. Use selected integral values $-5 < x < 5$.

EXAMPLE

$y = 3^x$

SOLUTION

Arbitrarily select integral values of x, say,

$(-2,\ \ ), (-1,\ \ ), (0,\ \ ), (1,\ \ ), (2,\ \ ), (3,\ \ )$.

Determine the y-components of each ordered pair.

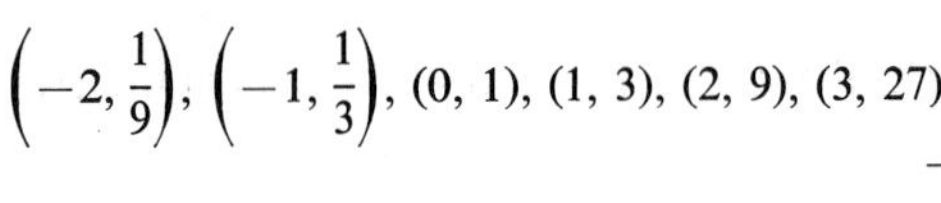

$\left(-2, \frac{1}{9}\right), \left(-1, \frac{1}{3}\right), (0, 1), (1, 3), (2, 9), (3, 27)$.

Plot the points and connect with a smooth curve.

9. $y = 4^x$ 10. $y = 5^x$ 11. $y = 10^x$ 12. $y = 2^{-x}$

13. $y = 3^{-x}$ 14. $y = \left(\frac{1}{3}\right)^x$ 15. $y = \left(\frac{1}{4}\right)^x$ 16. $y = \left(\frac{1}{10}\right)^x$

11.2 The logarithmic function

If the variables x and y in the equation

$$y = b^x, \quad b > 0, b \neq 1 \tag{1}$$

defining the exponential function are interchanged, we obtain

$$x = b^y, \quad b > 0, b \neq 1. \tag{2}$$

The functions defined by these equations are called **inverses** of each other. The graphs of functions defined by Equation (2) can be illustrated by the example

$$x = 10^y \quad (x > 0).$$

We assign arbitrary values of x, say, 0.01, 0.1, 1, 10, and 100, and obtain some ordered pairs that are solutions of the equation. We can then obtain their graphs and connect them with a smooth curve, as in Figure 11.2.

It is always useful to be able to express the variable y explicitly in terms of the variable x. To do this in equations such as (2), we use the notation

$$y = \log_b x \tag{3}$$

where $x > 0$, $b > 0$, $b \neq 1$ and the symbolism $\log_b x$ is read "logarithm to the base b of x," or "logarithm of x to the base b." The functions defined by such equations are called **logarithmic functions**.

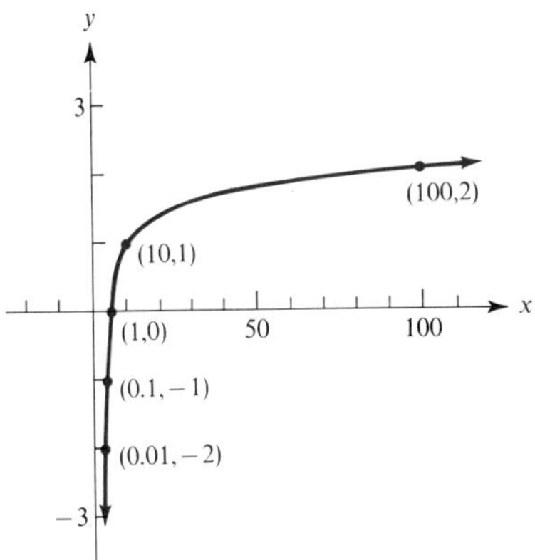

FIGURE **11.2**

It should be recognized that

$$x = b^y \quad \text{and} \quad y = \log_b x$$

are different forms of an equation defining the same function, in the same way that $x = y + 4$ and $y = x - 4$ define the same function, and we may use whichever equation suits our purpose. Thus, exponential statements may be written in logarithmic form; for example,

$$5^2 = 25 \quad \text{is equivalent to} \quad \log_5 25 = 2,$$

$$8^{1/3} = 2 \quad \text{is equivalent to} \quad \log_8 2 = \frac{1}{3},$$

$$3^{-2} = \frac{1}{9} \quad \text{is equivalent to} \quad \log_3 \frac{1}{9} = -2,$$

and so on. Also, logarithmic statements may be written in exponential form; for example,

$$\log_{10} 100 = 2 \quad \text{is equivalent to} \quad 10^2 = 100,$$

$$\log_3 81 = 4 \quad \text{is equivalent to} \quad 3^4 = 81,$$

$$\log_2 \frac{1}{2} = -1 \quad \text{is equivalent to} \quad 2^{-1} = \frac{1}{2}.$$

Notice that the logarithmic function associates with each number x the exponent on b ($\log_b x$) such that the power is equal to x. In other words, we can think of $\log_b x$ as an exponent on b. Thus, for positive real numbers b and x, we have the very important relationship

$$\boldsymbol{b^{\log_b x} = x.}$$

EXERCISE 11.2

Express each of the following in logarithmic notation.

EXAMPLES

a. $3^2 = 9$ b. $16^{1/4} = 2$ c. $64^{-1/3} = \frac{1}{4}$

SOLUTIONS

a. $\log_3 9 = 2$ b. $\log_{16} 2 = \frac{1}{4}$ c. $\log_{64} \frac{1}{4} = -\frac{1}{3}$

1. $4^2 = 16$ 2. $5^3 = 125$ 3. $3^3 = 27$ 4. $8^2 = 64$
5. $\left(\frac{1}{2}\right)^2 = \frac{1}{4}$ 6. $\left(\frac{1}{3}\right)^2 = \frac{1}{9}$ 7. $8^{-1/3} = \frac{1}{2}$ 8. $64^{-1/6} = \frac{1}{2}$
9. $10^2 = 100$ 10. $10^0 = 1$ 11. $10^{-1} = 0.1$ 12. $10^{-2} = 0.01$

Express each of the following logarithmic statements in exponential notation.

EXAMPLES

a. $\log_6 36 = 2$ b. $\log_{1/5} 125 = -3$ c. $\log_{10} 100 = 2$

SOLUTIONS

a. $6^2 = 36$ b. $\left(\frac{1}{5}\right)^{-3} = 125$ c. $10^2 = 100$

13. $\log_2 64 = 6$ 14. $\log_5 25 = 2$ 15. $\log_3 9 = 2$
16. $\log_{16} 256 = 2$ 17. $\log_{1/3} 9 = -2$ 18. $\log_{1/2} 8 = -3$
19. $\log_{10} 1000 = 3$ 20. $\log_{10} 1 = 0$ 21. $\log_{10} 0.01 = -2$

Find the value of each of the following.

EXAMPLES

a. $\log_4 16$ b. $\log_3 81$

SOLUTIONS

a. 4 raised to what power equals 16?
2

b. 3 raised to what power equals 81?
4

22. $\log_5 5$
23. $\log_7 49$
24. $\log_2 32$
25. $\log_4 64$
26. $\log_5 \sqrt{5}$
27. $\log_3 \sqrt{3}$
28. $\log_3 \frac{1}{3}$
29. $\log_5 \frac{1}{5}$
30. $\log_3 3$
31. $\log_2 2$
32. $\log_{10} 10$
33. $\log_{10} 100$
34. $\log_{10} 1$
35. $\log_{10} 0.1$
36. $\log_{10} 0.01$

Find an equivalent equation of the form $x = c$, $y = c$, or $b = c$, where c is a constant.

EXAMPLES

a. $\log_2 x = 3$

b. $\log_b 2 = \frac{1}{2}$

SOLUTIONS

Write in exponential form.

a. $2^3 = x$

b. $b^{1/2} = 2$

$(b^{1/2})^2 = 2^2$

Solve for the variable.

$\mathbf{x = 8}$

$\mathbf{b = 4}$

37. $\log_3 9 = y$
38. $\log_5 125 = y$
39. $\log_b 8 = 3$
40. $\log_b 625 = 4$
41. $\log_4 x = 3$
42. $\log_{1/2} x = -5$
43. $\log_2\left(\frac{1}{8}\right) = y$
44. $\log_5 5 = y$
45. $\log_b 10 = \frac{1}{2}$
46. $\log_b 0.1 = -1$
47. $\log_2 x = 2$
48. $\log_{10} x = -3$
49. What is $\log_2 1$? $\log_7 1$? $\log_a 1$, for $a > 0$?
50. Show that $\log_b b = 1$. *Hint*: Consider the definition of $\log_b x$.

11.3 Properties of logarithms

Because a logarithm by definition is an exponent, the following three laws are valid for positive numbers b, x_1, x_2, and all real numbers m:

1. $\mathbf{\log_b(x_1x_2) = \log_b x_1 + \log_b x_2}$.
2. $\mathbf{\log_b \frac{x_2}{x_1} = \log_b x_2 - \log_b x_1}$.
3. $\mathbf{\log_b(x_1)^m = m \log_b x_1}$.

The validity of law 1 is established as follows. Since

$$x_1 = b^{\log_b x_1} \quad \text{and} \quad x_2 = b^{\log_b x_2},$$

then

$$x_1 x_2 = b^{\log_b x_1} \cdot b^{\log_b x_2}$$
$$= b^{\log_b x_1 + \log_b x_2},$$

and, by the definition of a logarithm,

$$\log_b(x_1 x_2) = \log_b x_1 + \log_b x_2 .$$

The validity of Laws 2 and 3 can also be established.

EXERCISE 11.3

Express as the sum or difference of simpler logarithmic quantities. Assume that all variables denote positive real numbers.

EXAMPLE

$$\log_b\left(\frac{xy}{z}\right)^{1/2}$$

SOLUTION

Use the third law of logarithms.

$$\log_b\left(\frac{xy}{z}\right)^{1/2} = \frac{1}{2}\log_b\left(\frac{xy}{z}\right)$$

Use the first and second laws of logarithms.

$$\frac{1}{2}\log_b\left(\frac{xy}{z}\right) = \frac{1}{2}[\log_b x + \log_b y - \log_b z].$$

$$\boldsymbol{\log_b\left(\frac{xy}{z}\right)^{1/2} = \frac{1}{2}\log_b x + \frac{1}{2}\log_b y - \frac{1}{2}\log_b z.}$$

1. $\log_b(xy)$
2. $\log_b(xyz)$
3. $\log_b\left(\frac{x}{y}\right)$
4. $\log_b\left(\frac{xy}{z}\right)$
5. $\log_b x^5$
6. $\log_b x^{1/2}$
7. $\log_b \sqrt[3]{x}$
8. $\log_b \sqrt[3]{x^2}$
9. $\log_b x^2y^3$

10. $\log_b \dfrac{x^{1/2}y}{z^2}$

11. $\log_b \sqrt{\dfrac{x}{z}}$

12. $\log_b \sqrt{xy}$

13. $\log_{10} \sqrt[3]{\dfrac{xy^2}{z}}$

14. $\log_{10} \sqrt[5]{\dfrac{x^2 y}{z^3}}$

15. $\log_{10}(\sqrt{x}\sqrt[3]{y^2})$

16. $\log_{10} \dfrac{\sqrt{x}\sqrt[4]{y^3}}{z^2}$

17. $\log_{10}\sqrt{x(x-y)}$

18. $\log_{10} \sqrt[3]{(x-y)^2(x+y)}$

19. $\log_{10} 2\pi\sqrt{\dfrac{l}{g}}$

20. $\log_{10} \sqrt{s(s-a)(s-b)(s-c)}$

Express as a single logarithm with a coefficient of 1.

EXAMPLE

$\frac{1}{2}(\log_b x - \log_b y)$

SOLUTION

By the second law of logarithms,

$$\frac{1}{2}(\log_b x - \log, y) = \frac{1}{2}\log_b\left(\frac{x}{y}\right).$$

By the third law of logarithms,

$$\frac{1}{2}\left(\log_b\left(\frac{x}{y}\right)\right) = \log_b\left(\frac{x}{y}\right)^{1/2},$$

$$\mathbf{\frac{1}{2}(\log_b x - \log_b y) = \log_b\left(\frac{x}{y}\right)^{1/2}}.$$

21. $\log_b x + \log_b y$

22. $\log_b x - \log_b y$

23. $-\log_b x$

24. $-\frac{1}{2}\log_b x$

25. $2\log_b x + 3\log_b y$

26. $\frac{1}{4}\log_b x - \frac{3}{4}\log_b y$

27. $3\log_b x + \log_b y - 2\log_b z$

28. $\frac{1}{3}(\log_b x + \log_b y - 2\log_b z)$

29. $\log_{10}(x-2) + \log_{10} x - 2\log_{10} z$

30. $\frac{1}{2}(\log_{10} x - 3\log_{10} y - 5\log_{10} z)$

Verify that each of the following statements is true.

31. $\frac{1}{4}\log_{10} 8 + \frac{1}{4}\log_{10} 2 = \log_{10} 2$

32. $4 \log_{10} 3 - 2 \log_{10} 3 = \log_{10} 9$

33. $10^{2 \log_{10} x} = x^2$

34. $a^{2 \log_a 3} + b^{3 \log_b 2} = 17$

35. $\log_{10}[\log_3(\log_5 125)] = 0$

36. $\log_{10}[\log_2(\log_3 9)] = 0$

37. $\log_b b^2 + \log_b b^3 = \log_b b^5$

38. $\log_2 1 \cdot \log_2 3 = \log_2 1 + \log_3 1$

39. $\log_b 4 + \log_b 8 = \log_b 64 - \log_b 2$

40. $\log_2(\log_{10} 100) = \log_2 4 - 1$

11.4 Logarithms to the base 10

There are two logarithmic functions of special interest in mathematics; one is defined by

$$y = \log_{10} x, \tag{1}$$

and the other, by

$$y = \log_e x, \tag{2}$$

where e is an irrational number whose decimal approximation to eight digits is 2.7182818. Because these functions possess similar properties and because we are more familiar with the number 10 we shall confine our attention to Equation (1).

Values for $\log_{10} x$ are called **logarithms to the base 10** or **common logarithms.** From the definition of $\log_{10} x$,

$$10^{\log_{10} x} = x \quad (x > 0);$$

that is, $\log_{10} x$ is the exponent that must be placed on 10 so that the resulting power is x. The problem we are concerned with in this section is that of finding $\log_{10} x$ for each positive x. First, $\log_{10} x$ can easily be determined for all values of x that are integral powers of 10:

$$\begin{aligned} \log_{10} 10 &= 1, & \text{since } 10^1 &= 10; \\ \log_{10} 100 &= 2, & \text{since } 10^2 &= 100; \\ \log_{10} 1000 &= 3, & \text{since } 10^3 &= 1000; \end{aligned}$$

and

$$\begin{aligned} \log_{10} 1 &= 0, & \text{since } 10^0 &= 1; \\ \log_{10} 0.1 &= -1, & \text{since } 10^{-1} &= 0.1; \\ \log_{10} 0.01 &= -2, & \text{since } 10^{-2} &= 0.01; \\ \log_{10} 0.001 &= -3, & \text{since } 10^{-3} &= 0.001. \end{aligned}$$

Notice that the logarithm of a power of 10 is simply the exponent on the base 10 when the power is written in scientific notation. For example,

$$\log_{10} 100 = \log_{10} 10^2 = 2,$$
$$\log_{10} 0.01 = \log_{10} 10^{-2} = -2,$$

and so on.

A table of logarithms is used to find $\log_{10} x$ where $1 \leq x \leq 10$ (see page 314). Consider the excerpt from this table shown in Figure 11.3. Each number in the column headed x represents the first two significant digits of

x	0	1	2	3	4	5	6	7	8	9
3.8	.5798	.5809	.5821	.5832	.5843	.5855	.5866	.5877	.5888	.5899
3.9	.5911	.5922	.5933	.5944	.5955	.5966	.5977	.5988	.5999	.6010
4.0	.6021	.6031	.6042	.6053	.6064	.6075	.6085	.6096	.6107	.6117
4.1	.6128	.6138	.6149	.6160	.6170	.6180	.6191	.6201	.6212	.6222
4.2	.6232	.6243	.6253	.6263	.6274	.6284	.6294	.6304	.6314	.6325
4.3	.6335	.6345	.6355	.6365	.6375	.6385	.6395	.6405	.6415	.6425
4.4	.6435	.6444	.6454	.6464	.6474	.6484	.6493	.6503	.6513	.6522
4.5	.6532	.6542	.6551	.6561	.6571	.6580	.6590	.6599	.6609	.6618
4.6	.6628	.6637	.6646	.6656	.6665	.6675	.6684	.6693	.6702	.6712

FIGURE **11.3**

the numeral for x, while each number in the row containing x contains the third significant digit of the numeral for x. The digits located at the intersection of a row and a column form the logarithm of x. For example, to find $\log_{10} 4.25$, we look at the intersection of the row containing 4.2 under x and the column containing 5 to the right of x. Thus,

$$\log_{10} 4.25 = 0.6284.$$

Similarly,

$$\log_{10} 4.02 = 0.6042,$$
$$\log_{10} 4.49 = 0.6522,$$

and so on. The equals sign is being used here in a very loose sense. More properly $\log_{10} 4.25 \approx 0.6284$, $\log_{10} 4.02 \approx 0.6042$, and $\log_{10} 4.49 \approx 0.6522$, because these numbers are irrational and cannot be precisely represented by a decimal numeral. However, we shall follow customary usage and write $=$, instead of $\approx$, and leave the intent to the context.

Now suppose we wish to find $\log_{10} x$ for values of x outside the range of the table—that is, for $0 < x < 1$ or $x > 10$. This can be done quite readily by first representing the number in scientific notation—that is, as the product of a number between 1 and 10 and a power of 10—and applying the first law of logarithms. For example,

$$\begin{aligned}\log_{10} 42.5 = \log_{10}(4.25 \times 10^1) &= \log_{10} 4.25 + \log_{10} 10^1\\ &= 0.6284 + 1\\ &= 1.6284,\\ \log_{10} 425 = \log_{10}(4.25 \times 10^2) &= \log_{10} 4.25 + \log_{10} 10^2\\ &= 0.6284 + 2\\ &= 2.6284,\\ \log_{10} 4250 = \log_{10}(4.25 \times 10^3) &= \log_{10} 4.25 + \log_{10} 10^3\\ &= 0.6284 + 3\\ &= 3.6284.\end{aligned}$$

Observe that the decimal portion of the logarithm is always 0.6284, and *the integral portion is just the exponent on* 10 *when the number is written in scientific notation.*

This process can be reduced to a mechanical one by considering $\log_{10} x$ to consist of two parts, an integral part (called the **characteristic**) and a non-negative decimal fraction part (called the **mantissa**). Thus the table of values for $\log_{10} x$ for $1 < x < 10$ can be looked upon as a table of mantissas for $\log_{10} x$ for all $x > 0$.

To find $\log_{10} 4370$, we first write

$$\log_{10} 4370 = \log_{10}(4.37 \times 10^3).$$

Upon examining the table of logarithms, we find that $\log_{10} 4.37 = 0.6405$, so that

$$\log_{10} 4370 = 3.6405,$$

where we have prefixed the characteristic 3, the exponent on the base 10.

Now consider an example of the form $\log_{10} x$ for $0 < x < 1$. To find $\log_{10} 0.00402$, we write

$$\log_{10} 0.00402 = \log_{10}(4.02 \times 10^{-3}).$$

Examining the table, we find $\log_{10} 4.02$ is 0.6042. Upon adding 0.6042 to the characteristic -3, we obtain

$$\log_{10} 0.00402 = -2.3958,$$

where the decimal portion of the logarithm is no longer 0.6042 as it is in the case of all numbers $x > 1$ for which the first three significant digits of x are 402. To circumvent this situation, and thus provide access to the table, it is customary to write the logarithm in a form in which the decimal part is positive. In the foregoing example, we write

$$\begin{aligned}\log_{10} 0.00402 &= 0.6042 - 3\\ &= 0.6042 + (7 - 10)\\ &= 7.6042 - 10,\end{aligned}$$

and the decimal part is positive. The logarithms

$$6.6042 - 9 \quad \text{and} \quad 12.6042 - 15,$$

for example, are equally valid representations, but $7.6042 - 10$ is customary in most cases.

It is possible to reverse the process described in this section and, being given $\log_{10} x$, to find x. In this event, x is referred to as the **antilogarithm** (antilog_{10}) of $\log_{10} x$. For example, $\text{antilog}_{10}\ 1.6395$ can be obtained by locating the mantissa, 0.6395, in the body of the $\log_{10}$ tables and observing that the associated antilog_{10} is 4.36. Thus

$$\text{antilog}_{10}\ 1.6395 = 4.36 \times 10^1 = 43.6.$$

If we seek the common logarithm of a number that is not an entry in the table (for example, $\log_{10} 3712$) or if we seek x when $\log_{10} x$ is not an entry in the table, it is customary to use a procedure called **linear interpolation**, which is discussed in Section 11.5.

In some situations where logarithms to the base 10 are employed exclusively, the base is not shown. Thus, $\log_{10} x$ is simply written as $\log x$.

EXERCISE 11.4

Write the characteristic of each of the following.

EXAMPLES

a. $\log_{10} 248$ b. $\log_{10} 0.0057$

SOLUTIONS

Represent the number in scientific notation.

a. $\log_{10}(2.48 \times 10^2)$ b. $\log_{10}(5.7 \times 10^{-3})$

The exponent on the base 10 is the characteristic.

2 **−3**, or **7 − 10**

1. $\log_{10} 312$
2. $\log_{10} 8.12$
3. $\log_{10} 7912$
4. $\log_{10} 31$
5. $\log_{10} 0.02$
6. $\log_{10} 0.00851$
7. $\log_{10} 8.012$
8. $\log_{10} 752.31$
9. $\log_{10} 0.00031$
10. $\log_{10} 0.0004$
11. $\log_{10}(15 \times 10^3)$
12. $\log_{10}(820 \times 10^4)$

Find each logarithm.

EXAMPLES

a. $\log_{10} 16.8$ b. $\log_{10} 0.043$

SOLUTIONS

Represent the number in scientific notation.

a. $\log_{10}(1.68 \times 10^{1})$ b. $\log_{10}(4.3 \times 10^{-2})$

Determine the mantissa from the table of logarithms.

0.2253 0.6335

Add the characteristic as determined by the exponent on the base 10.

1.2253 **8.6335 − 10**

13. $\log_{10} 6.73$
14. $\log_{10} 891$
15. $\log_{10} 83.7$
16. $\log_{10} 21.4$
17. $\log_{10} 317$
18. $\log_{10} 219$
19. $\log_{10} 0.813$
20. $\log_{10} 0.00214$
21. $\log_{10} 0.08$
22. $\log_{10} 0.000413$
23. $\log_{10}(2.48 \times 10^{2})$
24. $\log_{10}(5.39 \times 10^{-3})$

Find each antilogarithm.

EXAMPLE

$\text{antilog}_{10} 2.7364$

SOLUTION

Locate the mantissa in the body of the table of mantissas and determine the associated antilog_{10} (a number between 1 and 10); write the characteristic as an exponent on the base 10.

$$\mathbf{5.45 \times 10^{2} = 545}$$

25. $\text{antilog}_{10} 0.6128$
26. $\text{antilog}_{10} 0.2504$
27. $\text{antilog}_{10} 1.5647$
28. $\text{antilog}_{10} 3.9258$
29. $\text{antilog}_{10}(8.8075 - 10)$
30. $\text{antilog}_{10}(3.9722 - 5)$
31. $\text{antilog}_{10} 1.2041$
32. $\text{antilog}_{10} 2.6590$
33. $\text{antilog}_{10} 3.7388$
34. $\text{antilog}_{10} 2.0086$
35. $\text{antilog}_{10}(6.8561 - 10)$
36. $\text{antilog}_{10}(1.8156 - 4)$

Find each of the following by means of the table of logarithms to the base 10.

EXAMPLE

$10^{0.6263}$

SOLUTION

The logarithmic function is the inverse of the exponential function. Hence, the element $10^{0.6263}$ in the range of the exponential function is also the element $\text{antilog}_{10}\ 0.6263$ in the domain of the logarithmic function. From the table of logarithms on page 314, we find that

$$10^{0.6263} = \text{antilog}_{10}\ 0.6263 = \mathbf{4.23}.$$

37. $10^{0.9590}$
38. $10^{0.8241}$
39. $10^{3.6990}$
40. $10^{2.3874}$
41. $10^{2.0531}$
42. $10^{1.7396}$

11.5 Linear interpolation

A table of common logarithms pairs each number x with an associated number $\log_{10} x$. Because of space limitations, only three digits for the number x and four for the number $\log_{10} x$ appear in the table. By means of a process called **linear interpolation**, however, the table can be used to find approximations to logarithms for numbers with four-digit numerals.

Let us examine geometrically the concepts involved. A portion of the graph of

$$y = \log_{10} x$$

is shown in Figure 11.4a (page 266). The curvature is exaggerated to illustrate the principle. We propose to use the straight line joining the points P_1 and P_2 as an approximation to the curve passing through the points. If an enlarged graph of $y = \log_{10} x$ were available, the value of $\log_{10} 4.257$ could be found by using the ordinate (RT) to the curve for $x = 4.257$. Since there is no way to accomplish this with a table of values only, we shall instead use the ordinate (RS) to the straight line as an approximation to the ordinate of the curve.

This can be accomplished directly from the set of numbers available in the table of logarithms. Consider Figure 11.4b, where the segments P_2P_3 and P_4P_5 are perpendicular to line segment P_1P_3. From geometry, we have

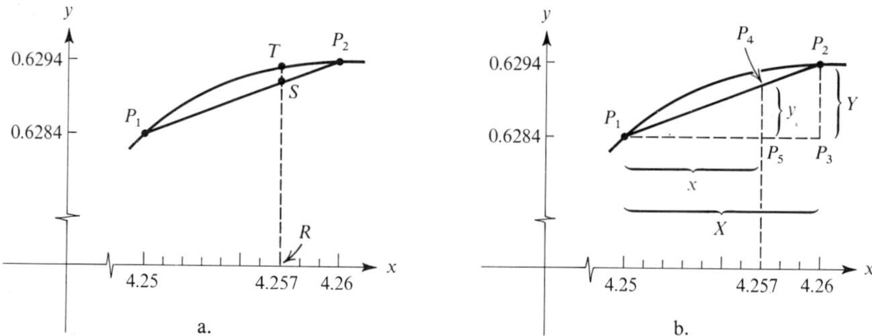

FIGURE **11.4**

$\Delta P_1P_4P_5$ is similar to $\Delta P_1P_2P_3$, where the corresponding lengths of the sides are proportional, and hence

$$\frac{x}{X} = \frac{y}{Y}. \tag{1}$$

If we know any three of these numbers, the fourth can be determined. For the purpose of interpolation, we assume that all of the members of the domain of the logarithmic function noted in the table on page 314 now have four-digit numerals; that is, we consider 4.250 instead of 4.25 and 4.260 instead of 4.26. We note that the number 4.257 falls on a point just $7/10$ of the distance between 4.250 and 4.260 and the distance Y (0.0010) is just the distance between the logarithms 0.6284 and 0.6294. It follows from (1) that

$$\frac{7}{10} = \frac{y}{0.0010}$$

and

$$y = \frac{7}{10}(0.0010) = 0.0007.$$

We can now add 0.0007 to 0.6284 and obtain a good approximation to the required logarithm. That is,

$$\log_{10} 4.257 = 0.6291.$$

The first example in Exercise 11.5 shows a convenient arrangement for the calculations involved in the example presented here. The antilogarithm of a number can be found by a similar procedure. The second example in Exercise 11.5 illustrates the process. However, with practice, it is possible to interpolate mentally in both procedures.

EXERCISE 11.5

Find each logarithm.

EXAMPLE

$\log_{10} 4.257$

SOLUTION

$$10\left\{\begin{array}{l} 7\left\{\begin{array}{l}4.250\\4.257\end{array}\right.\\ \quad 4.260\end{array}\right. \begin{array}{c|c} x & \log_{10} x \\ \hline 4.250 & 0.6284 \\ 4.257 & ? \\ 4.260 & 0.6294 \end{array} \left.\begin{array}{l}\left.\begin{array}{l}0.6284\\?\end{array}\right\}y\\ 0.6294\end{array}\right\}0.0010$$

Set up proportion and solve for y.

$$\frac{7}{10}=\frac{y}{0.0010}; \quad y=0.0007$$

Add this value of y to 0.6284.

$$\log_{10} 4.257=0.6284+0.0007=\mathbf{0.6291}$$

1. $\log_{10} 4.213$
2. $\log_{10} 8.184$
3. $\log_{10} 6.219$
4. $\log_{10} 10.31$
5. $\log_{10} 1522$
6. $\log_{10} 203.4$
7. $\log_{10} 37110$
8. $\log_{10} 72.36$
9. $\log_{10} 0.5123$
10. $\log_{10} 0.09142$
11. $\log_{10} 0.008351$
12. $\log_{10} 0.03741$

Find each antilogarithm.

EXAMPLE

$\text{antilog}_{10}\ 0.6446$

SOLUTION

$$0.0010\left\{\begin{array}{l} 0.0002\left\{\begin{array}{l}0.6444\\0.6446\end{array}\right.\\ \quad 0.6454\end{array}\right. \begin{array}{c|c} x & \text{antilog}_{10}\, x \\ \hline 0.6444 & 4.410 \\ 0.6446 & ? \\ 0.6454 & 4.420 \end{array} \left.\begin{array}{l}\left.\begin{array}{l}4.410\\?\end{array}\right\}y\\ 4.420\end{array}\right\}0.010$$

Set up proportion and solve for y.

$$\frac{0.0002}{0.0010}=\frac{y}{0.010}; \quad y=0.002$$

Solution continued overleaf.

Add this value of y to 4.410

$$\text{antilog}_{10}\, 0.6446 = 4.410 + 0.002 = \mathbf{4.412}$$

13. $\text{antilog}_{10}\, 0.5085$
14. $\text{antilog}_{10}\, 0.8087$
15. $\text{antilog}_{10}\, 1.9512$
16. $\text{antilog}_{10}\, 2.2620$
17. $\text{antilog}_{10}\, 1.0220$
18. $\text{antilog}_{10}\, 3.0759$
19. $\text{antilog}_{10}\, (8.7055 - 10)$
20. $\text{antilog}_{10}(3.6112 - 5)$
21. $\text{antilog}_{10}(9.8742 - 10)$
22. $\text{antilog}_{10}(20.9979 - 22)$
23. $\text{antilog}_{10}(2.8748 - 3)$
24. $\text{antilog}_{10}(7.7397 - 10)$

11.6 Computations with logarithms

The use of the slide rule and the advent of high-speed computing devices have almost removed the need to perform routine numerical computations with pencil and paper using logarithms. Nevertheless, we introduce the techniques involved in making such computations because the writing of the logarithmic equations involved sheds light on the properties of the logarithmic function and on the usefulness of the laws of logarithms which we reproduce here using the base 10.

If x_1 and x_2 are positive real numbers, then

1. $\log_{10}(x_1 x_2) = \log_{10} x_1 + \log_{10} x_2$,
2. $\log_{10} \dfrac{x_2}{x_1} = \log_{10} x_2 - \log_{10} x_1$,
3. $\log_{10}(x_1)^m = m \log_{10} x_1$.

We also state two assumptions that are useful in making computations using logarithms.

L–1 If $M = N$ $(M, N > 0)$, then $\log_{10} M = \log_{10} N$.

L–2 If $\log_{10} M = \log_{10} N$, then $M = N$.

These assumptions should seem plausible, because exponential and logarithmic functions are increasing functions.

Now, consider the product

$$(3.825)(0.00729).$$

If we set

$$N = (3.825)(0.00729),$$

then by assumption L–1,

$$\log_{10} N = \log_{10}[(3.825)(0.00729)].$$

Now, by the first law of logarithms,

$$\log_{10} N = \log_{10} 3.825 + \log_{10} 0.00729,$$

and by using the table we obtain,

$$\log_{10} 3.825 = 0.5826,$$
$$\log_{10} 0.00729 = 7.8627 - 10,$$

so that

$$\begin{aligned}\log_{10} N &= (0.5826) + (7.8627 - 10)\\ &= 8.4453 - 10.\end{aligned}$$

The computation is completed by referring to the table for

$$\begin{aligned}N = \text{antilog}_{10}(8.4453 - 10) &= 2.788 \times 10^{-2}\\ &= 0.02788.\end{aligned}$$

Hence,

$$N = (3.825)(0.00729) = 0.02788.$$

Actual computation shows the product to be 0.02788425 so that the result obtained by use of logarithms is correct to four significant digits. Some error should be expected, because we are using approximations to irrational numbers when we employ a table of logarithms.

Consider a more complicated example,

$$\frac{(8.21)^{1/2}(2.17)^{2/3}}{(3.14)^3}.$$

Setting

$$N = \frac{(8.21)^{1/2}(2.17)^{2/3}}{(3.14)^3},$$

we have

$$\begin{aligned}\log_{10} N &= \log_{10} \frac{(8.21)^{1/2}(2.17)^{2/3}}{(3.14)^3}\\ &= \log_{10}(8.21)^{1/2} + \log_{10}(2.17)^{2/3} - \log_{10}(3.14)^3,\\ \log_{10} N &= \frac{1}{2}\log_{10}(8.21) + \frac{2}{3}\log_{10}(2.17) - 3\log_{10}(3.14).\end{aligned}$$

The table provides values for the logarithms involved here, and the remainder of the computation is routine. In order to avoid confusion in computations of this sort, a systematic approach of some kind is desirable (see the example in Exercise 11.6).

You may wish to check your computations in the following exercise set by using a slide rule.

EXERCISE 11.6

Compute by means of logarithms.

EXAMPLE

$$\sqrt{\frac{(23.4)(0.681)}{4.13}}$$

SOLUTION

Let $P = \sqrt{\dfrac{(23.4)(0.681)}{4.13}}$.

Then $\log_{10} P = \dfrac{1}{2}(\log_{10} 23.4 + \log_{10} 0.681 - \log_{10} 4.13)$.

$$\left.\begin{aligned}\log_{10} 23.4 &= 1.3692\\ \log_{10} 0.681 &= \underline{9.8331 - 10}\end{aligned}\right\}\text{add}$$

$$\left.\begin{aligned}\log_{10}(23.4)(0.681) &= 11.2023 - 10\\ \log_{10} 4.13 &= \underline{\ \ 0.6160\qquad\ \ }\end{aligned}\right\}\text{subtract}$$

$$\log_{10} \frac{(23.4)(0.681)}{4.13} = 10.5863 - 10 = 0.5863$$

$$\frac{1}{2}\log_{10} \frac{(23.4)(0.681)}{4.13} = \frac{1}{2}(0.5863) = 0.2931$$

Hence, $\log_{10} P = 0.2931$, from which

$$P = \text{antilog}_{10}\ 0.2931 = \mathbf{1.964}.$$

1. $(2.32)(1.73)$
2. $(82.3)(6.12)$
3. $\dfrac{3.15}{1.37}$
4. $\dfrac{0.00214}{3.17}$
5. $(2.3)^5$
6. $(4.62)^3$
7. $\sqrt[3]{8.12}$
8. $\sqrt[5]{75}$
9. $(0.0128)^5$
10. $(0.0021)^6$
11. $\sqrt{0.0021}$
12. $\sqrt[5]{0.0471}$
13. $\dfrac{(8.12)(8.74)}{7.19}$
14. $\dfrac{(0.421)^2(84.3)}{\sqrt{21.7}}$
15. $\dfrac{(6.49)^2\sqrt[3]{8.21}}{17.9}$
16. $\dfrac{(2.61)^2(4.32)}{\sqrt{7.83}}$
17. $\dfrac{(0.3498)(27.16)}{6.814}$
18. $\dfrac{(4.813)^2(20.14)}{3.612}$

19. $\sqrt{\dfrac{(4.71)(0.00481)}{(0.0432)^2}}$

20. $\sqrt{\dfrac{(2.85)^3(0.97)}{(0.035)}}$

21. $\sqrt{25.1(25.1 - 18.7)(25.1 - 4.3)}$

22. $\sqrt{\dfrac{(4.17)^3(68.1 - 4.7)}{(68.1 - 52.9)}}$

23. $\dfrac{\sqrt{(23.4)^3(0.0064)}}{\sqrt[3]{69.1}}$

24. $\dfrac{\sqrt{38.7\sqrt[3]{491}}}{\sqrt[4]{9.21}}$

25. The period T of a simple pendulum is given by the formula $T = 2\pi\sqrt{L/g}$, where T is in seconds, L is the length of the pendulum in feet, and $g \approx 32$ ft/sec^2. Find the period of a pendulum 1 foot long.

26. The area A of a triangle in terms of the sides is given by the formula $A = \sqrt{s(s - a)(s - b)(s - c)}$, where a, b, and c are the lengths of the sides of the triangle and s equals one half of the perimeter. Find the area of a triangle in which the three sides are 2.314 inches, 4.217 inches, and 5.618 inches in length.

11.7 Exponential equations

An equation in which a variable occurs in an exponent is called an **exponential equation.** Solution sets of some such equations in one variable can be found by means of logarithms. Consider the equation

$$5^x = 7.$$

Because $5^x > 0$ for all x, we can apply assumption L–1 from the preceding section and write

$$\log_{10} 5^x = \log_{10} 7,$$

and, from the third law of logarithms,

$$x \log_{10} 5 = \log_{10} 7.$$

Multiplying each member by $1/\log_{10} 5$, we get

$$x = \frac{\log_{10} 7}{\log_{10} 5} = \frac{0.8451}{0.6990},$$

and the solution set is $\{1.209\}$. Note that when a numerical approximation to this solution is sought, the logarithms are *divided,* not subtracted. The quotient

$$\frac{\log_{10} 7}{\log_{10} 5} \neq \log_{10} \frac{7}{5},$$

from which we have

$$\frac{\log_{10} 7}{\log_{10} 5} \neq \log_{10} 7 - \log_{10} 5.$$

As another example of an exponential equation, consider

$$6^{3x-4} = 3.$$

We have from assumption L–1,

$$\log_{10} 6^{3x-4} = \log_{10} 3,$$

and, from the third law of logarithms,

$$(3x - 4)\log_{10} 6 = \log_{10} 3.$$

Multiplying each member by $1/\log_{10} 6$ gives

$$3x - 4 = \frac{\log_{10} 3}{\log_{10} 6},$$

from which

$$3x = \frac{\log_{10} 3}{\log_{10} 6} + 4,$$

$$x = \frac{\log_{10} 3}{3 \log_{10} 6} + \frac{4}{3}$$

$$= \frac{0.4771}{3(0.7782)} + \frac{4}{3}$$

$$= 0.204 + 1.333,$$

and the solution set is $\{1.537\}$.

EXERCISE 11.7

Solve. Express each answer to the nearest hundredth.

EXAMPLE

$3^{x-2} = 16$

SOLUTION

Use assumption L–1.

$$\log_{10} 3^{x-2} = \log_{10} 16$$

Use the third law of logarithms.

$$(x-2)\log_{10} 3 = \log_{10} 16$$

Multiply each member by $\dfrac{1}{\log_{10} 3}$.

$$x-2 = \frac{\log_{10} 16}{\log_{10} 3}$$

$$x = \frac{\log_{10} 16}{\log_{10} 3} + 2$$

$$= \frac{1.2041}{0.4771} + 2$$

The solution set is **{4.52}**.

1. $2^x = 7$
2. $3^x = 4$
3. $3^{x+1} = 8$
4. $2^{x-1} = 9$
5. $7^{2x-1} = 3$
6. $3^{x+2} = 10$
7. $4^{2x+3} = 15$
8. $3^{3x-1} = 21$
9. $3^{-x} = 10$
10. $2.13^{-x} = 8.1$
11. $3^{1-x} = 15$
12. $4^{2-x} = 10$

P dollars invested at an interest rate r compounded yearly yields an amount A after n years, given by $A = P(1+r)^n$. If the interest is compounded t times yearly, the amount is given by

$$A = P\left(1 + \frac{r}{t}\right)^{tn}.$$

In each of the following problems, solve for the variable n (nearest year), r (nearest ½%), or A (accuracy obtainable using 4-place table of mantissas).

EXAMPLE

$(1+r)^{12} = 1.127$

SOLUTION

Equate $\log_{10}$ of each member and apply the third law of logarithms.

$$12\log_{10}(1+r) = \log_{10} 1.127 = 0.0519$$

Multiply each member by $\dfrac{1}{12}$.

$$\log_{10}(1+r) = \frac{1}{12}(0.0519) = 0.0043$$

Solution continued overleaf.

Determine $\text{antilog}_{10}\ 0.0043$ and solve for r.

$$1 + r = \text{antilog}_{10}\ 0.0043 = 1.01$$

$$r = 0.01, \quad \text{or} \quad r = \mathbf{1\%}$$

EXAMPLE

$40(1 + 0.02)^n = 51.74$

SOLUTION

Multiply each member by $\frac{1}{40}$; equate $\log_{10}$ of each member and apply the second and third laws of logarithms. Solve for n.

$$n \log_{10}(1.02) = \log_{10} 51.74 - \log_{10} 40$$

$$n(0.0086) = 1.7138 - 1.6021$$

$$n = \frac{0.1117}{0.0086} = \mathbf{13}$$

13. $(1 + 0.03)^{10} = A$ 14. $(1 + 0.04)^8 = A$ 15. $(1 + r)^6 = 1.34$
16. $(1 + r)^{10} = 1.48$ 17. $(1 + 0.04)^n = 2.19$ 18. $(1 + 0.04)^n = 1.60$
19. $100\left(1 + \frac{r}{2}\right)^{10} = 113$ 20. $40\left(1 + \frac{r}{4}\right)^{12} = 50.9$
21. $150(1 + 0.01)^{4n} = 240$ 22. $60(1 + 0.02)^{2n} = 116$
23. Find the compounded amount of \$5000 invested at 4% for 10 years when compounded annually. When compounded semiannually.
24. Two men, A and B, each invested \$10,000 at 4% for 20 years with a bank that computed interest quarterly. A withdrew his interest at the end of each 3-month period but B let his investment be compounded. How much more did B earn over the period of 20 years?

The chemist defines the pH (hydrogen potential) of a solution by

$$p\text{H} = \log_{10} \frac{1}{[\text{H}^+]}$$

$$= \log_{10}[\text{H}^+]^{-1},$$

$$p\text{H} = -\log_{10}[\text{H}^+],$$

where $[\text{H}^+]$ is a numerical value for the concentration of hydrogen ions in aqueous solution in moles per liter. Calculate to the nearest tenth the pH of a solution whose hydrogen ion concentration is as given.

EXAMPLE

$[H^+] = 3.7 \times 10^{-6}$

SOLUTION

Substitute 3.7×10^{-6} for $[H^+]$ in the relationship $pH = -\log_{10}[H^+]$.

$$\begin{aligned} pH &= -\log_{10}(3.7 \times 10^{-6}) \\ &= -(\log_{10} 3.7 + \log_{10} 10^{-6}) \\ &= -(0.5682 - 6) \\ &\approx -(-5.4) \\ &\approx \mathbf{5.4} \end{aligned}$$

25. $[H^+] = 10^{-7}$
26. $[H^+] = 4.0 \times 10^{-5}$
27. $[H^+] = 2.0 \times 10^{-8}$
28. $[H^+] = 8.5 \times 10^{-3}$
29. $[H^+] = 6.3 \times 10^{-7}$
30. $[H^+] = 5.7 \times 10^{-7}$

Calculate the hydrogen ion concentration $[H^+]$ of a solution whose pH is as given.

EXAMPLE

$pH = 7.4$

SOLUTION

Substitute 7.4 for pH in the relationship $pH = \log_{10} \dfrac{1}{[H^+]}$.

$$\log_{10} \frac{1}{[H^+]} = 7.4$$

Find the antilog$_{10}$ 7.4 (7.4000) and solve for $[H^+]$.

$$\frac{1}{[H^+]} = \text{antilog}_{10}\, 7.4 = 2.5 \times 10^7$$

$$[H^+] = \frac{1}{2.5 \times 10^7} = 0.4 \times 10^{-7} = \mathbf{4.0 \times 10^{-8}}$$

31. $p\text{H} = 3.0$ 32. $p\text{H} = 4.2$ 33. $p\text{H} = 5.6$
34. $p\text{H} = 8.3$ 35. $p\text{H} = 7.2$ 36. $p\text{H} = 6.9$
37. The amount of a radioactive element available at any time (t) is given by $y = y_0 e^{-0.4t}$, where t is in seconds and y_0 is the amount present initially. How much of the element would remain after three seconds if 40 grams were present initially? Use: $e \approx 2.718$.
38. The number of bacteria present in a culture is related to time by the formula $N = N_0 e^{0.04t}$, where N_0 is the amount of bacteria present at time $t = 0$, and t is time in hours. If 10,000 bacteria are present 10 hours after the beginning of an experiment, how many are present when $t = 0$?
39. The atmospheric pressure p, in inches of mercury, is given approximately by $p = 30.0(10)^{-0.09a}$, where a is the altitude in miles above sea level. What is the atmospheric pressure at sea level? At 3 miles above sea level?
40. The intensity I (in lumens) of a light beam after passing through a thickness t (in centimeters) of a medium having an absorption coefficient of 0.1 is given by $I = 1000e^{-0.1t}$. How many centimeters of the material would reduce the illumination to 800 lumens?

Chapter review

[11.1] Sketch the graph of each equation.

1. $y = 5^x$ 2. $y = 5^{-x}$

[11.2] 3. Write each statement in logarithmic notation.

a. $9^{3/2} = 27$ b. $\left(\dfrac{4}{9}\right)^{1/2} = \dfrac{2}{3}$

4. Write each statement in exponential notation.

a. $\log_5 625 = 4$ b. $\log_{10} 0.0001 = -4$

5. Find x such that $\log_3 x = 3$.
6. Find b such that $\log_b 3 = 3$.

[11.3] 7. Express as the sum or difference of simpler logarithmic quantities.

a. $\log_b 3x^2y$ b. $\log_b \dfrac{y\sqrt{x}}{z^2}$

8. Show that $6(\log_2 8 - \log_2 4) = 6$.
9. Show that $\log_2[\log_4(\log_2 16)] = 0$.
10. Show that $\dfrac{1}{2} \log_2 16 + \log_4 16 = 8 \log_{16} 4$.

[11.4] 11. Find a value for each logarithm.

a. $\log_{10} 0.713$ b. $\log_{10} 1810$

12. Find a value for each antilogarithm.

a. $\text{antilog}_{10}\ 8.1761 - 10$ b. $\text{antilog}_{10}\ 3.7235$

13. Find each of the following.

a. $10^{0.8762}$ b. $10^{2.5490}$

[11.5] Use linear interpolation to find a value for each of the following.

14. $\log_{10} 27.93$ 15. $\text{antilog}_{10}\ 8.5000 - 10$

[11.6] Compute by means of logarithms.

16. $(3.17)(8.23)$ 17. $\dfrac{\sqrt{18.72}}{3.12}$ 18. $\sqrt[3]{\dfrac{(4.02)^2(27.1)}{14.3}}$

[11.7] 19. Solve $3^x = 15$.

20. Solve $10^{2x-1} = 7$.

12 Elements of geometry

12.1 Angles and their measure

Geometry is the branch of mathematics concerned with such things as points, lines, angles, and figures in planes and in space. In this chapter, we shall review some familiar geometric symbols and terms which are ordinarily introduced in earlier mathematics courses. Chart 12.1 is a brief summary of some symbols used to identify points, lines, and angles. Notice that in the chart we do not use different symbols for a line segment and its length or for an angle and its measure. This is because the meaning of the symbol is almost always clear from the context in which it is used.

In identifying angles, care must be taken to avoid ambiguity. Thus, in Figure 12.1(a), we can write either $\angle ABC$ or $\angle B$ to name the angle without confusion. However, we cannot use the symbol $\angle B$ to name one of the angles pictured in (b) because it is not clear whether $\angle ABC$, $\angle ABD$, or $\angle CBD$ is intended. For this reason, it is frequently convenient to use numbers to name angles as in (c).

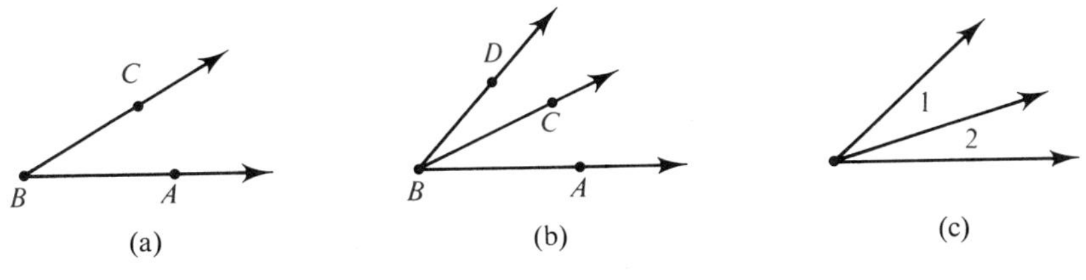

FIGURE **12.1**

CHART **12.1**

Geometric Object	Symbols	Examples
Point	$A, B, C, \ldots$ $P_1, P_2, \ldots$	
Line	$l_1, l_2, \ldots$ AB, where A and B are on the line; or $P_1P_2, \ldots$	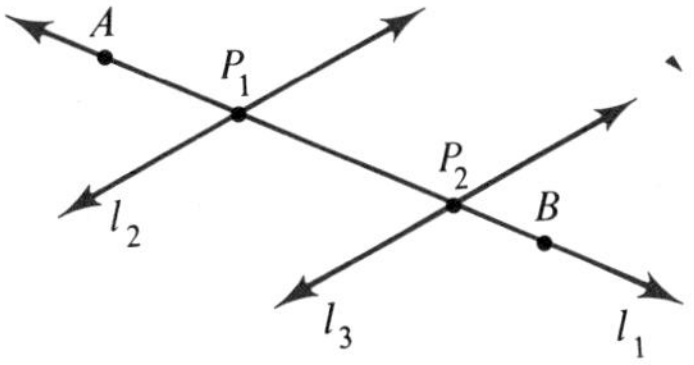
Ray, or **half-line**	AB, where A is the **initial** point and B is any other point on the ray.	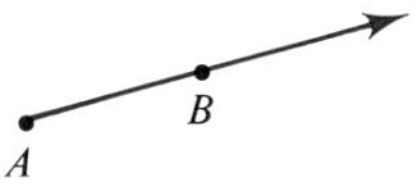
Line segment, or the **length of the segment**	AB or BA, where A and B are the end points; or $P_1P_2, \ldots$; or a, $b, \ldots$	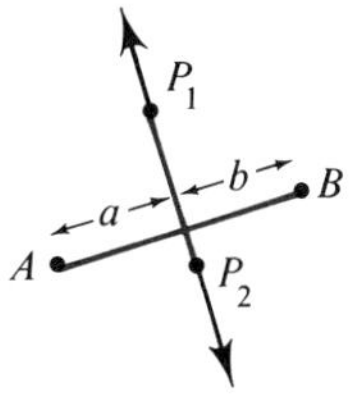
Angle, or the **measure of the angle**	$\angle ABC$, $\angle CBA$, or $\angle B$, where B is the vertex; or $\angle 1$, $\angle 2, \ldots$	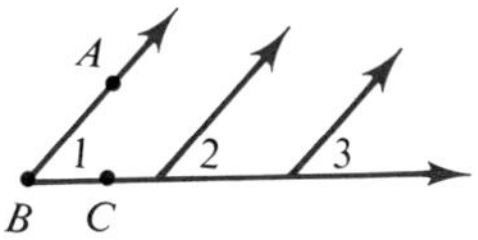

To measure an angle, we use a protractor such as that pictured in Figure 12.2 on page 280. The measure of $\angle ABC$ is 50°, while that of $\angle ABD$ is 134°. If the line AE is considered to be two sides of an angle with vertex at B, then $\angle ABE = 180°$.

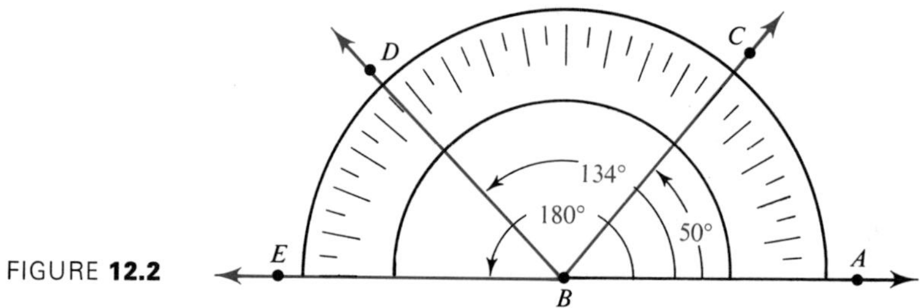

FIGURE **12.2**

For finer gradations of measure, the smaller units **minutes** and **seconds** are used. The symbol 12′ is read "twelve minutes" and 22″ is read "twenty-two seconds." There are 60 minutes in one degree and 60 seconds in one minute. In this book, we shall confine our attention to angles with measures in degrees and minutes only, or in decimal degree form. For example, 12° 48′ can also be written in decimal form as 12.8° because

$$48' = \frac{48}{60}(1)^\circ = \frac{4}{5}(1)^\circ = 0.8^\circ.$$

Angles are classified according to their measures in the following way.

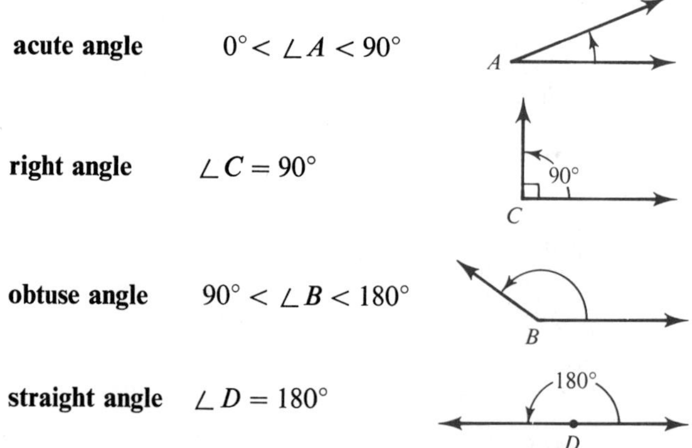

acute angle $0^\circ < \angle A < 90^\circ$

right angle $\angle C = 90^\circ$

obtuse angle $90^\circ < \angle B < 180^\circ$

straight angle $\angle D = 180^\circ$

Two angles such as $\angle 1$ and $\angle 2$ shown in Figure 12.3 are called **adjacent angles**. They share the common side BC, and we say that $\angle ABD$ is the sum

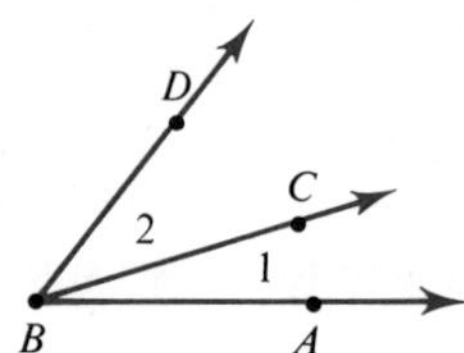

FIGURE **12.3**

of $\angle 1$ and $\angle 2$. That is, $\angle ABD = \angle 1 + \angle 2$. If $\angle 1 = \angle 2$, then ray BC is called the **bisector** of $\angle ABD$.

If the sum of two angles is 90°, the angles are called **complementary**, while if the sum of two angles is 180° they are called **supplementary**. For example, in Figure 12.4,

$\angle ADB$ and $\angle BDC$ are complementary,
$\angle EHF$ and $\angle FHG$ are supplementary,
$\angle J$ and $\angle L$ are complementary, and
$\angle K$ and $\angle L$ are supplementary.

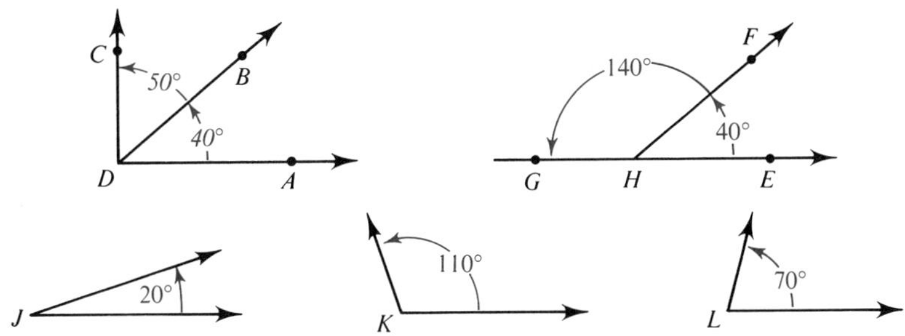

FIGURE **12.4**

Angles formed by two intersecting straight lines whose sides are rays in opposite directions have the same measure and are called **vertical angles**. Thus in Figure 12.5, $\angle 1$ and $\angle 3$ are vertical angles, as are $\angle 2$ and $\angle 4$. Since $\angle 1$

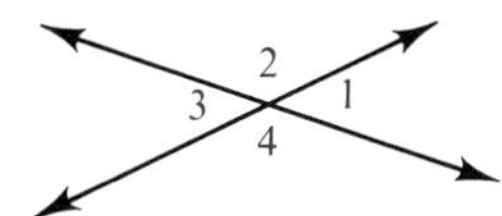

FIGURE **12.5**

and $\angle 2$ in this figure are supplementary, as are $\angle 3$ and $\angle 4$, it is obvious that the sum of the angles about a point is 360°. If two intersecting lines form equal adjacent angles, then their angles clearly measure 90°, and the lines are said to be **perpendicular**. To show that l_1 is perpendicular to l_2, we write $l_1 \perp l_2$.

EXERCISE 12.1

In the accompanying figure, lines DA and CE intersect at O, $OB \perp DA$ and $\angle 1 = 34°$. Find the measure of each angle.

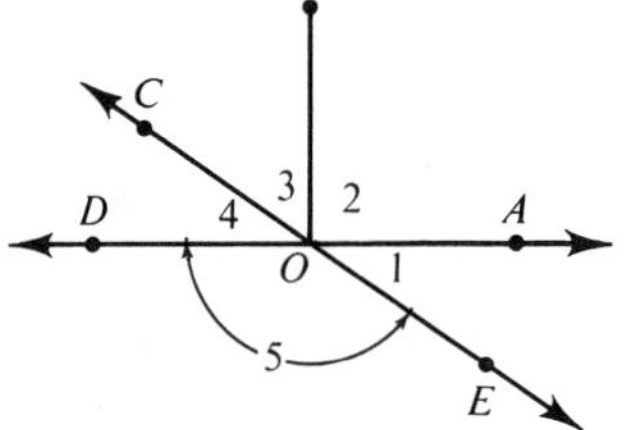

1. $\angle 4$ 2. $\angle 5$ 3. $\angle 2$ 4. $\angle 3$
5. Name the angle(s) supplementary to $\angle 5$.
6. Name the angle complementary to $\angle 4$.
7. Name all acute angles in the figure.
8. Name all right angles in the figure.

Name (a) the complement, and (b) the supplement of each angle.

9. 23°	10. 64°	11. 14° 10′
12. 28° 17′	13. 62° 04′	14. 19° 52′
15. 15.7°	16. 49.3°	17. 14.57°
18. 82.04°	19. 29.82°	20. 17.05°

Express each of the following in decimal form.

EXAMPLES

a. 3° 6′ b. 47° 24′

SOLUTIONS

a. $6' = \frac{6}{60}(1)° = \frac{1}{10}(1)° = 0.1°$; hence $3°\,6' = 3.1°$.

b. $24' = \frac{24}{60}(1)° = \frac{2}{5}(1)° = 0.4°$; hence $47°\,24' = \mathbf{47.4°}$.

21. 15° 30′	22. 127° 36′	23. 88° 12′
24. 54° 48′	25. 120° 45′	26. 107° 15′

Express each of the following in degrees and minutes.

EXAMPLE

52.6°

SOLUTION

$0.6° = \frac{6}{10}(1)° = \frac{6}{10}(60') = 36'$; hence $52.6° = \mathbf{52° \ 36'}$

27. 14.2°
28. 139.1°
29. 54.33 1/3°
30. 17.66 2/3°
31. 12.375°
32. 89.875°
33. Which angles in Exercises 21–26 are acute?
34. Which angles in Exercises 27–32 are obtuse?

EXAMPLE

The measure of an angle is equal to one-third that of its supplement. Find the measure of each angle.

SOLUTION

Let x = measure of the larger angle; then $\frac{1}{3}x$ = measure of the smaller angle.

Because the sum of the measures of supplementary angles equals 180°, we have

$$\begin{aligned} x + \frac{1}{3}x &= 180 \\ 3x + x &= 540 \\ 4x &= 540 \\ x &= 135; \\ \frac{1}{3}x &= 45. \end{aligned}$$

Hence, the measures of the angles are **135**° and **45**°.

35. The measure of an angle is equal to one-fifth that of its supplement. Find the measure of each angle.
36. The measure of an angle is equal to four-fifths that of its complement. Find the measure of each angle.
37. The measure of an angle is equal to 6° more than five times that of its complement. Find the measure of each angle.
38. The measure of an angle is equal to 52° less than three times that of its supplement. Find the measure of each angle.

12.2 Parallel and intersecting lines

Two different lines in a plane either intersect at a point as shown in Figure 12.6(a) or else do not intersect as suggested in (b). Lines that do not intersect are called **parallel lines**. To show that line l_1 is parallel to a line l_2 we write $l_1 \| l_2$.

A line that intersects two or more other lines is called a **transversal** of those lines. Thus, in Figure 12.7, l_1 is a transversal of lines l_2, l_3, and l_4.

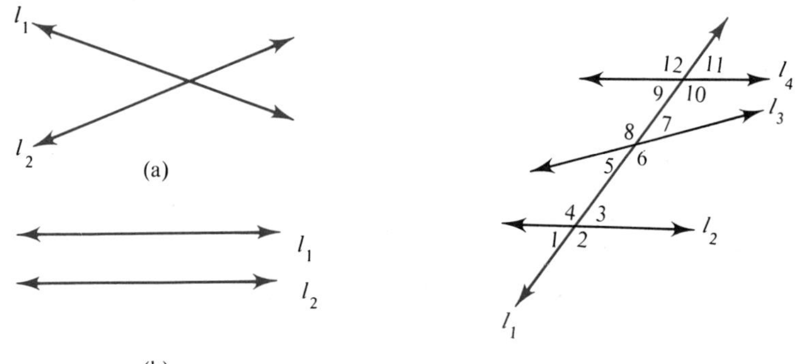

FIGURE **12.6** (b) FIGURE **12.7**

Certain angles formed by transversals and the lines they intersect are given special names. For example, referring to Figure 12.7, pairs of angles such as $\angle 1$ and $\angle 5$, $\angle 2$ and $\angle 6$, $\angle 8$ and $\angle 12$, and so forth, which are oriented in the same relative positions with respect to the transversal and the points of intersection of the transversal with two of the lines, are called **corresponding angles**. Angles such as $\angle 3$ and $\angle 5$, $\angle 8$ and $\angle 10$, and so forth, which are on opposite sides of the transversal, are called **alternate interior angles**. Similarly, angles such as $\angle 1$ and $\angle 7$, $\angle 6$ and $\angle 12$, are called **alternate exterior angles.**

When the lines cut by a transversal are **parallel**, any pairs of corresponding angles, alternate interior angles, or alternate exterior angles, respectively, have equal measures. Thus, in Figure 12.8, given that $l_2 \| l_3$, we know, for example, that $\angle 1 = \angle 5$, $\angle 4 = \angle 6$, and $\angle 2 = \angle 8$. Conversely, if two lines l_2 and l_3 are cut by a transversal, and if any pair of corresponding angles, alternate interior angles, or alternate exterior angles have equal measure, than $l_2 \| l_3$.

When parallel lines are cut by more than one transversal, as shown in Figure 12.9, the ratios of the lengths of the segments of the transversal included by the parallel lines are proportional. Thus from Figure 12.9,

$$\frac{AB}{BC} = \frac{DE}{EF} = \frac{GH}{HI}.$$

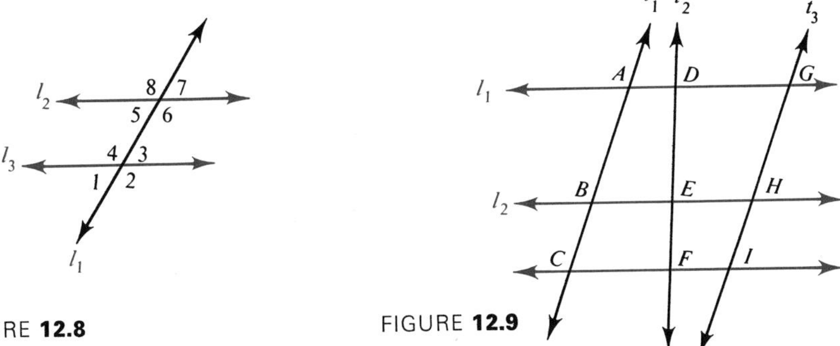

FIGURE **12.8** FIGURE **12.9**

Similar to the property of an angle bisector, a point, line, or line segment that separates another line segment into two equal parts is called a **bisector** of the given line segment. For example, in Figure 12.9, if $BE = EH$, l_2 is a bisector of BH.

EXERCISE 12.2

In the accompanying figure $l_1 \| l_2$ and $\angle 1 = 72°$. Find the measure of each of the following angles.

1. $\angle 5$ and $\angle 6$. (*Hint:* $\angle 1$ and $\angle 3$ are vertical angles and $\angle 3$ and $\angle 5$ are alternate interior angles.)
2. $\angle 7$ and $\angle 8$

In the figure $l_1 \| l_2 \| l_3$, $\angle 3 = 47°$, and $\angle 9 = 102°$. Find the measure of each of the following angles.

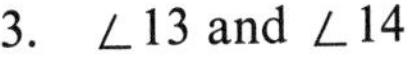

3. $\angle 13$ and $\angle 14$ 4. $\angle 15$ and 16
5. $\angle 7$ and $\angle 8$ 6. $\angle 5$ and $\angle 6$
7. $\angle 17$ and $\angle 18$ 8. $\angle 19$ and $\angle 20$

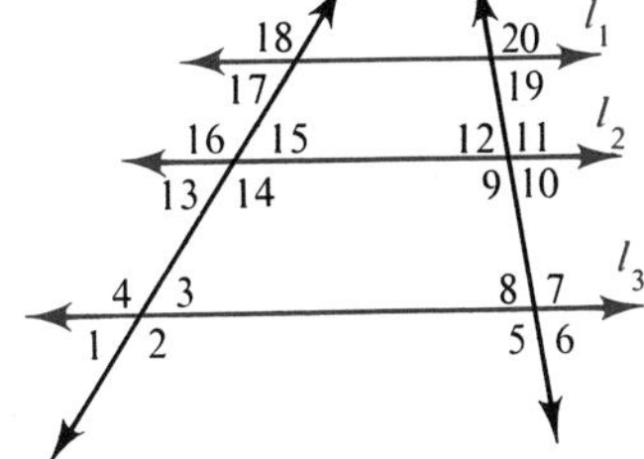

In the figure, $AD \| BF \| CG$, $AB = 7$, $BC = 3$, and $FG = 2$. Find each of the following.

9. The measure of $\angle ABF$
10. The measure of $\angle BCG$
11. The measure of $\angle BFD$
12. The measure of $\angle CGF$

13. The length of DF $\left(Hint: \dfrac{AB}{BC} = \dfrac{DF}{FG}.\right)$

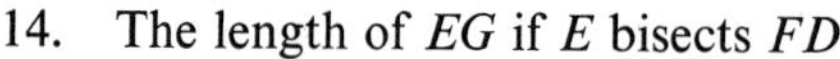

14. The length of EG if E bisects FD

In the figure, lines l_1, l_2, l_3, l_4, and l_5 are parallel, C bisects BD, D bisects BE, $AB = 5.6$, $BC = 3.2$, and $GH = 2.4$. Find each of the following lengths.

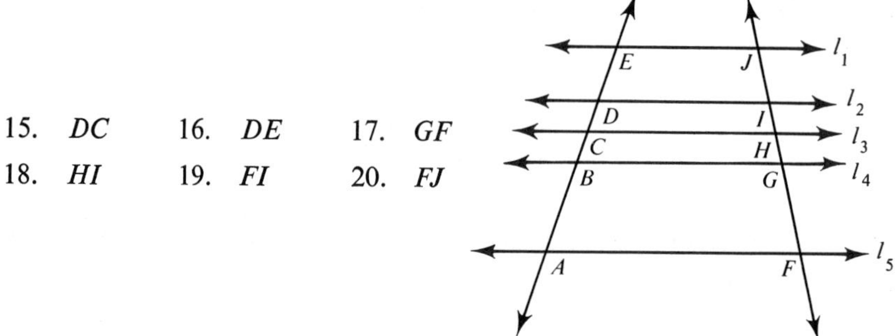

15. DC 16. DE 17. GF
18. HI 19. FI 20. FJ

12.3 Triangles

The triangle is the simplest closed figure having line segments as sides, yet it is also one of the most useful of such figures. We can classify triangles in various ways, one of which is in terms of the lengths of the sides. If the three sides of a triangle are equal in length, then the triangle is said to be **equilateral**. If any two sides of a triangle are equal in length, the triangle is **isosceles**, while if no two sides are equal in length, the triangle is **scalene**. Examples of these triangles are shown in Figure 12.10.

Another means of classifying triangles is by the measures of their angles. You may recall that the sum of the measures of the angles of any triangle is 180°. Thus for all triangles ABC,

$$\angle A + \angle B + \angle C = 180^\circ.$$

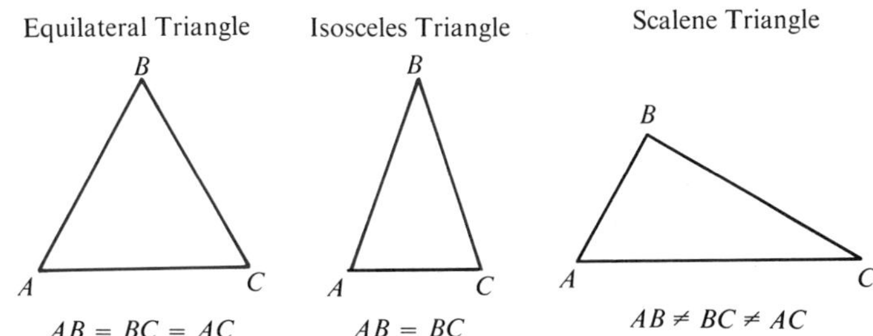

FIGURE **12.10**

If each of the angles of a triangle measures less than 90°, the triangle is an **acute triangle.** If one angle measures 90°, then the triangle is a **right triangle,** where the side opposite the 90° angle is called the **hypotenuse**. If one of the angles has a measure greater than 90°, the triangle is an **obtuse triangle.** Examples of these triangles are shown in Figure 12.11.

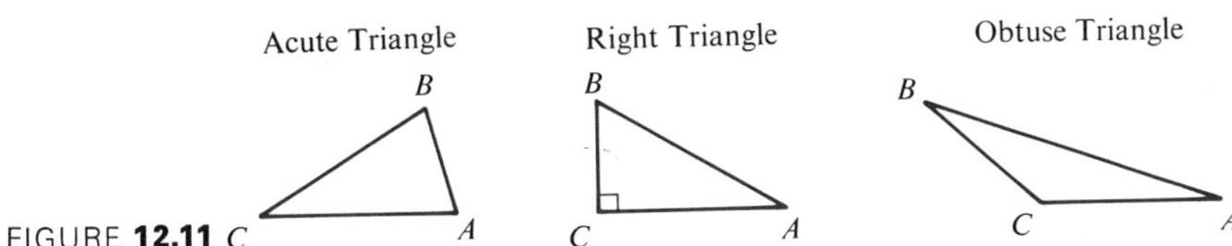

FIGURE **12.11**

In using letters to name the parts (angles and sides) of a triangle, it is customary to use capital letters to name the vertices of the angles, and lower case letters to specify the lengths of sides, and in general, if a given capital latter, say A, names the vertex of an angle, we use the lower-case version of the same letter, a, to represent the length of the side opposite that angle as shown in Figure 12.12. Using the symbol $\triangle ABC$ to represent the triangle whose vertices are A, B, and C, the **perimeter** p of $\triangle ABC$ is given by

$$p = a + b + c.$$

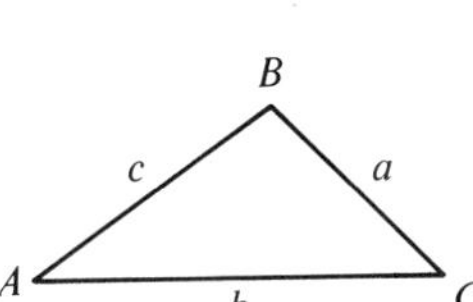

FIGURE **12.12**

The line segment from any vertex, perpendicular to the opposite side (or an extension of it) of a triangle is called an **altitude**, h, of the triangle, and the side to which the altitude is perpendicular then becomes the **base**. Figure 12.13 shows some altitudes of triangles. The line segment from the vertex of

FIGURE **12.13**

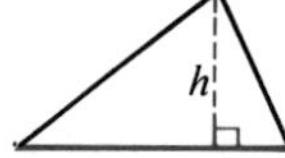

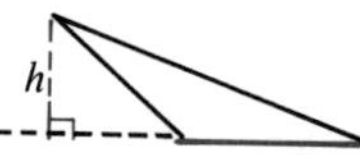

an angle to the midpoint of the side opposite the angle is called the **median** to the side. The three medians of a triangle are shown in Figure 12.14. The point of intersection of the three medians of a triangle is of special interest and is called the **centroid** of the triangle. It is the theoretical balance point of the triangular region.

CHART **12.2**

Equilateral triangle with sides s:

$\angle A = \angle B = \angle C = 60°$,

$h = \frac{s}{2}\sqrt{3}$, h bisects base,

$A = \frac{\sqrt{3}}{4} s^2$

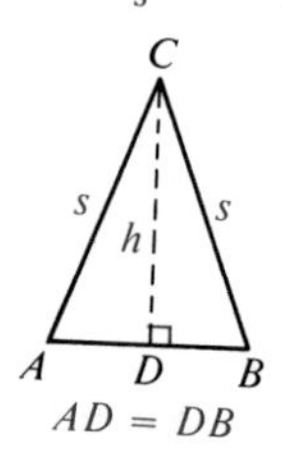

Isosceles triangle with equal sides s:

$\angle A = \angle B$, h bisects base

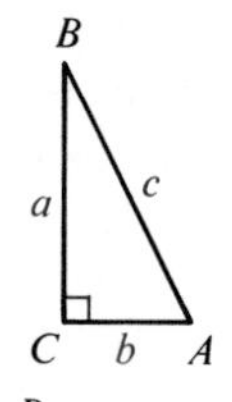

Right triangle with hypotenuse c:

$c^2 = a^2 + b^2$. Pythagorean theorem

a. 45°–45° right triangle;
Lengths of sides a, b, and c are in the ratio $1 : 1 : \sqrt{2}$

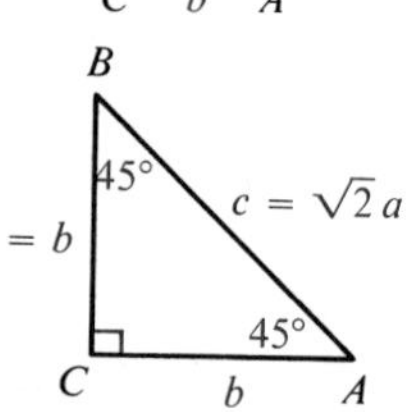

b. 30°–60° right triangle;
Lengths of sides a, c, and b are in the ratio $1 : 2 : \sqrt{3}$

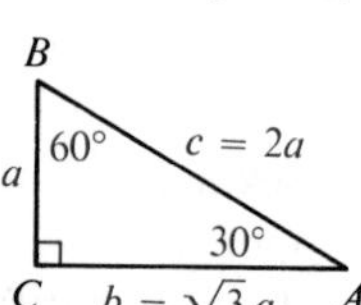

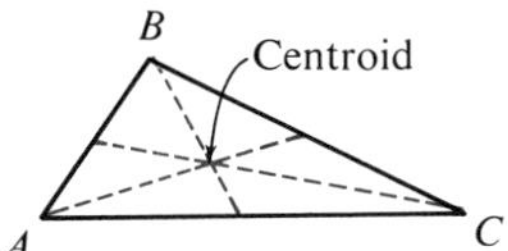

FIGURE **12.14**

In geometry we learn that the area A of a triangle is determined by

$$A = \frac{1}{2} \times \text{altitude} \times \text{base},$$

where the area measurement is given in square units such as square inches, square feet, and so forth. Using customary symbols, we express this by

$$\mathbf{A = \frac{1}{2}hb,}$$

where b is the length of the base.

Some special classes of triangles have special characteristics with which we should be familiar. These are listed in Chart 12.2.

EXERCISE 12.3

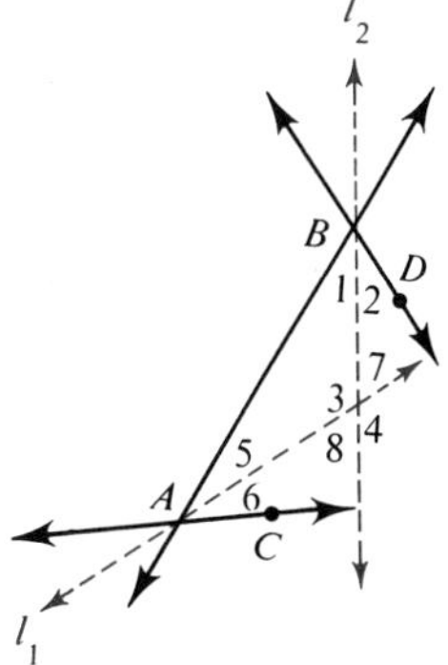

Given that l_1 bisects $\angle BAC$, l_2 bisects $\angle ABD$ $\angle 2 = 32°$, and $\angle 6 = 24°$. Find the measure of each of the following angles.

1. $\angle 1$
2. $\angle 5$
3. $\angle 3$
4. $\angle 4$
5. $\angle 7$
6. $\angle 8$

Use the information given in the figure to find measures of each angle.

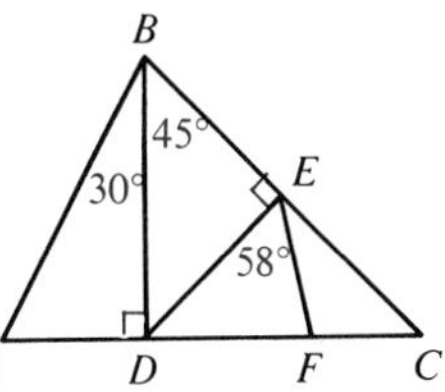

7. $\angle DAB$
8. $\angle CED$
9. $\angle BCD$
10. $\angle BDE$
11. $\angle CDE$
12. $\angle BDC$
13. $\angle FEC$
14. $\angle DFE$

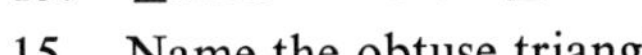

15. Name the obtuse triangle.
16. Name the 30°–60° right triangle.
17. Name the three 45°–45° right triangles.

18. Name the two acute triangles.
19. Name the altitude to base BC of $\triangle BDC$.
20. Name the altitude to base AC of $\triangle ABC$.

Find the lengths (to the nearest tenth of a unit) of the missing sides of each triangle as pictured in the examples below, where the length of one side is given.

EXAMPLES

a. $\triangle ABC$, $b = 6.2$ b. $\triangle DEF$, $e = 9.6$

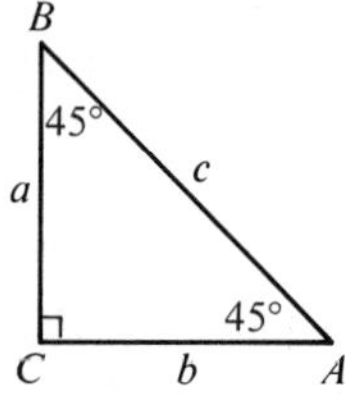

SOLUTIONS

a. Since the lengths of the sides of a 45°–45° right triangle are in the ratio $1:1:\sqrt{2}$ (see Chart 12.2), we have the relationship

$$\frac{6.2}{1} = \frac{a}{1} = \frac{c}{\sqrt{2}}.$$

Hence, $a = \mathbf{6.2}$ and, to the nearest tenth, $c = 6.2\sqrt{2} = 6.2(1.414) = \mathbf{8.8}$.

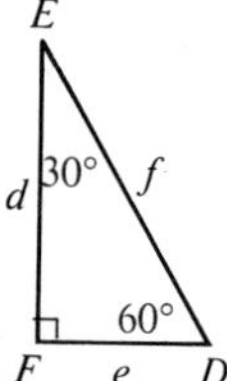

b. Since the lengths of the sides of a 30°–60° right triangle are in the ratio $1:\sqrt{3}:2$ (see Chart 12.2), we have

$$\frac{9.6}{1} = \frac{d}{\sqrt{3}} = \frac{f}{2}.$$

Hence, to the nearest tenth $d = 9.6\sqrt{3} = 9.6(1.732) = \mathbf{16.6}$, and $f = 2(9.6) = \mathbf{19.2}$.

21. $\triangle ABC$, $a = 4$
22. $\triangle ABC$, $c = 6$
23. $\triangle ABC$, $c = 8.5$
24. $\triangle ABC$, $b = 3.4$
25. $\triangle DEF$, $d = 4.4$
26. $\triangle DEF$, $f = 10.2$
27. $\triangle DEF$, $f = 0.48$
28. $\triangle DEF$, $e = 0.62$

Use the information in the figure to classify each triangle as equilateral, isosceles, or scalene.

29. $\triangle ABC$
30. $\triangle ABD$
31. $\triangle GFE$
32. $\triangle FHE$

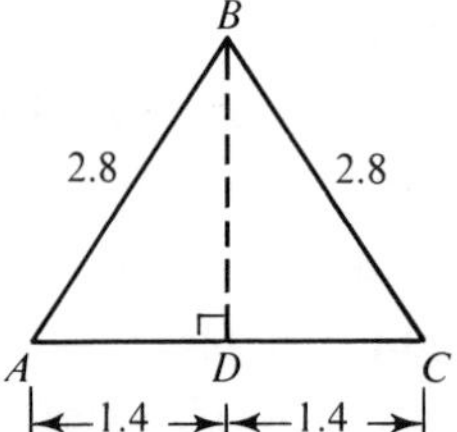

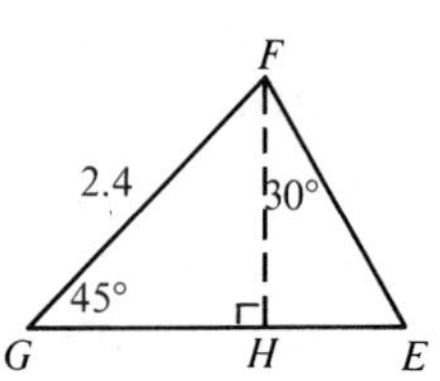

Ex. 29–42

33. What is the measure of $\angle DAB$?
34. What is the measure of $\angle ABD$?
35. What is the measure of $\angle GFH$?
36. What is the measure of $\angle GEF$?
37. Find the length of BD.
38. Find the lengths of GH and HF.
39. Find the perimeter of $\triangle ABD$.
40. Find the perimeter of $\triangle GFH$.
41. Find the area of $\triangle ABC$.
42. Find the area of $\triangle GFE$.

Use the information in the figure to find each of the following measures.

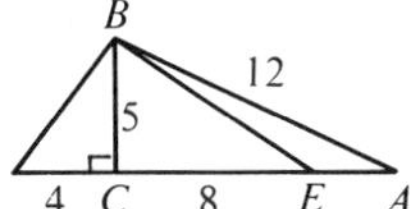

43. The length of CA. (*Hint:* Use the Pythagorean theorem.)
44. The length of EA.
45. The length of BD.
46. The length of BE.
47. The perimeter of $\triangle BDC$.
48. The perimeter of $\triangle BEA$.
49. The area of $\triangle BDC$.
50. The area of $\triangle BCE$.
51. The area of $\triangle DBA$.
52. The area of $\triangle BEA$.
53. Given that an altitude h of an equilateral triangle bisects its base, show that $h = \frac{\sqrt{3}}{2}s$, where s is the length of each side of the triangle.
54. Use the results obtained in Exercise 53 to show that the area A of an equilateral triangle with side s is given by $A = \frac{\sqrt{3}}{4}s^2$.

12.4 Quadrilaterals

Any closed four-sided figure, such as the one shown in Figure 12.15, is called a **quadrilateral.** If the vertices are named A, B, C, and D, the quadrilateral can be named $ABCD$. As in the case of the triangle, the perimeter of any quadrilateral is the sum of the lengths of its sides. Unlike a triangle, the sum of the measures of the angles of the quadrilateral is 360° rather than 180°. A line segment, other than a side, whose end points are vertices of a quadrilateral is a **diagonal** of that quadrilateral. In Figure 12.15, the diagonals d_1 and d_2 are shown by dashed lines.

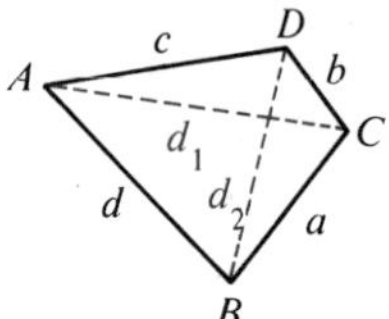

FIGURE **12.15**

You may recall from your previous mathematics courses the names and properties of some of the quadrilaterals that are of special interest. These are listed in Chart 12.3. You should refer to this chart as often as necessary to complete the exercises in Exercise set 12.4.

CHART **12.3**

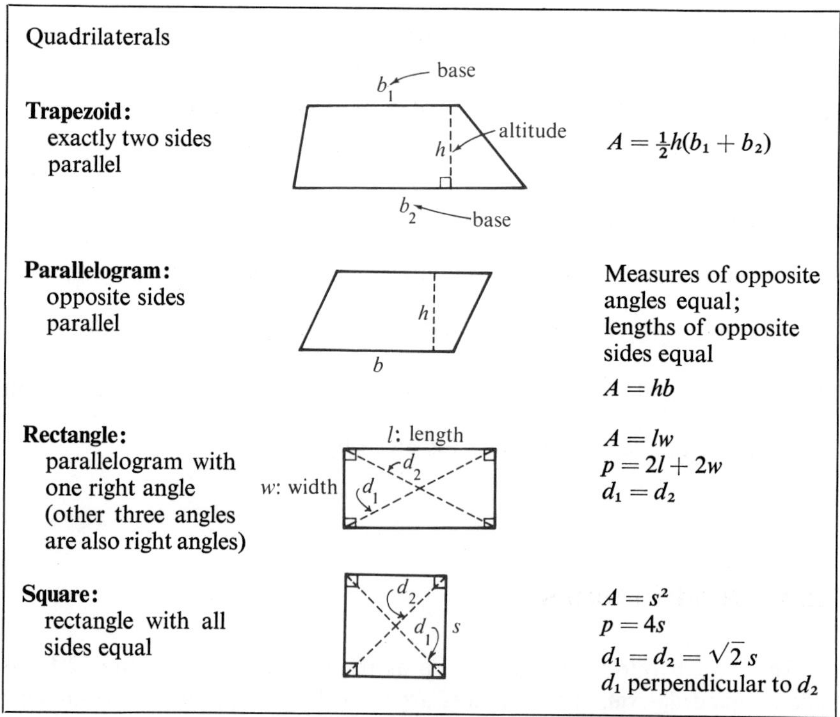

Note that more than one designation can be used to identify some quadrilaterals. For example, a square is a rectangle and also a parallelogram. However, a square has properties that not all rectangles and parallelograms have; therefore the word "square," being more descriptive, would be used to identify such a figure rather than either of the other terms. Similarly a non-square quadrilateral with opposite sides parallel and with four right angles is best described as a rectangle.

EXERCISE 12.4

(a) Give the most appropriate name for each figure. (b) Give all other possible names. (c) Find the measure of each angle whose measure is not shown.

EXAMPLE

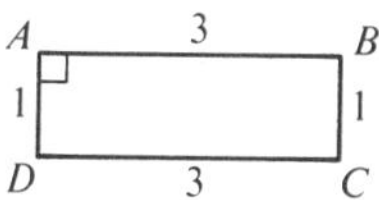

SOLUTION

(a) **Rectangle** (b) **Parallelogram, quadrilateral**

(c) $\angle B = \angle C = \angle D = 90°$

1. 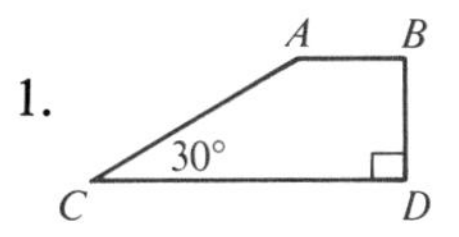

$AB \| CD$

2. 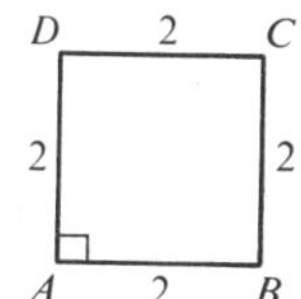

3. 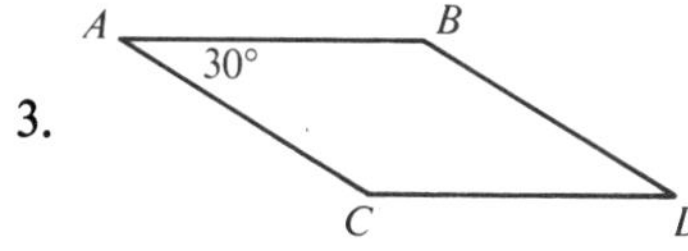

$AB \| CD$ and $AC \| BD$

4. 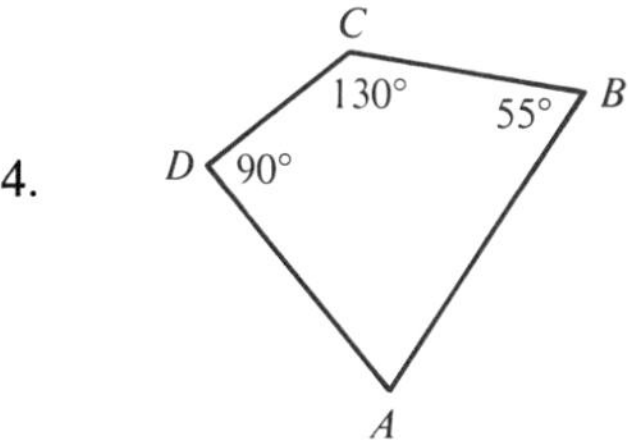

Use the information in the figure to find each of the following measures.

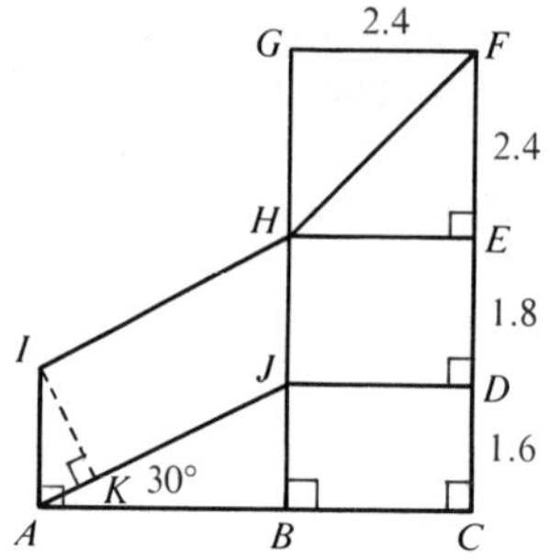

5. Area of $GHEF$.
6. Perimeter of $GHEF$.
7. Length of HF.
8. Area of $\triangle HEF$.
9. Area of $DEHJ$.
10. Area of $BCEH$.
11. Perimeter of $HECB$.
12. Perimeter of $GFDJ$.
13. Length of BJ.
14. Length of AJ.
15. Length of AB.
16. Area $\triangle ABJ$.
17. Measure of $\angle IAJ$.
18. Measure of $\angle IHJ$.
19. Measure of $\angle AJH$.
20. Measure of $\angle AIK$.
21. Length of AI.
22. Length of KI.
23. Perimeter of $AJHI$.
24. Area of $AJHI$.
25. The dimensions of a rectangle are 8 inches by 11 inches. Find the length (to the nearest tenth of an inch) of a side of a square that has the same area as the rectangle.
26. Find the area (to the nearest tenth of a square inch) of a square that has the same perimeter as the rectangle in Problem 25.
27. The area of a square with side s is doubled. Express the length of a side of the larger square in terms of s.
28. The dimensions x and y of a rectangle are decreased by 5 and increased by 8, respectively. (a) Find the area of the new rectangle in terms of x and y. (b) Find the length of a side in terms of x and y of a square with the same area as the original rectangle.
29. The dimensions of the floor of a room are 12 feet by 15 feet. The room has two doors 3 feet wide. How many square tiles, 9 inches on a side, would be required to cover the floor? How many feet of moulding would be required?
30. The inside dimensions of each of six window frames are 23 inches by 29 inches. How many square inches of screen would be required for all six frames if an additional one inch of screen is needed on each side under the moulding?
31. A gallon of paint will cover approximately 280 square feet. How much paint would be required to paint the walls and horizontal ceiling of a room whose dimensions are 12 feet by 15 feet by 8 feet if the room has two windows 2 feet by 3 feet and a door 3 feet by $6\frac{1}{2}$ feet?

32. A rectangular container is 4.2 inches deep and has a base that measures 6.8 inches by 9.4 inches. How many containers (without covers) can be made from 200 square feet of sheet metal if 10% waste is allowed for construction purposes?
33. Show that the length of a diagonal d of a square with sides of length s is given by $d = \sqrt{2}s$.
34. Show that the area A of a trapezoid with parallel sides of length b_1 and b_2, and whose altitude to these sides is h, is given by $A = \frac{1}{2}h\,(b_1 + b_2)$. (*Hint:* Separate the trapezoid into two triangles.)

12.5 Circles

A **circle** is usually defined as the set of all points equidistant from a given point, called the **center** of the circle. In geometry, we learn that a number of relationships of importance exist between parts of a circle and some auxiliary lines, line segments, and angles associated with the circle. First, since the distance from the center to a point on a circle is the **radius,** r, and the distance between two points on a circle on a line passing through the center is the **diameter,** d, we have that the diameter equals twice the radius; $d = 2r$. Next, we know that the perimeter of a circle, called the **circumference**, c, is determined by the equation

$$c = 2\pi r,$$

where π is an irrational number approximately equal to 3.1416. (In some cases, which should be clear from the context, the words "radius" and "circumference" refer to the respective line segments rather than their lengths.) It is sometimes convenient to use the alternative formula,

$$c = \pi d.$$

As in the case of the circumference, the area A involves π and is given by the formula

$$A = \pi r^2.$$

A **chord** is any line segment having its end points on a circle, a **secant** is a line that intersects a circle in two points, and a **tangent line** is a line that touches but does not pass through a circle. A tangent to a circle is perpendicular to the radius at the point of tangency, and the segment from the center of a circle to the midpoint of a chord is perpendicular to the chord.

An angle formed by two radii, $\angle AOC$, is a **central angle**. An **inscribed angle** has a vertex which is a point of a circle and its sides intercept two other points of the circle. If an inscribed angle and a central angle intercept the same arc, then the measure of the central angle is twice the measure of the inscribed angle.

An **arc** $\overset{\frown}{AB}$ is a part of a circle. If the end points of the arc determine a diameter, then the arc is called a **semicircle**. An angle inscribed in a semicircle is a right angle. The length of an arc can be determined by multiplying by $\pi r/180$ the measure of the central angle subtending (cutting off) the arc.

The foregoing facts are summarized in Chart 12.4.

CHART **12.4**

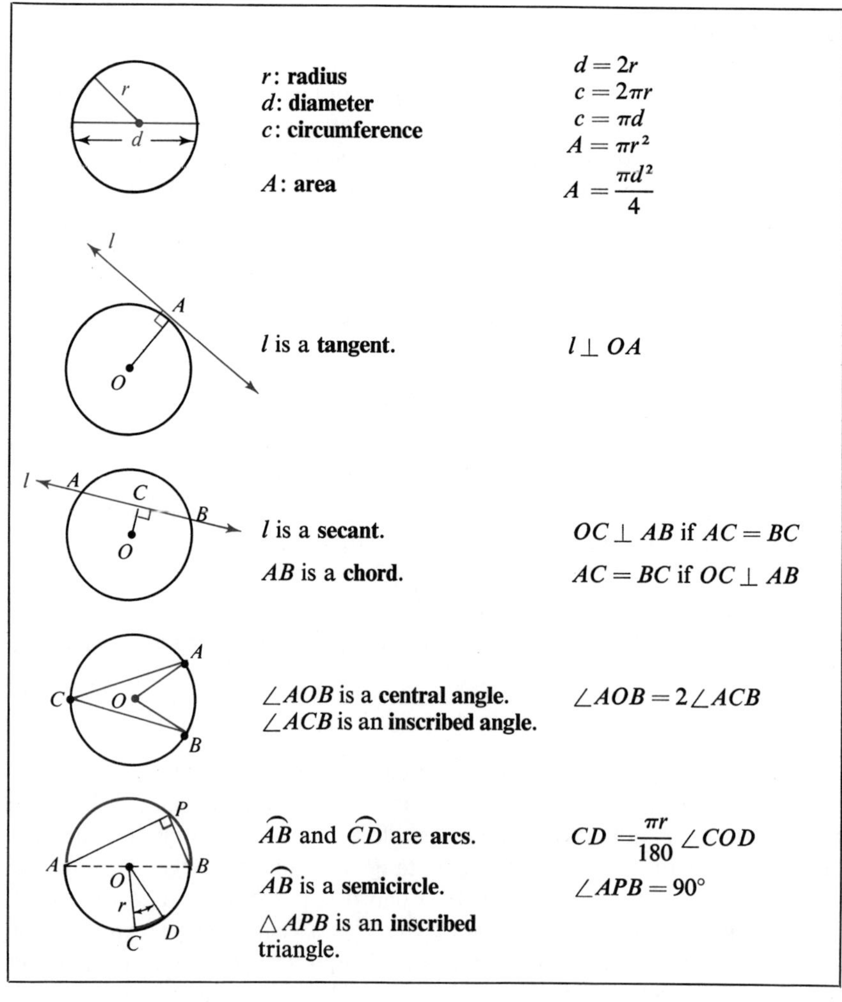

EXERCISE 12.5

In Exercises 1–8, name each part of the figure.

1. OA
2. AB
3. Segment AC
4. $\overparen{CB}$
5. l_1
6. l_2
7. $\angle COB$
8. $\angle CAB$

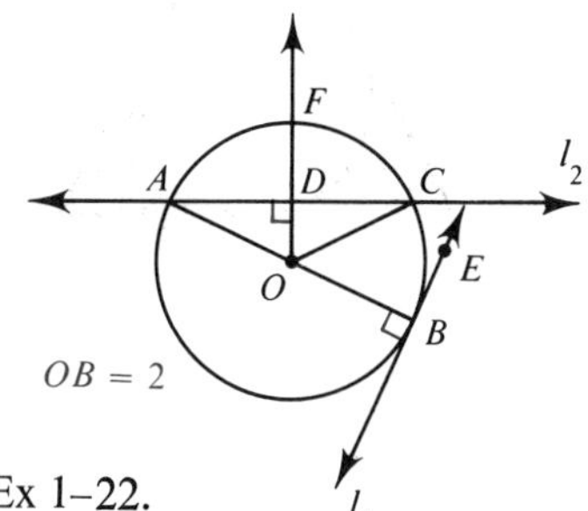

Ex 1–22.

In Exercises 9–20, assume that $\angle CAB = 30°$. Find the following measures of angles and lengths of line segments and arcs.

9. $\angle AOD$
10. $\angle ACO$
11. $\angle AOC$
12. $\angle COB$
13. $\angle ACB$
14. $\angle OBC$
15. $\angle ABC$
16. $\angle CBE$
17. $\overparen{AB}$
18. $\overparen{AF}$
19. $\overparen{FC}$
20. $\overparen{CB}$
21. Find the circumference of the circle.
22. Find the area of the circle.

In Exercises 23–30, find the area (to the nearest tenth of a unit) for each shaded region.

23. 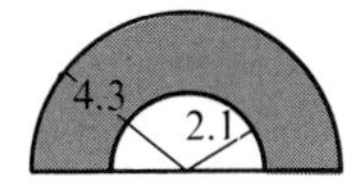

24. 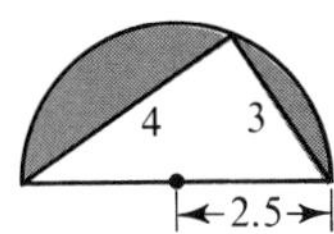

25. 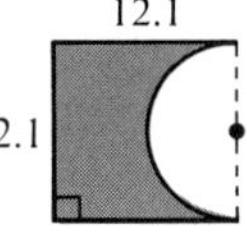

26. 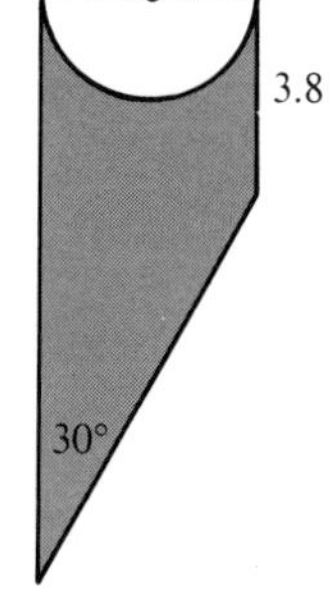

27. 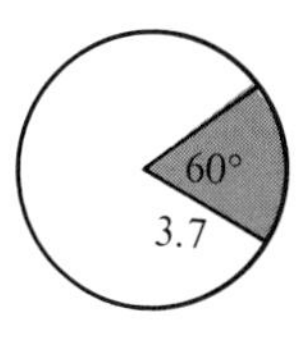

28. 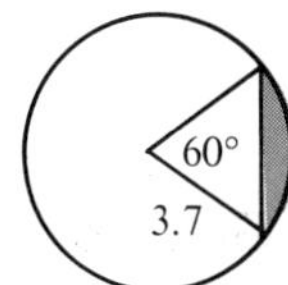

29. 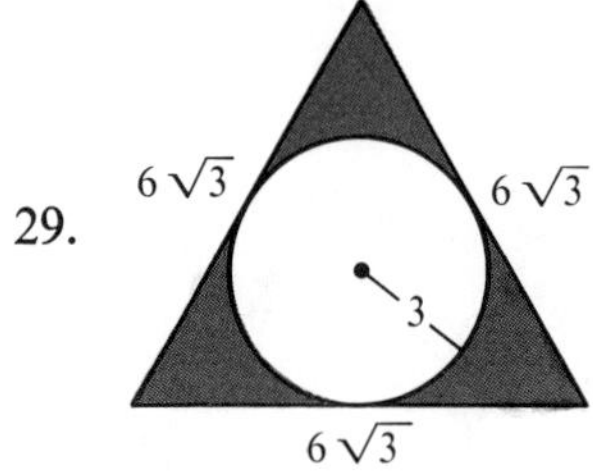

30. 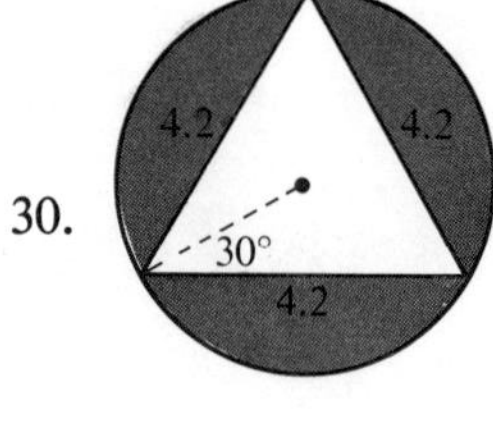

Find the perimeter of each shaded region to the nearest tenth of a unit.

31. Exercise 23 above
32. Exercise 24
33. Exercise 25
34. Exercise 26
35. Exercise 27
36. Exercise 28
37. A gallon of paint will cover 400 square feet. How many circular plates with diameter of 14.2 inches can be painted with 2 gallons of paint?
38. Two pulleys that are 8.6 inches in diameter have centers 26.4 inches apart, as shown in the accompanying figure. What is the length of a belt connecting the pulleys?

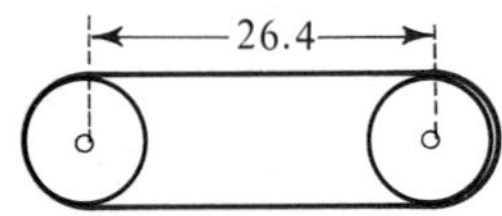

39. The outer surface of a tire is 15″ from the center of the wheel of a car. How many revolutions will the tire make if the car travels 10 miles?
40. Three circular pipes with diameters of 3.1, 4.6, and 7.2 inches are replaced with one circular pipe that has a cross sectional area equal to the sum of the cross-sectional areas of the three pipes. What is the diameter of the replacement pipe?
41. What is the radius of a circle if the circumference exceeds the radius by 38.4 inches?
42. What is the circumference of a circle if the circumference exceeds the diameter by 18.6 inches?

12.6 Geometric solids

There are a number of geometric solids that are important in a wide variety of practical applications. Many of these are encountered in earlier mathematics courses. We have listed the properties of some of the more important solids in Chart 12.5, which can be referred to as necessary to complete the exercises that follow.

CHART **12.5**

Solid	In all cases, B = area of base
Prism: Parallel bases; elements (line segments perpendicular to the bases) on the lateral surface are parallel.	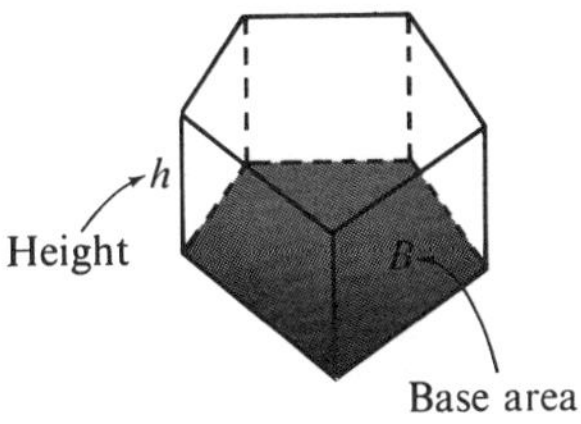 Volume: $V = Bh$
Right circular cylinder: Parallel circular bases; elements (line segments perpendicular to the bases) on the lateral surface are parallel.	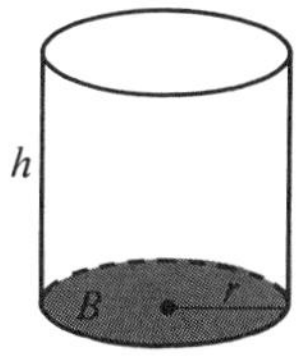 Volume: $V = Bh$ $= \pi r^2 h$ Lateral surface area: $S = 2\pi rh$
Pyramid: Base is closed geometric figure.	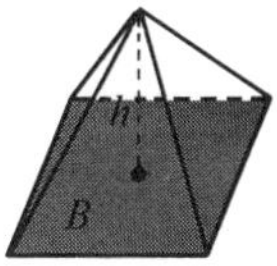 Volume: $V = \frac{1}{3}Bh$
Right circular cone: Base is a circle.	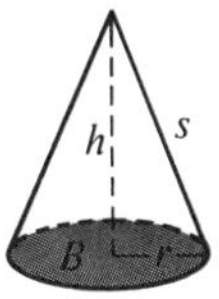 Volume: $V = \frac{1}{3}Bh$ $= \frac{1}{3}\pi r^2 h$ Lateral surface area: $S = \pi rs$
Sphere: Points are equidistant from a given point, the center.	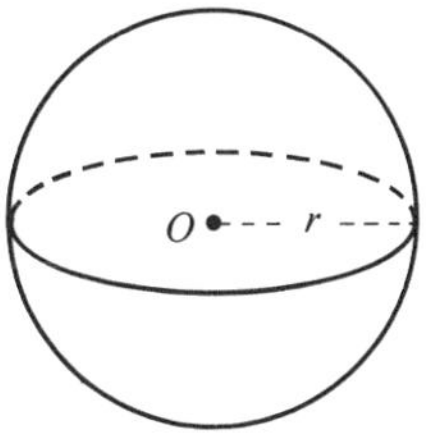 Volume: $V = \frac{4}{3}\pi r^3$ Surface area: $S = 4\pi r^2$

We have observed that lengths of line segments are in linear units, inches, feet, yards, and so forth, and that areas of regions are in square units, square inches, square feet, and so forth. You probably recall from your earlier mathematics courses, that volume is given in cubic units, such as cubic inches and cubic feet.

EXERCISE 12.6

In each of the following exercises, express your answers to the nearest tenth of a unit.

Find the volume and total surface area of each solid.

EXAMPLE

Prism: Base is an equilateral triangle with sides 2 inches in length, the height is 6 inches.

SOLUTION

A sketch is sometimes helpful. Area of the base (see Chart 12.2) is

$$A = \frac{\sqrt{3}}{4} s^2 = \frac{1.732}{4}(2)^2 = 1.732.$$

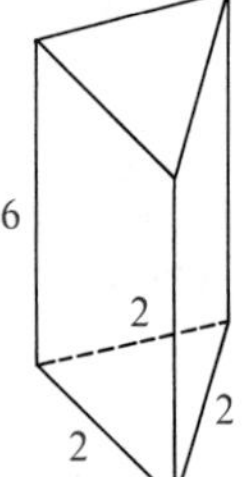

The volume (see Chart 12.5) is

$$V = Bh = 1.732(6) = \mathbf{10.392}.$$

Hence, the volume equals **10.4** cubic units to the nearest tenth of a unit.

The total surface area equals the sum of the areas of the two equilateral triangles and three rectangular sides, and is given by

$$\begin{aligned} S &= 2(1.732) + 3(6 \cdot 2) \\ &= 3.464 + 36.000 = 39.464. \end{aligned}$$

Hence, the total surface area equals **39.5** square units to the nearest tenth of a unit.

1. Prism: Base is a square with sides 3 inches in length, the height is 4 inches.
2. Prism: Base is a rectangle with dimensions 4 inches by 6 inches, the height is 10 inches.
3. Prism: Base is an isosceles triangle with sides 2.1, 2.1, and 1.8 inches in length, the height is 3.6 inches.
4. Prism: Base is an equilateral triangle with sides 1.4 feet in length, the height is 3.8 feet.

5. Right circular cylinder: Radius of the base is 2.3 inches, the height is 5.7 inches.
6. Right circular cylinder: Radius of the base is 1.71 feet, the height is 2.04 feet.
7. Circular cone: Radius of the base is 1.2 feet, the altitude is 4.3 feet.
8. Circular cone: Radius of the base is 0.82 feet, the altitude is 0.41 feet.
9. Sphere: Radius of 9.3 inches.
10. Sphere: Radius of 0.62 feet.

11. Pyramid: Square base with sides 2.5 inches in length, the altitude is 4.3 inches.

$$AE = BE = CE = DE$$

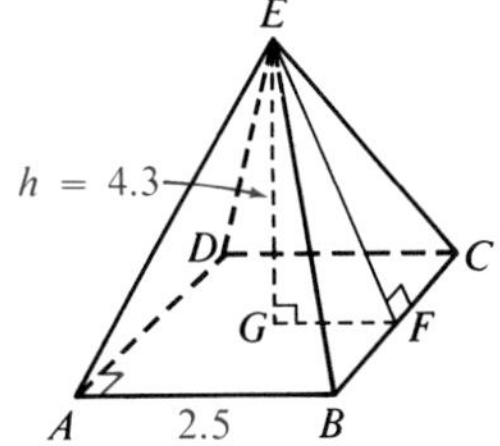

12. Pyramid: Base is an equilateral triangle with sides 2.8 inches in length, the altitude is 3.1 inches. (*Hint:* $EF = EG = EH = 0.8$.)

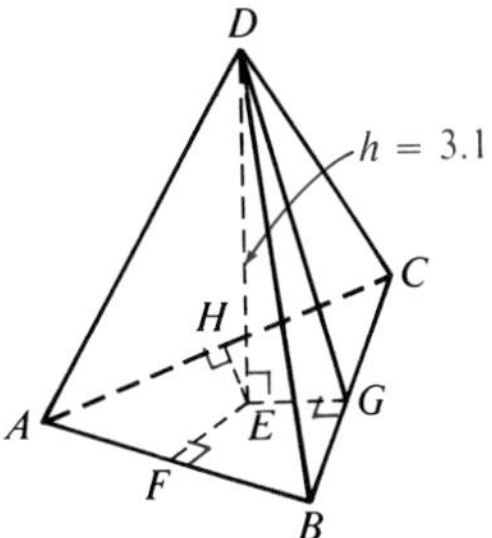

Find the volume of the following solids.

13.

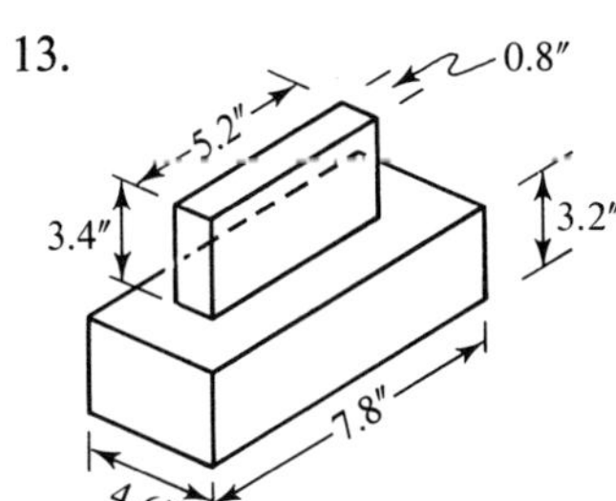

14.

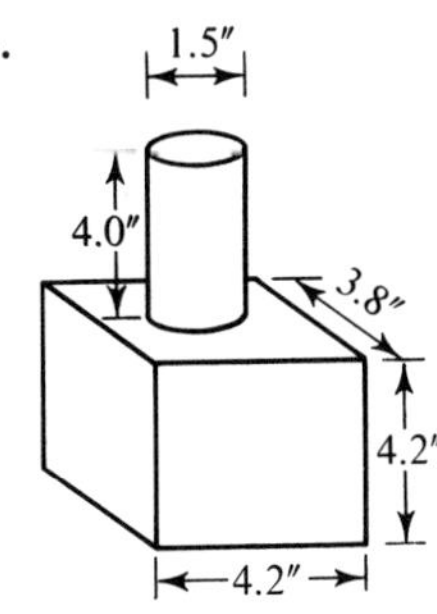

15.

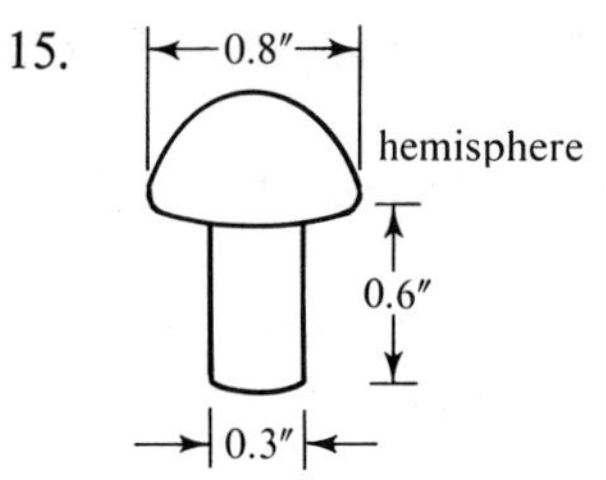

16.

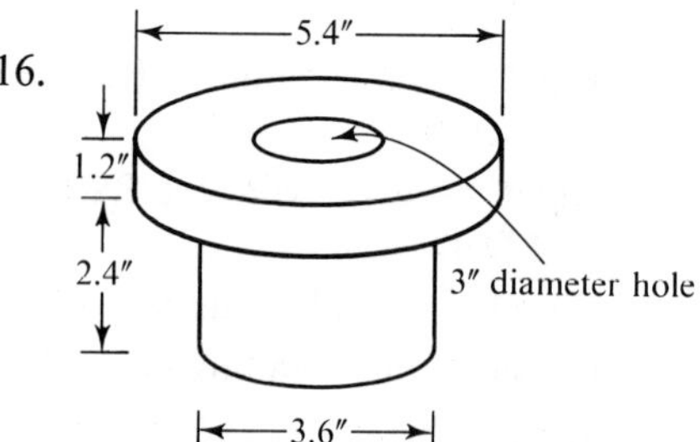

17.

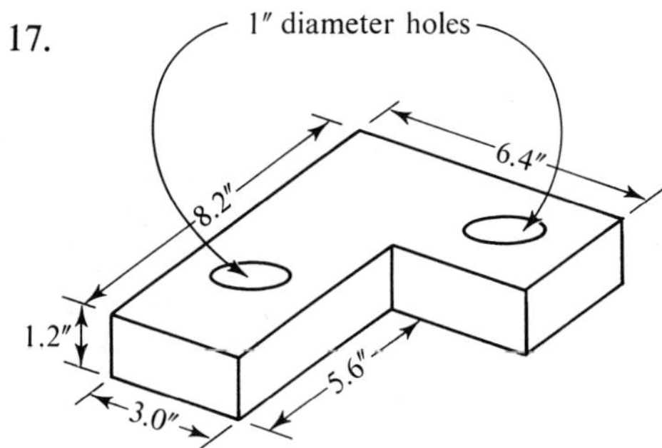

18.

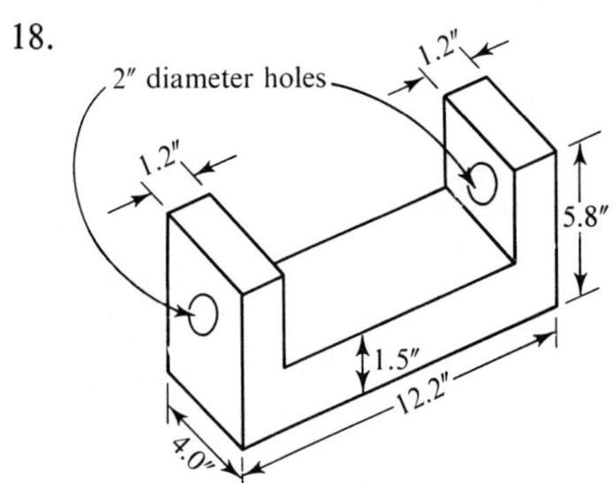

19.

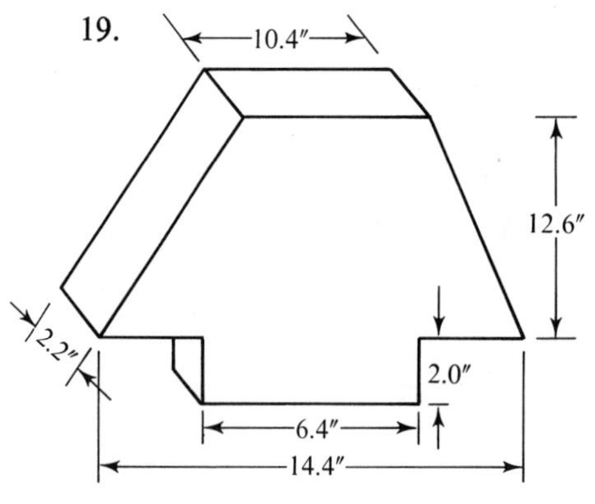

20.

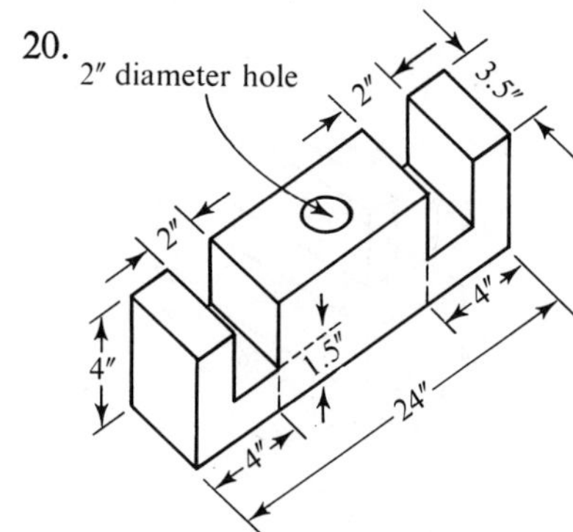

21. The length of a cylindrical tank is 24.6 feet and its diameter is 6.4 feet. How many cubic feet of water does it contain when it is half-full?

22. How much paint would be required to paint the entire outer surface of the tank in Problem 21 if one gallon of paint covers 400 square feet of surface?

23. Surfaces can be plated by a certain process at a cost of $0.10 per square inch. What is the total cost of plating 12 spheres if the diameter of each sphere is 14.4 inches?

24. What is the cost of plating the total surface of a circular cylinder with length 10 feet, whose volume equals the sum of the volumes of the twelve spheres in Problem 23?

25. A certain wire is made of metal which weighs 420 lb per cubic foot. How much does 1.8 miles of the wire weigh if it is 0.22 inches in diameter?

26. A tent in the form of a pyramid has a square base with sides 10.6 feet long and is 8.4 feet high. How many square feet of canvas is needed for the four sides? How many cubic feet of air does the tent contain?

Chapter review

[12.1] Name (a) the complement, and (b) the supplement of each angle.

1. a. 28° 14′ b. 34.6°

Express each of the following in decimal form.

2. a. 114° 24′ b. 67° 15′

Express each of the following in degrees and minutes.

3. a. 17.4° b. 18.6°

4. The measure of an angle is equal to three-fifths of its supplement. Find the measure of each angle.

[12.2] In the accompanying figure $AB \parallel CD \parallel EF$, and $AC = 6$, $DB = 8$, and $FD = 3$. Find each of the following.

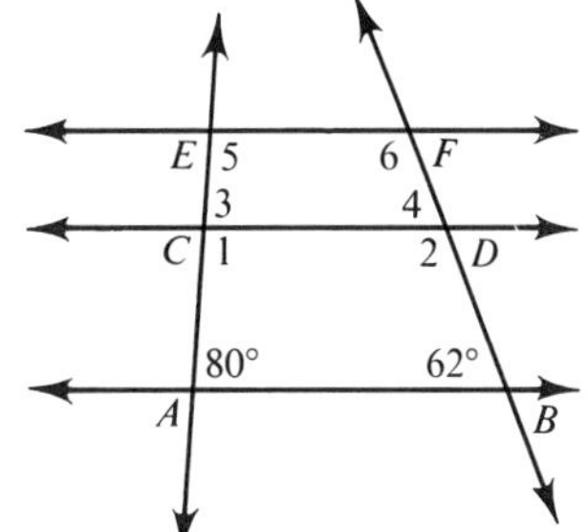

5. The measure of ∠3
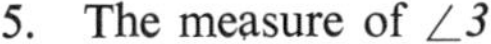
6. The measure of ∠5
7. The measure of ∠2
8. The length of EC

[12.3] Find each of the following lengths to the nearest tenth of a unit using the figure below.

9. *AB* 10. *DE* 11. *DF* 12. *HJ*

13. Perimeter and area of $\triangle DEF$ 14. Perimeter and area of $\triangle GHJ$

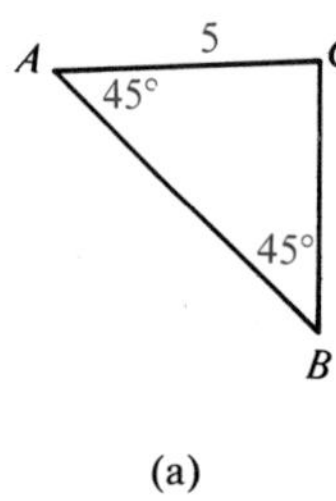

(a)

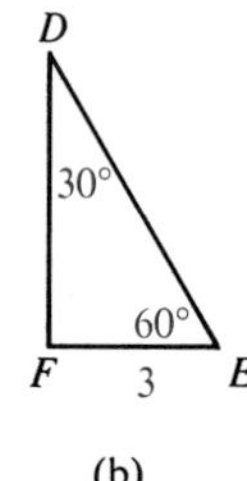

(b)

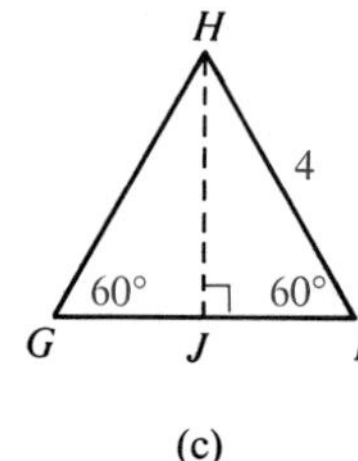

(c)

[12.4] 15. Find (a) the perimeter and (b) the length of the diagonal of a square whose area equals the area of a rectangle with length 9 units and width 4 units.

16. The length of a rectangle is 10 feet greater than its width. Find the dimensions if its perimeter is 168 feet.

[12.5] 17. What is (a) the length of the diameter, and (b) the area of a circle whose circumference is 56 inches?

18. What is the area of a semicircle if the circumference of the circle exceeds the diameter by 14.8 inches?

[12.6] 19. Find the volume and total surface area of a right circular cylinder whose base has a radius of 4.2 inches and whose height is 6.8 inches.

20. A box without a top is to be made from a square piece of tin by cutting a 3-inch square from each corner and folding up the sides. If the box will hold 147 cubic inches, what should be the length of the side of the original square piece of tin?

13 Trigonometry

13.1 Similar triangles

When comparing two (or more) triangles for various purposes, we frequently label their parts using the same basic letter and then distinguish between the parts of the two figures by affixing a raised mark ′ (read "prime") to the symbols applied to the parts of one of the triangles (see Figure 13.1).

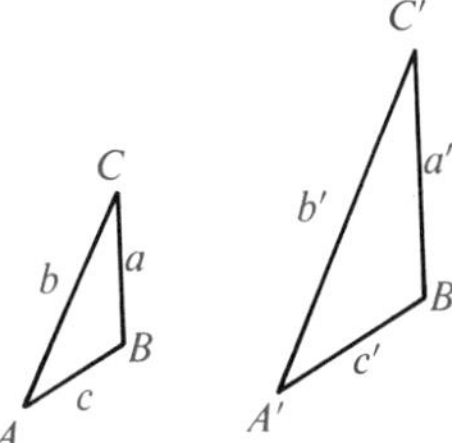

FIGURE **13.1**

Thus we use A' (read "A prime") to distinguish the vertex of an angle in one triangle from the vertex A of a comparable angle in the other.

If two triangles have the same size and shape, in the sense that one of them can be superimposed exactly upon the other, the triangles are said to be **congruent triangles**. The parts that coincide in this case are called **corresponding parts** of the triangles. Figure 13.2 shows congruent triangles ABC and

FIGURE **13.2**

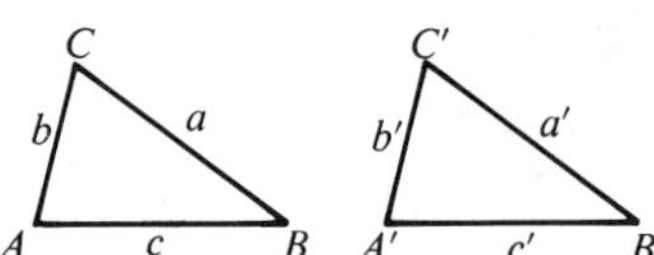

$A'B'C'$. We can represent this congruency by writing $\triangle ABC \cong \triangle A'B'C'$ (read "triangle ABC is congruent to triangle $A'B'C'$"). Clearly, if two triangles are congruent, then their corresponding parts have equal measures; that is, $\angle A = \angle A'$, $\angle B = \angle B'$, $\angle C = \angle C'$, $a = a'$, $b = b'$, and $c = c'$.

If two triangles have the same shape but are not necessarily the same size, that is, if their corresponding angles are equal in measure but the sides are not equal in length, then the triangles are called **similar triangles**. Thus in Figure 13.1, if $\angle A = \angle A'$, $\angle B = \angle B'$, and $\angle C = \angle C'$, we can write $\triangle ABC \sim \triangle A'B'C'$ (read "triangle ABC is similar to triangle $A'B'C'$). Of course, if two triangles are congruent, then they are also similar.

In geometry, it is shown that lengths of corresponding sides of similar triangles are proportional. From Figure 13.1, because $\triangle ABC \sim \triangle A'B'C'$, we know that

$$\frac{a}{a'} = \frac{b}{b'} = \frac{c}{c'}. \tag{1}$$

Since for any triangle the sum of the angle measures is 180°, it follows that if two angles of two triangles have equal measures, then so do the third angles. Thus, if two angles of one triangle have the same measures as two angles of another, the triangles must be similar.

EXERCISE 13.1

Given that $\triangle ABC \sim \triangle A'B'C'$.

EXAMPLE

If $\angle A = 42°$, $\angle C' = 108°$, $a = 12$, $a' = 4$, and $b = 9$, find $\angle B$ and b'.

SOLUTION

Make a sketch and label them with known values. To determine $\angle B$, we first observe that $\angle A = 42°$ and $\angle C = 108°$. Then, since $\angle A + \angle B + \angle C = 180°$, we have

$$42° + \angle B + 108° = 180,$$
$$\angle B = 180° - 150° = 30°.$$

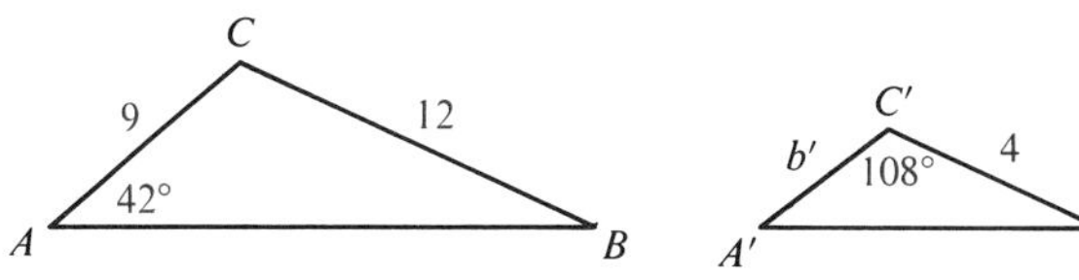

Next, to find b', we know that $a'/a = b'/b$. Using the given values for a, a', and b, we have

$$\frac{4}{12} = \frac{b'}{9},$$

from which

$$b' = 3.$$

Thus, $\angle B = \mathbf{30}°$ and $b' = \mathbf{3}$.

1. Given that $\angle A = 38°$, $\angle C' = 125°$, $a = 15$, $a' = 12$, and $c = 20$, find $\angle B'$ and c'.
2. Given that $\angle A = 25°$, $\angle C' = 95°$, $c = 32$, $c' = 24$, and $b' = 21$, find $\angle B'$ and b.
3. Given that $\angle B = 60°$, $\angle C' = 98°$, $b' = 40$, $b = 56$, and $a = 49$, find $\angle A$, and a'.
4. Given that $\angle C = 112°$, $\angle A' = 30°$, $b = 50$, $b' = 45$, and $c = 75$, find $\angle B'$ and c'.
5. Given that $\angle C = 3\angle B$, $\angle A' = 20°$, $a = 80$, $a' = 20$, and $c' = 26$, find $\angle B$, $\angle C$, and c.
6. Given that $\angle A = \angle B$, $\angle C' = 98°$, $a = 8$, $b' = 6$, and $c' = 9$, find $\angle A$, $\angle B'$, b and c.
7. Given that $\angle C = \angle A + 2\angle B$, $\angle A' = 3\angle B'$, $\angle C' = 100°$, $a = 3a'$, $b = 36$, and $c = 105$, find $\angle A$, $\angle B$, b', and c'.
8. Given that $\angle C - \angle B = 2\angle A$, $3\angle A = 4\angle B$, $\angle C' = 110°$, $b = 2b'$, $c = 12$, and $a = 8$, find $\angle A$, $\angle B$, $\angle C$, a', and c'.

In each Exercise 9–16, find x, y, and/or z to the nearest tenth.

9.

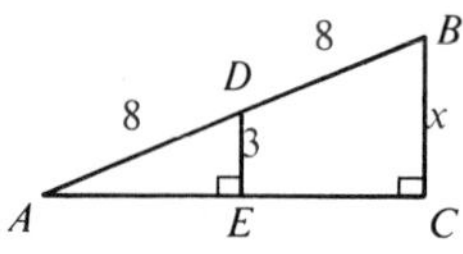

10.

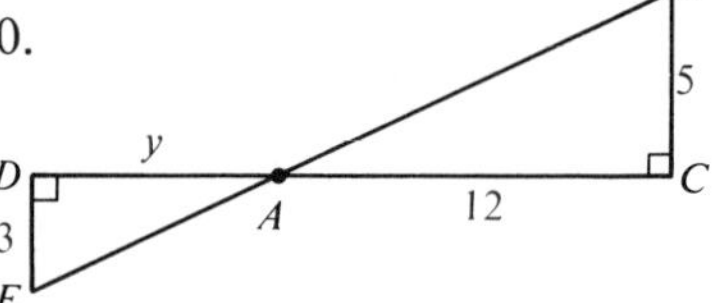

11.

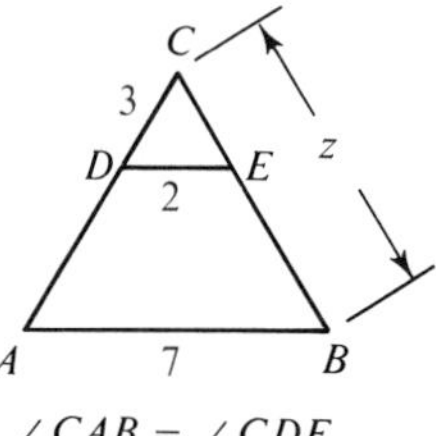

12.

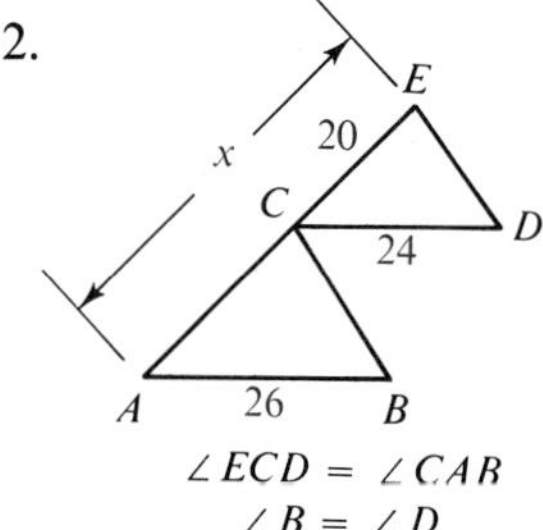

13.

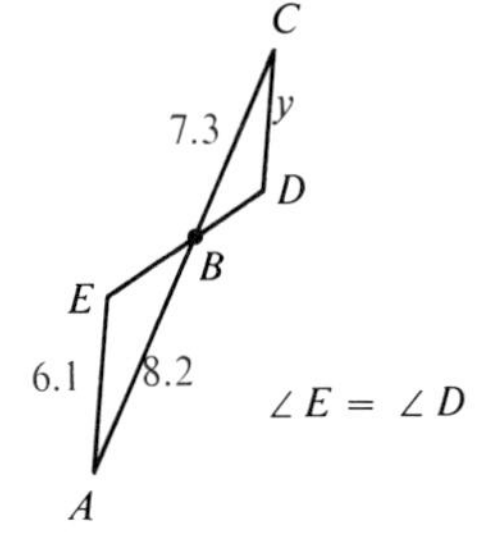

14.

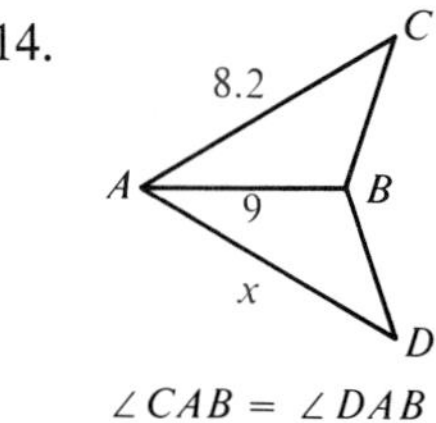

15.

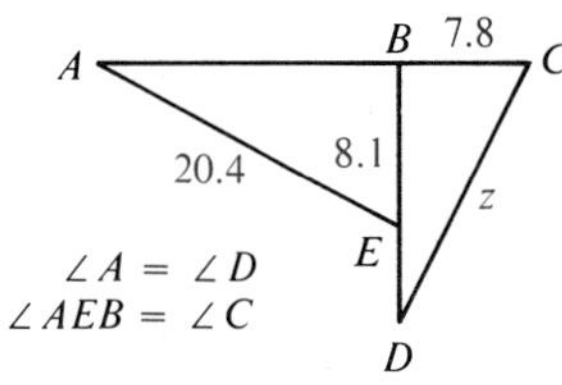

16.

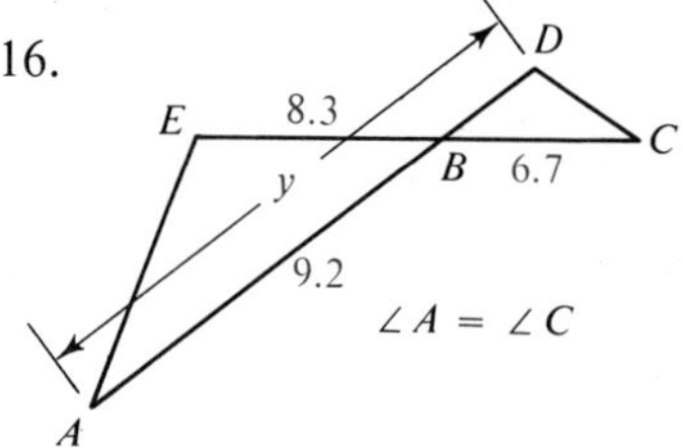

17. A 200-foot support wire is hooked to the top of a 140-foot TV antenna. How long should another support wire be if it is to be attached to the pole at a height of 110 feet and anchored so that it is parallel to the other support wire?

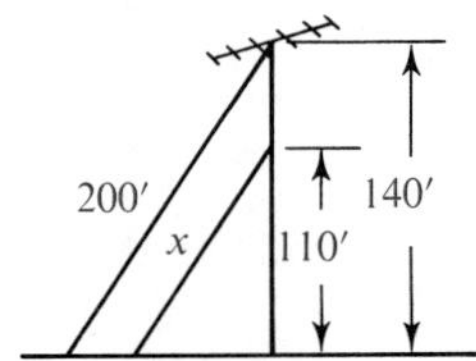

18. A steel ball rolled down an inclined plane for a distance of 14 inches continues to roll for 8 inches across a resistant medium. How far would the ball roll across the medium if it started 10 inches up the inclined plane?

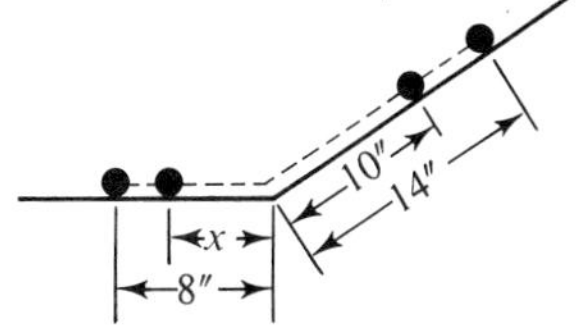

19. The figure at the right depicts a method of measuring the width x of a river at a certain point. If these measurements are taken as shown, how wide is the river at this point?

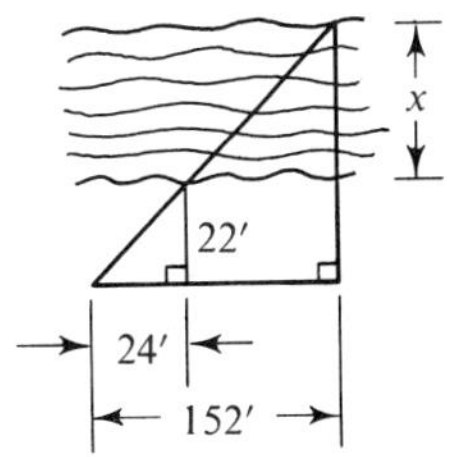

20. A man whose eyes are 5′ 8″ above the ground can just see a spot in the sea when he is standing 12′ from the edge of a 70-foot vertical cliff. How far is the spot from where he is standing?

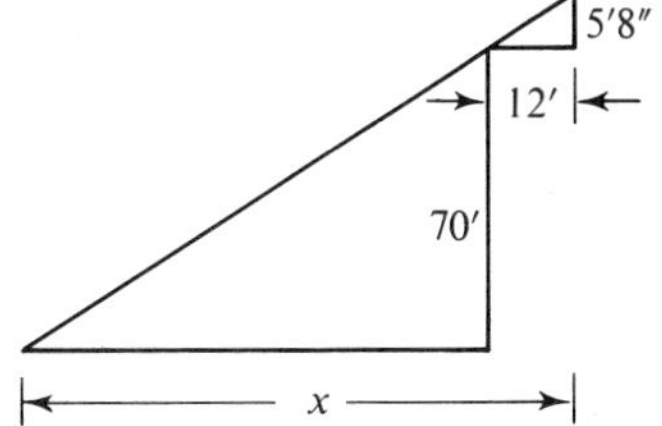

13.2 Trigonometric ratios and functions

An angle such as $\angle A$ or $\angle B$ shown in Figure 13.3 is said to be in **standard position** when one of its sides (its **initial side**) is in the positive x-axis and its vertex is the origin 0. By considering the similar triangles OPR and OQS, ($\angle A = \angle A$ and $\angle ORP = \angle OSQ$), we can see that for any points P and Q on the ray OQ (the **terminal side** of $\angle A$),

$$\frac{PR}{QS} = \frac{OP}{OQ} \quad \text{and} \quad \frac{OR}{OS} = \frac{OP}{OQ}.$$

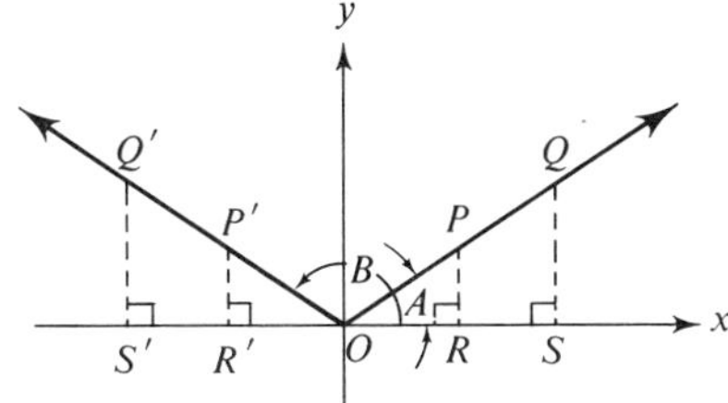

FIGURE **13.3**

which can be rewritten equivalently in the form

$$\frac{PR}{OP} = \frac{QS}{OQ} \quad \text{and} \quad \frac{OR}{OP} = \frac{OS}{OQ}.$$

By considering $\triangle OP'R' \sim \triangle OQ'S'$ in the same way, we have the ratios

$$\frac{P'R'}{OP'} = \frac{Q'S'}{OQ'} \quad \text{and} \quad \frac{OR'}{OP'} = \frac{OS'}{OQ'}.$$

Although we do not develop the idea here, similar conclusions apply for an angle in standard position whose terminal side lies in Quadrant III or Quadrant IV. What these facts suggest is that the ratios of the abscissa and ordinate of a point on the terminal side of an angle in standard position to the distance of the point from the vertex of the angle do not depend upon the particular point but, rather, depend entirely on the angle. Thus, we can define functions from the set of angles in standard position to the sets of all such ratios. These ratios we call the **sine** and **cosine** of the angle, respectively. From Figure 13.4,

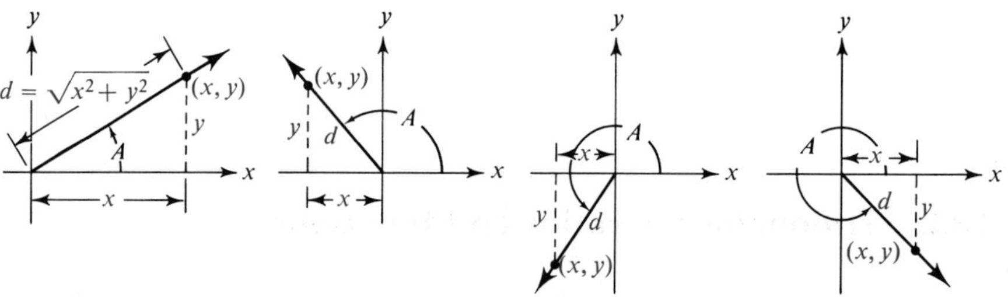

FIGURE **13.4**

since the distance d from the vertex to the point (x, y) on the terminal side of any angle A is $\sqrt{x^2 + y^2}$, we have

$$\sin A = \frac{y}{\sqrt{x^2 + y^2}} \quad \text{or} \quad \sin A = \frac{y}{d} \tag{1}$$

(sin A is read "the sine of A"), and

$$\cos A = \frac{x}{\sqrt{x^2 + y^2}} \quad \text{or} \quad \cos A = \frac{x}{d} \tag{2}$$

(cos A is read "the cosine of A").

The ratios for sin A and cos A are called **trigonometric ratios**, and the functions having these ratios as elements of the range are called **trigonometric functions**.

There are four other trigonometric functions of importance. These are defined as follows:

$$\textbf{tangent:} \quad \tan A = \frac{y}{x}, \quad (x \neq 0);$$

$$\textbf{cotangent:} \quad \cot A = \frac{x}{y}, \quad (y \neq 0);$$

$$\textbf{secant:} \quad \sec A = \frac{\sqrt{x^2 + y^2}}{x} = \frac{d}{x}, \quad (x \neq 0);$$

$$\textbf{cosecant:} \quad \csc A = \frac{\sqrt{x^2 + y^2}}{y} = \frac{d}{y}, \quad (y \neq 0).$$

Using the fact that $d = \sqrt{x^2 + y^2}$ is greater than 0 for any point (x, y) other than the origin, we can verify that the function values will be positive or negative when angle A has its terminal side in the respective quadrants according to Figure 13.5.

Quadrant II	Quadrant I
$\sin A > 0$, $\csc A > 0$ $\cos A < 0$, $\sec A < 0$ $\tan A < 0$, $\cot A < 0$	$\sin A > 0$, $\csc A > 0$ $\cos A > 0$, $\sec A > 0$ $\tan A > 0$, $\cot A > 0$
$\sin A < 0$, $\csc A < 0$ $\cos A < 0$, $\sec A < 0$ $\tan A > 0$, $\cot A > 0$	$\sin A < 0$, $\csc A < 0$ $\cos A > 0$, $\sec A > 0$ $\tan A < 0$, $\cot A < 0$

FIGURE **13.5**

For any angle in standard position whose terminal side lies on one of the coordinate axes, certain of the trigonometric ratios are not defined. Thus, if the terminal side of an angle A lies on the x-axis, so that for each point on the terminal side $y = 0$, then, for example, $\cot A = x/0$ is not defined.

Whether a denominator of a fraction is rationalized or not is usually determined by the way the fraction is to be used. For example,

$$\frac{1}{\sqrt{2}} = \frac{1}{\sqrt{2}} \cdot \frac{\sqrt{2}}{\sqrt{2}} = \frac{\sqrt{2}}{2}.$$

Hence, both $1/\sqrt{2}$ and $\sqrt{2}/2$ represent the same number. In this and the following sections, we shall use either form of such fractions as convenient.

EXERCISE 13.2

Given that the specified point is on the terminal side of an angle A in standard position, find the value of each of the six trigonometric functions of $\angle A$. If a particular value is not defined, so state.

EXAMPLE

(5, −1)

SOLUTION

A sketch may be helpful. In order to find values for all of the functions, we need values for x, y, and d. By inspection, $x = 5$ and $y = -1$. To find d, we have

$$d = \sqrt{x^2 + y^2} = \sqrt{5^2 + (-1)^2} = \sqrt{26}.$$

From the definitions of the functions, we have

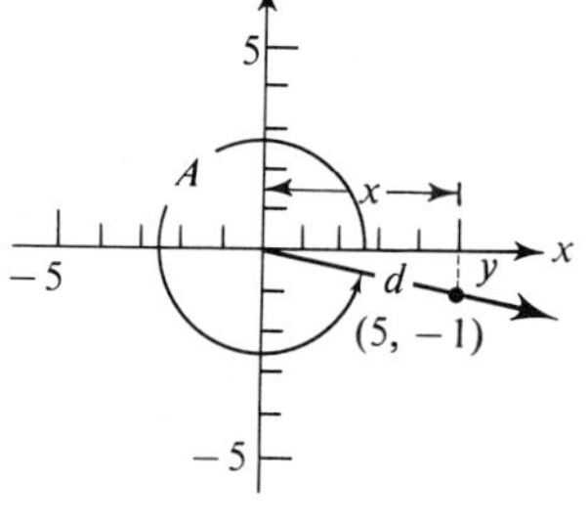

$$\sin A = \frac{\mathbf{-1}}{\sqrt{\mathbf{26}}} \qquad \csc A = \frac{\sqrt{26}}{-1} = \mathbf{-\sqrt{26}}$$

$$\cos A = \frac{\mathbf{5}}{\sqrt{\mathbf{26}}} \qquad \sec A = \frac{\sqrt{\mathbf{26}}}{\mathbf{5}}$$

$$\tan A = \frac{\mathbf{-1}}{\mathbf{5}} \qquad \cot A = \frac{5}{-1} = \mathbf{-5}$$

1. (3, 4)	2. (−4, 3)	3. (−4, 12)	4. (12, −5)
5. (1, 1)	6. (−1, −1)	7. $(1, \sqrt{3})$	8. $(-\sqrt{3}, 1)$
9. (6, −8)	10. $(5, 5\sqrt{3})$	11. $(3, \sqrt{7})$	12. $(-7, \sqrt{15})$
13. (0, 3)	14. (−2, 0)	15. (−5, 0)	16. (0, 3)

Given that the terminal side of $\angle A$ is in Quadrant I or on the positive x- or y-axis, and one trigonometric function value of $\angle A$ is as given, find the remaining function values. If a particular value is not defined, so state.

EXAMPLE

$$\sin A = \frac{2}{3}$$

SOLUTION

Make a sketch. As suggested by the sketch, the point on the terminal side of $\angle A$ located 3 units from the origin has a y-coordinate 2, because

$$\sin A = \frac{y}{d} = \frac{2}{3}.$$

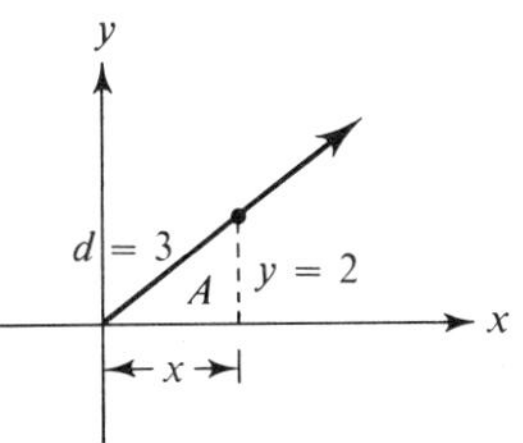

We can use the Pythagorean theorem to find the x-coordinate of this point. Since $x^2 + y^2 = d^2$,

$$x^2 + 2^2 = 3^2,$$
$$x^2 + 4 = 9,$$
$$x^2 = 5,$$
$$x = \sqrt{5} \quad \text{or} \quad x = -\sqrt{5}.$$

Because the terminal side of $\angle A$ is in Quadrant I, $x > 0$ and we select $\sqrt{5}$ for its value. Then, by definition,

$$\cos A = \frac{x}{d} = \frac{\sqrt{5}}{3}, \qquad \sec A = \frac{d}{x} = \frac{3}{\sqrt{5}},$$

$$\tan A = \frac{y}{x} = \frac{2}{\sqrt{5}}, \qquad \cot A = \frac{x}{y} = \frac{\sqrt{5}}{2},$$

$$\csc A = \frac{d}{y} = \frac{3}{2}.$$

17. $\sin A = \frac{1}{2}$
18. $\cos A = \frac{1}{2}$
19. $\tan A = \frac{3}{4}$
20. $\cot A = \frac{5}{12}$
21. $\sec A = \sqrt{2}$
22. $\csc A = \frac{5}{3}$
23. $\cos A = \frac{1}{\sqrt{10}}$
24. $\sin A = \frac{2}{\sqrt{5}}$
25. $\tan A = \frac{2}{3}$
26. $\cot A = \frac{5}{4}$
27. $\csc A = \frac{2}{\sqrt{3}}$
28. $\sec A = \frac{5}{\sqrt{3}}$
29. $\cos A = 0$
30. $\sin A = 0$
31. $\tan A = 0$
32. $\cot A = 0$
33. $\sec A = 0$
34. $\csc A = 0$

Use the definition of the appropriate trigonometric function(s) to show that each of the equations in Exercises 35–40 is an identity; that is, show that the equation is true for every angle A for which each function value is defined.

EXAMPLE

$$\csc A = \frac{1}{\sin A}.$$

SOLUTION

By definition, $\csc A = \frac{d}{y}$ and $\sin A = \frac{y}{d}$. Substituting in the given equation, we have

$$\frac{d}{y} = \frac{1}{\frac{y}{d}},$$

from which

$$\frac{d}{y} = \frac{d}{y}.$$

Since this equation is equivalent to the original equation for all values of y and d except 0, and since it is clearly an identity, the original equation is an identity.

35. $\sec A = \dfrac{1}{\cos A}$

36. $\cot A = \dfrac{1}{\tan A}$

37. $\sin^2 A + \cos^2 A = 1$

38. $\tan^2 A + 1 = \sec^2 A$

39. $\cot^2 A + 1 = \csc^2 A$

40. $\tan A + \cot A = \dfrac{\sin^2 A + \cos^2 A}{\sin A \cos A}$

State the quadrant in which the terminal side of $\angle A$ is located.

41. $\sin A > 0, \quad \cos A < 0$

42. $\sin A < 0, \quad \cos A < 0$

43. $\sin A < 0, \quad \cos A > 0$

44. $\sin A > 0, \quad \tan A > 0$

45. $\sin A < 0, \quad \tan A > 0$

46. $\sec A < 0, \quad \csc A > 0$

47. $\cos A < 0, \quad \cot A > 0$

48. $\sec A > 0, \quad \cot A < 0$

13.3 Values of trigonometric functions of special angles

In specifying function values for sine, cosine, tangent, and so forth, it is customary to use the symbolism $\sin A$, $\cos A$, $\tan A$, when A is identified by giving the degree measure of A. Thus, $\cos 25°$ represents the cosine of any angle measuring 25°.

From the definitions of the functions, we can see that because any point on the positive x-axis has coordinates of the form $(x, 0)$, with $x > 0$,

$$\sin 0^\circ = \frac{y}{\sqrt{x^2 + y^2}} = \frac{0}{\sqrt{x^2 + 0^2}} = \frac{0}{x} = 0,$$

$$\cos 0^\circ = \frac{x}{\sqrt{x^2 + y^2}} = \frac{x}{\sqrt{x^2 + 0^2}} = \frac{x}{x} = 1,$$

$$\tan 0^\circ = \frac{y}{x} = \frac{0}{x} = 0,$$

$$\csc 0^\circ = \frac{\sqrt{x^2 + y^2}}{y} = \frac{\sqrt{x^2 + 0^2}}{0} \quad \text{(not defined)},$$

$$\sec 0^\circ = \frac{\sqrt{x^2 + y^2}}{x} = \frac{\sqrt{x^2 + 0^2}}{x} = \frac{x}{x} = 1,$$

$$\cot 0^\circ = \frac{x}{y} = \frac{x}{0}, \quad \text{(not defined)}.$$

Similar considerations of angles measuring 90°, 180°, and 270° produce the values in the following table.

TABLE **13.1**

$\angle A$	$\sin A$	$\cos A$	$\tan A$	$\cot A$	$\sec A$	$\csc A$
0°	0	1	0	not defined	1	not defined
90°	1	0	not defined	0	not defined	1
180°	0	−1	0	not defined	−1	not defined
270°	−1	0	not defined	0	not defined	−1

To find values of the functions of some angles with other measures, consider Figure 13.6, which shows a familiar 30°–60° right triangle whose hypotenuse

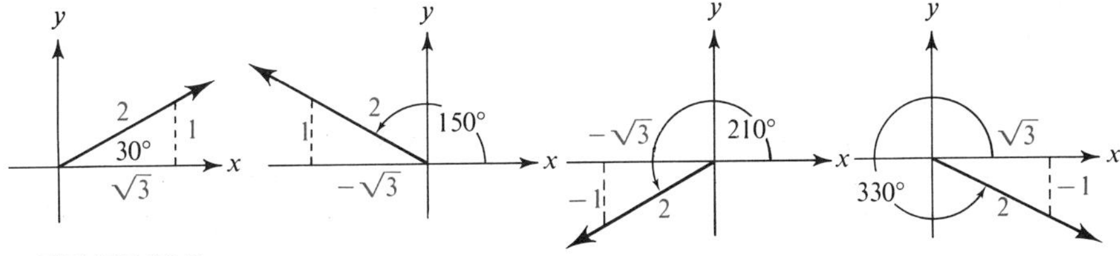

FIGURE **13.6**

lies on the terminal side of an angle A in standard position. Using the definitions of sin A and cos A, we have

$$\sin 30^\circ = \frac{1}{2}, \quad \sin 150^\circ = \frac{1}{2}, \quad \sin 210^\circ = -\frac{1}{2}, \quad \sin 330^\circ = -\frac{1}{2},$$

$$\cos 30^\circ = \frac{\sqrt{3}}{2}, \quad \cos 150^\circ = \frac{-\sqrt{3}}{2}, \quad \cos 210^\circ = \frac{-\sqrt{3}}{2}, \quad \cos 330^\circ = \frac{\sqrt{3}}{2}.$$

Values for these same angles can be determined for the remaining trigonometric functions. Then, by using the same triangle with the 60° angle in standard position (Figure 13.7), we can obtain function values for angles measuring 60°, 120°, 240°, and 300°.

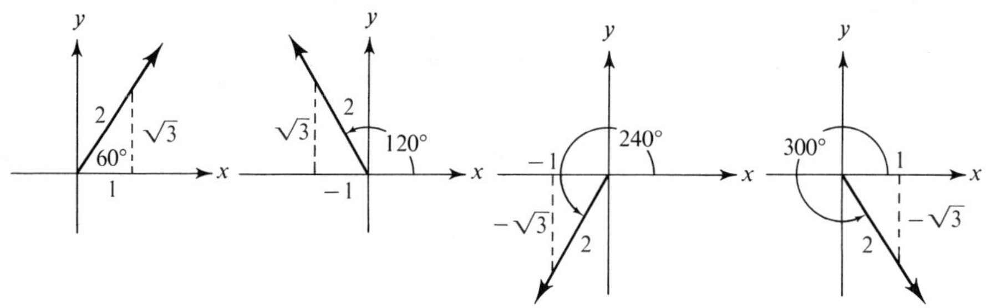

FIGURE **13.7**

Finally, using the fact that the sides of an isosceles 45°–45°–90° triangle are in the ratio $1 : 1 : \sqrt{2}$, we can obtain function values for angles measuring 45°, 135°, 225°, and 315° (Figure 13.8).

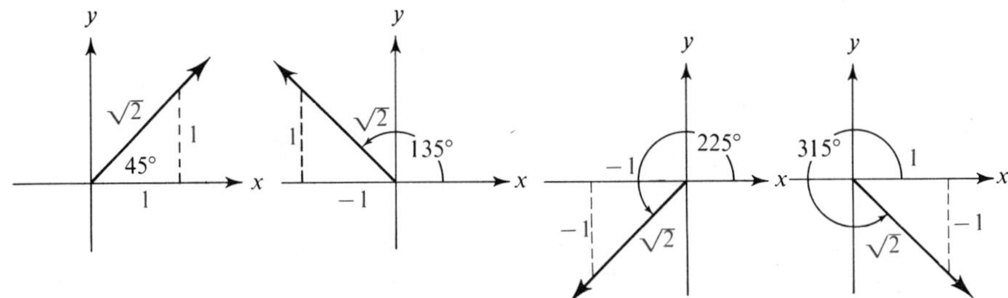

FIGURE **13.8**

The foregoing information about function values for special angles measuring between 0° and 180° is listed in Table 13.2. We have restricted our attention to angles with degree measures in this interval because we are

TABLE **13.2**

$\angle A$	$\sin A$	$\cos A$	$\tan A$	$\cot A$	$\sec A$	$\csc A$
0°	0	1	0	not defined	1	not defined
30°	$\frac{1}{2} = 0.50$	$\frac{\sqrt{3}}{2} = 0.87$	$\frac{\sqrt{3}}{3} = 0.58$	$\sqrt{3} = 1.73$	$\frac{2\sqrt{3}}{3} = 1.15$	2
45°	$\frac{\sqrt{2}}{2} = 0.71$	$\frac{\sqrt{2}}{2} = 0.71$	1	1	$\sqrt{2} = 1.41$	$\sqrt{2} = 1.41$
60°	$\frac{\sqrt{3}}{2} = 0.87$	$\frac{1}{2} = 0.50$	$\sqrt{3} = 1.73$	$\frac{\sqrt{3}}{3} = 0.58$	2	$\frac{2\sqrt{3}}{3} = 1.15$
90°	1	0	not defined	0	not defined	1
120°	$\frac{\sqrt{3}}{2} = 0.87$	$-\frac{1}{2} = -0.50$	$-\sqrt{3} = -1.73$	$-\frac{\sqrt{3}}{3} = -0.58$	-2	$\frac{2\sqrt{3}}{3} = 1.15$
135°	$\frac{\sqrt{2}}{2} = 0.71$	$-\frac{\sqrt{2}}{2} = -0.71$	-1	-1	$-\sqrt{2} = -1.41$	$\sqrt{2} = 1.41$
150°	$\frac{1}{2} = 0.50$	$-\frac{\sqrt{3}}{2} = -0.87$	$-\frac{\sqrt{3}}{3} = -0.58$	$-\sqrt{3} = -1.73$	$-\frac{2\sqrt{3}}{3} = -1.15$	2
180°	0	-1	0	not defined	-1	not defined

primarily interested in angles of triangles and the measure of any interior angle of a triangle must be less than 180°. Note that in some cases in Table 13.2 we are using the = symbol for an approximation as we did for approximations in Chapter 11 on logarithms.

EXERCISE 13.3

State each trigonometric function value. If the value is undefined, so state.

EXAMPLE

cos 135°

SOLUTION

From Table 13.2,

$$\cos 135° = -\frac{\sqrt{2}}{2} = \mathbf{-0.71}.$$

1. sin 60°	2. tan 30°	3. cos 45°	4. sec 0°
5. sin 90°	6. tan 90°	7. cos 120°	8. sec 120°
9. csc 150°	10. sin 0°	11. tan 180°	12. cot 90°
13. sec 150°	14. csc 0°	15. cot 45°	16. sin 135°

State all measures between 0° and 180° inclusive for the angle A with function value as given.

EXAMPLE

$\sec A = -\sqrt{2}$

SOLUTION

From Table 13.2,

$$\sec 135° = -\sqrt{2}, \quad \text{so that} \quad \angle A = \mathbf{135°}.$$

17. $\sin A = \frac{1}{2}$
18. $\cos A = -\frac{1}{2}$
19. $\tan A = 1$
20. $\cot A = -1$
21. $\sec A = \frac{2\sqrt{3}}{3}$
22. $\csc A = 1$
23. $\cos A = -\frac{\sqrt{3}}{2}$

24. $\sin A = \frac{\sqrt{3}}{2}$ 25. $\tan A = 0$ 26. $\cot A = 0$

27. $\sin A = 0.71$ 28. $\csc A = 1.41$ 29. $\cot A = -1.73$

30. $\sin A = 0.71$ 31. $\cos A = -0.87$ 32. $\sec A = -1$

Solve for x. Give answer to nearest tenth.

EXAMPLE

$2x + \sin 30° = \cos 30°$

SOLUTION

Since $\sin 30° = 0.50$ and $\cos 30° = 0.87$, we have

$$2x + 0.50 = 0.87,$$
$$2x = 0.87 - 0.50,$$
$$2x = 0.37,$$
$$x = \frac{0.37}{2} = 0.185.$$

Hence, $x = \mathbf{0.2}$.

33. $x = 2 \sec 30°$
34. $x = 5 \cos 60°$
35. $x = -4 \cos 120°$
36. $x = -3 \sin 150°$
37. $x = 2 \sin 30° + \csc 45°$
38. $x = \sin 135° - \cos 60°$
39. $x = \tan 30° + \tan 60°$
40. $x = 2 \cot 30° + \sec 45°$
41. $2x + \sin 45° = \sin 135°$
42. $2x + \tan 45° = \tan 135°$
43. $3x - 2 \sin 30° = 3 \tan 45°$
44. $5x - 3 \sin 150° = \cos 120°$

13.4 Trigonometric tables

To find values of trigonometric functions for angles other than those special angles discussed in Section 13.3, we use a table of function values. One such table is located on page 418, and an excerpt is shown on page 320.

This table gives the function values to four decimal digits for angles with measures between 0° and 90° at 10 minute intervals. It is used in much the same way as a table of logarithms. Thus, to find sin 32° 10′, for example, we locate 32° 10′ in the left-hand column and read sin 32° 10′ (equal to 0.5324) opposite this entry in the column headed "sin A." Unlike a table of logarithms, however, for values of A between 45° and 90° this table is read from

∠ A	sin A	csc A	tan A	cot A	sec A	cos A	
32° 00′	0.5299	1.887	0.6249	1.600	1.179	0.8480	58° 00′
10	.5324	.878	.6289	.590	.181	.8465	50
20	.5348	.870	.6330	.580	.184	.8450	40
30	.5373	.861	.6371	.570	.186	.8434	30
	cos A	sec A	cot A	tan A	csc A	sin A	∠ A

bottom to top. For example, to find tan 57° 40′, we locate 57° 40′ in the right-hand column and read tan 57° 40′ (equal to 1.580) directly opposite this entry in the column labeled "tan A" at the bottom.

The table can also be used to find the measure of an angle when a function value for this angle is known. Thus, given that sec $A = 1.186$, we locate 1.186 in the column labeled sec A and read $\angle A = 32°\ 30'$ directly across from this in the angle-measure column.

When interpolating between table entries in a table of trigonometric functions, it is necessary to be careful to observe whether successive entries are increasing or decreasing, and to add or subtract increments to or from table entries as necessary. An example in the exercises following illustrates the procedure.

Another table of trigonometric functions is found on page 424. This table, an excerpt of which is shown opposite, also gives function values for angles with measures between 0° and 90°, but at intervals of 0.1 of a degree instead of intervals of 10′. It is used in exactly the same way as Table III on page 418. For example, cos 24.7° = 0.9085 and sec 65.3° = 2.3931. Also, if we are given that tan $A = 0.4536$, then $\angle A = 24.4°$.

To find values for trigonometric functions of a given angle A whose terminal side lies in the second quadrant ($90° < A < 180°$), we can, as suggested by Figure 13.9, simply look up the values of the supplement of the given

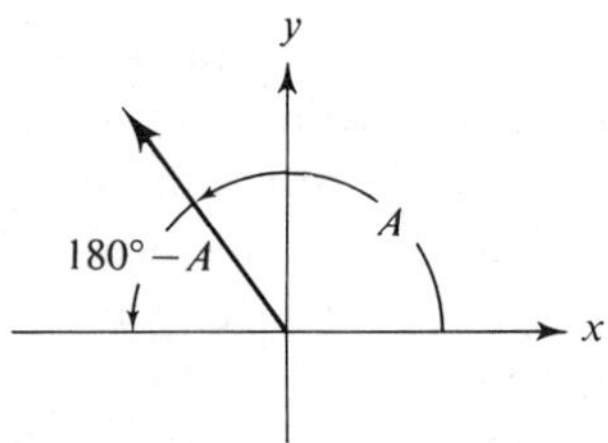

FIGURE **13.9**

∠ A	Sin A	Tan A	Cot A	Cos A	Sec A	Csc A	
24.0	0.4067	0.4452	2.246	0.9135	1.095	2.459	66.0
.1	.4083	.4473	2.236	.9128	1.096	2.449	.9
.2	.4099	.4494	2.225	.9121	1.096	2.440	.8
.3	.4115	.4515	2.215	.9114	1.097	2.430	.7
.4	.4131	.4536	2.204	.9107	1.098	2.421	.6
.5	.4147	.4557	2.194	.9100	1.099	2.411	.5
.6	.4163	.4578	2.184	.9092	1.100	2.402	.4
.7	.4179	.4599	2.174	.9085	1.101	2.393	.3
.8	.4195	.4621	2.164	.9078	1.102	2.384	.2
.9	.4210	.4642	2.154	.9070	1.102	2.375	.1
25.0	0.4226	0.4663	2.145	0.9063	1.103	2.366	65.0
	Cos A	Cot A	Tan A	Sin A	Csc A	Sec A	∠ A

angle and then attach the appropriate sign, + or −. For example,

$$\begin{aligned}\cot 155.1^\circ &= -\cot(180 - 155.1^\circ)\\ &= -\cot 24.9^\circ\\ &= -2.154,\end{aligned}$$

where the minus sign was attached because the cotangent of an angle in the second quadrant is negative.

EXERCISE 13.4

Use Table III or Table IV to find the given function value.

EXAMPLE

tan 122°

SOLUTION

Because the terminal side of an angle measuring 122° is in the second quadrant, it is not an entry in either table. Therefore, we look up the tangent of the supplement of 122°, namely, tan 58°. From either table,

$$\tan 58^\circ = 1.600.$$

Solution continued overleaf.

Then, because tan 122° < 0, we have

$$\tan 122° = \mathbf{-1.600.}$$

1.	sin 43° 40′	2.	cos 12° 20′	3.	cos 44° 50′	4.	sin 0° 40′
5.	cos 84° 10′	6.	sin 75° 30′	7.	sin 132°	8.	cos 154°
9.	tan 100° 10′	10.	cot 178° 50′	11.	tan 43.2°	12.	cot 31.7°
13.	sec 1.8°	14.	csc 37.5°	15.	tan 55.2°	16.	csc 73.4°
17.	cot 45.3°	18.	sec 89.1°	19.	csc 90.2°	20.	sec 111.6°

Use Table III to find the measures of all angles A between 0° and 180° for which the given equation is true.

EXAMPLE

$\cos A = 0.5831$

SOLUTION

Since cos A is given as positive, we know that A must be an angle whose terminal side is in Quadrant I. In the table, we find, in the column labeled "cos A" at the bottom, the entry 0.5831. Reading directly across to the right, we find that

$$\angle A = \mathbf{54° \ 20'.}$$

21. $\cos A = 0.9757$
22. $\sin A = 0.9886$
23. $\tan A = 0.8243$
24. $\cot A = 2.112$
25. $\csc A = 1.044$
26. $\sec A = 10.13$
27. $\sin A = 0.3529$
28. $\cos A = 0.7735$

Use Table IV to find the measuers of all angles A between 0° and 180° for which the given equation is true.

29. $\sin A = 0.5255$
30. $\tan A = 0.7080$
31. $\sec A = 1.1512$
32. $\csc A = 1.4192$
33. $\cos A = 0.9839$
34. $\cot A = 0.07168$
35. $\cos A = 0.7059$
36. $\sin A = 0.00349$

Use Table III and linear interpolation to find each function value to four significant digits.

EXAMPLE

tan 42° 17′

SOLUTION

We first find tan 42° 10′ and tan 42° 20′. The use of a tabular arrangement is helpful.

A	tan A
42° 10′	0.9057
42° 17′	?
42° 20′	0.9110

(The 7′ difference from 42° 10′ to 42° 17′ corresponds to d; the 10′ difference from 42° 10′ to 42° 20′ corresponds to 0.0053.)

Using linear interpolation we have

$$\frac{d}{0.0053} = \frac{7}{10},$$

from which

$$d = \frac{7}{10}(0.0053) = 0.0037.$$

Then, adding 0.0037 to 0.9057, we have

$$\tan 42° 17' = 0.9057 + 0.0037 = \mathbf{0.9094}.$$

37. sin 22° 18′ 38. cos 54° 43′ 39. tan 115° 45′
40. cot 121° 6′ 41. sec 63° 59′ 42. csc 77° 27′

Use Table IV and linear interpolation to find each function value to four significant digits.

43. cos 27.23° 44. sin 71.34° 45. cot 20.35°
46. tan 68.19° 47. csc 98.83° 48. sec 142.67°

Use Table III and linear interpolation to find to the nearest minute the measures of all angles A between 0° and 180° for which the given equation is true.

EXAMPLE

$\sin A = 0.3125$

SOLUTION

Since $\sin A$ is positive in both Quadrant I and Quadrant II, we have two solutions to the equation. In the sine column in Table III, we can determine that

Solution continued overleaf.

$\sin 18^\circ 10' = 0.3118$ and $\sin 18^\circ 20' = 0.3145$. We can use these values to set up the diagram as shown.

$$\begin{array}{c|c} A & \sin A \\ \hline 10' \left\{ \begin{array}{l} d \left\{ \begin{array}{l} 18^\circ 10' \\ ? \end{array} \right. \\ \quad 18^\circ 20' \end{array} \right. & \left. \begin{array}{l} \left. \begin{array}{l} 0.3118 \\ 0.3125 \end{array} \right\} 7 \\ 0.3145 \end{array} \right\} 27 \end{array}$$

We wish to find the value d such that

$$\frac{d}{10} = \frac{7}{27},$$

from which

$$d = \frac{7}{27}(10) = \frac{70}{27} \approx 3.$$

Adding $3'$ to $18^\circ 10'$, we have $\angle A = \mathbf{18^\circ 13'}$. In Quadrant II, the angle we seek is

$$180^\circ - 18^\circ 13' = \mathbf{161^\circ 47'}.$$

49. $\cos A = 0.2935$
50. $\sin A = 0.5530$
51. $\cot A = 2.382$
52. $\tan A = 1.415$
53. $\sin A = 1.362$
54. $\csc A = 1.234$

Use Table IV and linear interpolation to find to the nearest 0.01° angles A between 0° and 180° for which the given equation is true.

55. $\sin A = 0.8150$
56. $\cos A = 0.5183$
57. $\sec A = 3.0270$
58. $\cot A = 0.12550$
59. $\tan A = 0.3222$
60. $\csc A = 1.9800$

13.5 Right triangles

As shown in Figure 13.10a, any point (b, a) on the terminal side of an angle A in standard position determines a right triangle with sides of length b and a, and with hypotenuse of length c. Therefore, in the special case where A is an acute angle in a right triangle:

$$\sin A = \frac{\text{length of side opposite } \angle A}{\text{length of hypotenuse}} = \frac{a}{c},$$

$$\cos A = \frac{\text{length of side adjacent to } \angle A}{\text{length of hypotenuse}} = \frac{b}{c},$$

$$\tan A = \frac{\text{length of side opposite } \angle A}{\text{length of side adjacent to } \angle A} = \frac{a}{b},$$

$$\cot A = \frac{\text{length of side adjacent to } \angle A}{\text{length of side opposite } \angle A} = \frac{b}{a},$$

$$\sec A = \frac{\text{length of hypotenuse}}{\text{length of side adjacent to } \angle A} = \frac{c}{b},$$

$$\csc A = \frac{\text{length of hypotenuse}}{\text{length of side opposite } \angle A} = \frac{c}{a}.$$

The above relationships are equally applicable for right triangles in any position. For example, they are also valid for the six trigonometric ratios of

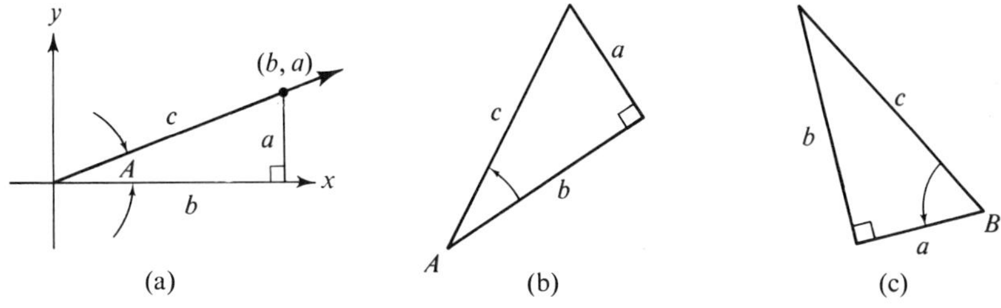

FIGURE **13.10**

angle A in (b) of the figure. In (c), the trigonometric ratios of angle B are:

$$\sin B = \frac{\text{length of side opposite } \angle B}{\text{length of hypotenuse}} = \frac{b}{c},$$

$$\cos B = \frac{\text{length of side adjacent to } \angle B}{\text{length of hypotenuse}} = \frac{a}{c},$$

$$\tan B = \frac{\text{length of side opposite } \angle B}{\text{length of side adjacent to } \angle B} = \frac{b}{a},$$

$$\cot B = \frac{\text{length of side adjacent to } \angle B}{\text{length of side opposite } \angle B} = \frac{a}{b},$$

$$\sec B = \frac{\text{length of hypotenuse}}{\text{length of side adjacent to } \angle B} = \frac{c}{a},$$

$$\csc B = \frac{\text{length of hypotenuse}}{\text{length of side opposite } \angle B} = \frac{c}{b}.$$

As we can see, each trigonometric function of an acute angle in a right triangle can be expressed in terms of ratios involving the lengths of the hypotenuse and the sides opposite and adjacent to the angle.

We can use the foregoing trigonometric ratios to solve a right triangle, that is, to determine missing parts of the triangle when given some information about certain other parts (length of sides or measures of angles). For example, to solve $\triangle ABC$ in Figure 13.11, given that $c = 40$ and $\angle A = 28°$, we can use

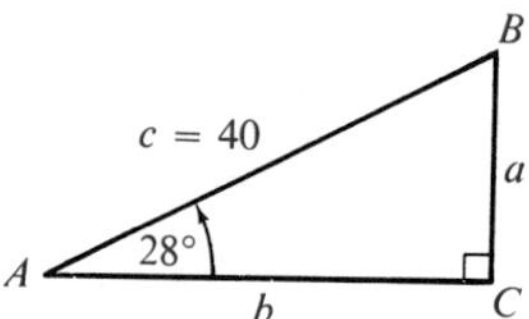

FIGURE **13.11**

the fact that

$$\sin A = \frac{a}{c} \quad \text{and} \quad \cos A = \frac{b}{c}$$

to determine a and b. Thus,

$$\sin 28° = \frac{a}{40} \quad \text{and} \quad \cos 28° = \frac{b}{40},$$

or

$$a = 40 \sin 28° \quad \text{and} \quad b = 40 \cos 28°.$$

Using Table III or Table IV, we have (to the nearest tenth of a unit),

$$a = 40(0.4695) = 18.8 \quad \text{and} \quad b = 40(0.8829) = 35.3.$$

Since the acute angles of a right triangle are complements of each other, we know that

$$\angle B = 90° - \angle A = 90° - 28° = 62°.$$

In solving a right triangle, it is simplest to use functions in which the numerator of the trigonometric ratio is the unknown, because it is easier to multiply by a function value than to divide by the reciprocal of the same value. For example, to find c in the right triangle in which we are given $b = 25$ and $\angle A = 52°$ it is easier to use the fact that

$$\sec A = \frac{c}{b}$$

than it would be to employ the cosine ratio,

$$\cos A = \frac{b}{c}.$$

This is because, on the one hand,

$$c = b \sec A,$$

while on the other,

$$c = \frac{b}{\cos A}.$$

EXERCISE 13.5

Solve the right triangle which has the given parts. Use Table IV and give angle measures and lengths to the nearest tenth of a unit.

EXAMPLE

$a = 56, \quad b = 67$

SOLUTION

Make a sketch and label the known parts.

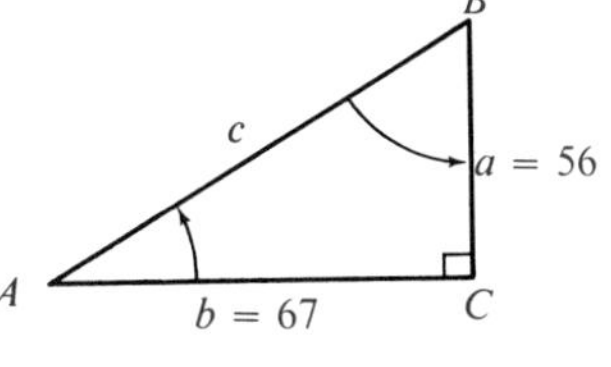

To find $\angle A$, we use the fact that

$$\tan A = \frac{56}{67} = 0.8358.$$

From Table IV, $\angle A = 39.9°$. Since $\angle C = 90°$, to find $\angle B$ we have

$$\angle B = 90° - 39.9° = 50.1°.$$

Finally, to find c, we can use the fact that

$$\sec A = \frac{c}{67},$$

so that

$$c = 67 \sec A = 67 \sec 39.9°$$
$$= 67(1.304) = 87.4.$$

The missing parts of the triangle, then, are

$$A = \mathbf{39.9°}, \quad B = \mathbf{50.1°}, \quad \text{and} \quad c = \mathbf{87.4}.$$

1. $a = 15, b = 23$
2. $a = 17, c = 31$
3. $b = 24, c = 27$
4. $a = 112, b = 92$
5. $c = 28, \angle A = 43°$
6. $c = 72, \angle B = 65°$
7. $\angle A = 14°, a = 24$
8. $\angle B = 81°, b = 73$
9. $\angle B = 27.3°, a = 214$
10. $\angle A = 67.7°, b = 165$
11. $a = 28.3, c = 37.9$
12. $a = 121.3, b = 203.4$
13. $\angle A = 27.5°, c = 93.8$
14. $\angle B = 77.7°, a = 112.5$
15. $\angle B = 47.3°, b = 81.1$
16. $\angle A = 19.8°, c = 427.3$

Using Table III, solve the right triangle whose parts are given. Give angle measures to the nearest minute and lengths to the nearest hundredth of a unit.

17. $a = 38, c = 114$
18. $b = 18, c = 40$
19. $a = 84, b = 112$
20. $a = 98, b = 52$
21. $\angle A = 27°, c = 112$
22. $\angle B = 53°, a = 227$
23. $\angle A = 15°, b = 36$
24. $\angle B = 45°, a = 84$
25. $\angle A = 18° 20', c = 315$
26. $\angle B = 77° 40', b = 248$
27. $a = 83.7, b = 106.5$
28. $c = 153.8, a = 42.3$
29. $\angle A = 28° 18', c = 84.3$
30. $\angle B = 62° 27', a = 61.2$
31. $\angle B = 54° 54', b = 27.81$
32. $\angle A = 47° 36', b = 84.28$

The relationships given in Exercises 33–38 are applicable for all angles. Use the information in the figure to show that they are valid for acute angles in a right triangle.

33. $\sin A = \cos B$
34. $\cos A = \sin B$
35. $\tan A = \cot B$
36. $\cot A = \tan B$
37. $\sec A = \csc B$
38. $\csc A = \sec B$

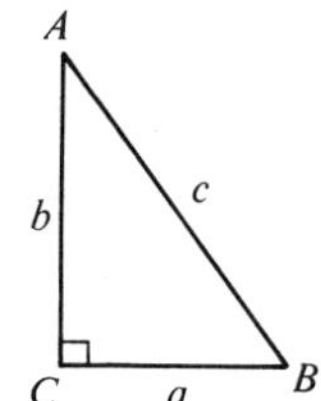

13.6 Applications of right triangles

We can obtain solutions to many real-life mathematics problems more easily by using trigonometric ratios than by any other mathematical means. The exercise set that follows provides several examples of such problems.

EXERCISE 13.6

In Exercises 1–10 give all angle measures to the nearest 10 minutes and all lengths to the nearest tenth of a unit.

EXAMPLE

The arm of a crane makes an angle measuring 67° with the level ground and is 42 feet in length. If the base of the arm is 4 feet above ground level, how far above ground level is the tip of the crane's arm?

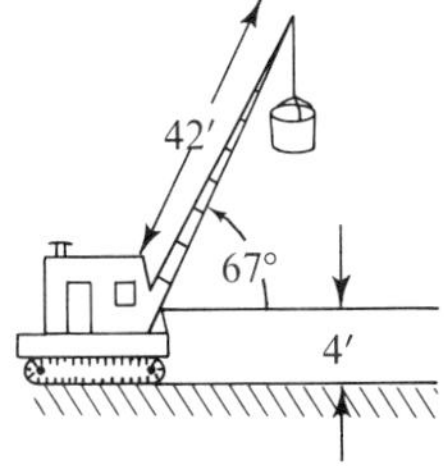

SOLUTION

Make a simple sketch and label the known parts. Let x represent the height of the tip of the crane above its base. Then,

$$\frac{x}{42} = \sin 67°$$

so that

$$x = 42 \sin 67°$$
$$= 42(0.9205)$$
$$= 38.66.$$

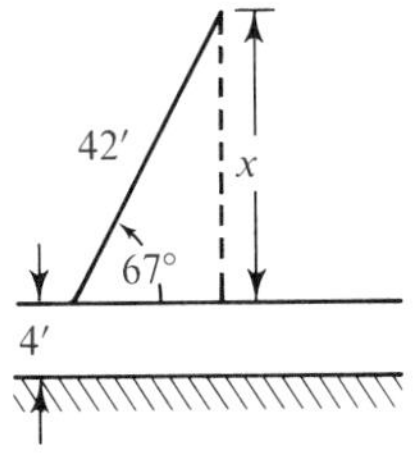

Since the base of the crane is 4 feet above ground level, the tip of the crane to the nearest tenth of a foot is $38.7 + 4 = $ **42.7 feet** above ground level.

1. How long should an escalator be if it is to make an angle measuring 33° with the horizontal and carry persons through a vertical distance of 35 feet between floors?

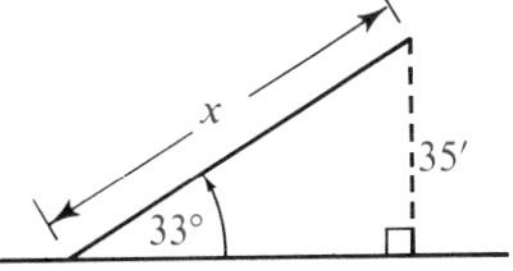

2. An airplane descends on a path making a constant angle measuring 12° with the level ground. How far from its point of impact with the ground will it be when it is 3000 feet above the ground?

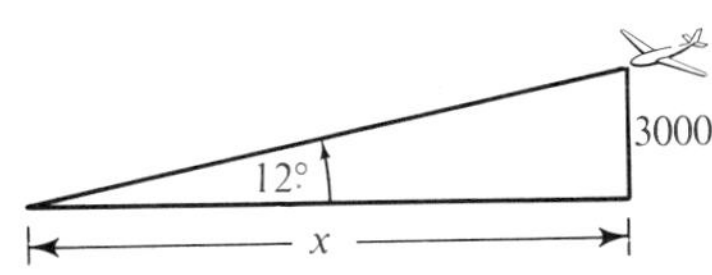

3. What angle should the bottom of a swimming pool 40 feet in length make with the horizontal if it is desired to have a steady increase in depth from 3 feet to 9 feet?

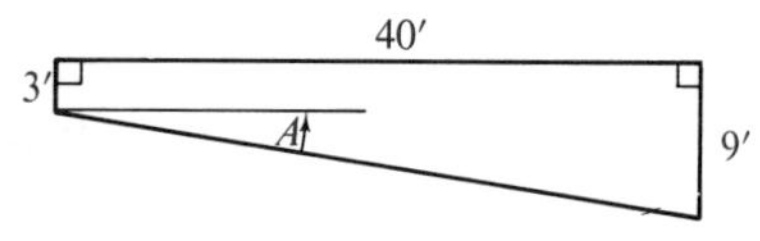

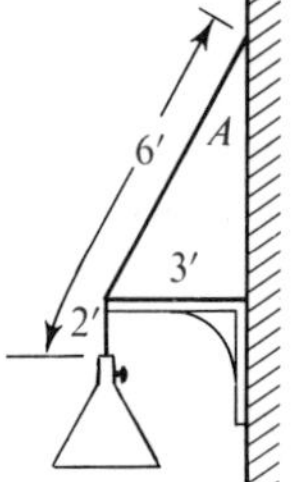

4. A lamp is hung from a wall as pictured at the right. If the support bracket is 3 feet in length and the cord from the wall to the lamp is 6 feet long, at what distance above the support bracket will the cord meet the wall? What angle will the cord make with the wall?

5. How long should a rafter be cut if it is to extend 2 feet beyond the wall of a building and rise 8 feet over a run of 15 feet? What angle will the rafter make with the vertical?

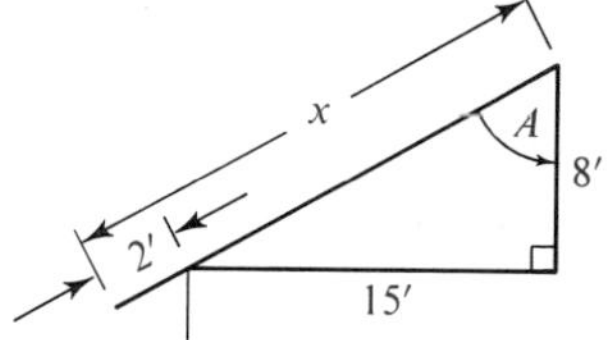

6. A sailor observes that the measure of the angle of elevation of the top of a 120-foot lighthouse to the ship is 6°. How far, in feet, is the ship from the lighthouse?

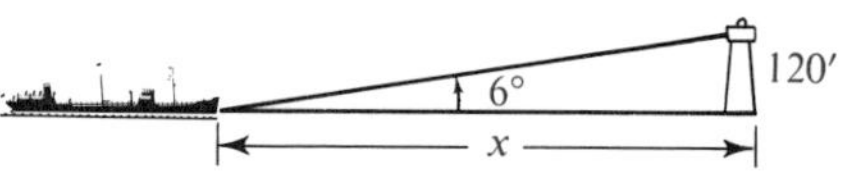

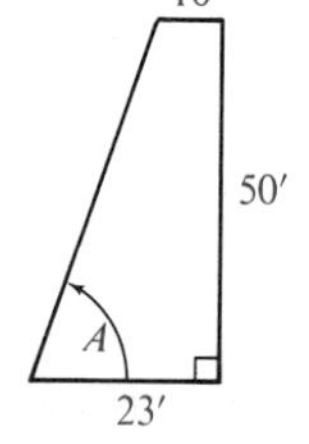

7. What angle will the slanting face of a trapezoidal dam make with the surface of the ground if the dam is 50 feet tall with a top 10 feet in width and a base 23 feet in width?

8. A 25-foot ladder is leaning against a wall so that the foot of the ladder is 7 feet from the base of the wall. How high up the wall does the ladder touch the wall? What angle does the ladder make with the ground?

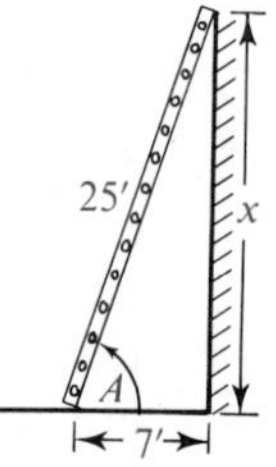

9. The canopy of a patio is 14 feet long and forms an angle measuring 78° 10′ with the wall of a house. If the canopy is attached to the house at a height of 12 feet, how long should a support post at the extremity of the canopy be?

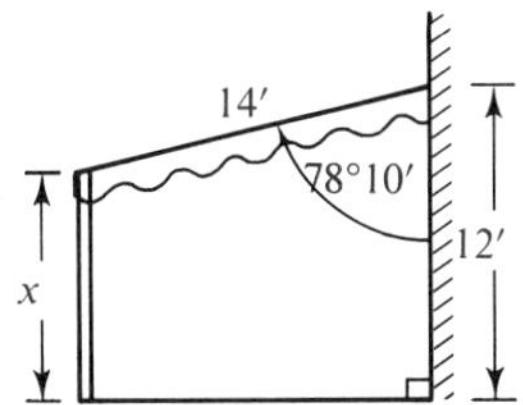

10. If the slant-height of a truncated pyramid is 108 feet and if the side forms an angle measuring 72° 20′ with the ground, what is the height of the pyramid?

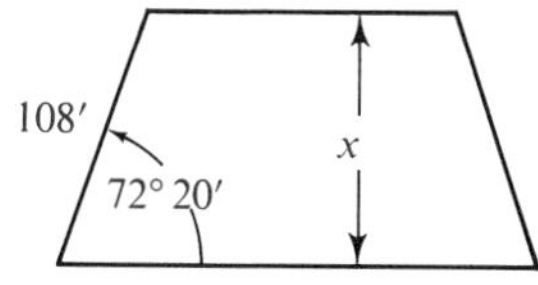

In the following exercises give angle measures and lengths to the nearest tenth of a unit.

11. A circle has a radius of 22.4 inches. What is the measure of the central angle that subtends a chord 16.8 inches in length? (*Hint:* Form two right triangles.)

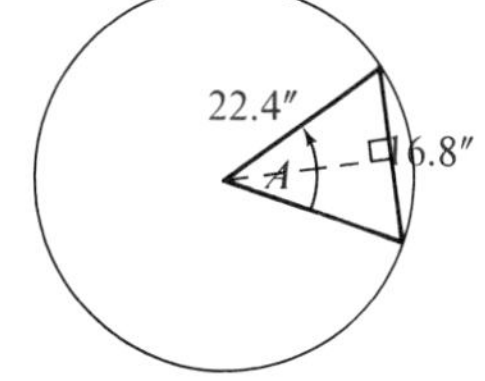

12. A ship leaves harbor sailing at 13.2 knots on a course forming an angle measuring 42.6° with a straight coastline. How many nautical miles is the ship from the shore after 2 hours?

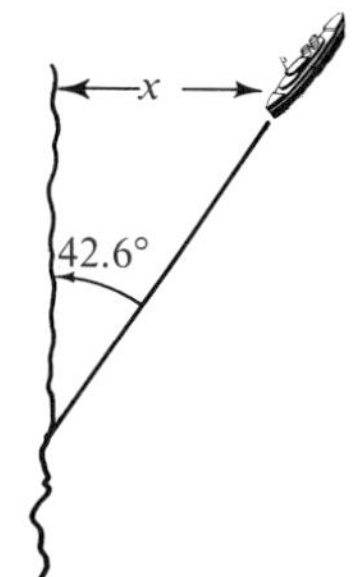

13. From a window 84.3 feet above the ground, the angle of depression to a window 47.1 feet above the ground in another building measures 37°. How far apart are the buildings?

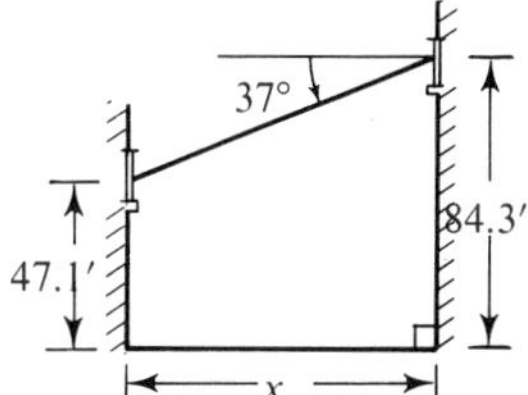

14. From a point 250 feet from the bottom of a building, the angle of elevation to the base of a flagpole on top of the building measures 28.3°. From the same point, the angle of elevation to the top of the flagpole measures 37.7°. How tall is the flagpole?

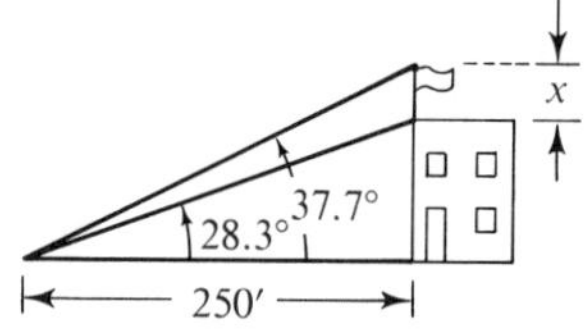

15. A ship is sailing directly away from an observer who is at a height of 80 feet. If the angle of depression to one end of the ship measures 5.9° and the angle of depression to the other end measures 6.1°, how long is the ship?

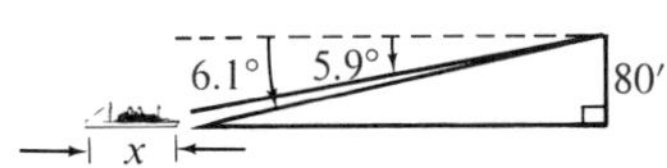

16. From a window 35.7′ above the ground, the angle of elevation to the top of a building across the street measures 42.3°, and the angle of depression to the front of the building measures 49.5°. How tall is the building across the street?

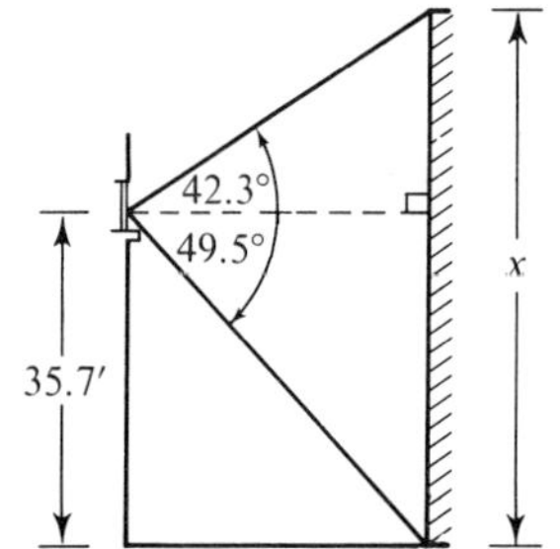

17. From a point 1284′ from an end of a lake, the angles measured to the ends of the lake are 52.3° and 61.5°, respectively, as shown in the figure. How long is the lake?

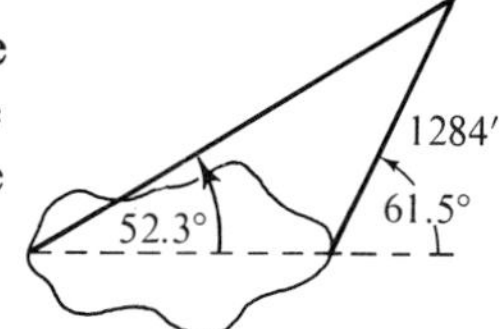

18. An isosceles triangle is made from a piece of wire 16.8 inches long. One of the angles formed is 96°. What is the length of each side of the triangle? (*Hint:* $\angle A = \angle B$; form two right triangles.)

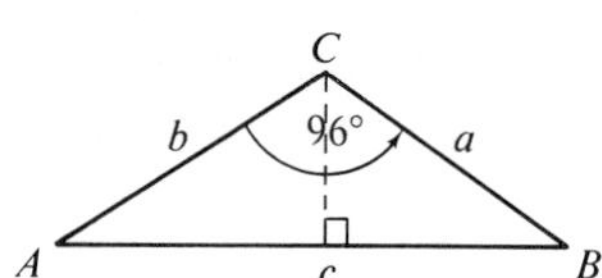

13.7 The law of sines

In addition to their application to right triangles, trigonometric functions are also useful in working with oblique triangles.

One such application of trigonometric functions is the **law of sines**. In more advanced courses, it is shown that for any triangle ABC,

$$\frac{\sin A}{a} = \frac{\sin B}{b} = \frac{\sin C}{c},$$

FIGURE **13.12**

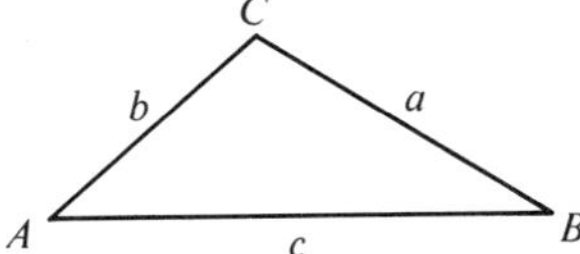

or

$$\frac{a}{\sin A} = \frac{b}{\sin B} = \frac{c}{\sin C}.$$

Thus, if we are given the measures of two sides and an angle opposite one of these sides, or of two angles and a side, we can use the law of sines to find the remaining measures for the triangle. For example, if we know that in triangle ABC in Figure 13.12, $b = 10$, $\angle B = 32°$, and $\angle A = 40°$, we can determine $\angle C$ from the fact that

$$\angle A + \angle B + \angle C = 180°,$$
$$40° + 32° + \angle C = 180°,$$

from which

$$\angle C = 108°.$$

Then, to find a and c, we use the law of sines to obtain

$$\frac{a}{\sin 40°} = \frac{10}{\sin 32°} = \frac{c}{\sin 108°},$$

from which

$$a = \frac{10 \sin 40°}{\sin 32°} \quad \text{and} \quad c = \frac{10 \sin 108°}{\sin 32°},$$

Using Table III or Table IV, and that $\sin 108° = \sin(180° - 108°) = \sin 72°$, we have,

$$a = \frac{10(0.6428)}{0.5299} \quad \text{and} \quad c = \frac{10(0.9511)}{0.5299},$$

or, to the nearest tenth,

$$a = 12.1 \quad \text{and} \quad b = 17.9.$$

In a complete course in trigonometry, it is shown, that, under certain given conditions, there may be two triangles with a given set of parts. We do not consider such conditions in this book.

EXERCISE 13.7

Use Table III to solve the triangle having the given parts. State angle measures to nearest ten minutes and lengths to the nearest tenth of a unit.

EXAMPLE

$A = 52° 10', C = 92° 30', c = 36$

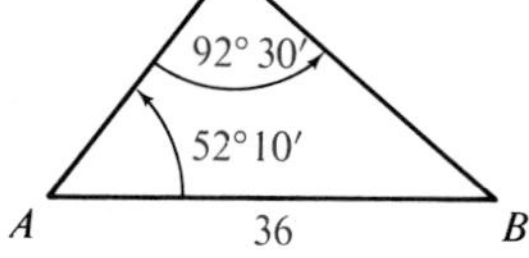

SOLUTION

Make a sketch and label the known parts. Since

$$\angle A + \angle B + \angle C = 180°,$$

we have

$$52° 10' + \angle B + 92° 30' = 180°,$$

from which

$$\angle B = 180° - 144° 40' = 35° 20'.$$

Then, using the law of sines, we have

$$\frac{a}{\sin 52° 10'} = \frac{b}{\sin 35° 20'} = \frac{36}{\sin 92° 30'},$$

or

$$\frac{a}{\sin 52° 10'} = \frac{36}{\sin 92° 30'} \quad \text{and} \quad \frac{b}{\sin 35° 20'} = \frac{36}{\sin 92° 30'}.$$

Solving for a and b yields

$$a = \frac{36 \sin 52° 10'}{\sin 92° 30'} \quad \text{and} \quad b = \frac{36 \sin 35° 20'}{\sin 92° 30'}.$$

Using Table III and the fact that

$$\sin 92° 30' = \sin (180° - 92° 30') = \sin 87° 30',$$

we have

$$a = \frac{36(0.7898)}{0.9990} \quad \text{and} \quad b = \frac{36(0.5783)}{0.9990},$$

from which

$$a = 28.4 \quad \text{and} \quad b = 20.8.$$

The missing parts of the triangle are $\angle B = \mathbf{35° 20'}$, $a = \mathbf{28.4}$, and $b = \mathbf{20.8}$.

1. $a = 10$, $\angle A = 40°$, $\angle B = 70°$
2. $b = 20$, $\angle C = 12°$, $\angle B = 84°$
3. $c = 30$, $\angle A = 60°$, $\angle B = 45°$
4. $a = 25$, $\angle A = 70°$, $\angle C = 36°$

5. $c = 14.1$, $\angle A = 36° 20'$, $\angle B = 44° 40'$
6. $a = 27.3$, $\angle B = 48° 40'$, $\angle C = 57° 10'$
7. $a = 24$, $b = 40$, $\angle B = 43° 10'$
8. $b = 20$, $c = 40$, $\angle C = 82° 10'$
9. $a = 18.4$, $c = 24.3$, $\angle C = 70° 50'$
10. $b = 28.7$, $a = 12$, $\angle B = 100° 20'$

Use Table IV to solve the triangle having the given parts. State angle measures and lengths to nearest tenth of a unit.

11. $a = 10$, $\angle A = 61.4°$, $\angle B = 74°$
12. $b = 20$, $\angle B = 81.3°$, $\angle C = 15.9°$
13. $c = 31.2$, $\angle C = 15.7°$, $\angle A = 47.2°$
14. $b = 102.3$, $\angle A = 42.6°$, $\angle B = 38.8°$
15. $a = 10$, $b = 20$, $\angle B = 62.5°$
16. $b = 20$, $c = 22$, $\angle C = 82.3°$
17. $c = 40$, $b = 52.1$, $\angle B = 26.5°$
18. $a = 12.3$, $c = 6.9$, $\angle A = 82.6°$

13.8 Applications of the law of sines

In many instances, the law of sines can be used to solve practical problems involving oblique triangles.

EXERCISE 13.8

In each exercise, give lengths to the nearest tenth of a unit.

EXAMPLE

From the top of a cabin 42′ high, the angle of elevation to the top of a nearby tree measures 14.1°. From the ground level at the base of the cabin the angle of elevation to the top of the same tree measures 28.3°. How tall is the tree?

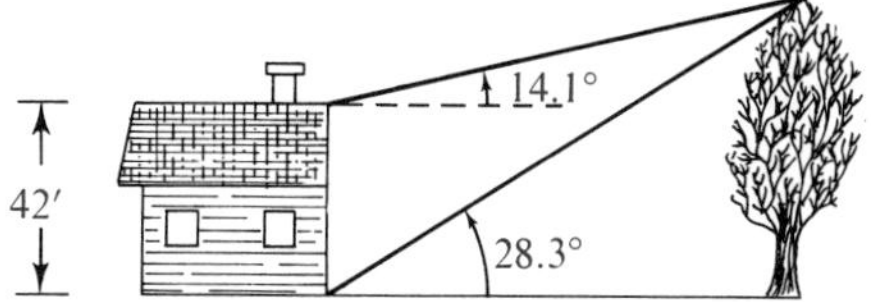

Solution overleaf.

SOLUTION

Make a sketch and label the known parts. To determine x, we need to know the length of either AC or AB. We can determine the length of AC by using the law of sines as follows.

$$\angle ADC = 14.1^\circ + 90^\circ = 104.1^\circ,$$
$$\angle DAC = 90^\circ - 28.3^\circ = 61.7^\circ.$$

Since,

$$\angle DCA + 104.1^\circ + 61.7^\circ = 180^\circ,$$

we have

$$\angle DCA = 14.2^\circ.$$

Then, from the law of sines,

$$\frac{AC}{\sin 104.1^\circ} = \frac{42}{\sin 14.2^\circ}.$$

Using Table IV and the fact that $\sin 104.1^\circ = \sin(180^\circ - 104.1^\circ) = \sin 75.9^\circ$, we have

$$\frac{AC}{0.9699} = \frac{42}{0.2453},$$

$$AC = \frac{42(0.9699)}{0.2453} \approx 166.1.$$

CB, the height, of the tree, can then be determined by observing that

$$\frac{CB}{AC} = \sin 28.3^\circ,$$

or

$$\begin{aligned} CB &= AC(\sin 28.3^\circ) \\ &= 166.1(0.4741) \\ &= 78.7 \end{aligned}$$

The tree is approximately **78.7 feet** in height.

1. The angle of elevation from a window 37 feet above the ground to the top of the building across the street measures 51.3°. From the ground directly beneath the window the angle of elevation to the top of the building measures 81.3°. Find the height of the building.

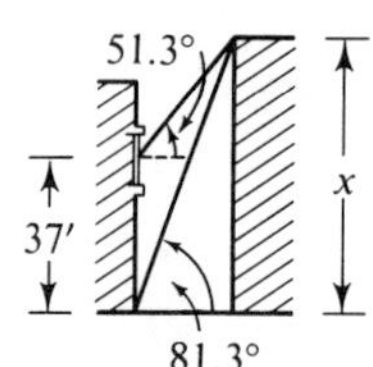

2. The angle of elevation from a point on the ground to the top of a 130′ television tower on top of a building measures 36.2°. From the same spot on the ground, the angle of elevation to the base of the tower measures 27.3°. Find the height of the building.

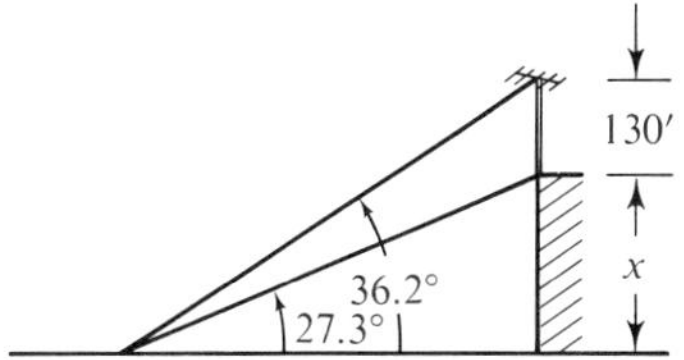

3. A slanted gravel screen forms an angle measuring 44.7° with the ground. At a distance of 6.3 feet from the bottom of the screen, a 7.8-foot support brace is attached to the screen. How far from the base of the screen does the support brace touch the ground?

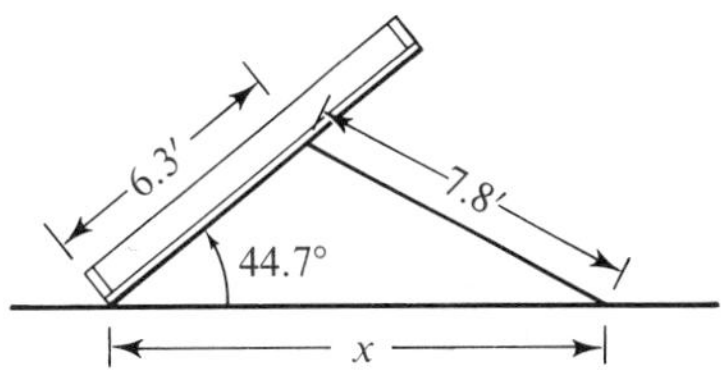

4. The bed of a dump truck forms an angle of 78.3° with the frame when in a fully raised position. If the 6.3′ dumping arm is attached to the bed 5.7′ from the frame, how far from this point of attachment does the dumping arm meet the frame?

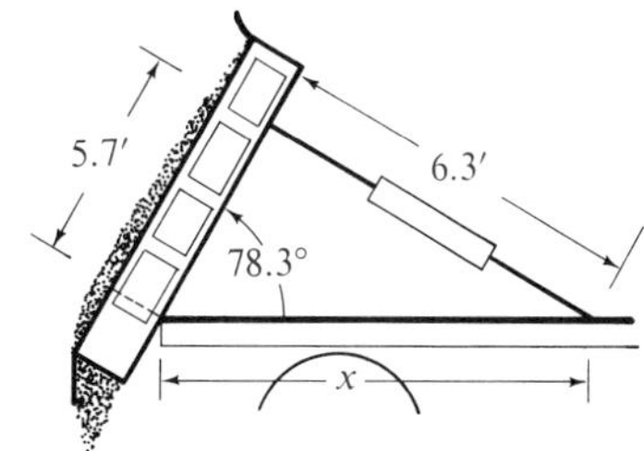

5. A ship sails on a course forming an angle measuring 23.7° with a straight coastline. After sailing 74.3 nautical miles on this course, the ship changes heading and sails back toward the coast, meeting the coast after sailing another 98.7 miles. How far from its starting point is the point at which the ship again reaches the coast?

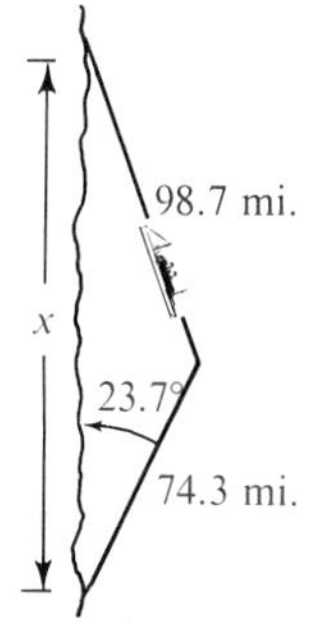

6. A sloping dam face forms an angle measuring 107.3° with the level ground. How far up the face will a 70-foot ladder touch the face if the base of the ladder rests on the ground 8.3 feet from the base of the dam?

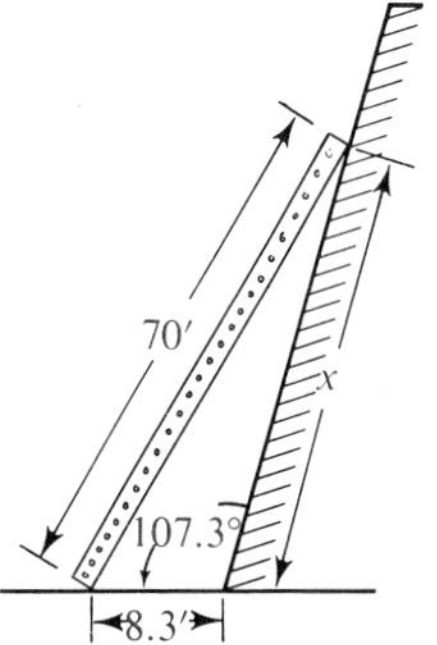

7. From harbors 103 miles apart on a straight coastline, two ships sail courses measuring 43.1° and 27.9°, respectively, with the coastline as pictured. At what distance from each harbor will their courses intersect?

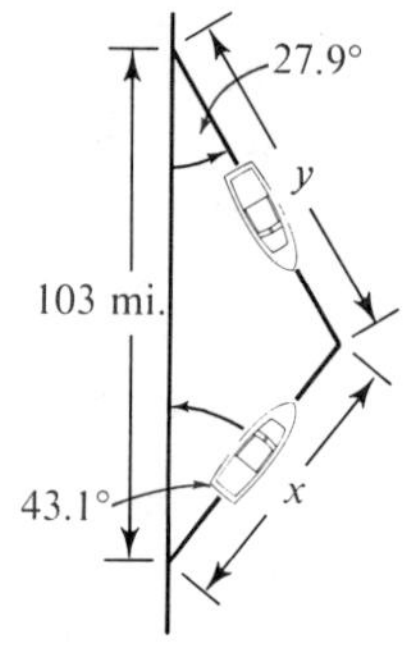

8. One side of a triangular plot of land is 423′ long. How long are the other two sides if they form angles measuring 36.3° and 38.7°, respectively, with the first side?

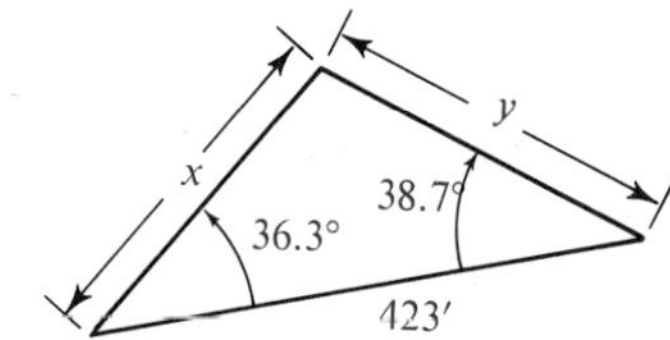

9. Two beacon lights are 1.4 miles apart along a straight coastline. If the lines of sight to the beacons from a ship form an angle of 104° 50′ when the ship is 0.8 miles from one light, how far is the ship from the other light?

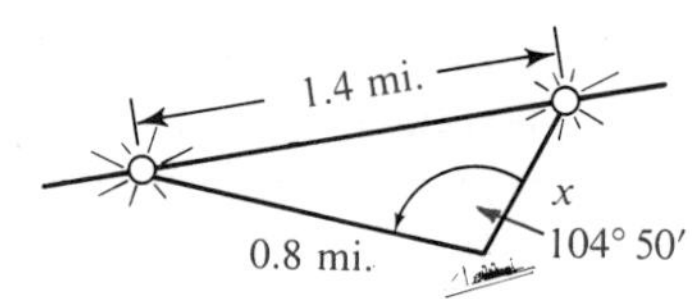

10. In Exercise 9, how far is the ship from the coastline?

11. What is the area of a triangular lot if one side of the lot measures 228.3 feet in length and forms angles measuring 28° 20′ and 97° 40′, respectively, with the other two sides?

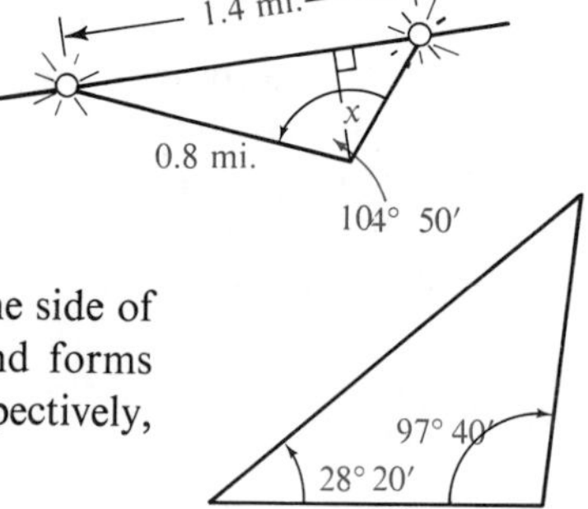

12. What is the perimeter of the lot described in Exercise 11?

13.9 The law of cosines

Another relationship existing between the sides and angles of a triangle relates the lengths of two sides and their included angle with the length of the side opposite this angle. The relationship, called the **law of cosines**, can be given in symbols by

$$\mathbf{a^2 = b^2 + c^2 - 2bc \cos A,}$$
$$\mathbf{b^2 = a^2 + c^2 - 2ac \cos B,}$$
$$\mathbf{c^2 = a^2 + b^2 - 2ab \cos C.}$$

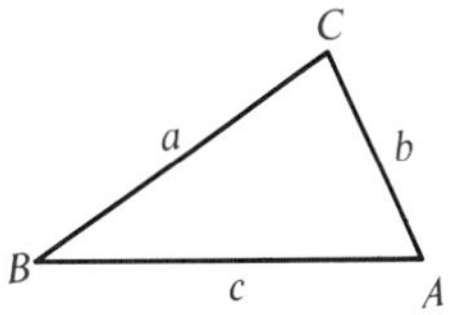

For example, given that $a = 12$, $b = 10$, and $\angle C = 72°$, we can determine c by using

$$c^2 = a^2 + b^2 - 2ab \cos C.$$

Thus,

$$\begin{aligned} c^2 &= (12)^2 + (10)^2 - 2(12)(10) \cos 72° \\ &= 144 + 100 - 240(0.3090) \\ &= 244 - 74.16 = 169.84. \end{aligned}$$

Hence,

$$c = \sqrt{169.84}.$$

or approximately 13.0 units. If required to solve the triangle completely, that is, to find the measure of the two remaining angles, we could use the law of sines very conveniently.

If we know only the lengths of three sides of a triangle, the law of cosines can be used to determine the measure of one or more of angles of the triangle. This can be accomplished by using one or another of the following equivalent forms of the law of cosines.

$$\cos A = \frac{b^2 + c^2 - a^2}{2bc},$$

$$\cos B = \frac{a^2 + c^2 - b^2}{2ac},$$

$$\cos C = \frac{a^2 + b^2 - c^2}{2ab}.$$

Thus, for a triangle whose sides have lengths $a = 20$, $b = 30$, and $c = 40$ we can determine the measure of $\angle C$ by using

$$\cos C = \frac{a^2 + b^2 - c^2}{2ab}.$$

We have

$$\cos C = \frac{(20)^2 + (30)^2 - (40)^2}{2(20)(30)}$$

$$= \frac{400 + 900 - 1600}{1200}$$

$$= -\frac{300}{1200} = -0.2500.$$

Since $\cos C$ is negative, $\angle C$ must be an obtuse angle. From Table III, we find that, to the nearest 10 minutes, the supplement of $\angle C$ measures $75° \, 30'$. It follows that

$$c = 180° - 75° \, 30' = 104° \, 30'.$$

If the measure of $\angle A$ or $\angle B$ is then desired, the law of sines offers perhaps the easiest means of computation.

EXERCISE 13.9

Solve each triangle giving angle measures and lengths to the nearest tenth of a unit.

EXAMPLE

$a = 2.3$, $c = 6.2$, $\angle B = 51.8°$

SOLUTION

We first find b by using $\cos 51.8° = 0.6184$ and

$$b^2 = a^2 + c^2 - 2ac \cos B;$$

$$b^2 = (2.3)^2 + (6.2)^2 - 2(2.3)(6.2)(0.6184)$$

$$= 5.29 + 38.44 - 17.64$$

$$= 26.09.$$

From Table I, we see that $(5.1)^2 = 26.01$ and $(5.2)^2 = 27.04$, so that, to the nearest tenth, $b = 5.1$. To find the measures of $\angle A$ and $\angle C$, we use

$$\frac{\sin A}{a} = \frac{\sin B}{b}.$$

Therefore,

$$\frac{\sin A}{2.3} = \frac{\sin 51.8°}{5.1},$$

from which

$$\sin A = \frac{2.3(0.7859)}{5.1}$$
$$= 0.3544.$$

From Table IV, we see that, to the nearest tenth of a degree, $\angle A = 20.8°$. Then we obtain the measure of $\angle C$ using

$$\angle A + \angle B + \angle C = 180°;$$
$$20.8° + 51.8° + \angle C = 180°,$$
$$\angle C = 107.4°.$$

The missing parts of the triangle are

$$b = \mathbf{5.1}, \quad \angle A = \mathbf{20.8°}, \quad \text{and} \quad \angle C = \mathbf{107.4°}.$$

1. $a = 5,\ b = 8,\ \angle C = 30°$
2. $a = 3,\ c = 10,\ \angle B = 60°$
3. $b = 4,\ c = 5,\ \angle A = 120°$ (*Hint:* Recall that $\cos 120° < 0$.)
4. $b = 10,\ c = 8,\ \angle A = 100°$
5. $a = 1.4,\ b = 7.8,\ \angle C = 44.3°$
6. $a = 3.9,\ b = 8.1,\ \angle C = 92.4°$
7. $b = 2.3,\ c = 2.3,\ \angle A = 27.8°$
8. $b = 14.4,\ c = 11.4,\ \angle A = 52.4°$

Solve each triangle giving angle measures to the nearest 10 minutes and lengths to the nearest tenth of a unit.

EXAMPLE

$a = 8,\ b = 5,\ c = 7$

SOLUTION

We first find the measure of $\angle C$ using

$$\cos C = \frac{a^2 + b^2 - c^2}{2ab};$$

$$\cos C = \frac{8^2 + 5^2 - 7^2}{2(8)(5)}$$
$$= \frac{64 + 25 - 49}{80}$$
$$= \frac{40}{80} = \frac{1}{2}.$$

Thus, $\angle C = 60°$.

Solution continued overleaf.

Then, using the law of sines in the form

$$\frac{\sin A}{a} = \frac{\sin C}{c},$$

we have

$$\frac{\sin A}{8} = \frac{\sin 60^\circ}{7},$$

or

$$\sin A = \frac{8(0.8660)}{7} = 0.9897.$$

From Table III, we find that, to the nearest 10 minutes, $\angle A = 81^\circ\, 50'$. Using the fact that

$$\angle A + \angle B + \angle C = 180^\circ,$$

we have

$$\angle B = 180^\circ - 81^\circ\, 50' - 60^\circ = 38^\circ\, 10'.$$

The missing parts of the triangle are

$$\angle A = \mathbf{81^\circ\, 50'}, \quad \angle B = \mathbf{38^\circ\, 10'}, \quad \text{and} \quad \angle C = \mathbf{60^\circ}.$$

9. $a = 7,\ b = 9,\ c = 12$
10. $a = 3,\ b = 8,\ c = 10$
11. $a = 7,\ b = 8,\ c = 13$
12. $a = 6,\ b = 9,\ c = 12$
13. $a = 5.4,\ b = 5.2,\ c = 8.3$
14. $a = 8.9,\ b = 8.0,\ c = 8.5$
15. $a = 2.4,\ b = 7.5,\ c = 7.0$
16. $a = 12.3,\ b = 14.2,\ c = 22.7$
17. Show that the Pythagorean theorem is a special case of the law of cosines, where one angle of a triangle is a right angle.

13.10 Applications of the law of cosines

In many practical situations, the law of cosines offers the easiest means of solving problems.

EXERCISE 13.10

Solve each problem giving lengths to the nearest tenth of a unit and angle measures to the nearest 10 minutes.

EXAMPLE

Two ships leave a harbor at the same time sailing on courses forming an angle of 104° with each other. If one ship sails at 10 knots and the other at 12 knots, how many nautical miles apart are the ships after 2 hours?

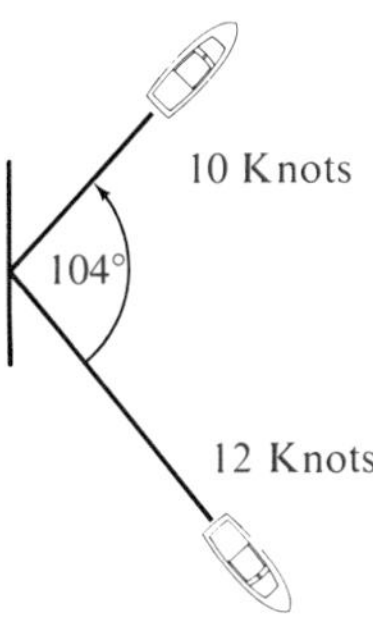

SOLUTION

Make a sketch and label all known parts. Since the ships are sailing at rates of 10 and 12 knots, after 2 hours they will have sailed 20 and 24 nautical miles, respectively. Using the law of cosines, we find that the distance x between the ships is given by

$$x^2 = (20)^2 + (24)^2 - 2(20)(24)\cos 104°.$$

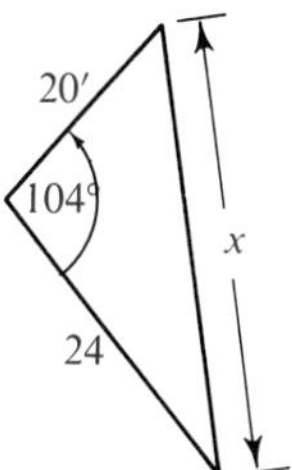

Since 104° is not an entry in Table III, we use the fact that

$$\cos 104° = -\cos(180° - 104°) = -\cos 76°.$$

Then,

$$x^2 = 400 + 576 - 960(-0.2419)$$
$$= 976 + 232.2 = 1208.2.$$

To find $\sqrt{1208.2}$ we can either use logarithms, or else reason as follows:

$$\sqrt{1208.2} \approx \sqrt{1200} = 20\sqrt{3} \approx 34.64.$$

However, because $\sqrt{1208.2} > \sqrt{1200}$, we round upward, to obtain 34.7. Hence, after 2 hours, the ships are approximately **34.7** nautical miles apart.

1. A telephone pole is set into a hillside so that an angle measuring 110° is determined. How long should a support wire be if it is to be hooked to the pole at a height of 32 feet and anchored 48 feet down the hill?

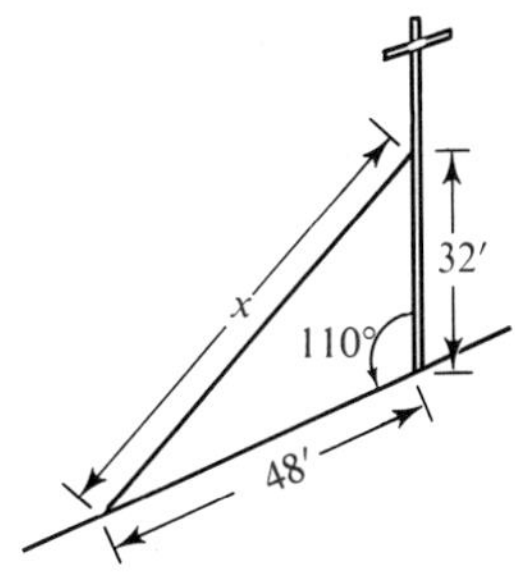

2. A cross brace is mounted 6.3′ down each leg of a tall stepladder. If the legs of the ladder form an angle measuring 28° 40′, how long is the cross brace?

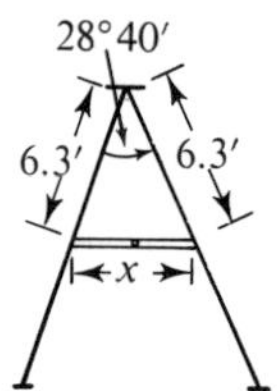

3. Two airplanes depart from the same airport at the same time on courses forming an angle measuring 57°. If the planes fly at a ground speed of 150 and 180 miles per hour, respectively, how far apart are the planes at the end of 1/2 hour?

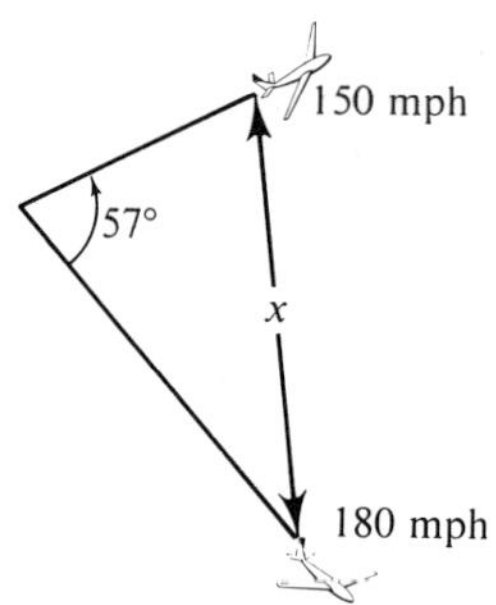

4. A ship sailing at a rate of 14 knots sails from a harbor on a straight course. Two hours later the ship changes course and sails another hour. If the second course forms an angle measuring 116° with the original course, how far is the ship from the harbor after 3 hours?

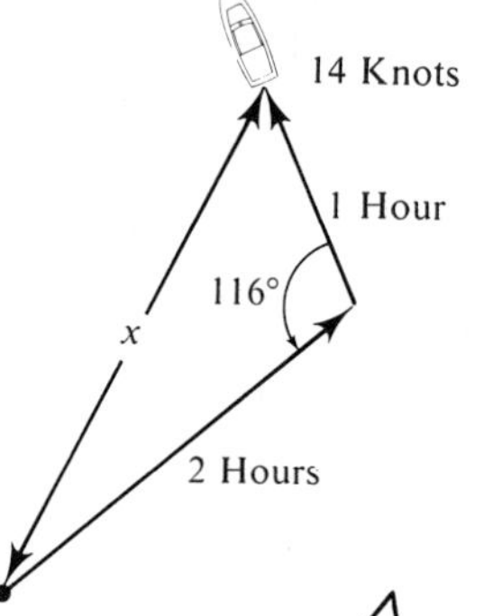

5. The sides of a triangular lot measure 120′, 143′, and 150′. What is the measure of the angle determined by the sides measuring 120′ and 150′?

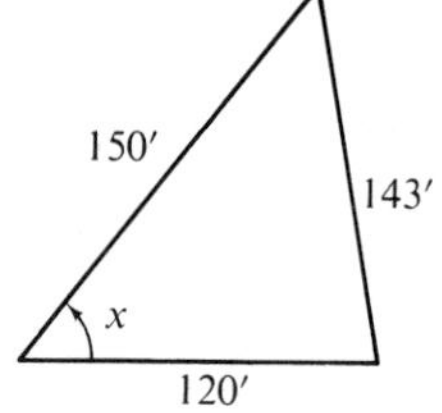

6. Find the area of the triangular lot described in Exercise 5. *Hint*: First find the altitude to any side.

7. A bracket is to be constructed having two arms of length 3.2″ and 5.1″. If the ends of the arms are to be attached to a wall at a distance 4.2″ apart, what must be the measure of the angle determined by the arms?

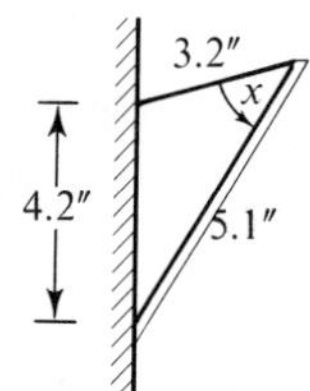

8. A 16.2′ beam supports a roof with rafters each 12.3′ long. What is the measure of the angle determined by the rafters?

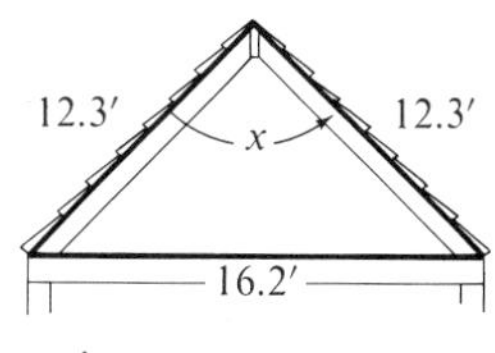

9. A ship leaves a harbor sailing on a course east of north at a rate of 14 knots. From another harbor 200 nautical miles to the north of the first harbor another ship leaves at the same time sailing at a rate of 16 knots. What angle should the second ship's course make with due south if the ships are to meet in 8 hours?

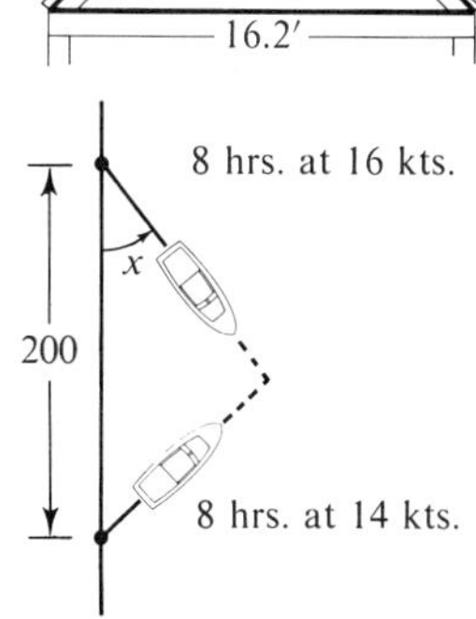

10. In Exercise 9, what is the measure of the angle the first ship's course formed with due north?

11. A ladder 28.5′ in length touches the ground 12.3′ down a slope from the base of a wall. The other end of the ladder just reaches the top of the wall which is 18.2′ high. What angle does the ground make with the wall?

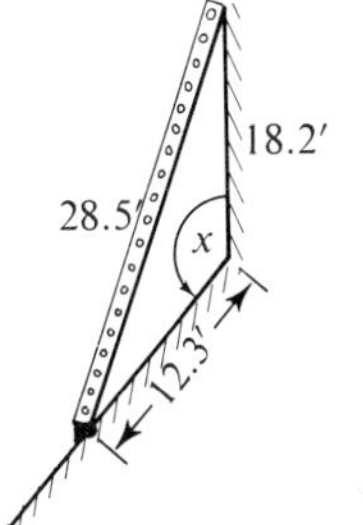

12. In Exercise 11, what angle does the ladder form with the ground?

13. Forces of 8 and 14 pounds act on a point at an angle of 45° with respect to each other. Find the magnitude of the resultant force.

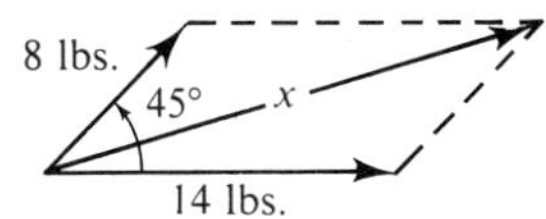

14. Find the measure of the angles that the resultant force makes with the given forces in Exercise 13.

Chapter review

[13.1] 1. Given that $\triangle ABC \sim \triangle A'B'C'$ and $a' = 38$, $a = 46$, and $b = 24$, find b' to the nearest tenth of a unit.

2. Find x to the nearest tenth of a unit.

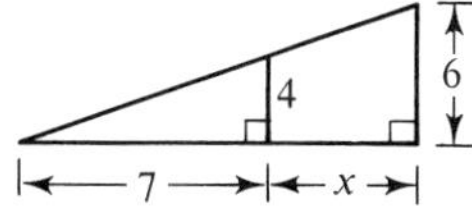

[13.2] 3. Given that the point $(-2, 3)$ is on the terminal side of an angle in standard position, find the value of each of the trigonometric functions of the angle.

4. Given that the terminal side of $\angle A$ is in Quadrant I and $\cos A = 4/7$, find the remaining function values.

[13.3] 5. Find each trigonometric function value.

a. $\sec 135°$ b. $\cos 180°$

6. Find all measures between 0° and 180°, inclusive, for the angle A with function value as given.

a. $\cos A = \frac{1}{2}$ b. $\tan A = -1$

[13.4] 7. Find each function value.

a. $\sin 68°\ 10'$. b. $\tan 122.3°$

8. Find the measures in degrees and minutes of all angles A between 0° and 180° for which each equation is true.

a. $\cos A = -0.7934$ b. $\sin A = 0.7173$

9. Use linear interpolation to find each function value to four significant digits.

a. $\cos 47°\ 13'$ b. $\tan 122.45°$

10. Use linear interpolation to find to the nearest minute the measures of all angles A between 0° and 180° for which each equation is true.

a. $\tan A = 0.8765$ b. $\cos A = -0.7425$

[13.5–13.6] 11. Solve the right triangle in which $\angle C = 90°$, $\angle A = 14°\ 10'$, and $c = 6.4$.

12. A 30-foot ladder is used to reach the top of a 14 foot wall. Find the measure of the angle that the ladder forms with the horizontal ground if it extends 8 feet past the top of the wall.

[13.7–13.8] 13. Solve the triangle in which $b = 14$, $\angle B = 62°$, and $\angle C = 47°$. Give angle measures to the nearest 10 minutes and lengths to the nearest tenth of a unit.

14. Solve the triangle in which $a = 6.8$, $b = 14.7$, and $\angle B = 104.7°$. Give angle measures to the nearest tenth of a degree and lengths to the nearest tenth of a unit.

15. Two sides of a triangular lot form angles measuring 29° 06′ and 33° 48′ with the third side which is 481′ long. How much will it cost to fence the lot if fencing is done at the rate of \$3.28 per foot?

16. What is the value of the lot described in Exercise 15 if the land is worth \$1.12 per square foot?

[13.9–13.10] 17. Given $\triangle ABC$ with $b = 4.6$, $c = 6.8$, and $\angle A = 48.6°$. Solve the triangle giving angle measures to the nearest tenth of a degree and lengths to the nearest tenth of a unit.

18. Given $\triangle ABC$ with $a = 4.6$, $b = 5.2$, and $c = 7.4$. Solve the triangle giving angle measures to the nearest 10 minutes.

19. Forces of 6 and 9 pounds act on a point at an angle of 36° with respect to each other. Find the magnitude of the resultant force.

20. Find the measures of the angles that the resultant force makes with the given forces in Exercise 19.

Appendix

The slide rule

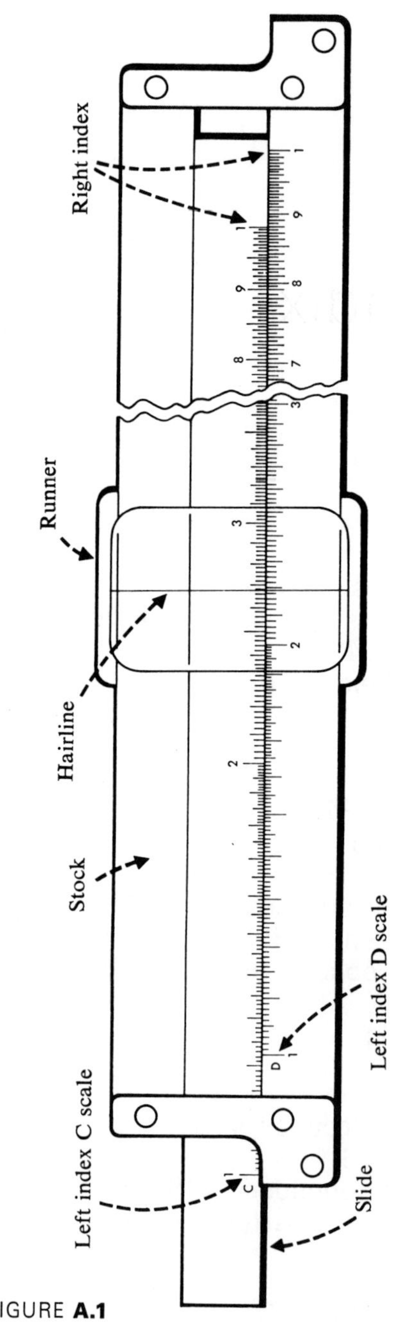

FIGURE **A.1**

A.1 Reading the C and D scales

The slide rule consists of three major parts: a **stock**, a **slide,** and a **runner**. The runner is transparent and contains a vertical **hairline** to aid in aligning numbers. Both the slide and the stock contain several **scales**; the markings at the extremities of the scales are called the **left index** and the **right index**, respectively (see Figure A.1).

The C and D scales are divided into ten main divisions, called **primary divisions**, with the numbers 1, 2, 3, ... marking their boundaries (see Figure A.2). These divisions determine the first significant figure of a number, *the location of a number on the slide rule being independent of the position of the decimal point in the number.*

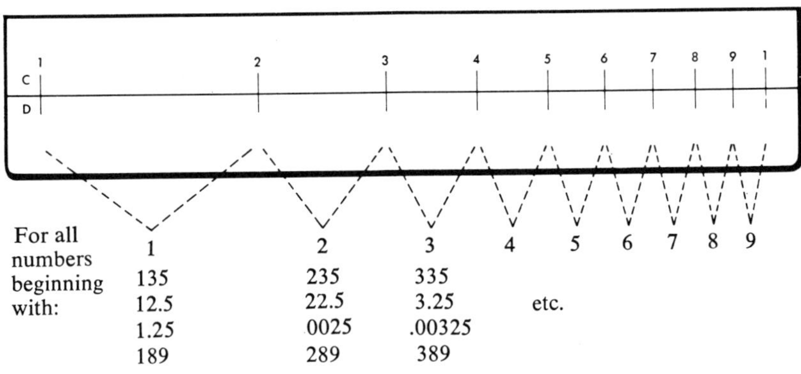

FIGURE **A.2**

Each primary division is divided into ten **secondary divisions**. These are individually numbered in the first primary division and simply indicated with a line in the other primary divisions. The secondary divisions determine the second significant digit of a number. Any number beginning with 13*d*—for example, 13.5, 1.36, 0.138, 0.00132, etc.—will be located between the numbers 1 and 2 in the primary division and between the numbers 3 and 4 in the secondary division. The symbol *d* here simply holds a place for an unspecified digit. This usage is not to be confused with algebraic symbolism in which 13*d* indicates a product. Numbers beginning with 26*d* will be found between the numbers 2 and 3 in the primary division and between the numbers 6 and 7 in the secondary division (see Figure A.3).

In the primary division bounded by the numbers 1 and 2, each secondary division is divided into ten *tertiary* divisions. These divisions determine the third significant digit of a number. In the primary divisions bounded by the

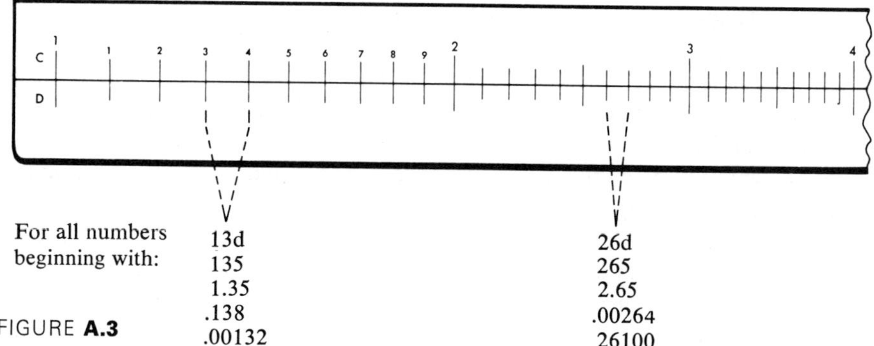

FIGURE **A.3**

numbers 2 and 3 and the numbers 3 and 4, there are only five tertiary divisions. However, since each secondary division represents ten possible third-place digits (1, 2, 3, 4, 5, 6, 7, 8, 9, 0), each tertiary division between 2 and 4 represents two digits; or each one-half of a division represents one digit (see Figure A.4).

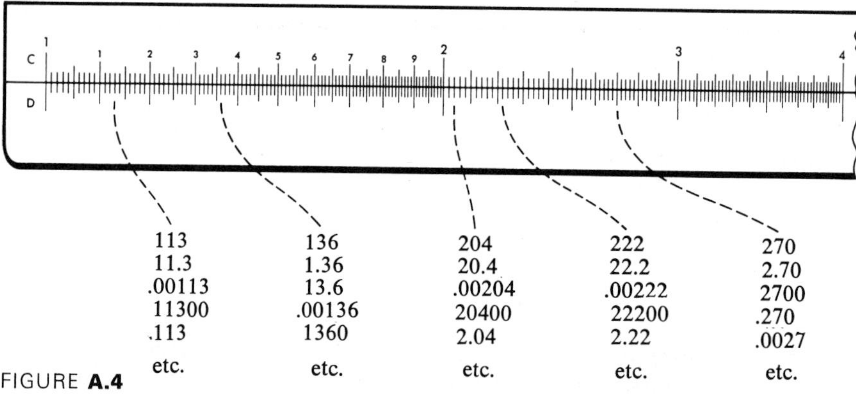

FIGURE **A.4**

In the remaining primary divisions—that is, from the fourth to the tenth—there are only two tertiary divisions in each secondary division. Thus in this section of the scale each tertiary division represents five digits or each one-fifth of a division represents one digit. Numbers whose third significant digit is different from 0 or 5 are estimated (see Figure A.5).

(A) 464 is located a little to the left of 465.
(B) 481 is located a little to the right of 480.
(C) 533 is located about halfway between 530 and 535, but slightly closer to 535.

In the first primary division, three digits can be read directly and a fourth estimated with an accuracy of approximately ± 2. Thus 1352 can be read

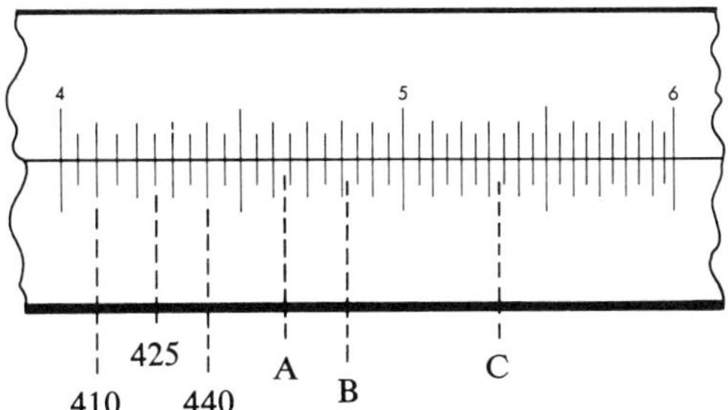

FIGURE **A.5**

with the assurance that the number lies between 1350 and 1354. In primary divisions bounded by 2 and 4, two digits can be read and a third estimated with an accuracy of approximately ± 1. The number 0.0241 can be read with the assurance that it lies between 0.0240 and 0.0242. In the remainder of the primary divisions, two digits can be read and a third estimated with an accuracy of approximately ± 2. The number 5.72 can be read with the assurance that it lies between 5.70 and 5.74. With practice, you can obtain accuracy to three significant figures without difficulty.

You should make a practice of reading numbers by digits. That is, the number 427 should be read "four–two–seven" rather than "four hundred twenty-seven"; the number 3.28 should be read "three–point–two–eight" rather than "three and twenty-eight hundredths." It is very important that you establish this habit.

We have used the terms "primary divisions," "secondary divisions," and "tertiary divisions" to refer to the intervals bounded by the respective markings on the rule. We shall also have occasion to refer to the markings themselves as divisions on the rule.

EXERCISE A.1

Read the indicated settings.

A.

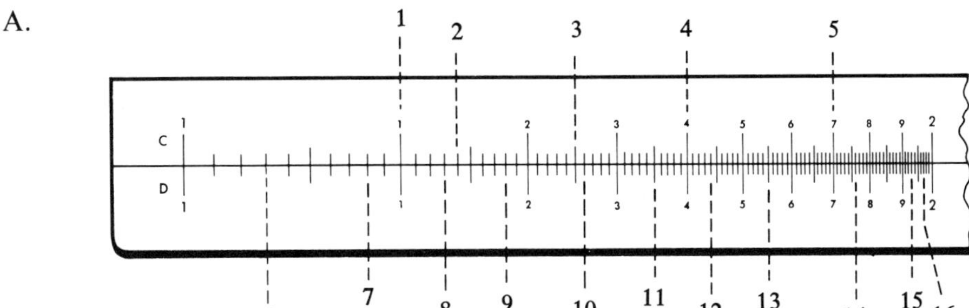

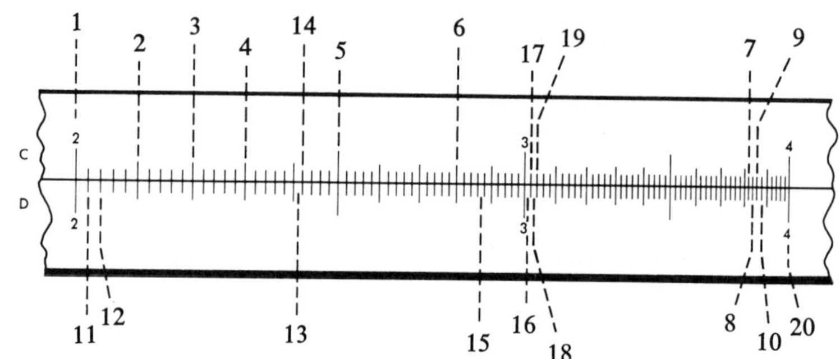

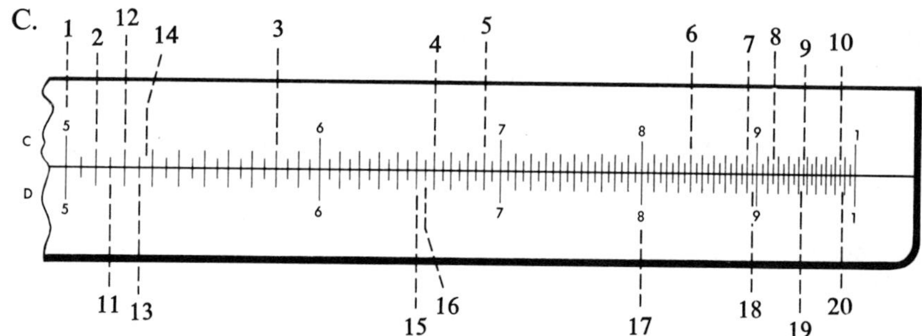

A.2 The product of two numbers; placing the decimal point by approximation

In a sense, all mathematical operations may be considered variations of addition, which in turn is simply counting. If two rulers, A and B, are laid side by side with the left end of ruler A opposite the number 2 on ruler B, we can find the sum of 2 and 3 by counting to the number 3 on ruler A and observing that the sum, 5, lies directly opposite this point on ruler B (see Figure A.6).

The C and D scales on a slide rule are laid out on a logarithmic scale with the distance between the indexes representing one unit; that is, the number 2 is in the position corresponding to the mantissa of the logarithm of 2, the number 3 is in the position corresponding to the mantissa of the logarithm of 3, etc. (see Figure A.7). You should not be overly concerned here if the

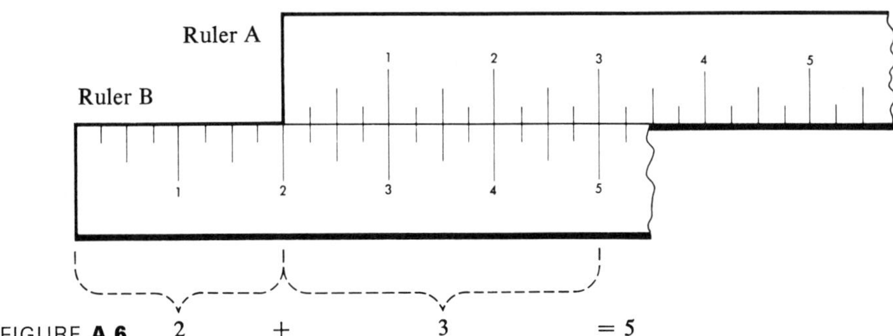

FIGURE **A.6**

nature and properties of logarithms which are introduced in Chapter 11 have not been considered at this time. Although such knowledge is necessary if you are to understand the principles underlying the construction of the scales on a rule, you can learn to use the slide rule very effectively for elementary computations without it. You should reread this section after you have studied Chapter 11.

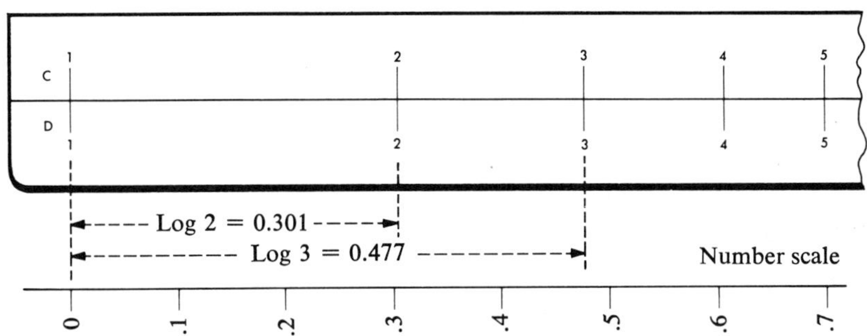

FIGURE **A.7**

From the first law of logarithms,

$$\log M + \log N = \log (MN).$$

Thus,

$$\log 2 + \log 3 = \log (2 \cdot 3)$$

or

$$0.301 + 0.477 = 0.778.$$

Note that in this appendix when we use notation such as log M we mean $\log_{10} M$.

When the rulers are arranged as in Figure A.8, the *product* of 2 and 3 is located on the D scale in a position corresponding to the *sum* of their logarithms, which correspond to lengths along the C and D scales.

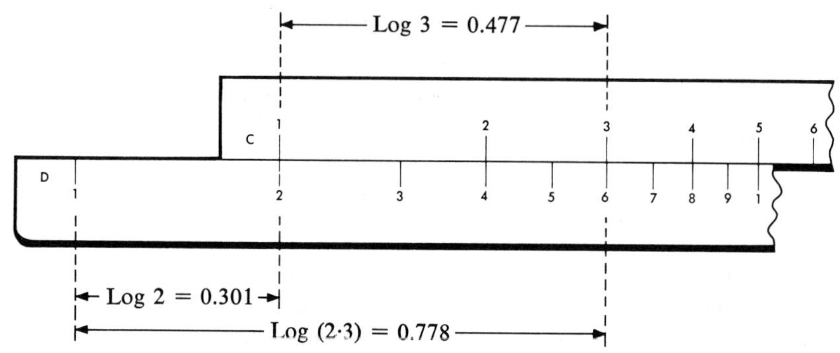

FIGURE **A.8**

To multiply two numbers:

1. *Opposite one factor on D, set the index of C.*
2. *Opposite the second factor on C, read answer on D.*

EXAMPLE 1

$2 \times 4 = ?$

SOLUTION

(1) Opposite 2 on D, set index of C.
(2) Opposite 4 on C, read **8** on D.

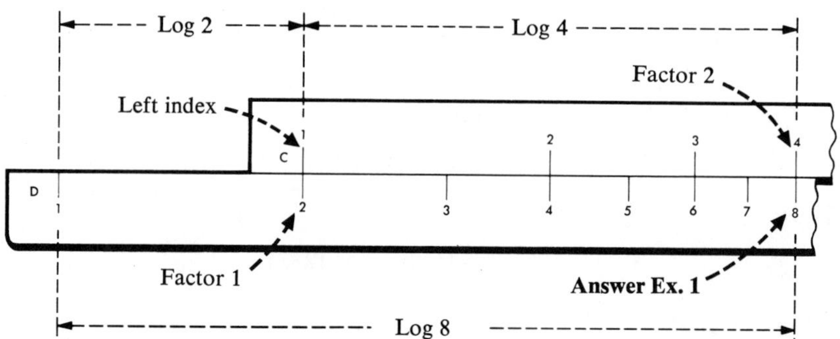

In multiplying two numbers, it is possible that the procedure will cause the second factor, which is located on the C scale, to be to the right of the stock. If this occurs, simply *reverse the index; use the right index of the C scale instead of the left index and continue with the computation.*

EXAMPLE 2

$2 \times 6 = ?$

SOLUTION

(1) Opposite 2 on D, set right index of C.
(2) Opposite 6 on C, read **12** on D.

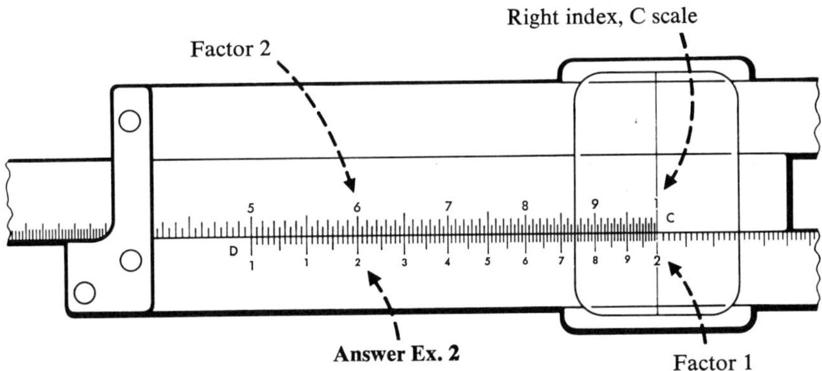

In Example 2, placing the decimal point presented no difficulty. Since this is not always the case, we need some means of determining where the decimal point in a product should be placed. We will, for the present, use an approximation method for placing the decimal point. This is accomplished by rounding off the factors to one significant digit and then multiplying these factors to obtain an approximate product where the decimal point can be placed by inspection, and then placed in the actual product by comparing it with the approximate product.

Recall that numbers are rounded to one significant digit by first examining the second significant digit of the number. If the second significant digit is less than 5, the first significant digit is used as it is and the remaining digits are replaced with zeros. If the second significant digit is 5 or greater, the first significant digit is increased by 1 and the remaining digits are replaced with zeros.

EXAMPLE 3

$2.3 \times 49 = ?$

SOLUTION

(1) Opposite 23 on D, set right index of C.
(2) Opposite 49 on C, read **1127** on D.

To place the decimal point, round off 2.3 and 49 to 2 and 50, respectively, and multiply.

$$2.3 \times 49 \approx 2 \times 50 = 100.$$

The answer is approximately 100 and the decimal point is placed accordingly. **112.7**

EXAMPLE 4

$24.5 \times 3.6 = ?$

SOLUTION

(1) Opposite 245 on D, set left index of C.
(2) Opposite 36 on C, read **882** on D.

To place the decimal point, round off 24.5 and 3.6 to 20 and 4, respectively, and multiply.

$$24.5 \times 3.6 \approx 20 \times 4 = 80.$$

The answer is approximately 80 and the decimal point is placed accordingly. **88.2**

If both factors in a product lie near the center of an interval—for example, 35 and 36 or 24 and 24—it is usually more accurate to round them alternately regardless of the value of the second digit. In the case of 35 and 36 we would use 30 and 40, while with 24 and 24, 20 and 30 would give a better estimate than 20 and 20.

EXAMPLE 5

$2.58 \times 55.3 = ?$

SOLUTION

(1) Opposite 258 on D, set right index of C.
(2) Opposite 553 on C, read **1427** on D.

To place the decimal point, round off 2.58 and 55.3 to 3 and 50, respectively, and multiply.

$$2.58 \times 55.3 \approx 3 \times 50 = 150.$$

The answer is approximately 150 and the decimal point is placed accordingly. **142.7**

EXAMPLE 6

106.5 × 67.2 = ?

SOLUTION

(1) Opposite 1065 on D, set left index of C.
(2) Opposite 672 on C, read **716** on D.

To place the decimal point, round off 106.5 and 67.2 to 100 and 70, respectively, and multiply.

$$106.5 \times 67.2 \approx 100 \times 70 = 7000.$$

The anwer is approximately 7000 and the decimal point is placed accordingly. **7160**

EXAMPLE 7

11.44 × 16.25 = ?

SOLUTION

(1) Opposite 1144 on D, set left index of C.
(2) Opposite 1625 on C, read **1859** on D.

To place the decimal point, round off 11.44 and 16.25 to 10 and 20, respectively, and multiply.

$$11.44 \times 16.25 \approx 10 \times 20 = 200.$$

The answer is approximately 200 and the decimal point is placed accordingly. **185.9**

EXAMPLE 8

9400 × 1.845 = ?

SOLUTION

(1) Opposite 94 on D, set right index of C.
(2) Opposite 1845 on C, read **1734** on D.

To place the decimal point, round off 9400 and 1.845 to 9000 and 2, respectively, and multiply.

$$9400 \times 1.845 \approx 9000 \times 2 = 18{,}000.$$

The answer is approximately 18,000 and the decimal point is placed accordingly. **17,340**

Approximating an answer by rounding off factors to one significant digit offers the easiest means of placing the decimal point in simple problems. Since this method becomes less practical in more complicated computations, we shall discuss another method of placing the decimal point in Section A.4 of this appendix.

EXERCISE A.2

Compute:

1. 2×4	2. 3×20	3. 12×3
4. 2×47	5. 12×15	6. 7×12
7. 21×6	8. 6×24	9. 8×155
10. 364×4	11. 3×2.5	12. 3.5×18
13. 20×3.1	14. 0.4×17	15. 6.5×0.5
16. 0.8×135	17. 0.6×23.5	18. 3.5×6.4
19. 32.5×0.8	20. 3.6×0.95	21. 2.32×4.15
22. 46.4×1.7	23. 4.5×12.3	24. 103×0.81
25. 45.2×13.3	26. 3.09×8.83	27. 8.50×2.33
28. 5.31×595	29. 307×9.25	30. 301×0.835
31. 10.23×75.3	32. 19.05×4.42	33. 0.577×110.2
34. 0.365×1907	35. 0.845×100	36. 88.6×2.45
37. 0.996×888	38. 4.72×625	39. 864×12.25
40. 37.6×950	41. 2.34×121.5	42. 1625×8.95
43. 1055×0.733	44. 49.1×121.5	45. 6.82×174.5
46. 1100×84.5	47. 16.45×12.55	48. 1.005×1945
49. 1.408×164.6	50. 180.5×14.65	51. 0.843×1840
52. 10.25×64.2	53. 1250×3.08	54. 7.43×1620
55. 155.5×1.283	56. 1900×12.45	

A.3 The quotient of two numbers

In dividing numbers on a slide rule, we use the logarithmic principle

$$\log\left(\frac{M}{N}\right) = \log M - \log N.$$

That is, the logarithm of a quotient equals the logarithm of the dividend minus the logarithm of the divisor, each of which corresponds to a length along the scale.

To divide one number by another:

1. Opposite numerator on D, set denominator on C.
2. Opposite index on C, read answer on D.

Since one index of C is always opposite some point on D, the answer will always be on the stock.

EXAMPLE 1

$6 \div 2 = \frac{6}{2} = ?$

SOLUTION

(1) Opposite 6 on D, set 2 on C.
(2) Opposite index on C, read **3** on D.

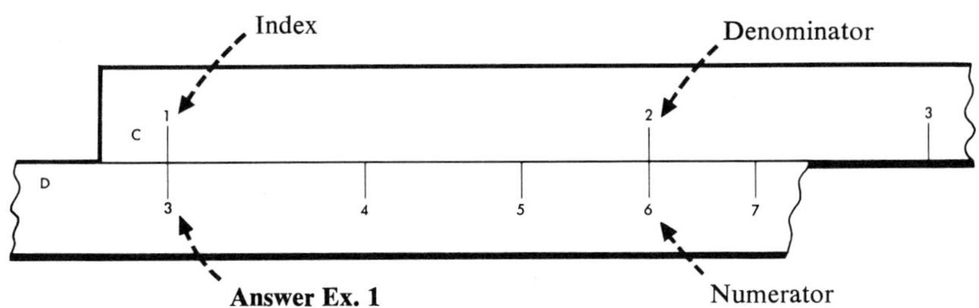

EXAMPLE 2

$\frac{279}{17} = ?$

SOLUTION

(1) Opposite 279 on D, set 17 on C.
(2) Opposite index on C, read **1641** on D.

Solution continued overleaf.

To place the decimal point, round off 279 and 17 to 300 and 20, respectively, and divide. Since

$$\frac{279}{17} \approx \frac{300}{20} = 15,$$

the decimal point is placed accordingly. **16.41**

EXAMPLE 3

$$\frac{582}{275} = ?$$

SOLUTION

(1) Opposite 582 on D, set 275 on C.
(2) Opposite index on C, read **212** on D.

Since

$$\frac{582}{275} \approx \frac{600}{300} = 2,$$

the decimal point is placed accordingly. **2.12**

EXAMPLE 4

$$\frac{58{,}900}{230} = ?$$

SOLUTION

(1) Opposite 589 on D, set 230 on C.
(2) Opposite index on C, read **256** on D.

Since

$$\frac{58{,}900}{230} \approx \frac{60{,}000}{200} = 300,$$

the decimal point is placed accordingly. **256**

In more complicated computations the decimal point can be placed by rounding off the numbers and changing the fraction to an equivalent fraction with a denominator between 1 and 10. This can be accomplished by moving the decimal point in the numerator and the decimal point in the denominator the same number of places to the right or the left, which is equivalent to multiplying or dividing the numerator and the denominator of the fraction by 10 or 100 or 1000, and so forth.

EXAMPLE 5

$$\frac{57}{385} = ?$$

SOLUTION

(1) Opposite 57 on D, set 385 on C.
(2) Opposite index on C, read **148** on D.

$$\frac{57}{385} \approx \frac{60}{400}.$$

Move decimal point in numerator and decimal point in denominator two places to the left and divide.

$$\frac{0.60}{4.00} = 0.15.$$

The decimal point is placed accordingly. **0.148**

EXAMPLE 6

$$\frac{3.6}{6400} = ?$$

SOLUTION

(1) Opposite 36 on D, set 640 on C.
(2) Opposite index on C, read **563** on D.

$$\frac{3.6}{6400} \approx \frac{4}{6000}.$$

Move decimal points three places to the left and divide.

$$\frac{0.004}{6.000} \approx 0.00066.$$

The decimal point is placed accordingly. **0.000563**

EXAMPLE 7

$$\frac{0.6340}{28.1} = ?$$

SOLUTION

(1) Opposite 634 on D, set 281 on C.
(2) Opposite index on C, read **226** on D.

$$\frac{0.6340}{28.1} \approx \frac{0.6000}{30}.$$

Solution continued overleaf.

Move decimal points one place to the left and divide.

$$\frac{0.06000}{3.0} = 0.02.$$

The decimal point is placed accordingly. **0.0226**

Recall that the reciprocal of a number n ($n \neq 0$) is the number $1/n$. There are several ways of finding reciprocals on the slide rule. One way is simply to divide the number into 1.

EXAMPLE 8

$\frac{1}{8} = ?$

SOLUTION

Since we want the reciprocal of 8,

(1) Opposite either index on D, set 8 on C.
(2) Opposite index on C, read **125** on D.

Placement of the decimal point by approximation yields **0.125**.

Other methods of determining reciprocals will be discussed in several of the following sections.

EXERCISE A.3

Compute:

1. $\frac{8}{4}$	2. $\frac{6}{3}$	3. $\frac{12}{2}$	4. $\frac{48}{6}$
5. $\frac{82}{41}$	6. $\frac{52}{13}$	7. $\frac{27}{4}$	8. $\frac{25}{6}$
9. $\frac{450}{4.5}$	10. $\frac{220}{40}$	11. $\frac{31}{4}$	12. $\frac{29}{19.3}$
13. $\frac{52.5}{2.3}$	14. $\frac{48.1}{3.5}$	15. $\frac{105}{50.5}$	16. $\frac{122}{48.3}$
17. $\frac{23}{21}$	18. $\frac{29}{27}$	19. $\frac{1045}{23.1}$	20. $\frac{2073}{25.7}$

21. $\frac{27.3}{290}$ 22. $\frac{25.6}{186}$ 23. $\frac{231}{305}$ 24. $\frac{5.07}{16.3}$

25. $\frac{1224}{1321}$ 26. $\frac{1074}{985}$ 27. $\frac{4.53}{0.92}$ 28. $\frac{5.33}{0.86}$

29. $\frac{110.5}{1.83}$ 30. $\frac{147.2}{2.87}$ 31. $\frac{1}{2}$ 32. $\frac{1}{5}$

33. $\frac{1}{7}$ 34. $\frac{1}{9}$ 35. $\frac{1}{32}$ 36. $\frac{1}{28}$

37. $\frac{1}{0.8}$ 38. $\frac{1}{0.7}$ 39. $\frac{1}{235}$ 40. $\frac{1}{407}$

A.4 Multiple operations and powers of ten

In Sections A.2 and A.3, we discussed the fundamental operations of multiplication and division using only C and D scales. These operations, when accomplished on a slide rule, involve the addition and subtraction of the logarithms that correspond to lengths along the scales. It is possible, in some cases, to perform an addition and a subtraction of line segments successively with one setting on the rule. Thus, where a computation involves both a multiplication and a division the number of movements of the slide can be reduced by means of the following:

To compute $\frac{p \times q}{r}$:

1. *Opposite p on D, set r on C.*

   (*The quotient* $\frac{p}{r}$ *will appear on D opposite the index on C.*)

2. *Move the hairline to q on C and read answer opposite on D.*

In computations of this type, it does not matter in which order the operations are performed, but note that, by dividing before multiplying, only one setting of the slide was necessary. In general, it is usually advisable to begin with the operation that leaves most of the slide in the stock. As before, the decimal point in the result can be placed by approximation.

EXAMPLE 1

$$\frac{37 \times 42}{54} = ?$$

SOLUTION

(1) Opposite 37 on D, set 54 on C.
(2) Opposite 42 on C, read **288** on D.

Since

$$\frac{37 \times 42}{54} \approx \frac{40 \times 40}{50} = 32,$$

the decimal point is placed accordingly. **28.8**

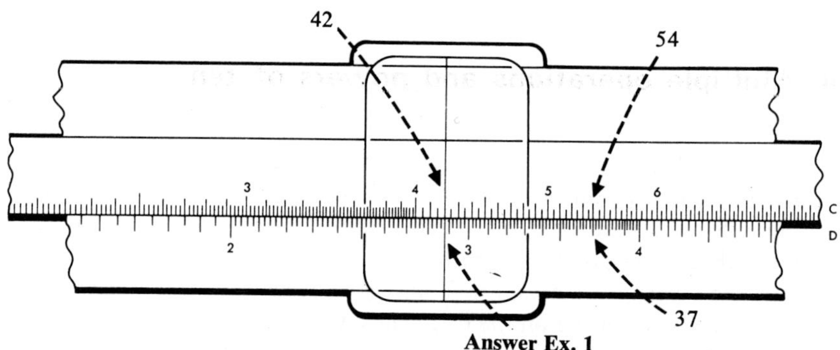

EXAMPLE 2

$$\frac{54.2 \times 2.31}{5.6} = ?$$

SOLUTION

(1) Opposite 542 on D, set 56 on C.
(2) Opposite 231 on C, read **224** on D.

Since

$$\frac{54.2 \times 2.31}{5.6} \approx \frac{50 \times 2}{6} \approx 17,$$

the decimal point is placed accordingly. **22.4**

If there are a number of factors involved in a computation of this type, for instance $\frac{a \times b \times c}{d \times e}$, it is generally advisable to divide a by d, multiply by b, divide by e, and then multiply by c.

EXAMPLE 3

$$\frac{25 \times 35 \times 45}{55 \times 65} = ?$$

SOLUTION

(1) Opposite 25 on D, set 55 on C.
(2) Set hairline on 35 on C.
(3) Set 65 on C under hairline.
(4) Opposite 45 on C, read **110** on D.

Since

$$\frac{25 \times 35 \times 45}{55 \times 65} \approx \frac{30 \times 30 \times 50}{60 \times 60} \approx 12,$$

the decimal point is placed accordingly. **11.0**

Sometimes, after a multiplication operation, the result is off the stock. If this is the case, simply reverse the index and continue the computations.

EXAMPLE 4

$$\frac{35 \times 95 \times 32}{25 \times 21} = ?$$

SOLUTION

(1) Opposite 35 on D, set 25 on C.
(2) Since 95 on C is off the stock, reverse the index and set hairline on 95 on C.
(3) Set 21 on C under hairline.
(4) Opposite 32 on C, read **203** on D.

Since

$$\frac{35 \times 95 \times 32}{25 \times 21} \approx \frac{40 \times 100 \times 30}{30 \times 20} = 200,$$

the decimal point is placed accordingly. **203**

When there are additional factors in the numerator, we proceed as far as we can by alternating the operations and then multiply the result by the additional factor or factors. Similarly, if there are additional factors in the denominator, we proceed as far as we can by alternating the operations and then divide the result by the additional factor or factors.

EXAMPLE 5

$$\frac{25 \times 22 \times 42 \times 12}{46 \times 31} = ?$$

SOLUTION

(1) Opposite 25 on D, set 46 on C.
(2) Set hairline on 22 on C.
(3) Set 31 on C under hairline.
(4) Set hairline on 42 on C.
(5) Set index on C under hairline.
(6) Opposite 12 on C, read **1944** on D.

Since

$$\frac{25 \times 22 \times 42 \times 12}{46 \times 31} \approx \frac{30 \times 20 \times 40 \times 10}{50 \times 30} = 160,$$

the decimal point is placed accordingly. **194.4**

EXAMPLE 6

$$\frac{35 \times 20}{42 \times 16 \times 15} = ?$$

SOLUTION

(1) Opposite 35 on D, set 42 on C.
(2) Set hairline on 20 on C.
(3) Set 16 on C under hairline.
(4) Set hairline on index.
(5) Set 15 on C under hairline.
(6) Opposite index on C, read **694** on D.

Since

$$\frac{35 \times 20}{42 \times 16 \times 15} \approx \frac{40 \times 20}{40 \times 20 \times 20} = 0.05,$$

the decimal point is placed accordingly. **0.0694**

If it is difficult to place the decimal point by approximation, the "powers of ten" method can be used. This method is based on the fact that any real number can be expressed in scientific notation as the product of a number between 1 and 10 and a power of 10, the exponent of the power being equal to the number of places the decimal point is moved to obtain the first factor (see Section 5.5). The exponent is positive if the original number is larger than 10 and negative if the original number is less than 1.

EXAMPLE 7

(a) $379 = 3.79 \times 10^2$,
(b) $30.2 = 3.02 \times 10^1$,
(c) $0.0214 = 2.14 \times 10^{-2}$,
(d) $0.000301 = 3.01 \times 10^{-4}$.

A number given in scientific notation can be written in standard form by moving the decimal point the number of places indicated by the exponent on 10 (to the left if negative, to the right if positive).

EXAMPLE 8

(a) $3.75 \times 10^4 = 37{,}500$,
(b) $2.03 \times 10^1 = 20.3$,
(c) $7.34 \times 10^{-4} = 0.000734$,
(d) $2.98 \times 10^{-1} = 0.298$.

In the following examples we shall use the first two laws of exponents, which are discussed in Section 5.1.

$$10^m \cdot 10^n = 10^{m+n}$$

and

$$\frac{10^m}{10^n} = 10^{m-n}.$$

Now to apply the "powers of ten" method:

1. Round off each factor to its first significant digit and then express each as a single digit times a power of 10.
2. Multiply or divide the single digits as applicable.
3. Multiply this result by the result of multiplying and dividing the powers of 10.

For example,

$$\frac{2180 \times 6392}{319} = \frac{(2 \times 10^3)(6 \times 10^3)}{(3 \times 10^2)}$$

$$= \frac{2 \times 6}{3} \times 10^{3+3-2}$$

$$= 4 \times 10^4 = 40{,}000.$$

EXAMPLE 9

$0.0356 \times 4.79 \times 0.0264 = ?$

SOLUTION

(1) Opposite 356 on D, set right index on C.
(2) Set hairline on 479 on C.

Solution continued overleaf.

(3) Set left index on C under hairline.
(4) Opposite 264 on C, read **450** on D.

Since

$$0.0356 \times 4.79 \times 0.0264 \approx 0.04 \times 5 \times 0.03$$
$$(0.04 = 4 \times 10^{-2};\ 5 = 5 \times 10^{0};\ 0.03 = 3 \times 10^{-2}),$$

we have

$$4 \times 10^{-2} \times 5 \times 10^{0} \times 3 \times 10^{-2} = 60 \times 10^{-4},$$

and the decimal point is placed accordingly. **0.00450**

EXAMPLE 10

$$\frac{275 \times 82.9}{0.0066} = ?$$

SOLUTION

(1) Opposite 275 on D, set 66 on C.
(2) Opposite 829 on C, read **345** on D.

Since

$$\frac{275 \times 82.9}{0.0066} \approx \frac{300 \times 80.0}{0.007}$$
$$(300 = 3 \times 10^{2};\ 80.0 = 8 \times 10^{1};\ 0.007 = 7 \times 10^{-3}),$$
$$\frac{3 \times 10^{2} \times 8 \times 10^{1}}{7 \times 10^{-3}} = \frac{3 \times 8}{7} \times 10^{2+1-(-3)} \approx 3 \times 10^{6},$$

and the decimal point is placed accordingly. **3,450,000**

EXERCISE A.4

Compute:

1. $\dfrac{6 \times 2}{4}$
2. $\dfrac{5 \times 6}{4}$
3. $\dfrac{8 \times 1.8}{16}$
4. $\dfrac{7 \times 12.4}{5}$
5. $\dfrac{3.5 \times 4.6}{14}$
6. $\dfrac{4.7 \times 6.2}{3.1}$
7. $\dfrac{573}{28.1 \times 9.1}$
8. $\dfrac{641}{38.2 \times 5.9}$
9. $\dfrac{0.21}{843 \times 0.0076}$
10. $\dfrac{3.7}{2.86 \times 0.97}$
11. $\dfrac{12 \times 75}{5 \times 30}$
12. $\dfrac{5.9 \times 3.7}{0.051}$

13. $\dfrac{0.0034 \times 462}{1.74 \times 0.732}$

14. $\dfrac{0.0031 \times 64}{0.00118 \times 47}$

15. $\dfrac{234 \times 0.0021}{0.0411 \times 521}$

16. $\dfrac{258 \times 0.791}{0.0031 \times 0.002}$

17. $0.025 \times 34 \times 61$

18. $73 \times 0.86 \times 40$

19. $54.1 \times 62.4 \times 1.05$

20. $0.003 \times 4.1 \times 6.4$

21. $795 \times 0.037 \times 1.054$

22. $0.587 \times 0.37 \times 493$

23. $\dfrac{249 \times 312 \times 110}{905}$

24. $\dfrac{7.3 \times 68 \times 0.069}{21}$

25. $\dfrac{8 \times 9 \times 7}{6 \times 5}$

26. $\dfrac{30 \times 81 \times 69}{6 \times 17}$

27. $\dfrac{2.3 \times 0.0067 \times 7.5}{0.072 \times 5.3}$

28. $\dfrac{1.2 \times 5.4 \times 0.017}{92 \times 221}$

29. $\dfrac{7 \times 12}{3 \times 5 \times 9}$

30. $\dfrac{34 \times 14}{1.9 \times 3.8 \times 16}$

31. $\dfrac{2.1 \times 0.34}{0.17 \times 0.052 \times 1.3}$

32. $\dfrac{0.058 \times 248}{7.87 \times 351 \times 0.03}$

33. $\dfrac{81 \times 62 \times 34}{57 \times 98 \times 14}$

34. $\dfrac{59 \times 63 \times 87}{21 \times 41 \times 51}$

35. $\dfrac{231 \times 356 \times 32.7}{121 \times 40 \times 13}$

36. $\dfrac{87 \times 69 \times 135}{0.41 \times 0.71 \times 0.032}$

37. $\dfrac{1}{17 \times 21}$

38. $\dfrac{1}{3.8 \times 9.64}$

39. $\dfrac{1}{0.51 \times 0.73 \times 0.84}$

40. $\dfrac{1}{174 \times 286 \times 14.3}$

A.5 Ratio and proportion

The slide rule is particularly useful in solving for a missing term in a proportion, since at any position of the slide the ratios determined by corresponding numbers on the C and D scales are equal. In Figure A.9, observe

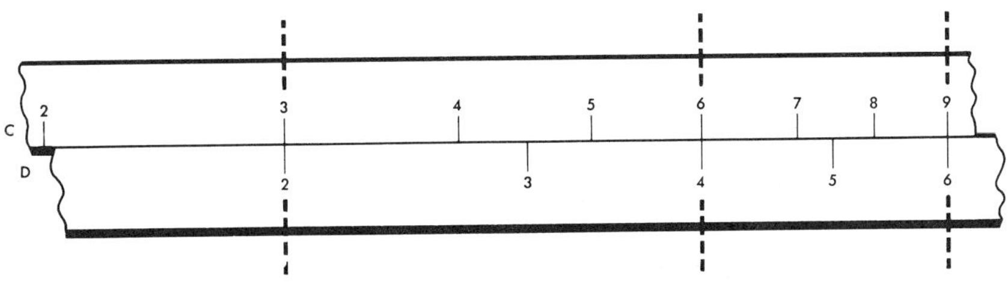

FIGURE **A.9**

that if 3 on C is placed opposite 2 on D, which as a ratio would appear as $\frac{3}{2}$, all other ratios exhibited by the C and D scales are equal to $\frac{3}{2}$, e.g., $\frac{6}{4}$ or $\frac{9}{6}$.

To solve a proportion $\frac{a}{b} = \frac{x}{c}$, the slide rule can be set exactly as the proportion appears when written; that is, both numerators will appear on the C scale and both denominators on D scale. The unknown can then be determined on the appropriate scale.

To solve a proportion of the form $\frac{a}{b} = \frac{x}{c}$:

1. *Opposite b on D, set a on C.*
2. *Opposite c on D, read x on C.*

To solve a proportion of the form $\frac{a}{b} = \frac{c}{x}$:

1. *Opposite b on D, set a on C.*
2. *Opposite c on C, read x on D.*

(*Note that in each case the hairline is first set on the denominator of the fraction not containing the unknown quantity.*)

EXAMPLE 1

$$\frac{5}{20} = \frac{4}{x}$$

SOLUTION

(1) Opposite 20 on D, set 5 on C.
(2) Opposite 4 on C, read **16** on D.

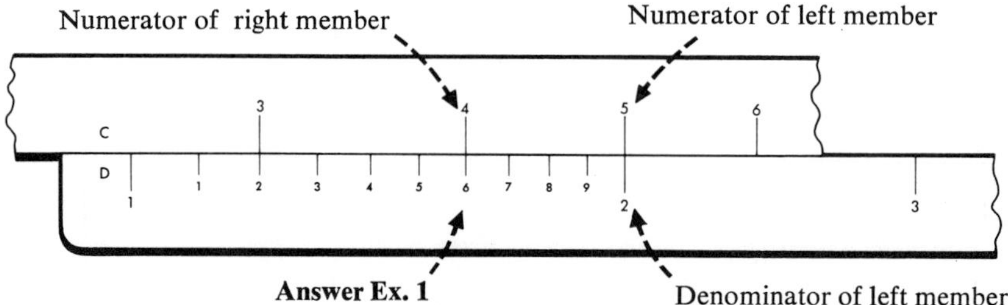

EXAMPLE 2

$$\frac{2}{5} = \frac{6}{x}$$

SOLUTION

(1) Opposite 5 on D, set 2 on C.
(2) Since 6 on C is off the stock, reverse the index and, opposite 6 on C, read **15** on D.

The position of the decimal point can be approximated by noting that the same ratio exists between a and b as between c and x. Thus, in Example 2 above, 5 is between two and three times as large as 2 and therefore x must be between two and three times as large as 6. Another approach is to note that 6 is three times as large as 2, and therefore x will be three times as large as 5. If you wish, you may shift the decimal points in the numerator and denominator of one member of the proportion to assist in estimating the position of the decimal point in the result.

EXAMPLE 3

$$\frac{x}{2.47} = \frac{0.000622}{0.00125}$$

SOLUTION

(1) Opposite 125 on D, set 622 on C.
(2) Since 247 on D is off the slide, reverse the index and, opposite 247 on D, read **123** on C.

Since

$$\frac{0.000622}{0.00125} = \frac{0.622}{1.25} \approx \frac{1}{2},$$

the fraction $\frac{x}{2.47} \approx \frac{1}{2}$; the numerator equals one-half of the denominator and the decimal point is placed accordingly. **1.23**

The technique of solving for a missing term in a proportion can be used to find the reciprocal of a number. To find the reciprocal of a number n, observe that $\frac{1}{n} = \frac{x}{1}$, where x is the desired reciprocal. This procedure has an advantage over that discussed in Section A.3 in that the reciprocal of a number already located on the D scale can be found without having to reset the number on the C scale.

EXAMPLE 4

$\frac{1}{3.2} = ?$ $\left(\frac{1}{3.2} \text{ is the reciprocal of } 3.2\right)$

SOLUTION

(1) Opposite 32 on D, set index on C.
(2) Opposite index on D, read **313** on C.

Since

$$\frac{1}{3.2} \approx \frac{1}{3},$$

the decimal point is placed accordingly. **0.313**

EXAMPLE 5

$\frac{1}{12.2} = ?$

SOLUTION

(1) Opposite 122 on D, set index of C.
(2) Opposite index on D, read **820** on C.

Since

$$\frac{1}{12.2} \approx \frac{1}{10} = 0.1,$$

the decimal point is placed accordingly. **0.0820**

EXAMPLE 6

$\frac{1}{1.8 \times 4.32} = ?$

SOLUTION

(1) Opposite 18 on D, set left index of C.
(2) Opposite 432 on C, read 778 on D.
(3) Opposite 778 on D, set index of C.
(4) Opposite index on D, read **1286** on C.

Since

$$\frac{1}{1.8 \times 4.32} \approx \frac{1}{2 \times 4} = \frac{1}{8} \approx \frac{1}{10} = 0.1,$$

the decimal point is placed accordingly. **0.1286**

EXERCISE A.5

Solve for x.

1. $\frac{2}{4} = \frac{x}{8}$
2. $\frac{3}{6} = \frac{x}{4}$
3. $\frac{2}{14} = \frac{x}{28}$
4. $\frac{3}{5} = \frac{x}{25}$
5. $\frac{x}{14} = \frac{4}{28}$
6. $\frac{x}{12.5} = \frac{6}{25}$
7. $\frac{5}{8} = \frac{2.5}{x}$
8. $\frac{2}{3.1} = \frac{6}{x}$
9. $\frac{82.5}{x} = \frac{55}{4}$
10. $\frac{0.9}{x} = \frac{30}{11}$
11. $\frac{2}{4} = \frac{x}{16}$
12. $\frac{6}{x} = \frac{2}{6}$
13. $\frac{9}{2.5} = \frac{18}{x}$
14. $\frac{41}{91} = \frac{5.5}{x}$
15. $\frac{x}{19.5} = \frac{1.71}{3.91}$
16. $\frac{3.24}{5.86} = \frac{x}{14.2}$
17. $\frac{x}{34.3} = \frac{62.4}{12.2}$
18. $\frac{107.5}{x} = \frac{48.3}{2.68}$
19. $\frac{281}{100} = \frac{57.3}{x}$
20. $\frac{x}{1.48} = \frac{3.87}{2.05}$
21. $\frac{1.00}{1.05} = \frac{98.2}{x}$
22. $\frac{0.94}{9.2} = \frac{0.63}{x}$
23. $\frac{x}{6.92} = \frac{0.041}{0.032}$
24. $\frac{43}{32.1} = \frac{0.67}{x}$
25. $\frac{0.76}{0.22} = \frac{16.1}{x}$
26. $\frac{x}{4.71} = \frac{68}{48.3}$
27. $\frac{2.1}{x} = \frac{42}{2.6}$
28. $\frac{8.62}{28.4} = \frac{x}{0.762}$
29. $\frac{12.5}{9.91} = \frac{0.721}{x}$
30. $\frac{0.431}{4.7} = \frac{x}{8.62}$

Solve for x and y.

31. $\frac{0.623}{8.21} = \frac{0.421}{x} = \frac{0.743}{y}$
32. $\frac{1.89}{x} = \frac{483}{21.7} = \frac{y}{0.432}$
33. $\frac{0.00614}{0.00721} = \frac{32.1}{x} = \frac{y}{0.074}$
34. $\frac{0.408}{0.064} = \frac{6.42}{x} = \frac{y}{74}$

Compute.

35. $\frac{1}{7}$

36. $\frac{1}{4.3 \times 6.8}$

37. $\frac{1}{6.2 \times 3.1 \times 0.95}$

38. $\frac{1}{0.87 \times 2.1 \times 38}$

A.6 Percentage

Problems involving percentage can be solved efficiently with a slide rule by using the methods developed in Section A.5 to solve for a missing term in proportion.

Recall that percent simply means hundredths. For example,

$$100\% = \frac{100}{100} = 1,$$

$$63\% = \frac{63}{100} = 0.63,$$

$$0.4\% = \frac{0.4}{100} = 0.004,$$

and

$$152\% = \frac{152}{100} = 1.52.$$

Thus, many problems involving percent can be set up as a proportion. Consider the following examples.

EXAMPLE 1

18% of 130 = ?

SOLUTION

First view the left-hand member as a number, say N. Next, write the equation in the form

$$\frac{18}{100} \times 130 = N,$$

and then as the proportion

$$\frac{18}{100} = \frac{N}{130}.$$

(1) Opposite left index on D, set 18 on C.
(2) Opposite 130 on D, read **234** on C.

Since

$$\frac{18}{100} \approx \frac{20}{100} = \frac{1}{5},$$

the numerator is one-fifth of the denominator and the decimal point is placed accordingly. **23.4**

EXAMPLE 2

6 is what percent of 45?

SOLUTION

First write the equation

$$6 = \frac{N}{100} \times 45,$$

and then the proportion

$$\frac{6}{45} = \frac{N}{100}.$$

(1) Opposite 45 on D, set 6 on C.
(2) Opposite the left index on D, read **1333** on C.

Since

$$\frac{6}{45} \approx \frac{0.6}{4} = 0.15 = 15\%,$$

the decimal point is placed accordingly. **13.33%**

EXAMPLE 3

12% of what number equals 32?

SOLUTION

First write the equation

$$\frac{12}{100} \times N = 32,$$

and then the proportion

$$\frac{12}{100} = \frac{32}{N}.$$

Solution continued overleaf.

(1) Opposite left index on D, set 12 on C.
(2) Opposite 32 on C, read **267** on D.

Since

$$\frac{12}{100} \approx \frac{10}{100} = \frac{1}{10},$$

the denominator is ten times the numerator, and the decimal point is placed accordingly. **267**

EXERCISE A.6

Compute:

1. 20% of 60
2. 30% of 60
3. 12% of 50
4. 80% of 7
5. 43% of 60
6. 35% of 80
7. 54% of 450
8. 61% of 360
9. 25% of 57
10. 37.5% of 80
11. 62.5% of 400
12. 84.4% of 600
13. 140% of 20
14. 180% of 30
15. 15 is what percent of 60?
16. 12 is what percent of 60?
17. 60 is what percent of 70?
18. 20 is what percent of 35?
19. 16 is what percent of 350?
20. 21 is what percent of 380?
21. 8 is what percent of 3600?
22. 6 is what percent of 3200?
23. 86 is what percent of 50?
24. 22 is what percent of 6?
25. 874 is what percent of 900?
26. 634 is what percent of 738?
27. 0.07 is what percent of 1.54?
28. 0.23 is what percent of 3.69?
29. 20% of what number equals 15?
30. 40% of what number equals 30?
31. 5.8% of what number equals 240?
32. 4.7% of what number equals 360?
33. 2.5% of what number equals 170?
34. 3.5% of what number equals 240?
35. $2\frac{1}{4}$% of what number equals 600?
36. $3\frac{3}{4}$% of what number equals 400?

37. $5\frac{1}{2}\%$ of what number equals 1200?

38. $4\frac{1}{4}\%$ of what number equals 1600?

39. 150% of what number equals 430?

40. 210% of what number equals 170?

A.7 Squares and square roots

The A and B scales are double scales and, insofar as significant digits are concerned, are read in the same fashion as the C and D scales.

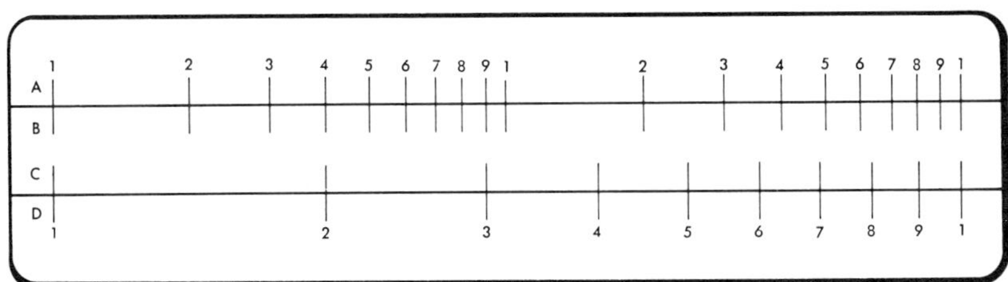

FIGURE **A.10**

Since the scale on A or B is exactly half of that on C or D and since all scales are logarithmic, the square of a number on C or D can be read directly opposite on B or A. We shall use the A and D scales. Thus:

To square a number N:

Opposite N on D, read answer on A.
Place decimal point as in multiplication.

The procedure is shown in Examples 1, 2, and 3.

EXAMPLE 1

$(34.6)^2 = ?$

SOLUTION

Opposite 34.6 on D, read **1197** on A.

Since

$$(34.6)^2 \approx (30)^2 = 900,$$

the decimal point is placed accordingly. **1197**

EXAMPLE 2

$(2.6)^2 = ?$

SOLUTION

Opposite 2.6 on D, read **676** on A.

Since

$$(2.6)^2 \approx (3)^2 = 9,$$

the decimal point is placed accordingly. **6.76**

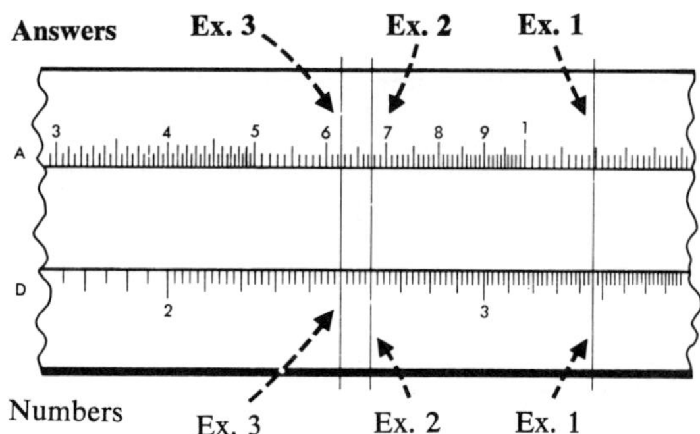

EXAMPLE 3

$(0.0025)^2 = ?$

SOLUTION

Opposite 25 on D, read **625** on A.

Since

$$(0.0025)^2 = (2.5 \times 10^{-3})^2 \approx 6 \times 10^{-6},$$

the decimal point is placed accordingly. **0.00000625**

The use of the A or B scale reduces the degree of accuracy obtainable. Numbers between primary divisions 1 and 2 on each section can be read to approximately ± 1 in the third significant digit; numbers between primary divisions 2 and 5 can be read to approximately ± 2 in the third significant digit; and numbers over the remainder of the scale can be read with an accuracy of approximately ± 3 in the third significant digit. Considering the decreased accuracy obtainable on the A and B scales, it is usually desirable to use the C and D scales to square a number, treating the process as a standard multiplication.

Finding the square root of a number is accomplished by reversing the process used to square a number. However, since the A and B scales contain two sections, you must decide whether N should be set on the left or the right section. There is a relatively straightforward means to reach this decision. If N is first expressed as a "power of ten" (Section A.4), you may examine the exponent on 10; if the exponent is even (or zero), the number is set on the left section; if the exponent is odd, the number is set on the right section. You can also decide whether N should be set on the left or right section as follows:

1. *Mark off the digits in N in groups of two to the left and right of the decimal point.*
2. *If the group containing the first significant digit of N contains only one significant digit, set N on the left section of A or B; if it contains two significant digits, set N on the right section of A or B.*
3. *Read $\sqrt{N}$ directly opposite on C or D.*

If each significant digit in the square root of N is paired with its corresponding group of two digits in N beginning with the group containing the first significant digit, the decimal point in the square root will appear directly opposite the decimal point in N.

The procedure is illustrated in Examples 4 to 7.

EXAMPLE 4

$\sqrt{235} = ?$

SOLUTION

(1) Mark 235 in groups of two digits from decimal point: $\sqrt{2\ \underset{\smile}{35}.}$
(2) Opposite 235 on left section of A read **153** on D.

Pair one digit in answer with each group of two digits in the number: $\overset{1\ \ 5\ .\ 3}{\sqrt{2\ 35.\underset{\smile}{00}}}$.
15.3

EXAMPLE 5

$\sqrt{0.00235} = ?$

Solution overleaf.

SOLUTION

(1) Mark 0.00235 in groups of two digits from decimal point: $\sqrt{0.00\ 23\ 50}$
(2) Opposite 235 on right section of A, read **485** on D.

Pair one digit in answer with each group of two digits in the number, the first digit in the answer being paired with the first group that contains a significant digit:

$$\overset{0.0\ \ 4\ \ 8\ \ 5}{\sqrt{0.00\ 23\ 50\ 00}}. \qquad \mathbf{0.0485}$$

EXAMPLE 6

$\sqrt{48500} = ?$

SOLUTION

(1) Mark 48500 in groups of two digits from decimal point: $\sqrt{4\ 85\ 00}$
(2) Opposite 485 on left section of A, read **220** on D.

Pair one digit in answer with each group of two digits in the number:

$$\overset{2\ \ 2\ \ 0}{\sqrt{4\ 85\ 00}}. \qquad \mathbf{220}$$

EXAMPLE 7

$\sqrt{0.0000463} = ?$

SOLUTION

(1) Mark 0.0000463 in groups of two digits from decimal point: $\sqrt{0.00\ 00\ 46\ 30}$
(2) Opposite 463 on right section of A, read **680** on D.

Pair one digit in answer with each group of two digits in the number:

$$\overset{0.\ 0\ \ 0\ \ 6\ \ 8\ \ 0}{\sqrt{0.00\ 00\ 46\ 30\ 00}}. \qquad \mathbf{0.00680}$$

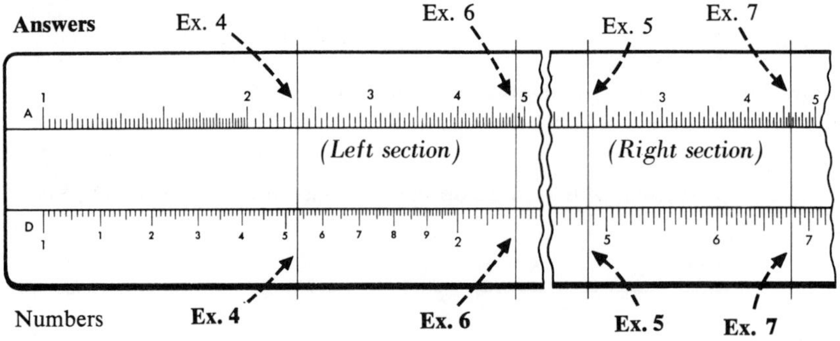

You can multiply or divide by a square root as illustrated in Examples 8, 9, and 10.

EXAMPLE 8

$3\sqrt{8} = ?$

SOLUTION

(1) Opposite 3 on D, set index of C.
(2) Opposite 8 on left section of B, read **848** on D. ($\sqrt{8}$ is located on C opposite 8 on B.)

Since

$$3\sqrt{8} \approx 3 \times 3 = 9,$$

the decimal point is placed accordingly. **8.48**

EXAMPLE 9

$\dfrac{17}{\sqrt{125}} = ?$

SOLUTION

(1) Opposite 17 on D, set 125 on left section of B. ($\sqrt{125}$ is located on C opposite 125 on B.)
(2) Opposite index on C, read **152** on D.

Since

$$\frac{17}{\sqrt{125}} \approx \frac{17}{11} \approx 1.6,$$

the decimal point is placed accordingly. **1.52**

EXAMPLE 10

$\dfrac{7.1\sqrt{42}}{\sqrt{3}} = ?$

SOLUTION

(1) Opposite 42 on right section of A, set 3 on left section of B.

$\left(\dfrac{\sqrt{42}}{\sqrt{3}}\right.$ is located on D opposite index on C.$\left.\right)$

Solution continued overleaf.

(2) Set hairline on left index of C and reverse index.
(3) Opposite 71 on C, read **266** on D.

Since

$$\frac{7.1\sqrt{42}}{\sqrt{3}} \approx \frac{7 \times 6}{2} = 21,$$

the decimal point is placed accordingly. **26.6**

EXAMPLE 11

$$\frac{\sqrt{15}\sqrt{740}}{\sqrt{8}} = ?$$

SOLUTION

(1) Opposite 15 on right section of A, set 8 on left section of B.

$\left(\frac{\sqrt{15}}{\sqrt{8}}\right.$ is located on D opposite index on C.$\left.\right)$

(2) Opposite 740 on left section of B, read **372** on D.

Since

$$\frac{\sqrt{15}\sqrt{740}}{\sqrt{8}} \approx \frac{4 \times 30}{3} = 40,$$

the decimal point is placed accordingly. **37.2**

EXERCISE A.7

Compute.

1. $(3)^2$
2. $(8.3)^2$
3. $(0.5)^2$
4. $(4.78)^2$
5. $(6.7)^2$
6. $(0.0061)^2$
7. $(35.7)^2$
8. $(4.53)^2$
9. $(2.41 \times 0.98)^2$
10. $(0.031 \times 4.7)^2$
11. $\left(\frac{27.9}{3.6}\right)^2$
12. $\left(\frac{2.5 \times 4.07}{28.1}\right)^2$
13. $\left(\frac{1}{4.8}\right)^2$
14. $\left(\frac{1}{3.52}\right)^2$
15. $\sqrt{4}$

16. $\sqrt{9}$
17. $\sqrt{8}$
18. $\sqrt{80}$
19. $\sqrt{800}$
20. $\sqrt{0.008}$
21. $\sqrt{3}$
22. $30\sqrt{3}$
23. $\dfrac{\sqrt{3}}{0.962}$
24. $\dfrac{\sqrt{7}}{22.8}$
25. $0.0062\sqrt{34}$
26. $\dfrac{0.073\sqrt{39}}{1.24}$
27. $\dfrac{0.21\sqrt{0.7}}{0.43}$
28. $\sqrt{7}\sqrt{29}$
29. $\sqrt{(0.7)(39.3)}$
30. $\dfrac{\sqrt{43}}{\sqrt{5}}$
31. $\sqrt{\dfrac{39.5}{2.1}}$
32. $3.8\sqrt{3}\sqrt{21}$
33. $\dfrac{\sqrt{4.7}\sqrt{0.9}}{3.72}$
34. $\dfrac{45\sqrt{62.1}}{\sqrt{83}}$
35. $\dfrac{23.4\sqrt{0.002}}{\sqrt{0.16}}$
36. $\dfrac{1}{\sqrt{3.92}}$
37. $\dfrac{1}{7\sqrt{3.7}}$
38. $\dfrac{1}{\sqrt{0.75}\sqrt{0.085}}$
39. $\dfrac{\sqrt{23.1}}{3\sqrt{4.7}}$
40. $\dfrac{62.3}{\sqrt{41}\sqrt{26}}$
41. $\dfrac{\sqrt{6.1}\sqrt{29}}{\sqrt{0.05}}$
42. $\dfrac{\sqrt{5.49}\sqrt{0.041}}{\sqrt{0.84}}$
43. $\dfrac{\sqrt{69.1}}{\sqrt{220}\sqrt{0.62}}$
44. $\dfrac{\sqrt{2.04}}{\sqrt{0.42}\sqrt{0.08}}$
45. $\dfrac{1}{\sqrt{2}} + \dfrac{1}{\sqrt{3}}$
46. $\dfrac{1}{2\sqrt{2}} + \dfrac{1}{3\sqrt{3}} + \dfrac{1}{4\sqrt{4}}$

A.8 Cubes and cube roots

The K and C or D scales are used to find the cube of a number or the cube root of a number. The K scale consists of three sections, each of which is similar to the C or D scale but one-third as long. Numbers whose logarithms are three times the logarithms of numbers on the C and D scales are located directly opposite on the K scale.

To cube N:

Opposite N on D, read N^3 on K.
The decimal point may be placed by approximation.

EXAMPLE 1

$(3.25)^3 = ?$

SOLUTION

Opposite 325 on D, read **343** on K.
Since

$$(3.25)^3 \approx 3^3 = 27,$$

the decimal point is placed accordingly. **34.3**

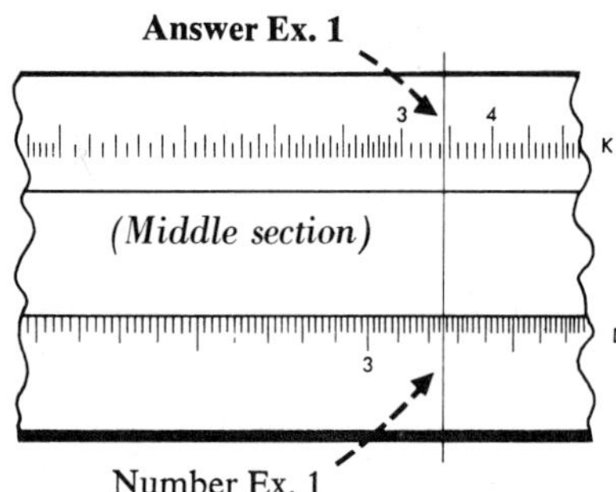

The use of the K scale reduces the degree of accuracy obtainable; numbers between primary divisions 1 and 3 on each section can be read to approximately ± 2 in the third significant digit; numbers between primary divisions 3 and 6 on each section can be read to approximately ± 3 in the third significant digit; and numbers over the remainder of the scale can be read with an accuracy of approximately ± 5 in the third significant digit.

The process of cubing N is reversed to find $\sqrt[3]{N}$. That is, N is set on the appropriate section of the K scale and $\sqrt[3]{N}$ is read directly opposite on the D scale. The procedure used to determine which of the three sections of the K scale to use for a given number N is comparable to the procedure used to determine the appropriate section of the A or B scale to use in finding the square root of N. The only difference is in the method of grouping digits.

To find $\sqrt[3]{N}$:

1. *Group the digits of N in groups of three from the decimal point.*
2. *If there is one significant digit in the group containing the first significant digit of N, set N on the left section of K; if there are two significant digits, set N on the center section of K; and if there are three significant digits, set N on the right section of K.*
3. *Read $\sqrt[3]{N}$ direcly opposite on D.*

The decimal point can be placed by a procedure similar to that used in the computation of square roots; however, one digit in $\sqrt[3]{N}$ corresponds to each group of three digits in the number N. The procedure is illustrated in Examples 2, 3, and 4.

EXAMPLE 2

$\sqrt[3]{1230} = ?$

SOLUTION

(1) Mark 1230 in groups of three digits from the decimal point: $\sqrt[3]{1\ 230}$

(2) Since left group contains one significant digit, opposite 123 on left section of K, read **107** on D.

To place the decimal point, pair one digit in the answer with each group of three digits in the number: $\sqrt[3]{1\ 230.000}$, with 1 0 . 7 written above. **10.7**

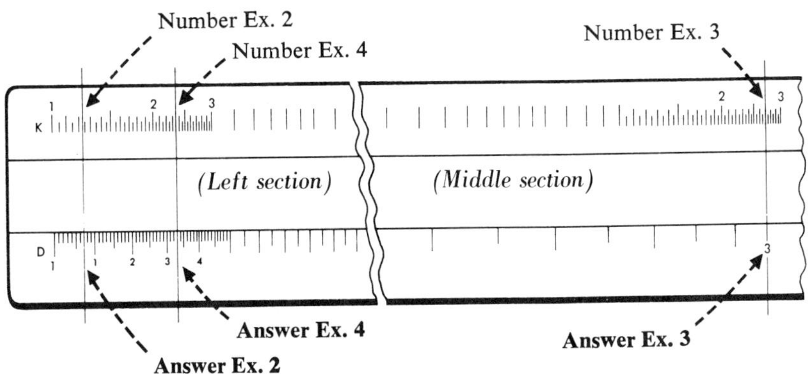

EXAMPLE 3

$\sqrt[3]{27} = ?$

SOLUTION

(1) Mark 27 in groups of three digits from the decimal point: $\sqrt[3]{27}$

(2) Since left group contains two significant digits, opposite 27 on center section of K, read **300** on D.

To place the decimal point, pair one digit in the answer with each group of three digits in the number: $\sqrt[3]{27.000\ 000}$, with 3 . 0 0 written above. **3.00**

EXAMPLE 4

$\sqrt[3]{0.00000235} = ?$

SOLUTION

(1) Mark 0.00000235 in groups of three digits from the decimal point:
$\sqrt{0.000\ 002\ 35}$
(2) Since left group containing a significant digit contains one digit, opposite 235 on left section of K, read **133** on D.

To place the decimal point, pair one digit in the answer with each group of three digits in the number: $\overset{0.\ 0\ \ 1\ \ 3\ \ 3}{\sqrt[3]{0.000\ 002\ 350\ 000}}$. **0.0133**

Computations involving cube roots as factors can be accomplished in the same way as those made in the preceding lesson using square roots, except that where the K scale is located on the stock, this setting must be made first.

In the remaining examples in this section, the details of determining the section of the K scale to be used are not shown.

EXAMPLE 5

$3\sqrt[3]{7} = ?$

SOLUTION

(1) Opposite 7 on left section of K, set index on C. ($\sqrt[3]{7}$ is now located on the D scale opposite index on C.)
(2) Opposite 3 on C, read **575** on D.

Since

$$3\sqrt[3]{7} \approx 3\sqrt[3]{8} = 3(2) = 6,$$

the decimal point is placed accordingly. **5.75**

EXAMPLE 6

$\dfrac{\sqrt[3]{17}}{2} = ?$

SOLUTION

(1) Opposite 17 on center section of K, set 2 on C. ($\sqrt[3]{17}$ is now located on the D scale opposite 2 on C.)
(2) Opposite index on C, read **129** on D.

Since

$$\frac{\sqrt[3]{17}}{2} \approx \frac{2}{2} = 1,$$

the decimal point is placed accordingly. **1.29**

EXERCISE A.8

Compute.

1. 2^3
2. 4^3
3. $(2.6)^3$
4. $(21)^3$
5. $(0.026)^3$
6. $(0.0034)^3$
7. $(0.62 \times 36.8)^3$
8. $[(0.421)(48.9)]^3$
9. $\left(\frac{26.3}{3.87}\right)^3$
10. $\left[\frac{(1007)(234.2)}{12.1}\right]^3$
11. $(\sqrt{16})^3$
12. $\left(\frac{\sqrt{54}}{3.7}\right)^3$
13. $(7\sqrt{31})^3$
14. $\left(\frac{1}{\sqrt{3.7}}\right)^3$
15. $\left[\frac{1}{(3)(3.14)}\right]^3$
16. $\left(\frac{1}{(4)(7.83)}\right)^3$
17. $\sqrt[3]{8}$
18. $\sqrt[3]{64}$
19. $\sqrt[3]{216}$
20. $\sqrt[3]{3.4}$
21. $\sqrt[3]{34}$
22. $\sqrt[3]{34000}$
23. $\sqrt[3]{0.34}$
24. $\sqrt[3]{0.00034}$
25. $\sqrt[3]{7}$
26. $\sqrt[3]{385}$
27. $\sqrt[3]{12}$
28. $6\sqrt[3]{15}$
29. $\frac{\sqrt[3]{12}}{7}$
30. $\frac{\sqrt[3]{17 \times 6.2}}{0.9}$
31. $\sqrt{10}\,\sqrt[3]{10}$
32. $\sqrt[3]{6}\,\sqrt{0.8}$
33. $\frac{\sqrt[3]{2.3} \times 400}{\sqrt{3}}$
34. $\frac{\sqrt[3]{12}\,\sqrt{5}}{17}$
35. $\frac{\sqrt{0.2}\,\sqrt[3]{0.05}}{0.072}$
36. $\frac{\sqrt[3]{241}\,\sqrt{0.9}}{5\sqrt{21}}$
37. $\frac{\sqrt{20.7}\,\sqrt[3]{13.3}}{\sqrt{0.062}}$
38. $\frac{\sqrt[3]{42.2}\,\sqrt{7.2}}{3.5\sqrt{9.3}}$
39. $\frac{4\sqrt{3}\,\sqrt[3]{21}}{1.7\sqrt{12.2}}$

40. $\dfrac{2.3\sqrt{2}\sqrt[3]{84}}{\sqrt{3}\sqrt{5}}$

41. $\dfrac{\sqrt[3]{0.094}\sqrt{0.043}}{16\sqrt{0.005}}$

42. $\dfrac{\sqrt[3]{0.0024}\sqrt{42.1}}{1.7\sqrt{6.3}}$

43. $\dfrac{1}{\sqrt[3]{37.3}}$

44. $\dfrac{1}{\sqrt[3]{0.89}}$

45. $\dfrac{1}{\sqrt{5}\sqrt[3]{33}}$

46. $\dfrac{1}{\sqrt{3}\sqrt[3]{7}}$

A.9 Logarithms to the base 10

The L scale can be used in conjunction with the D (or C) scale to find the logarithm of a number or the antilogarithm of a number in the same direct way that a table of logarithms is used.

To find the logarithm of a number x on the slide rule:

1. *Write x as the product of a number between* 1 *and* 10 *and a power of* 10; *the exponent on* 10 *will be the characteristic of the logarithm of x.*
2. *Opposite x on D, read mantissa of the logarithm of x on L.*

EXAMPLE 1

$\log_{10} 75 = ?$

SOLUTION

Opposite 75 on D, read **875** on L. (0.875 is the mantissa of $\log_{10} 75$.) Since

$$75 = 7.5 \times 10^1,$$

the characteristic of $\log_{10} 75$ is 1. **1.875** **Answer Ex. 1**

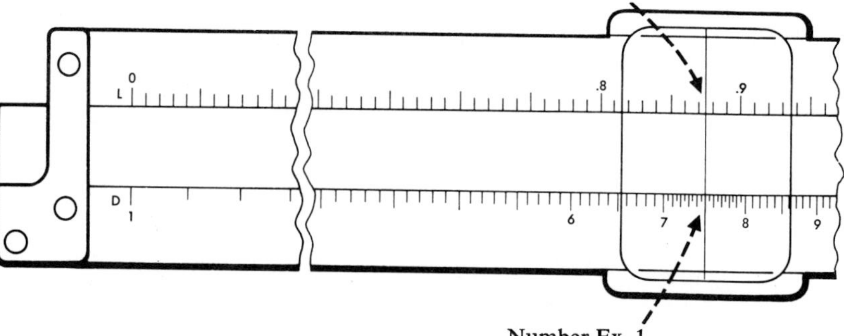

Number Ex. 1

EXAMPLE 2

$\log_{10} 0.047 = ?$

SOLUTION

Opposite 47 on D, read **672** on L. (0.672 is the mantissa of $\log_{10} 0.047$.)

Since

$$0.047 = 4.7 \times 10^{-2},$$

the characteristic of $\log_{10} 0.047$ is -2. **8.672 – 10**

To find the antilogarithm of a number x:

1. *Opposite mantissa of x on L, read antilog$_{10}$ x on D.*
2. *Write antilog$_{10}$ x as a number between 1 and 10 times a power of 10, with the characteristic of x as the exponent on 10. Antilog$_{10}$ x can be written in standard form.*

This procedure is illustrated in Examples 3 and 4.

EXAMPLE 3

$\text{antilog}_{10} 4.658 = ?$

SOLUTION

(1) Opposite 0.658 on L, read **455** on D.
(2) Write 4.55×10^4, then **45,500.**

EXAMPLE 4

antilog $8.554 - 10 = ?$

SOLUTION

(1) Opposite 0.554 on L, read **358** on D.
(2) Write 3.58×10^{-2}, then **0.0358.**

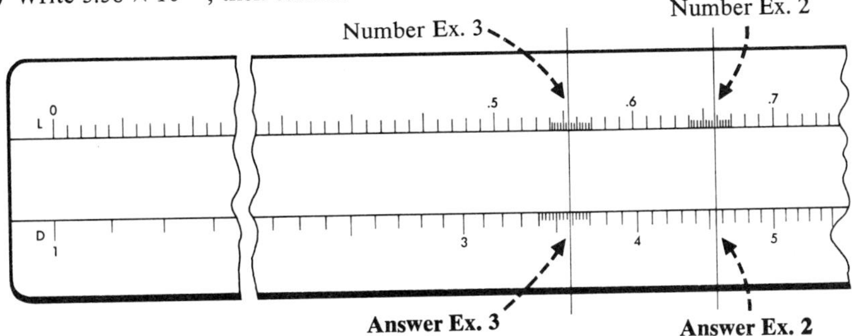

EXAMPLE 5

antilog$_{10}$ 0.432 = ?

SOLUTION

(1) Opposite 0.432 on L, read **270** on D.
(2) The characteristic of 0.432 is 0. Write 2.70×10^0, then **2.70.**

EXAMPLE 6

antilog$_{10}$ 1.082 = ?

SOLUTION

(1) Opposite 0.082 on L, read **1208** on D.
(2) Write 1.208×10^1, then **12.08.**

EXERCISE A.9

Find an approximation for each of the following.

1. $\log_{10} 3$
2. $\log_{10} 5$
3. $\log_{10} 1.45$
4. $\log_{10} 7.1$
5. $\log_{10} 9.2$
6. $\log_{10} 8.3$
7. antilog$_{10}$ 0.372
8. antilog$_{10}$ 0.845
9. antilog$_{10}$ 0.103
10. antilog$_{10}$ 0.391
11. $\log_{10} 30$
12. $\log_{10} 91$
13. $\log_{10} 300$
14. $\log_{10} 710$
15. $\log_{10} 1403.5$
16. $\log_{10} 1731.5$
17. antilog$_{10}$ 1.372
18. antilog$_{10}$ 2.845
19. antilog$_{10}$ 2.103
20. antilog$_{10}$ 3.542
21. $\log_{10} 0.3$
22. $\log_{10} 0.91$
23. $\log_{10} 0.03$
24. $\log_{10} 0.071$
25. $\log_{10} 0.0064$
26. $\log_{10} 0.00042$
27. antilog$_{10}$ 9.372 − 10
28. antilog$_{10}$ 8.845 − 10
29. antilog$_{10}$ 8.103 − 10
30. antilog$_{10}$ 7.380 − 10

A.10 Trigonometric scales: sines and cosines

Trigonometric scales on slide rules may be labeled in degrees and decimal fractions of degrees (the deci-trig rule) or in degrees and minutes (the trig rule). In this lesson we shall concentrate on the deci-trig rule, although you may convert quite easily from one type of measurement to the other by noting that 0.1 degree equals 6 minutes. Also, we shall restrict our attention to rules where the trigonometric scales S, T, and ST are used in conjunction with the C and D scales.

In this and the following lesson we will find values for the trigonometric functions, sine, cosine, and tangent, for given angles, and we will find angles for given function values. Chart I, which displays the range of values of the trigonometric functions, will be helpful in placing the decimal point, when reading specific function values.

CHART I

Measure of the angle	Range of the function
$0.575° \leq \angle A \leq 5.75°$	$0.01 \leq \sin A \leq 0.1$
$5.75° \leq \angle A \leq 90°$	$0.1 \leq \sin A \leq 1$
$0° \leq \angle A \leq 84.25°$	$1 \geq \cos A \geq 0.1$
$84.25° \leq \angle A \leq 89.425°$	$0.1 \geq \cos A \geq 0.01$
$0.575° \leq \angle A \leq 5.75°$	$0.01 \leq \tan A \leq 0.1$
$5.75° \leq \angle A \leq 45°$	$0.1 \leq \tan A \leq 1$
$45° \leq \angle A \leq 84.25°$	$1 \leq \tan A \leq 10$
$84.25° \leq \angle A \leq 89.425°$	$10 \leq \tan A \leq 100$

As an illustration of the use of Chart I, consider an angle of 35°. Note that since 35° is between 5.75° and 90°, the sine must be between 0.1 and 1. After you have found the digits for sin 35° on the slide rule, the decimal point can be placed accordingly.

The slide rule does not offer a direct means of determining the values for the trigonometric functions of very small angles ($A < 0.575°$) or angles near 90° ($A > 89.425°$). We shall not be concerned here with trigonometric functions in this domain.

We find the sine of an angle by using either the ST scale or the S scale in conjunction with the C scale. There are generally two sets of numbers on the S scale. One set increases in value from left to right and the other from right to left. To make the examples easier to follow, we shall use the letter S to mean the S scale reading from left to right and a subscript arrow on the letter $S_{\leftarrow}$ to mean the S scale reading from right to left. In our examples, the S, T, and ST scales are located on the slide. Similar procedures are applicable when these scales are on the stock.

To find sin *A*:

If $0.575° \leq \angle A \leq 5.75°$:

Opposite $\angle A$ *on ST, read* sin *A on C.* ($0.01 \leq \sin A \leq 0.1$)

If $5.75° \leq \angle A \leq 90°$:

Opposite $\angle A$ *on S, read* sin *A on C.* ($0.1 \leq \sin A \leq 1$)

EXAMPLE 1

$\sin 4° = ?$

SOLUTION

Opposite 4° on ST, read **698** on C.
Since $0.575° < 4° < 5.75°$, then $0.01 < \sin 4° < 0.1$, and the function value is **0.0698.**

The inequalities appearing in Example 1 and all following examples are taken from Chart I. Since 4° is between 0.575° and 5.75°, sin 4° must lie between 0.01 and 0.1, and the decimal point is placed accordingly.

EXAMPLE 2

$\sin 24° = ?$

SOLUTION

Opposite 24° on S, read **407** on C.
Since $5.75° < 24° < 90°$, then $0.1 < \sin 24° < 1$, and the function value is **0.407.**

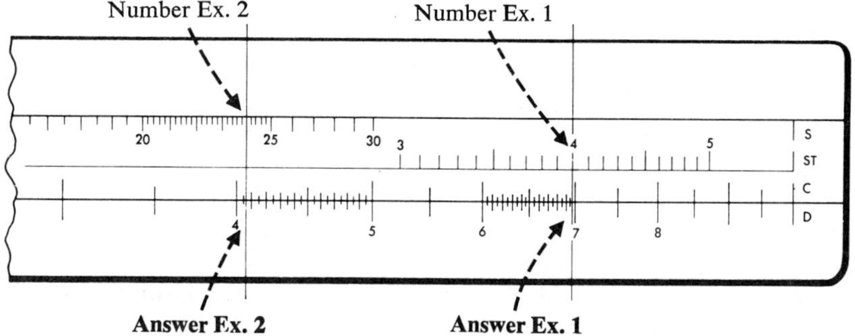

The angle A may be found when $\sin A$ is given by reversing the above process.

To find $\angle A$:

If $0.01 \leq \sin A \leq 0.1$:
Opposite $\sin A$ *on C, read* $\angle A$ *on ST.*

If $0.1 \leq \sin A \leq 1$:
Opposite $\sin A$ *on C, read* $\angle A$ *on S.*

EXAMPLE 3

If $\sin A = 0.05$, then $\angle A = ?$.

SOLUTION

Opposite 5 on C, read **2.86°** on ST.

EXAMPLE 4

If $\sin A = 0.551$, then $\angle A = ?$.

SOLUTION

Opposite 551 on C, read **33.5°** on S.

On most slide rules, the S scale carries a separate set of numbers reading from right to left. These numbers are used when working with the cosines of angles. If you are working with a rule that does not have these angles printed on it, you can use the trigonometric identity

$$\cos A = \sin (90° - A)$$

See Exercise 34, page 328. For example, to find cos 35°, we first subtract 35° from 90° to obtain 55° and then set 55° on the S scale. Cos 35° will then appear opposite on the C scale. In our examples, we shall assume the S scale carries the printed scale (which we have designated $S_{\leftarrow}$) for use with the cosine. However, the ST scale usually does not show measures of A for $\cos A$.

To find $\cos A$*:*

If $0° \leq \angle A \leq 84.25°$:
Opposite $\angle A$ *on* $S_{\leftarrow}$, *read* $\cos A$ *on* C. $(1 \geq \cos A \geq 0.1)$

If $84.25° \leq \angle A \leq 89.425°$:
Opposite $90° - \angle A$ *on* ST, *read* $\cos A$ *on* C. $(0.1 \geq \cos A \geq 0.01.)$

EXAMPLE 5

$\cos 72° = ?$

SOLUTION

Opposite 72° on $S_{\leftarrow}$, read **309** on C.
Since $0° < 72° < 84.25°$, then $1 > \cos 72° > 0.1$, and the function value is **0.309**.

EXAMPLE 6

cos 86.2° = ?

SOLUTION

Opposite 3.8° on ST, read **664** on C.
Since 84.25° < 86.2° < 89.425°, then 0.1 > cos 86.2° < 0.01, and the function value is **0.0664**.

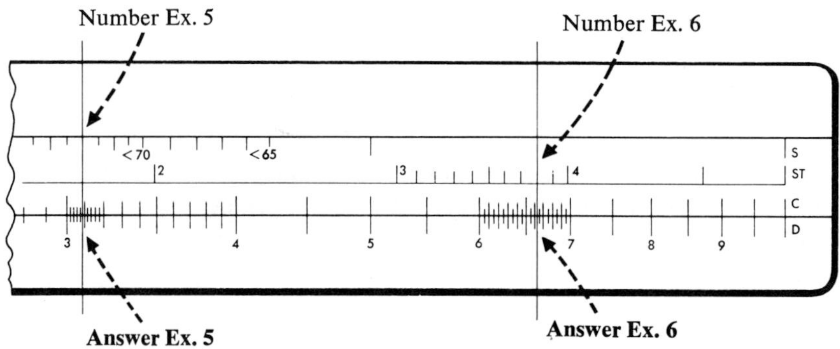

To find ∠*A* when cos *A* is given, the above process is reversed.

To find ∠*A*:

If 0.1 ≤ cos *A* ≤ 1:
Opposite cos *A on C, read* ∠*A on* $S_{\leftarrow}$.

If 0.01 ≤ cos *A* ≤ 0.1:
Opposite cos *A on C, read* 90° − ∠*A on ST.* (*Subtract from* 90° *to obtain* ∠*A*.)

EXAMPLE 7

If cos *A* = 0.866, then ∠*A* = ?.

SOLUTION

Opposite 866 on C, read **30°** on $S_{\leftarrow}$.

EXAMPLE 8

If $\cos A = 0.472$, then $\angle A = ?$.

SOLUTION

Opposite 472 on C, read **61.8**° on S←.

EXAMPLE 9

If $\cos A = 0.0410$, then $\angle A = ?$.

SOLUTION

Opposite 41 on C, read **2.35**° on ST. Take 90° − 2.35° = **87.65**° for $\angle A$.

EXAMPLE 10

If $\cos A = 0.0234$, then $\angle A = ?$.

SOLUTION

Opposite 234 on C, read **1.34** on ST. Take 90° − 1.34° = **88.66**° for $\angle A$.

Accuracy on the ST scale varies from approximately ±0.01 degree on the left to approximately ±0.05 degree on the right; on the S scale from ±0.05 degree on the left to ±5 degrees on the right.

EXERCISE A.10

Find a decimal approximation for each function value.

1. sin 30°
2. sin 40°
3. sin 22°
4. sin 54°
5. sin 73°
6. sin 70°
7. sin 10.5°
8. sin 7.3°
9. sin 15.6°
10. sin 7.25°
11. sin 4°
12. sin 3.4°
13. sin 2.4°
14. sin 1.6°
15. sin 1.25°
16. sin 1.85°

Find an approximation for each $\angle A$.

17. $\sin A = 0.628$
18. $\sin A = 0.615$
19. $\sin A = 0.431$
20. $\sin A = 0.188$
21. $\sin A = 0.500$
22. $\sin A = 0.584$
23. $\sin A = 0.941$
24. $\sin A = 0.742$
25. $\sin A = 0.234$
26. $\sin A = 0.0463$
27. $\sin A = 0.0621$
28. $\sin A = 0.0573$
29. $\sin A = 0.0125$
30. $\sin A = 0.0813$

Find a decimal approximation for each function value.

31. cos 60°
32. cos 40°
33. cos 30°
34. cos 10°
35. cos 27°
36. cos 47°
37. cos 75.5°
38. cos 58.3°
39. cos 80.5°
40. cos 47.75°
41. cos 88°
42. cos 86.2°
43. cos 87.8°
44. cos 88.9°
45. cos 87.3°
46. cos 88.4°

Find an approximation for each $\angle A$.

47. $\cos A = 0.809$
48. $\cos A = 0.604$
49. $\cos A = 0.452$
50. $\cos A = 0.314$
51. $\cos A = 0.368$
52. $\cos A = 0.702$
53. $\cos A = 0.115$
54. $\cos A = 0.0524$
55. $\cos A = 0.500$
56. $\cos A = 0.0742$

A.11 Trigonometric scales: tangents

To find the tangent of an angle, the ST or T scales are used in conjunction with the C scale. The T scale on most slide rules contains two sets of numbers, one increasing from left to right and the other from right to left. Let us first look only at the scale that increases from left to right (T), and consider tangents of angles less than 45°.

To find tan A:

If $0.575° \leq \angle A \leq 5.75°$:
Opposite $\angle A$ *on ST, read* tan A *on* C. ($0.01 \leq \tan A \leq 0.1$)

If $5.75° \leq \angle A \leq 45°$:
Opposite $\angle A$ *on T, read* tan A *on* C. ($0.1 \leq \tan A \leq 1$)

Note that the tangent of an angle between 0.575° and 5.75° is read from the ST scale in the same manner as the sine of an angle. This accounts for the label ST. Within this domain tan A and sin A do not differ enough to be distinguishable within the limits of accuracy of the slide rule.

EXAMPLE 1

$\tan 4° = ?$

SOLUTION

Opposite 4° on ST, read **698** on C.
Since $0.575° < 4° < 5.75°$, then $0.01 < \tan 4° < 0.1$, and the function value equals **0.0698**.

EXAMPLE 2

$\tan 21° = ?$

SOLUTION

Opposite 21° on T, read **384** on C.
Since $5.75° < 21° < 45°$, then $0.1 < \tan 21° < 1$, and the function value equals **0.384**.

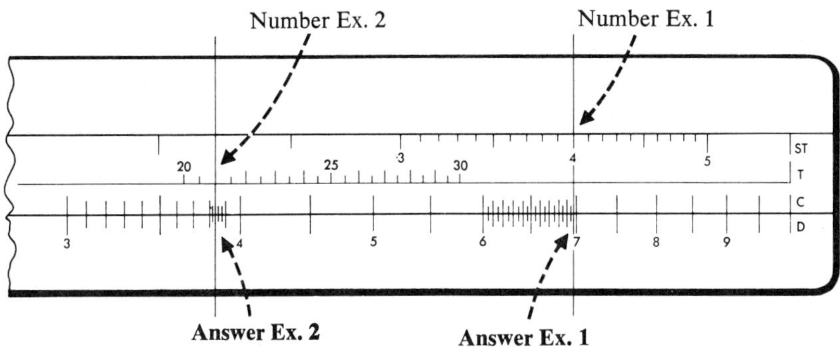

Conversely, the procedures above can be reversed to find $\angle A$ when $\tan A$ is given.

To find $\angle A$:

If $0.01 \leq \tan A \leq 0.1$:
Opposite $\tan A$ *on C, read* $\angle A$ *on ST.* $(0.575° \leq \angle A \leq 5.75°)$

If $0.1 \leq \tan A \leq 1$:
Opposite $\tan A$ *on C, read* $\angle A$ *on T.* $(5.75° \leq \angle A \leq 45°)$

EXAMPLE 3

If $\tan A = 0.0244$, then $\angle A = ?$.

SOLUTION

Since $0.01 < 0.0244 < 0.1$, then opposite 244 on C, read **1.40°** on ST.

EXAMPLE 4

If $\tan A = 0.528$, then $\angle A = ?$.

SOLUTION

Since $0.1 < 0.528 < 1$, then opposite 528 on C, read **27.8**° on T.

Since

$$\tan A = \frac{1}{\tan(90^\circ - \angle A)},$$

the tangent of an angle A greater than 45° can be found by finding the reciprocal of the tangent of $(90^\circ - \angle A)$. The set of numbers on the T scale which increase in measure from right to left ($T_{\leftarrow}$) is used in conjunction with the CI scale to find the tangents of angles between 45° and 84.25°. Since the ST scale does not have the dual labeling, the tangent of an angle A between 84.25° and 89.425° can be found by setting $90^\circ - \angle A$ on the ST scale.

To find $\tan A$:

If $45^\circ \leq \angle A \leq 84.25^\circ$:
Opposite $\angle A$ *on* $T_{\leftarrow}$, *read* $\tan A$ *on CI.* $(1 \leq \tan A \leq 10)$

If $84.25^\circ \leq \angle A \leq 89.425^\circ$:
Opposite $90^\circ - \angle A$ *on ST, read* $\tan A$ *on CI.* $(10 \leq \tan A \leq 100)$

EXAMPLE 5

$\tan 71.4^\circ = ?$

SOLUTION

Opposite 71.4° on $T_{\leftarrow}$, read **297** on CI.
Since $45^\circ < 71.4^\circ < 84.25^\circ$, then $1 < \tan 71.4^\circ < 10$, and the function value equals **2.97**.

EXAMPLE 6

$\tan 87^\circ = ?$

SOLUTION

Opposite 3° on ST, read **191** on CI.
Since $84.25^\circ < 87^\circ < 89.425^\circ$, then $10 < \tan 87^\circ < 100$, and the function value equals **19.1**.

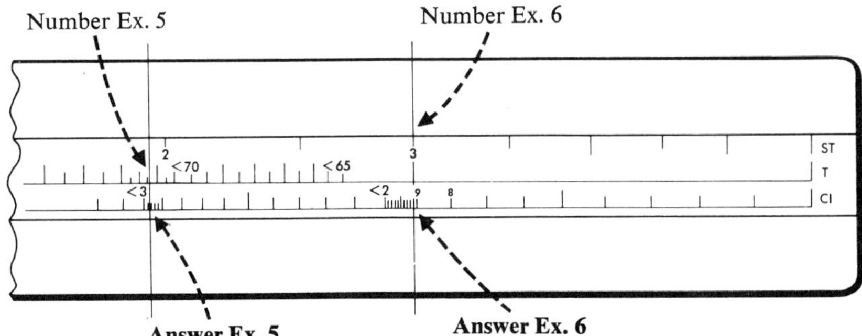

Conversely, the procedures above can be reversed to find $\angle A$ when tan A is given.

To find $\angle A$:

If $1 \leq \tan A \leq 10$:
Opposite tan A *on CI, read* $\angle A$ *on* $T_{\leftarrow}$.

If $10 \leq \tan A \leq 100$:
Opposite tan A *on CI, read* $90° - \angle A$ *on ST*. (*Subtract reading from* 90° *to obtain* $\angle A$.)

EXAMPLE 7

If $\tan A = 3.84$, then $\angle A = ?$.

SOLUTION

Since $1 < 3.84 < 10$, opposite 384 on CI, read **75.4**° on $T_{\leftarrow}$.

EXAMPLE 8

If $\tan A = 40.4$, then $\angle A = ?$.

SOLUTION

Since $10 < 40.4 < 100$, opposite 404 on CI, read **1.42**° on ST. Take $90° - 1.42° =$ **88.58**° for $\angle A$.

Accuracy on the T scale varies from ± 0.05 degree on the left to ± 0.3 degree on the right.

EXERCISE A.11

Find a decimal approximation for each function value.

1. tan 30°	2. tan 20°	3. tan 15.5°	4. tan 24.3°
5. tan 18°	6. tan 10.5°	7. tan 3°	8. tan 3.5°
9. tan 2.2°	10. tan 1.8°	11. tan 60°	12. tan 70°
13. tan 82.5°	14. tan 86.2°	15. tan 88.5°	16. tan 87.4°
17. tan 14.8°	18. tan 11.4°	19. tan 2.6°	20. tan 1.7°
21. tan 62.5°	22. tan 84.3°	23. tan 74.1°	24. tan 0.9°
25. tan 41.3°	26. tan 67.4°		

Find an approximation for each $\angle A$.

27. $\tan A = 0.194$	28. $\tan A = 0.506$	29. $\tan A = 0.0681$
30. $\tan A = 0.0346$	31. $\tan A = 0.0233$	32. $\tan A = 0.866$
33. $\tan A = 0.500$	34. $\tan A = 0.614$	35. $\tan A = 0.0623$
36. $\tan A = 1.73$	37. $\tan A = 4.681$	38. $\tan A = 3.46$
39. $\tan A = 6.201$	40. $\tan A = 6.48$	41. $\tan A = 3.20$
42. $\tan A = 5.07$	43. $\tan A = 7.95$	44. $\tan A = 8.14$
45. $\tan A = 1.03$	46. $\tan A = 1.07$	47. $\tan A = 12.25$
48. $\tan A = 15.74$	49. $\tan A = 31.2$	50. $\tan A = 47.9$
51. $\tan A = 52.0$	52. $\tan A = 54.5$	

A.12 Solution of triangles

The slide rule offers a rapid means of solving certain types of problems arising in trigonometry. We shall restrict our treatment here to the solution of triangles that can be performed efficiently on the slide rule.

From the definitions of the trigonometric functions, we have the following (see Figure A.11):

$$\sin A = \frac{a}{c} \qquad \sin B = \frac{b}{c}$$

$$\cos A = \frac{b}{c} \qquad \cos B = \frac{a}{c}$$

$$\tan A = \frac{a}{b} \qquad \tan B = \frac{b}{a}$$

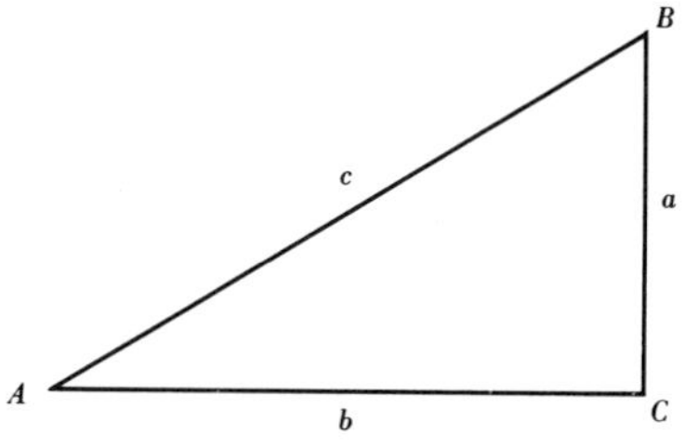

FIGURE **A.11**

As you have seen in Section 13.5, if two parts of a right triangle (other than the right angle) are known, any of the remaining parts can be determined by the use of one or more of the three trigonometric functions noted.

We can approximate the position of the decimal point by considering the range of the trigonometric function as listed in Chart I, Section A.9. For simplicity, we shall not include units of measurement for the length of the sides of the triangles in the examples and in the exercises.

EXAMPLE 1

Solve for a, given $b = 6.42$ and $\angle A = 20°$.

SOLUTION

$$\tan 20° = \frac{a}{6.42}; \quad a = 6.42 \tan 20°.$$

(1) Opposite 642 on D, set index on C.
(2) Opposite 20° on T, read **234** on D.

Since $b = 6.42$, we take $a =$ **2.34**.

EXAMPLE 2

Solve for c, given $a = 62$ and $\angle B = 41°$.

SOLUTION

$$\cos 41° = \frac{62}{c}; \quad c = \frac{62}{\cos 41°}$$

(1) Opposite 62 on D, set 41° on $S_{\leftarrow}$.
(2) Opposite index on C, read **823** on D.

Since $a = 62$, we take $c =$ **82.3**.

EXAMPLE 3

Solve for a, given $b = 23.5$ and $c = 41.7$.

SOLUTION

$$\sin B = \frac{23.5}{41.7}.$$

(1) Opposite 235 on D, set 417 on C.
(2) Set hairline on index on C.
(3) Opposite index on D, set index on C.
(4) Under hairline, read **34.3°** on S.

$$\cos 34.3° = \frac{a}{41.7}; \quad a = 41.7 \cos 34.3°.$$

Solution continued overleaf.

(5) Opposite 417 on D, set index on C.
(6) Opposite 34.3 on $S_{\leftarrow}$, read **344** on D.

Since $b = 23.5$, we take $a =$ **34.4**.

When one side and one acute angle of a right triangle are known, an alternative method of solution is available. This method has the advantage of reducing the computation to a single setting on the slide rule, and is based on the law of sines in the form

$$\frac{\sin A}{a} = \frac{\sin B}{b} = \frac{\sin C}{c}.$$

To apply the law of sines given the length of one side and one acute angle A, first determine the remaining acute angle B from the relationship $B = 90° - \angle A$. Then apply the law of sines.

1. *Opposite the length of the given side on D, set the appropriate angle on S.*
2. *Read lengths of remaining sides on D opposite remaining angles on S. If either angle is off the stock, reverse the index. The decimal point is placed by approximation, the smallest side being opposite the smallest angle.*

EXAMPLE 4

Solve for b and c, given $a = 3$ and $\angle A = 31°$.

SOLUTION

$$\frac{\sin 31°}{3} = \frac{\sin 59°}{b} = \frac{\sin 90°}{c}$$

(1) Opposite 3 on D, set 31° on S.
(2) Opposite 59° on S, read **500** on D.
(3) Opposite 90° on S, read **583** on D.

Since $a = 3$, we take $b =$ **5.00** and $c =$ **5.83**.

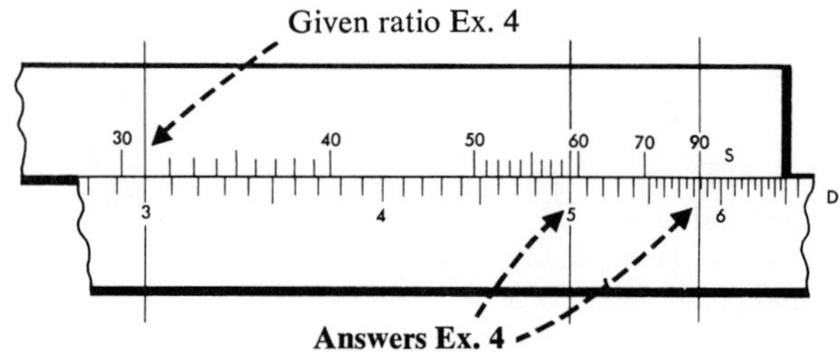

In the event that one of the angles involved is less than 5.75°, the ST scale is used for the initial setting instead of the S scale.

EXAMPLE 5

Solve for b and c, given $a = 4$ and $\angle A = 3°$.

SOLUTION

$$\frac{\sin 3°}{4} = \frac{\sin 87°}{b} = \frac{\sin 90°}{c}$$

(1) Opposite 4 on D, set 3° on ST.
(2) Opposite 87° on S, read **763** on D.
(3) Opposite 90° on S, read **765** on D.

Since $a = 4$, we take $b =$ **76.3** and $c =$ **76.5**

Given the two sides of a right angle, the tangent of one of the acute angles and the angle itself can be determined. The remaining parts of the triangle can then be found as above.

EXAMPLE 6

Solve for $\angle A$, $\angle B$, and c, given $a = 15$ and $b = 17$.

SOLUTION

(1) Opposite 17 on D, set 15 on C.
(2) Opposite index on D, read **41.4°** on T.

$$(\angle B = 90° - 41.4° = \mathbf{48.6°})$$

$$\frac{\sin 41.4°}{15} = \frac{\sin 48.6°}{17} = \frac{\sin 90°}{c}$$

(3) Opposite 15 on D, set 41.4° on S.
(4) Opposite 90° on S, read **227** on D.

Hence, $\angle A =$ **41.4°**, $\angle B =$ **48.6°**, and $c =$ **22.7**.

As we observed in Section 13.7, given two angles and one side of an oblique triangle, the third angle can be found immediately by subtracting the sum of the two given angles from 180°. The two unknown sides can then be found with one setting of the rule by use of the law of sines.

EXAMPLE 7

Solve for $\angle A$, b and c, given $a = 40$, $\angle B = 25°$, and $\angle C = 80°$.

Solution overleaf.

SOLUTION

(1) $\angle A = 180° - (80° + 25°) = \mathbf{75}°$.

$$\frac{\sin 75°}{40} = \frac{\sin 25°}{b} = \frac{\sin 80°}{c}$$

(2) Opposite 40 on D, set 75° on S.
(3) Opposite 25° on S, read **175** on D.
(4) Opposite 80° on S, read **408** on D.

Hence, $\angle A = \mathbf{75}°$, $b = \mathbf{17.5}$, and $c = \mathbf{40.8}$.

Given two sides and an angle opposite one of the given sides, the law of sines can also be used, but this time an intermediate step is needed to determine the third angle.

EXAMPLE 8

Find $\angle B$, a, and c if $b = 2.4$, $\angle A = 13°$, and $\angle C = 110°$.

SOLUTION

$\angle B = 180° - (13° + 110°) = \mathbf{57}°$.

$$\frac{\sin 57°}{2.4} = \frac{\sin (180° - 110°)}{c} = \frac{\sin 13°}{a}$$

(1) Opposite 24 on D, set 57° on S.
(2) Opposite 70° on S, read **269** on D.
(3) Reverse index; opposite 13° on S, read **644** on D.

Hence, $\angle B = \mathbf{57}°$, $a = \mathbf{0.644}$, and $c = \mathbf{2.69}$.

EXAMPLE 9

Find c, $\angle A$, and $\angle C$ if $a = 12$, $b = 15$, and $\angle B = 55°$.

SOLUTION

$$\frac{\sin 55°}{15} = \frac{\sin A}{12} = \frac{\sin C}{c}.$$

(1) Opposite 15 on D, set 55° on S.
(2) Opposite 12 on D, read **40.9°** on S.
(3) $\angle C = 180° - (55° + 40.9°) = 84.1°$.
(4) Opposite 84.1° on S, read **1822** on D.

Hence, $c = \mathbf{18.22}$, $\angle A = \mathbf{40.9}°$, and $\angle C = \mathbf{84.1}°$.

EXAMPLE 10

Find a, $\angle A$, and $\angle C$, if $b = 8.5$, $c = 6.3$, and $\angle B = 42.5°$.

SOLUTION

$$\frac{\sin 42.5°}{8.5} = \frac{\sin C}{6.3} = \frac{\sin A}{a}.$$

(1) Opposite 85 on D, set 42.5° on S.
(2) Opposite 6.3 on D, read **30.0**° on S.
(3) $\angle A = 180° - (42.5° + 30.0°) = 107.5°$.
(4) Reverse index; opposite 72.5° (180° − 107.5°) on S, read **1200** on D.

Therefore $a = \mathbf{12.0}$, $\angle A = \mathbf{107.5}°$, and $\angle C = \mathbf{30.0}°$.

An analysis of the ambiguity involved in problems of the type in Examples 9 and 10, in which two sides and the angle opposite one of these sides are known, will not be undertaken here. Neither do we discuss the case where two sides and the included angle are given or the case where three sides are given. These cases are best solved using the law of cosines, which involves the operations of addition and subtraction. These operations are not readily performed on the slide rule.

EXERCISE A.12

Given *right* triangle ABC, find each of the following angles and sides.

1. $c = 20$ and $\angle A = 30°$; $a = ?$
2. $a = 21.2$ and $\angle A = 20.5°$; $c = ?$
3. $a = 0.601$ and $\angle A = 10.5°$; $b = ?$
4. $b = 12.5$ and $\angle A = 22°$; $a = ?$
5. $c = 0.84$ and $\angle A = 54°$; $b = ?$
6. $a = 48$ and $c = 81$; $\angle A = ?$
7. $a = 2.3$ and $c = 4.6$; $\angle A = ?$
8. $b = 142$ and $c = 200$; $\angle A = ?$
9. $b = 12.3$ and $c = 45.0$; $\angle B = ?$
10. $\angle B = 69°$ and $c = 63$; $a = ?$
11. $a = 0.072$ and $c = 0.193$; $\angle A = ?$
12. $\angle A = 45°$ and $b = 0.0024$; $c = ?$
13. $\angle A = 31°$ and $a = 17.2$; $\angle B = ?$, $b = ?$, $c = ?$
14. $\angle A = 56.5°$ and $b = 25.8$; $\angle B = ?$, $a = ?$, $c = ?$
15. $a = 2.3$ and $c = 4.6$; $\angle A = ?$, $\angle B = ?$, $b = ?$
16. $b = 14.3$ and $c = 26.4$; $\angle B = ?$, $\angle A = ?$, $a = ?$

17. $a = 17$ and $b = 19$;
$\angle B = ?$, $\angle A = ?$, $c = ?$

18. $a = 15.2$ and $b = 12.4$;
$\angle B = ?$, $\angle A = ?$, $c = ?$

19. $\angle A = 32.4°$ and $c = 27.4$,
$\angle B = ?$, $a = ?$, $b = ?$

20. $\angle B = 22.3°$ and $a = 14.6$;
$\angle A = ?$, $b = ?$, $c = ?$

Given *oblique* triangle ABC. Find each of the following measures.

21. $a = 12$, $\angle A = 30°$, and $\angle B = 80°$;
$b = ?$, $c = ?$, $\angle C = ?$

22. $c = 234$, $\angle C = 64.45°$, and $\angle A = 53.63°$;
$\angle B = ?$, $a = ?$, $b = ?$

23. $b = 49.5$, $\angle B = 50.83°$, and $\angle C = 28.6°$;
$\angle A = ?$, $c = ?$, $a = ?$

24. $a = 21.6$, $b = 26.7$, and $\angle A = 34.25°$;
$\angle B = ?$, $\angle C = ?$, $c = ?$

25. $a = 540$, $c = 600$, and $\angle C = 30°$;
$\angle A = ?$, $b = ?$, $\angle B = ?$

26. $b = 5.21$, $c = 4.64$, and $\angle B = 42.4°$;
$\angle C = ?$, $\angle A = ?$, $a = ?$

27. $a = 17.3$, $c = 24.4$, and $\angle C = 37.7°$;
$\angle A = ?$, $\angle B = ?$, $b = ?$

28. $a = 0.79$, $b = 0.63$, and $\angle A = 21.8°$;
$\angle B = ?$, $\angle C = ?$, $c = ?$

29. $b = 3.21$, $c = 1.71$, and $\angle B = 54.3°$;
$\angle C = ?$, $\angle A = ?$, $a = ?$

30. $a = 217$, $c = 494.4$, and $\angle C = 76.5°$;
$\angle A = ?$, $\angle B = ?$, $b = ?$

Some properties of numbers

[1.3] If $a = b$, then $a + c = b + c$ and $c + a = c + b$.

If $a = b$, then $ac = bc$ and $ca = cb$.

For every $a \in R$, $a \cdot 0 = 0$.

For each $a \in R$, $-(-a) = a$.

$$|a| = \begin{cases} a & \text{if } a \geq 0, \\ -a & \text{if } a < 0. \end{cases}$$

[1.4] $(-a) + (-b) = -(a + b)$.

$a - b = c$ if and only if $b + c = a$.

$a - b = a + (-b)$.

[1.5] $a(-b) = -ab$.

$(-a)(b) = -ab$.

$(-a)(-b) = ab$.

$\frac{a}{b} = q$ if and only if $bq = a$; $b \neq 0$.

$\frac{a}{b} = a\left(\frac{1}{b}\right)$; $b \neq 0$.

[2.3] $a^m \cdot a^n = a^{m+n}$.

[2.4] $\frac{a^m}{a^n} = a^{m-n}$; $a \neq 0$.

[2.7] $x^3 + a^3 = (x + a)(x^2 - ax + a^2)$

$x^3 - a^3 = (x - a)(x^2 + ax + a^2)$.

[3.1] $\frac{a}{b} = \frac{c}{d}$ if and only if $ad = bc$; $b, d \neq 0$.

$\frac{a}{b} = \frac{ac}{bc}$; $b, c \neq 0$.

$\frac{a}{b} = \frac{-a}{-b} = -\frac{a}{-b} = -\frac{-a}{b}$; $b \neq 0$.

$\frac{-a}{b} = \frac{a}{-b} = -\frac{a}{b} = -\frac{-a}{-b}$; $b \neq 0$.

[3.5] $\frac{a}{c} + \frac{b}{c} = \frac{a+b}{c}$; $c \neq 0$.

$\frac{a}{b} + \frac{c}{d} = \frac{ad+bc}{bd}$; $b, d \neq 0$.

$\frac{a}{c} - \frac{b}{c} = \frac{a-b}{c}$; $c \neq 0$.

$\frac{a}{b} - \frac{c}{d} = \frac{ad-bc}{bd}$; $b, d \neq 0$.

[3.6] $\frac{a}{b} \cdot \frac{c}{d} = \frac{ac}{bd}$; $b, d \neq 0$.

$\frac{a}{b} \div \frac{c}{d} = \frac{ad}{bc}$; $b, c, d \neq 0$.

[5.1] $a^m \cdot a^n = a^{m+n}$.

$\frac{a^m}{a^n} = a^{m-n} = \frac{1}{a^{n-m}}$; $a \neq 0$.

$(a^m)^n = a^{mn}$.

$(ab)^n = a^n b^n$.

$\left(\frac{a}{b}\right)^n = \frac{a^n}{b^n}$; $b \neq 0$.

[5.2] $a^{1/n}$ denotes a number such that $(a^{1/n})^n = a$; $a \geq 0, n \in N$.

$a^{1/n}$ denotes a number such that $(a^{1/n})^n = a$; $a < 0$, n an odd natural number.

$a^{m/n} = (a^{1/n})^m = (a^m)^{1/n}$.

[5.3] $a^0 = 1$; $a \neq 0$.

$a^{-n} = \frac{1}{a^n}$; $a \neq 0, n \in N$.

[6.1] $\sqrt[n]{a} = a^{1/n}$; $a^{1/n} \in R$.

$\sqrt[n]{a^n} = |a|$; n an even natural number.

$\sqrt[n]{a^n} = a$; n an odd natural number.

[6.2] $\sqrt[n]{ab} = \sqrt[n]{a}\sqrt[n]{b}$; $\sqrt[n]{a}, \sqrt[n]{b} \in R$.

$$\sqrt[n]{\frac{a}{b}} = \frac{\sqrt[n]{a}}{\sqrt[n]{b}}; \quad \sqrt[n]{a}, \sqrt[n]{b} \in R.$$

[6.7] $\sqrt{-b} = i\sqrt{b}, \quad b > 0$.

[7.1] $ab = 0$ if and only if $a = 0$ or $b = 0$.

[7.4] If $ax^2 + bx + c = 0$, $a \neq 0$, then

$$x = \frac{-b + \sqrt{b^2 - 4ac}}{2a} \quad \text{or} \quad x = \frac{-b - \sqrt{b^2 - 4ac}}{2a}.$$

[8.5] For any two points in a geometric plane (x_1, y_1) and (x_2, y_2), the distance d between the points is given by

$$d = \sqrt{(x_2 - x_1)^2 + (y_2 - y_1)^2},$$

and the slope m of the line containing the points is given by

$$m = \frac{y_2 - y_1}{x_2 - x_1}; \quad x_2 \neq x_1.$$

[8.6] The point slope of an equation of a line is

$$y - y_1 = m(x - x_1),$$

where the line has slope m and contains the point corresponding to (x_1, y_1). The slope-intercept form for an equation of a line is

$$y = mx + b,$$

where the line has slope m and y-intercept b.

[9.4] $y = kx$. Direct variation.

$y = \dfrac{k}{x}$. Indirect variation

[10.3] $\begin{vmatrix} a_1 & b_1 \\ a_2 & b_2 \end{vmatrix} = a_1b_2 - a_2b_1$.

If $a_1x + b_1y = c_1$
$a_2x + b_2y = c_2$, then

$$x = \frac{D_x}{D} = \frac{\begin{vmatrix} c_1 & b_1 \\ c_2 & b_2 \end{vmatrix}}{\begin{vmatrix} a_1 & b_1 \\ a_2 & b_2 \end{vmatrix}} \quad \text{and} \quad y = \frac{D_y}{D} = \frac{\begin{vmatrix} a_1 & c_1 \\ a_2 & c_2 \end{vmatrix}}{\begin{vmatrix} a_1 & b_1 \\ a_2 & b_2 \end{vmatrix}}.$$

[10.4] $\begin{vmatrix} a_1 & b_1 & c_1 \\ a_2 & b_2 & c_2 \\ a_3 & b_3 & c_3 \end{vmatrix} = a_1 b_2 c_3 - a_1 b_3 c_2 + a_3 b_1 c_2 - a_2 b_1 c_3 + a_2 b_3 c_1 - a_3 b_2 c_1.$

[10.6] If $a_1 x + b_1 y + c_1 z = d_1$
$a_2 x + b_2 y + c_2 z = d_2$
$a_3 x + b_3 y + c_3 z = d_3$, then $x = \frac{D_x}{D}$, $y = \frac{D_y}{D}$, and $z = \frac{D_z}{D}$.

[11.3] For all $b > 0$, $b \neq 1$, and $x > 0$:

$$b^{\log_b x} = x,$$

$$\log_b(x_1 x_2) = \log_b x_1 + \log_b x_2,$$

$$\log_b \frac{x_2}{x_1} = \log_b x_2 - \log_b x_1,$$

$$\log_b(x_1)^m = m \log_b x_1.$$

[11.6] If $M = N$ $(M, N > 0)$, then $\log_b M = \log_b N$.
If $\log_b M = \log_b N$, then $M = N$.

[12.3] For special properties of triangles see Chart 12.2 (page 288).

$$\angle A + \angle B + \angle C = 180°$$

[12.4] For special properties of quadrilaterals see Chart 12.3 (page 292).

[12.5] For special properties of circles see Chart 12.4 (page 296).

[12.6] For special properties of geometric solids see Chart 12.5 (page 299).

[13.2] $\sin A = \frac{y}{\sqrt{x^2 + y^2}}$ $\qquad \csc A = \frac{\sqrt{x^2 + y^2}}{y}$ $(y \neq 0)$

$\cos A = \frac{x}{\sqrt{x^2 + y^2}}$ $\qquad \sec A = \frac{\sqrt{x^2 + y^2}}{x}$ $(x \neq 0)$

$\tan A = \frac{y}{x}$ $(x \neq 0)$ $\qquad \cot A = \frac{x}{y}$ $(y \neq 0)$

[13.7] Law of sines:

$$\frac{\sin A}{a} = \frac{\sin B}{b} = \frac{\sin C}{c}$$

[13.9] Law of cosines:

$$a^2 = b^2 + c^2 - 2bc \cos A$$

$$b^2 = a^2 + c^2 - 2ac \cos B$$

$$c^2 = a^2 + b^2 - 2ab \cos C$$

Table of measures

Liquid measure

16 fluid ounces = 1 pint
2 pints = 1 quart
4 quarts = 1 gallon
1.06 quarts ≈ 1 liter

Length

39.37 inches = 1 meter
2.54 centimeters ≈ 1 inch
12 inches = 1 foot
3 feet = 1 yard
5280 feet = 1 mile
6,080 feet ≈ 1 nautical mile

Weight

16 ounces = 1 pound
2000 pounds = 1 ton

Area

144 square inches = 1 square foot
9 square feet = 1 square yard

Volume

1728 cubic inches = 1 cubic foot
27 cubic feet = 1 cubic yard

Table I Squares, square roots, and prime factors

No.	Sq.	Sq. Root	Prime Factors	No.	Sq.	Sq. Root	Prime Factors
1	1	1.000		51	2,601	7.141	$3 \cdot 17$
2	4	1.414	2	52	2,704	7.211	$2^2 \cdot 13$
3	9	1.732	3	53	2,809	7.280	53
4	16	2.000	2^2	54	2,916	7.348	$2 \cdot 3^3$
5	25	2.236	5	55	3,025	7.416	$5 \cdot 11$
6	36	2.449	$2 \cdot 3$	56	3,136	7.483	$2^3 \cdot 7$
7	49	2.646	7	57	3,249	7.550	$3 \cdot 19$
8	64	2.828	2^3	58	3,364	7.616	$2 \cdot 29$
9	81	3.000	3^2	59	3,481	7.681	59
10	100	3.162	$2 \cdot 5$	60	3,600	7.746	$2^2 \cdot 3 \cdot 5$
11	121	3.317	11	61	3,721	7.810	61
12	144	3.464	$2^2 \cdot 3$	62	3,844	7.874	$2 \cdot 31$
13	169	3.606	13	63	3,969	7.937	$3^2 \cdot 7$
14	196	3.742	$2 \cdot 7$	64	4,096	8.000	2^6
15	225	3.873	$3 \cdot 5$	65	4,225	8.062	$5 \cdot 13$
16	256	4.000	2^4	66	4,356	8.124	$2 \cdot 3 \cdot 11$
17	289	4.123	17	67	4,489	8.185	67
18	324	4.243	$2 \cdot 3^2$	68	4,624	8.246	$2^2 \cdot 17$
19	361	4.359	19	69	4,761	8.307	$3 \cdot 23$
20	400	4.472	$2^2 \cdot 5$	70	4,900	8.367	$2 \cdot 5 \cdot 7$
21	441	4.583	$3 \cdot 7$	71	5,041	8.426	71
22	484	4.690	$2 \cdot 11$	72	5,184	8.485	$2^3 \cdot 3^2$
23	529	4.796	23	73	5,329	8.544	73
24	576	4.899	$2^3 \cdot 3$	74	5,476	8.602	$2 \cdot 37$
25	625	5.000	5^2	75	5,625	8.660	$3 \cdot 5^2$
26	676	5.099	$2 \cdot 13$	76	5,776	8.718	$2^2 \cdot 19$
27	729	5.196	3^3	77	5,929	8.775	$7 \cdot 11$
28	784	5.292	$2^2 \cdot 7$	78	6,084	8.832	$2 \cdot 3 \cdot 13$
29	841	5.385	29	79	6,241	8.888	79
30	900	5.477	$2 \cdot 3 \cdot 5$	80	6,400	8.944	$2^4 \cdot 5$
31	961	5.568	31	81	6,561	9.000	3^4
32	1,024	5.657	2^5	82	6,724	9.055	$2 \cdot 41$
33	1,089	5.745	$3 \cdot 11$	83	6,889	9.110	83
34	1,156	5.831	$2 \cdot 17$	84	7,056	9.165	$2^2 \cdot 3 \cdot 7$
35	1,225	5.916	$5 \cdot 7$	85	7,225	9.220	$5 \cdot 17$
36	1,296	6.000	$2^2 \cdot 3^2$	86	7,396	9.274	$2 \cdot 43$
37	1,369	6.083	37	87	7,569	9.327	$3 \cdot 29$
38	1,444	6.164	$2 \cdot 19$	88	7,744	9.381	$2^3 \cdot 11$
39	1,521	6.245	$3 \cdot 13$	89	7,921	9.434	89
40	1,600	6.325	$2^3 \cdot 5$	90	8,100	9.487	$2 \cdot 3^2 \cdot 5$
41	1,681	6.403	41	91	8,281	9.539	$7 \cdot 13$
42	1,764	6.481	$2 \cdot 3 \cdot 7$	92	8,464	9.592	$2^2 \cdot 23$
43	1,849	6.557	43	93	8,649	9.644	$3 \cdot 31$
44	1,936	6.633	$2^2 \cdot 11$	94	8,836	9.695	$2 \cdot 47$
45	2,025	6.708	$3^2 \cdot 5$	95	9,025	9.747	$5 \cdot 19$
46	2,116	6.782	$2 \cdot 23$	96	9,216	9.798	$2^5 \cdot 3$
47	2,209	6.856	47	97	9,409	9.849	97
48	2,304	6.928	$2^4 \cdot 3$	98	9,604	9.899	$2 \cdot 7^2$
49	2,401	7.000	7^2	99	9,801	9.950	$3^2 \cdot 11$
50	2,500	7.071	$2 \cdot 5^2$	100	10,000	10.000	$2^2 \cdot 5^2$

Table II Common logarithms

x	0	1	2	3	4	5	6	7	8	9
1.0	.0000	.0043	.0086	.0128	.0170	.0212	.0253	.0294	.0334	.0374
1.1	.0414	.0453	.0492	.0531	.0569	.0607	.0645	.0682	.0719	.0755
1.2	.0792	.0828	.0864	.0899	.0934	.0969	.1004	.1038	.1072	.1106
1.3	.1139	.1173	.1206	.1239	.1271	.1303	.1335	.1367	.1399	.1430
1.4	.1461	.1492	.1523	.1553	.1584	.1614	.1644	.1673	.1703	.1732
1.5	.1761	.1790	.1818	.1847	.1875	.1903	.1931	.1959	.1987	.2014
1.6	.2041	.2068	.2095	.2122	.2148	.2175	.2201	.2227	.2253	.2279
1.7	.2304	.2330	.2355	.2380	.2405	.2430	.2455	.2480	.2504	.2529
1.8	.2553	.2577	.2601	.2625	.2648	.2672	.2695	.2718	.2742	.2765
1.9	.2788	.2810	.2833	.2856	.2878	.2900	.2923	.2945	.2967	.2989
2.0	.3010	.3032	.3054	.3075	.3096	.3118	.3139	.3160	.3181	.3201
2.1	.3222	.3243	.3263	.3284	.3304	.3324	.3345	.3365	.3385	.3404
2.2	.3424	.3444	.3464	.3483	.3502	.3522	.3541	.3560	.3579	.3598
2.3	.3617	.3636	.3655	.3674	.3692	.3711	.3729	.3747	.3766	.3784
2.4	.3802	.3820	.3838	.3856	.3874	.3892	.3909	.3927	.3945	.3962
2.5	.3979	.3997	.4014	.4031	.4048	.4065	.4082	.4099	.4116	.4133
2.6	.4150	.4166	.4183	.4200	.4216	.4232	.4249	.4265	.4281	.4298
2.7	.4314	.4330	.4346	.4362	.4378	.4393	.4409	.4425	.4440	.4456
2.8	.4472	.4487	.4502	.4518	.4533	.4548	.4564	.4579	.4594	.4609
2.9	.4624	.4639	.4654	.4669	.4683	.4698	.4713	.4728	.4742	.4757
3.0	.4771	.4786	.4800	.4814	.4829	.4843	.4857	.4871	.4886	.4900
3.1	.4914	.4928	.4942	.4955	.4969	.4983	.4997	.5011	.5024	.5038
3.2	.5051	.5065	.5079	.5092	.5105	.5119	.5132	.5145	.5159	.5172
3.3	.5185	.5198	.5211	.5224	.5237	.5250	.5263	.5276	.5289	.5302
3.4	.5315	.5328	.5340	.5353	.5366	.5378	.5391	.5403	.5416	.5428
3.5	.5441	.5453	.5465	.5478	.5490	.5502	.5514	.5527	.5539	.5551
3.6	.5563	.5575	.5587	.5599	.5611	.5623	.5635	.5647	.5658	.5670
3.7	.5682	.5694	.5705	.5717	.5729	.5740	.5752	.5763	.5775	.5786
3.8	.5798	.5809	.5821	.5832	.5843	.5855	.5866	.5877	.5888	.5899
3.9	.5911	.5922	.5933	.5944	.5955	.5966	.5977	.5988	.5999	.6010
4.0	.6021	.6031	.6042	.6053	.6064	.6075	.6085	.6096	.6107	.6117
4.1	.6128	.6138	.6149	.6160	.6170	.6180	.6191	.6201	.6212	.6222
4.2	.6232	.6243	.6253	.6263	.6274	.6284	.6294	.6304	.6314	.6325
4.3	.6335	.6345	.6355	.6365	.6375	.6385	.6395	.6405	.6415	.6425
4.4	.6435	.6444	.6454	.6464	.6474	.6484	.6493	.6503	.6513	.6522
4.5	.6532	.6542	.6551	.6561	.6571	.6580	.6590	.6599	.6609	.6618
4.6	.6628	.6637	.6646	.6656	.6665	.6675	.6684	.6693	.6702	.6712
4.7	.6721	.6730	.6739	.6749	.6758	.6767	.6776	.6785	.6794	.6803
4.8	.6812	.6821	.6830	.6839	.6848	.6857	.6866	.6875	.6884	.6893
4.9	.6902	.6911	.6920	.6928	.6937	.6946	.6955	.6964	.6972	.6981
5.0	.6990	.6998	.7007	.7016	.7024	.7033	.7042	.7050	.7059	.7067
5.1	.7076	.7084	.7093	.7101	.7110	.7118	.7126	.7135	.7143	.7152
5.2	.7160	.7168	.7177	.7185	.7193	.7202	.7210	.7218	.7226	.7235
5.3	.7243	.7251	.7259	.7267	.7275	.7284	.7292	.7300	.7308	.7316
5.4	.7324	.7332	.7340	.7348	.7356	.7364	.7372	.7380	.7388	.7396
x	0	1	2	3	4	5	6	7	8	9

Table II (*continued*)

x	0	1	2	3	4	5	6	7	8	9
5.5	.7404	.7412	.7419	.7427	.7435	.7443	.7451	.7459	.7466	.7474
5.6	.7482	.7490	.7497	.7505	.7513	.7520	.7528	.7536	.7543	.7551
5.7	.7559	.7566	.7574	.7582	.7589	.7597	.7604	.7612	.7619	.7627
5.8	.7634	.7642	.7649	.7657	.7664	.7672	.7679	.7686	.7694	.7701
5.9	.7709	.7716	.7723	.7731	.7738	.7745	.7752	.7760	.7767	.7774
6.0	.7782	.7789	.7796	.7803	.7810	.7818	.7825	.7832	.7839	.7846
6.1	.7853	.7860	.7868	.7875	.7882	.7889	.7896	.7903	.7910	.7917
6.2	.7924	.7931	.7938	.7945	.7952	.7959	.7966	.7973	.7980	.7987
6.3	.7993	.8000	.8007	.8014	.8021	.8028	.8035	.8041	.8048	.8055
6.4	.8062	.8069	.8075	.8082	.8089	.8096	.8102	.8109	.8116	.8122
6.5	.8129	.8136	.8142	.8149	.8156	.8162	.8169	.8176	.8182	.8189
6.6	.8195	.8202	.8209	.8215	.8222	.8228	.8235	.8241	.8248	.8254
6.7	.8261	.8267	.8274	.8280	.8287	.8293	.8299	.8306	.8312	.8319
6.8	.8325	.8331	.8338	.8344	.8351	.8357	.8363	.8370	.8376	.8382
6.9	.8388	.8395	.8401	.8407	.8414	.8420	.8426	.8432	.8439	.8445
7.0	.8451	.8457	.8463	.8470	.8476	.8482	.8488	.8494	.8500	.8506
7.1	.8513	.8519	.8525	.8531	.8537	.8543	.8549	.8555	.8561	.8567
7.2	.8573	.8579	.8585	.8591	.8597	.8603	.8609	.8615	.8621	.8627
7.3	.8633	.8639	.8645	.8651	.8657	.8663	.8669	.8675	.8681	.8686
7.4	.8692	.8698	.8704	.8710	.8716	.8722	.8727	.8733	.8739	.8745
7.5	.8751	.8756	.8762	.8768	.8774	.8779	.8785	.8791	.8797	.8802
7.6	.8808	.8814	.8820	.8825	.8831	.8837	.8842	.8848	.8854	.8859
7.7	.8865	.8871	.8876	.8882	.8887	.8893	.8899	.8904	.8910	.8915
7.8	.8921	.8927	.8932	.8938	.8943	.8949	.8954	.8960	.8965	.8971
7.9	.8976	.8982	.8987	.8993	.8998	.9004	.9009	.9015	.9020	.9025
8.0	.9031	.9036	.9042	.9047	.9053	.9058	.9063	.9069	.9074	.9079
8.1	.9085	.9090	.9096	.9101	.9106	.9112	.9117	.9122	.9128	.9133
8.2	.9138	.9143	.9149	.9154	.9159	.9165	.9170	.9175	.9180	.9186
8.3	.9191	.9196	.9201	.9206	.9212	.9217	.9222	.9227	.9232	.9238
8.4	.9243	.9248	.9253	.9258	.9263	.9269	.9274	.9279	.9284	.9289
8.5	.9294	.9299	.9304	.9309	.9315	.9320	.9325	.9330	.9335	.9340
8.6	.9345	.9350	.9355	.9360	.9365	.9370	.9375	.9380	.9385	.9390
8.7	.9395	.9400	.9405	.9410	.9415	.9420	.9425	.9430	.9435	.9440
8.8	.9445	.9450	.9455	.9460	.9465	.9469	.9474	.9479	.9484	.9489
8.9	.9494	.9499	.9504	.9509	.9513	.9518	.9523	.9528	.9533	.9538
9.0	.9542	.9547	.9552	.9557	.9562	.9566	.9571	.9576	.9581	.9586
9.1	.9590	.9595	.9600	.9605	.9609	.9614	.9619	.9624	.9628	.9633
9.2	.9638	.9643	.9647	.9652	.9657	.9661	.9666	.9671	.9675	.9680
9.3	.9685	.9689	.9694	.9699	.9703	.9708	.9713	.9717	.9722	.9727
9.4	.9731	.9736	.9741	.9745	.9750	.9754	.9759	.9763	.9768	.9773
9.5	.9777	.9782	.9786	.9791	.9795	.9800	.9805	.9809	.9814	.9818
9.6	.9823	.9827	.9832	.9836	.9841	.9845	.9850	.9854	.9859	.9863
9.7	.9868	.9872	.9877	.9881	.9886	.9890	.9894	.9899	.9903	.9908
9.8	.9912	.9917	.9921	.9926	.9930	.9934	.9939	.9943	.9948	.9952
9.9	.9956	.9961	.9965	.9969	.9974	.9978	.9983	.9987	.9991	.9996
x	0	1	2	3	4	5	6	7	8	9

Table III Values of trigonometric functions

∠ A	sin A	csc A	tan A	cot A	sec A	cos A	
0° 00′	0.0000	No value	0.0000	No value	1.000	1.0000	90° 00′
10	.0029	343.8	.0029	343.8	.000	.0000	50
20	.0058	171.9	.0058	171.9	.000	.0000	40
30	.0087	114.6	.0087	114.6	.000	1.0000	30
40	.0116	85.95	.0116	85.94	.000	.9999	20
50	.0145	68.76	.0145	68.75	.000	.9999	10
1° 00′	.0175	57.30	.0175	57.29	1.000	.9998	89° 00′
10	.0204	49.11	.0204	49.10	.000	.9998	50
20	.0233	42.98	.0233	42.96	.000	.9997	40
30	.0262	38.20	.0262	38.19	.000	.9997	30
40	.0291	34.38	.0291	34.37	.000	.9996	20
50	.0320	31.26	.0320	31.24	.001	.9995	10
2° 00′	.0349	28.65	.0349	28.64	1.001	.9994	88° 00′
10	.0378	26.45	.0378	26.43	.001	.9993	50
20	.0407	24.56	.0407	24.54	.001	.9992	40
30	.0436	22.93	.0437	22.90	.001	.9990	30
40	.0465	21.49	.0466	21.47	.001	.9989	20
50	.0494	20.23	.0495	20.21	.001	.9988	10
3° 00′	.0523	19.11	.0524	19.08	1.001	.9986	87° 00′
10	.0552	18.10	.0553	18.07	.002	.9985	50
20	.0581	17.20	.0582	17.17	.002	.9983	40
30	.0610	16.38	.0612	16.35	.002	.9981	30
40	.0640	15.64	.0641	15.60	.002	.9980	20
50	.0669	14.96	.0670	14.92	.002	.9978	10
4° 00′	.0698	14.34	.0699	14.30	1.002	.9976	86° 00′
10	.0727	13.76	.0729	13.73	.003	.9974	50
20	.0765	13.23	.0758	13.20	.003	.9971	40
30	.0785	12.75	.0787	12.71	.003	.9969	30
40	.0814	12.29	.0816	12.25	.003	.9967	20
50	.0843	11.87	.0846	11.83	.004	.9964	10
5° 00′	.0872	11.47	.0875	11.43	1.004	.9962	85° 00′
10	.0901	11.10	.0904	11.06	.004	.9959	50
20	.0929	10.76	.0934	10.71	.004	.9957	40
30	.0958	10.43	.0963	10.39	.005	.9954	30
40	.0987	10.13	.0992	10.08	.005	.9951	20
50	.1016	9.839	.1022	9.788	.005	.9948	10
6° 00′	.1045	9.567	.1051	9.514	1.006	.9945	84° 00′
10	.1074	9.309	.1080	9.255	.006	.9942	50
20	.1103	9.065	.1110	9.010	.006	.9939	40
30	.1132	8.834	.1139	8.777	.006	.9936	30
40	.1161	8.614	.1169	8.556	.007	.9932	20
50	.1190	8.405	.1198	8.345	.007	.9929	10
7° 00′	.1219	8.206	.1228	8.144	1.008	.9925	83° 00′
10	.1248	8.016	.1257	7.953	.008	.9922	50
20	.1276	7.834	.1287	7.770	.008	.9918	40
30	.1305	7.661	.1317	7.596	.009	.9914	30
40	.1334	7.496	.1346	7.429	.009	.9911	20
50	.1363	7.337	.1376	7.269	.009	.9907	10
8° 00′	0.1392	7.185	0.1405	7.115	1.010	0.9903	82° 00′
	cos A	sec A	cot A	tan A	csc A	sin A	∠ A

Table III (*continued*)

∠ A	sin A	csc A	tan A	cot A	sec A	cos A	∠ A
8° 00′	0.1392	7.185	0.1405	7.115	1.010	0.9903	82° 00′
10	.1421	7.040	.1435	6.968	.010	.9899	50
20	.1449	6.900	.1465	.827	.011	.9894	40
30	.1478	.765	.1495	.691	.011	.9890	30
40	.1507	.636	.1524	.561	.012	.9886	20
50	.1536	.512	.1554	.435	.012	.9881	10
9° 00′	.1564	6.392	.1584	6.314	1.012	.9877	81° 00′
10	.1593	.277	.1614	.197	.013	.9872	50
20	.1622	.166	.1644	6.084	.013	.9868	40
30	.1650	6.059	.1673	5.976	.014	.9863	30
40	.1679	5.955	.1703	.871	.014	.9858	20
50	.1708	.855	.1733	.769	.015	.9853	10
10° 00′	.1736	5.759	.1763	5.671	1.015	.9848	80° 00′
10	.1765	.665	.1793	.576	.016	.9843	50
20	.1794	.575	.1823	.485	.016	.9838	40
30	.1822	.487	.1853	.396	.017	.9833	30
40	.1851	.403	.1883	.309	.018	.9827	20
50	.1880	.320	.1914	.226	.018	.9822	10
11° 00′	.1908	5.241	.1944	5.145	1.019	.9816	79° 00′
10	.1937	.164	.1974	5.066	.019	.9811	50
20	.1965	.089	.2004	4.989	.020	.9805	40
30	.1994	5.016	.2035	.915	.020	.9799	30
40	.2022	4.945	.2065	.843	.021	.9793	20
50	.2051	.876	.2095	.773	.022	.9787	10
12° 00′	.2079	4.810	.2126	4.705	1.022	.9781	78° 00′
10	.2108	.745	.2156	.638	.023	.9775	50
20	.2136	.682	.2186	.574	.024	.9769	40
30	.2164	.620	.2217	.511	.024	.9763	30
40	.2193	.560	.2247	.449	.025	.9757	20
50	.2221	.502	.2278	.390	.026	.9750	10
13° 00′	.2250	4.445	.2309	4.331	1.026	.9744	77° 00′
10	.2278	.390	.2339	.275	.027	.9737	50
20	.2306	.336	.2370	.219	.028	.9730	40
30	.2334	.284	.2401	.165	.028	.9724	30
40	.2363	.232	.2432	.113	.029	.9717	20
50	.2391	.182	.2462	.061	.030	.9710	10
14° 00′	.2419	4.134	.2493	4.011	1.031	.9713	76° 00′
10	.2447	.086	.2524	3.962	.031	.9696	50
20	.2476	4.039	.2555	.914	.032	.9689	40
30	.2504	3.994	.2586	.867	.033	.9681	30
40	.2532	.950	.2617	.821	.034	.9674	20
50	.2560	.906	.2648	.776	.034	.9667	10
15° 00′	.2588	3.864	.2679	3.732	1.035	.9659	75° 00′
10	.2616	.822	.2711	.689	.036	.9652	50
20	.2644	.782	.2742	.647	.037	.9644	40
30	.2672	.742	.2773	.606	.038	.9636	30
40	.2700	.703	.2805	.566	.039	.9628	20
50	.2728	.665	.2836	.526	.039	.9621	10
16° 00′	0.2756	3.628	0.2867	3.487	1.040	0.9613	74° 00′
	cos A	sec A	cot A	tan A	csc A	sin A	∠ A

Table III (*continued*)

∠ A	sin A	csc A	tan A	cot A	sec A	cos A	
16° 00′	0.2756	3.628	0.2867	3.487	1.040	0.9613	74° 00′
10	.2784	.592	.2899	.450	.041	.9605	50
20	.2812	.556	.2931	.412	.042	.9596	40
30	.2840	.521	.2962	.376	.043	.9588	30
40	.2868	.487	.2944	.340	.044	.9580	20
50	.2896	.453	.3026	.305	.045	.9572	10
17° 00′	.2924	3.420	.3057	3.271	1.046	.9563	73° 00′
10	.2952	.388	.3089	.237	.047	.9555	50
20	.2979	.357	.3121	.204	.048	.9546	40
30	.3007	.326	.3153	.172	.048	.9537	30
40	.3035	.295	.3185	.140	.049	.9528	20
50	.3062	.265	.3217	.108	.050	.9520	10
18° 00′	.3090	3.236	.3249	3.078	1.051	.9511	72° 00′
10	.3118	.207	.3281	.047	.052	.9502	50
20	.3145	.179	.3314	3.018	.053	.9492	40
30	.3173	.152	.3346	2.989	.054	.9483	30
40	.3201	.124	.3378	.960	.056	.9474	20
50	.3228	.098	.3411	.932	.057	.9465	10
19° 00′	.3256	3.072	.3443	2.904	1.058	.9455	71° 00
10	.3283	.046	.3476	.877	.059	.9446	50
20	.3311	3.021	.3508	.850	.060	.9436	40
30	.3338	2.996	.3541	.824	.061	.9426	30
40	.3365	.971	.3574	.798	.062	.9417	20
50	.3393	.947	.3607	.773	.063	.9407	10
20° 00′	.3420	2.924	.3640	2.747	1.064	.9397	70° 00′
10	.3448	.901	.3673	.723	.065	.9387	50
20	.3475	.878	.3706	.699	.066	.9377	40
30	.3502	.855	.3739	.675	.068	.9367	30
40	.3529	.833	.3772	.651	.069	.9356	20
50	.3557	.812	.3805	.628	.070	.9346	10
21° 00′	.3584	2.790	.3839	2.605	1.071	.9336	69° 00′
10	.3611	.769	.3872	.583	.072	.9325	50
20	.3638	.749	.3906	.560	.074	.9315	40
30	.3665	.729	.3939	.539	.075	.9304	30
40	.3692	.709	.3973	.517	.076	.9293	20
50	.3719	.689	.4006	.496	.077	.9283	10
22° 00′	.3746	2.669	.4040	2.475	1.079	.9272	68° 00′
10	.3773	.650	.4074	.455	.080	.9261	50
20	.3800	.632	.4108	.434	.081	.9250	40
30	.3827	.613	.4142	.414	.082	.9239	30
40	.3854	.595	.4176	.394	.084	.9228	20
50	.3881	.577	.4210	.375	.085	.9216	10
23° 00′	.3907	2.559	.4245	2.356	1.086	.9205	67° 00′
10	.3934	.542	.4279	.337	.088	.9194	50
20	.3961	.525	.4314	.318	.089	.9182	40
30	.3987	.508	.4348	.300	.090	.9171	30
40	.4014	.491	.4383	.282	.092	.9159	20
50	.4041	.475	.4417	.264	.093	.9147	10
24° 00′	0.4067	2.459	0.4452	2.246	1.095	0.9135	66° 00′
	cos A	sec A	cot A	tan A	csc A	sin A	∠ A

Table III (*continued*)

∠ A	sin A	csc A	tan A	cot A	sec A	cos A	
24° 00′	0.4067	2.459	0.4452	2.246	1.095	0.9135	66° 00′
10	.4094	.443	.4487	.229	.096	.9124	50
20	.4120	.427	.4522	.211	.097	.9112	40
30	.4147	.411	.4557	.194	.099	.9100	30
40	.4173	.396	.4592	.177	.100	.9088	20
50	.4200	.381	.4628	.161	.102	.9075	10
25° 00′	.4226	2.336	.4663	2.145	1.103	.9063	65° 00′
10	.4253	.352	.4699	.128	.105	.9051	50
20	.4279	.337	.4734	.112	.106	.9038	40
30	.4305	.323	.4770	.097	.108	.9026	30
40	.4331	.309	.4806	.081	.109	.9013	20
50	.4358	.295	.4841	.066	.111	.9001	10
26° 00′	.4384	2.281	.4877	2.050	1.113	.8988	64° 00′
10	.4410	.268	.4913	.035	.114	.8975	50
20	.4436	.254	.4950	.020	.116	.8962	40
30	.4462	.241	.4986	2.006	.117	.8949	30
40	.4488	.228	.5022	1.991	.119	.8936	20
50	.4514	.215	.5059	.977	.121	.8923	10
27° 00′	.4540	2.203	.5095	1.963	1.122	.8910	63° 00′
10	.4566	.190	.5132	.949	.124	.8897	50
20	.4592	.178	.5169	.935	.126	.8884	40
30	.4617	.166	.5206	.921	.127	.8870	30
40	.4643	.154	.5243	.907	.129	.8857	20
50	.4669	.142	.5280	.894	.131	.8843	10
28° 00′	.4695	2.130	.5317	1.881	1.133	.8829	62° 00′
10	.4720	.118	.5354	.868	.134	.8816	50
20	.4746	.107	.5392	.855	.136	.8802	40
30	.4772	.096	.5430	.842	.138	.8788	30
40	.4797	.085	.5467	.829	.140	.8774	20
50	.4823	.074	.5505	.816	.142	.8760	10
29° 00′	.4848	2.063	.5543	1.804	1.143	.8746	61° 00′
10	.4874	.052	.5581	.792	.145	.8732	50
20	.4899	.041	.5619	.780	.147	.8718	40
30	.4924	.031	.5658	.767	.149	.8704	30
40	.4950	.020	.5696	.756	.151	.8689	20
50	.4975	.010	.5735	.744	.153	.8675	10
30° 00′	.5000	2.000	.5774	1.732	1.155	.8660	60° 00′
10	.5025	1.990	.5812	.720	.157	.8646	50
20	.5050	.980	.5851	.709	.159	.8631	40
30	.5075	.970	.5890	.698	.161	.8616	30
40	.5100	.961	.5930	.686	.163	.8601	20
50	.5125	.951	.5969	.675	.165	.8587	10
31° 00′	.5150	1.942	.6009	1.664	1.167	.8572	59° 00′
10	.5175	.932	.6048	.653	.169	.8557	50
20	.5200	.923	.6088	.643	.171	.8542	40
30	.5225	.914	.6128	.632	.173	.8526	30
40	.5250	.905	.6168	.621	.175	.8511	20
50	.5275	.896	.6208	.611	.177	.8496	10
32° 00′	0.5299	1.887	0.6249	1.600	1.179	0.8480	
	cos A	sec A	cot A	tan A	csc A	sin A	∠ A

Table III (*continued*)

∠ A	sin A	csc A	tan A	cot A	sec A	cos A	
32° 00′	0.5299	1.887	0.6249	1.600	1.179	0.8480	58° 00′
10	.5324	.878	.6289	.590	.181	.8465	50
20	.5348	.870	.6330	.580	.184	.8450	40
30	.5373	.861	.6371	.570	.186	.8434	30
40	.5398	.853	.6412	.560	.188	.8418	20
50	.5422	.844	.6453	.550	.190	.8403	10
33° 00′	.5446	1.836	.6494	1.540	1.192	.8387	57° 00′
10	.5471	.828	.6536	.530	.195	.8371	50
20	.5495	.820	.6577	.520	.197	.8355	40
30	.5519	.812	.6619	.511	.199	.8339	30
40	.5544	.804	.6661	.501	.202	.8323	20
50	.5568	.796	.6703	.492	.204	.8307	10
34° 00′	.5592	1.788	.6745	1.483	1.206	.8290	56° 00′
10	.5616	.781	.6787	.473	.209	.8274	50
20	.5640	.773	.6830	.464	.211	.8258	40
30	.5664	.766	.6873	.455	.213	.8241	30
40	.5688	.758	.6916	.446	.216	.8225	20
50	.5712	.751	.6959	.437	.218	.8208	10
35° 00′	.5736	1.743	.7002	1.428	1.221	.8192	55° 00′
10	.5760	.736	.7046	.419	.223	.8175	50
20	.5783	.729	.7089	.411	.226	.8158	40
30	.5807	.722	.7133	.402	.228	.8141	30
40	.5831	.715	.7177	.393	.231	.8124	20
50	.5854	.708	.7221	.385	.233	.8107	10
36° 00′	.5878	1.701	.7265	1.376	1.236	.8090	54° 00′
10	.5901	.695	.7310	.368	.239	.8073	50
20	.5925	.688	.7355	.360	.241	.8056	40
30	.5948	.681	.7400	.351	.244	.8039	30
40	.5972	.675	.7445	.343	.247	.8021	20
50	.5995	.668	.7490	.335	.249	.8004	10
37° 00′	.6018	1.662	.7536	1.327	1.252	.7986	53° 00′
10	.6041	.655	.7581	.319	.255	.7966	50
20	.6065	.649	.7627	.311	.258	.7951	40
30	.6088	.643	.7673	.303	.260	.7934	30
40	.6111	.636	.7720	.295	.263	.7916	20
50	.6134	.630	.7766	.288	.266	.7898	10
38° 00′	.6157	1.624	.7813	1.280	1.269	.7880	52° 00′
10	.6180	.618	.7860	.272	.272	.7862	50
20	.6202	.612	.7907	.265	.275	.7844	40
30	.6225	.606	.7954	.257	.278	.7826	30
40	.6248	.601	.8002	.250	.281	.7808	20
50	.6271	.595	.8050	.242	.284	.7790	10
39° 00′	.6293	1.589	.8098	1.235	1.287	.7771	51° 00′
10	.6316	.583	.8146	.228	.290	.7753	50
20	.6338	.578	.8195	.220	.293	.7735	40
30	.6361	.572	.8243	.213	.296	.7716	30
40	.6383	.567	.8292	.206	.299	.7698	20
50	.6406	.561	.8342	.199	.302	.7679	10
40° 00′	0.6428	1.556	0.8391	1.192	1.305	0.7660	50° 00′
	cos A	sec A	cot A	tan A	csc A	sin A	∠ A

Table III (*continued*)

∠ A	sin A	csc A	tan A	cot A	sec A	cos A	
40° 00′	0.6428	1.556	0.8391	1.192	1.305	0.7660	50° 00′
10	.6450	.550	.8441	.185	.309	.7642	50
20	.6472	.545	.8491	.178	.312	.7623	40
30	.6494	.540	.8541	.171	.315	.7604	30
40	.6517	.535	.8591	.164	.318	.7585	20
50	.6539	.529	.8642	.157	.322	.7566	10
41° 00′	.6561	1.524	.8693	1.150	1.325	.7547	49° 00′
10	.6583	.519	.8744	.144	.328	.7528	50
20	.6604	.514	.8796	.137	.332	.7509	40
30	.6626	.509	.8847	.130	.335	.7490	30
40	.6648	.504	.8899	.124	.339	.7470	20
50	.6670	.499	.8952	.117	.342	.7451	10
42° 00′	.6691	1.494	.9004	1.111	1.346	.7431	48° 00′
10	.6713	.490	.9057	.104	.349	.7412	50
20	.6734	.485	.9110	.098	.353	.7392	40
30	.6756	.480	.9163	.091	.356	.7373	30
40	.6777	.476	.9217	.085	.360	.7353	20
50	.6799	.471	.9271	.079	.364	.7333	10
43° 00′	.6820	1.466	.9325	1.072	1.367	.7314	47° 00′
10	.6841	.462	.9380	.066	.371	.7294	50
20	.6862	.457	.9435	.060	.375	.7274	40
30	.6884	.453	.9490	.054	.379	.7254	30
40	.6905	.448	.9545	.048	.382	.7234	20
50	.6926	.444	.9601	.042	.386	.7214	10
44° 00′	.6947	1.440	.9657	1.036	1.390	.7193	46° 00′
10	.6967	.435	.9713	.030	.394	.7173	50
20	.6988	.431	.9770	.024	.398	.7153	40
30	.7009	.427	.9827	.018	.402	.7133	30
40	.7030	.423	.9884	.012	.406	.7112	20
50	.7050	.418	.9942	.006	.410	.7092	10
45° 00′	0.7071	1.414	1.0000	1.000	1.414	0.7071	45° 00′
	cos A	sec A	cot A	tan A	csc A	sin A	∠ A

Table IV Values of trigonometric functions for decimal fractions of a degree

∠ A	Sin A	Tan A	Cot A	Cos A	Sec A	Csc A	
0.0	0.0000	0.0000		1.0000	1.000		90.0
.1	.0018	.0018	573.0	1.0000	1.000	573.0	.9
.2	.0035	.0035	286.5	1.0000	1.000	286.5	.8
.3	.0052	.0052	191.0	1.0000	1.000	191.0	.7
.4	.0070	.0070	143.2	1.0000	1.000	143.2	.6
.5	.0087	.0087	114.6	1.0000	1.000	114.6	.5
.6	.0105	.0105	95.49	0.9999	1.000	95.50	.4
.7	.0122	.0122	81.85	.9999	1.000	81.85	.3
.8	.0140	.0140	71.62	.9999	1.000	71.62	.2
.9	.0157	.0157	63.66	.9999	1.000	63.66	.1
1.0	0.0174	0.0175	57.29	0.9998	1.000	57.30	89.0
.1	.0192	.0192	52.08	.9998	1.000	52.09	.9
.2	.0209	.0210	47.74	.9998	1.000	47.75	.8
.3	.0227	.0227	44.07	.9997	1.000	44.08	.7
.4	.0244	.0244	40.92	.9997	1.000	40.93	.6
.5	.0262	.0262	38.19	.9997	1.000	38.20	.5
.6	.0279	.0279	35.80	.9996	1.000	38.82	.4
.7	.0297	.0297	33.69	.9996	1.000	33.71	.3
.8	.0314	.0314	31.82	.9995	1.001	31.84	.2
.9	.0332	.0332	30.14	.9995	1.001	30.16	.1
2.0	0.0349	0.0349	28.64	0.9994	1.001	28.65	88.0
.1	.0366	.0367	27.27	.9993	1.001	27.29	.9
.2	.0384	.0384	26.03	.9993	1.001	26.05	.8
.3	.0401	.0402	24.90	.9992	1.001	24.92	.7
.4	.0419	.0419	23.86	.9991	1.001	23.88	.6
.5	.0436	.0437	22.90	.9990	1.001	22.93	.5
.6	.0454	.0454	22.02	.9990	1.001	22.04	.4
.7	.0471	.0472	21.20	.9989	1.001	21.23	.3
.8	.0488	.0489	20.45	.9988	1.001	20.47	.2
.9	.0506	.0507	19.74	.9987	1.001	19.77	.1
3.0	0.0523	0.0524	19.08	0.9986	1.001	19.11	87.0
.1	.0541	.0542	18.46	.9985	1.002	18.49	.9
.2	.0558	.0559	17.89	.9984	1.002	17.91	.8
.3	.0576	.0577	17.34	.9983	1.002	17.37	.7
.4	.0593	.0594	16.83	.9982	1.002	16.86	.6
.5	.0610	.0612	16.35	.9981	1.002	16.38	.5
.6	.0628	.0629	15.90	.9980	1.002	15.93	.4
.7	.0645	.0647	15.46	.9979	1.002	15.50	.3
.8	.0663	.0664	15.06	.9978	1.002	15.09	.2
.9	.0680	.0682	14.67	.9977	1.002	14.70	.1
4.0	0.0698	0.0699	14.30	0.9976	1.002	14.34	86.0
.1	.0715	.0717	13.95	.9974	1.003	13.99	.9
.2	.0732	.0734	13.62	.9973	1.003	13.65	.8
.3	.0750	.0752	13.30	.9972	1.003	13.34	.7
.4	.0767	.0770	13.00	.9971	1.003	13.04	.6
	Cos A	Cot A	Tan A	Sin A	Csc A	Sec A	∠ A

Table IV (*continued*)

∠ A	Sin A	Tan A	Cot A	Cos A	Sec A	Csc A	
.5	.0785	.0787	12.71	.9969	1.003	12.74	.5
.6	.0802	.0805	12.43	.9968	1.003	12.47	.4
.7	.0819	.0822	12.16	.9966	1.003	12.20	.3
.8	.0837	.0840	11.91	.9965	1.004	11.95	.2
.9	.0854	.0857	11.66	.9963	1.004	11.71	.1
5.0	0.0872	0.0875	11.43	0.9962	1.004	11.47	85.0
.1	.0889	.0892	11.20	.9960	1.004	11.25	.9
.2	.0906	.0910	10.99	.9959	1.004	11.03	.8
.3	.0924	.0928	10.78	.9957	1.004	10.83	.7
.4	.0941	.0945	10.58	.9956	1.004	10.63	.6
.5	.0958	.0963	10.38	.9954	1.005	10.43	.5
.6	.0976	.0981	10.20	.9952	1.005	10.25	.4
.7	.0993	.0998	10.02	.9951	1.005	10.07	.3
.8	.1011	.1016	9.845	.9949	1.005	9.896	.2
.9	.1028	.1033	9.677	.9947	1.005	9.728	.1
6.0	0.1045	0.1051	9.514	0.9945	1.006	9.567	84.0
.1	.1063	.1069	9.357	.9943	1.006	9.410	.9
.2	.1080	.1086	9.205	.9942	1.006	9.259	.8
.3	.1097	.1104	9.058	.9940	1.006	9.113	.7
.4	.1115	.1122	8.915	.9938	1.006	8.971	.6
.5	.1132	.1139	8.777	.9936	1.006	8.834	.5
.6	.1149	.1157	8.643	.9934	1.007	8.700	.4
.7	.1167	.1175	8.513	.9932	1.007	8.571	.3
.8	.1184	.1192	8.386	.9930	1.007	8.446	.2
.9	.1201	.1210	8.264	.9928	1.007	8.324	.1
7.0	0.1219	0.1228	8.144	0.9925	1.008	8.206	83.0
.1	.1236	.1246	8.028	.9923	1.008	8.090	.9
.2	.1253	.1263	7.916	.9921	1.008	7.979	.8
.3	.1271	.1281	7.806	.9919	1.008	7.870	.7
.4	.1288	.1299	7.700	.9917	1.008	7.764	.6
.5	.1305	.1316	7.596	.9914	1.009	7.661	.5
.6	.1323	.1334	7.495	.9912	1.009	7.561	.4
.7	.1340	.1352	7.396	.9910	1.009	7.464	.3
.8	.1357	.1370	7.300	.9907	1.009	7.368	.2
.9	.1374	.1388	7.207	.9905	1.010	7.276	.1
8.0	0.1392	0.1405	7.115	0.9903	1.010	7.185	82.0
.1	.1409	.1423	7.026	.9900	1.010	7.097	.9
.2	.1426	.1441	6.940	.9898	1.010	7.011	.8
.3	.1444	.1459	6.855	.9805	1.011	6.927	.7
.4	.1461	.1477	6.772	.9893	1.011	6.845	.6
.5	.1478	.1494	6.691	.9890	1.011	6.766	.5
.6	.1495	.1512	6.612	.9888	1.011	6.687	.4
.7	.1513	.1530	6.535	.9985	1.012	6.611	.3
.8	.1530	.1548	6.460	.9882	1.012	6.537	.2
.9	.1547	.1566	6.386	.9880	1.012	6.464	.1
	Cos A	Cot A	Tan A	Sin A	Csc A	Sec A	∠ A

Table IV (*continued*)

∠ A	Sin A	Tan A	Cot A	Cos A	Sec A	Csc A	
9.0	0.1564	0.1584	6.314	0.9877	1.012	6.392	81.0
.1	.1582	.1602	6.243	.9874	1.013	6.323	.9
.2	.1599	.1620	6.174	.9871	1.013	6.255	.8
.3	.1616	.1638	6.107	.9869	1.013	6.188	.7
.4	.1633	.1656	6.041	.9866	1.014	6.123	.6
.5	.1651	.1673	5.976	.9863	1.014	6.059	.5
.6	.1668	.1691	5.912	.9860	1.014	5.996	.4
.7	.1685	.1709	5.850	.9857	1.014	5.935	.3
.8	.1702	.1727	5.789	.9854	1.015	5.875	.2
.9	.1719	.1745	5.730	.9851	1.015	5.816	.1
10.0	0.1736	0.1763	5.671	0.9848	1.015	5.759	80.0
.1	.1754	.1781	5.614	.9845	1.016	5.702	.9
.2	.1771	.1799	5.558	.9842	1.016	5.647	.8
.3	.1788	.1817	5.503	.9839	1.016	5.593	.7
.4	.1805	.1835	5.449	.9836	1.017	5.540	.6
.5	.1822	.1853	5.396	.9833	1.017	5.487	.5
.5	.1840	.1871	5.343	.9829	1.017	5.436	.4
.7	.1857	.1870	5.292	.9826	1.018	5.386	.3
.8	.1874	.1908	5.242	.9823	1.018	5.337	.2
.9	.1891	.1926	5.193	.9820	1.018	5.288	.1
11.0	0.1908	0.1944	5.145	0.9816	1.019	5.241	79.0
.1	.1925	.1962	5.097	.9813	1.019	5.194	.9
.2	.1942	.1980	5.050	.9810	1.019	5.148	.8
.3	.1959	.1998	5.005	.9806	1.020	5.103	.7
.4	.1977	.2016	4.959	.9803	1.020	5.059	.6
.5	.1994	.2035	4.915	.9799	1.020	5.016	.5
.6	.2011	.2053	4.872	.9796	1.021	4.973	.4
.7	.2028	.2071	4.829	.9792	1.021	4.931	.3
.8	.2045	.2089	4.787	.9789	1.022	4.890	.2
.9	.2062	.2107	4.745	.9785	1.022	4.850	.1
12.0	0.2079	0.2126	4.705	0.9781	1.022	4.810	78.0
.1	.2096	.2144	4.665	.9778	1.023	4.771	.9
.2	.2113	.2162	4.625	.9774	1.023	4.732	.8
.3	.2130	.2180	4.586	.9770	1.024	4.694	.7
.4	.2147	.2199	4.548	.9767	1.024	4.657	.6
.5	.2164	.2217	4.511	.9763	1.024	4.620	.5
.6	.2181	.2235	4.474	.9759	1.025	4.584	.4
.7	.2198	.2254	4.437	.9755	1.025	4.549	.3
.8	.2215	.2272	4.402	.9751	1.026	4.514	.2
.9	.2233	.2290	4.366	.9748	1.026	4.479	.1
13.0	0.2250	0.2309	4.331	0.9744	1.026	4.445	77.0
.1	.2267	.2327	4.297	.9740	1.027	4.412	.9
.2	.2284	.2345	4.264	.9736	1.027	4.379	.8
.3	.2300	.2364	4.230	.9732	1.028	4.347	.7
.4	.2317	.2382	4.198	.9728	1.028	4.315	.6
	Cos A	Cot A	Tan A	Sin A	Csc A	Sec A	∠ A

Table IV (*continued*)

∠ A	Sin A	Tan A	Cot A	Cos A	Sec A	Csc A	
.5	.2334	.2401	4.165	.9724	1.028	4.284	.5
.6	.2351	.2419	4.134	.9720	1.029	4.253	.4
.7	.2368	.2438	4.102	.9715	1.029	4.222	.3
.8	.2385	.2456	4.071	.9711	1.030	4.192	.2
.9	.2402	.2475	4.041	.9707	1.030	4.163	.1
14.0	0.2419	0.2493	4.011	0.9703	1.031	4.134	76.0
.1	.2436	.2512	3.981	.9699	1.031	4.105	.9
.2	.2453	.2530	3.952	.9694	1.032	4.076	.8
.3	.2470	.2549	3.923	.9690	1.032	4.049	.7
.4	.2487	.2568	3.895	.9686	1.032	4.021	.6
.5	.2504	.2586	3.867	.9681	1.033	3.994	.5
.6	.2521	.2605	3.839	.9677	1.033	3.967	.4
.7	.2538	.2623	3.812	.9673	1.034	3.941	.3
.8	.2554	.2642	3.785	.9668	1.034	3.915	.2
.9	.2571	.2661	3.758	.9664	1.035	3.889	.1
15.0	0.2605	0.2679	3.732	0.9659	1.035	3.864	75.0
.1	.2505	.2698	3.706	.9655	1.036	3.839	.9
.2	.2622	.2717	3.681	.9650	1.036	3.814	.8
.3	.2639	.2736	3.655	.9646	1.037	3.790	.7
.4	.2656	.2754	3.630	.9641	1.037	3.766	.6
.5	.2672	.2773	3.606	.9636	1.038	3.742	.5
.6	.2689	.2792	3.582	.9632	1.038	3.719	.4
.7	.2706	.2811	3.558	.9627	1.039	3.696	.3
.8	.2723	.2830	3.534	.9622	1.039	3.673	.2
.9	.2740	.2849	3.511	.9617	1.040	3.650	.1
16.0	0.2756	0.2867	3.487	0.9613	1.040	3.628	74.0
.1	.2773	.2886	3.465	.9608	1.041	3.606	.9
.2	.2790	.2905	3.442	.9603	1.041	3.584	.8
.3	.2807	.2924	3.420	.9598	1.042	3.563	.7
.4	.2823	.2943	3.398	.9593	1.042	3.542	.6
.5	.2840	.2962	3.376	.9588	1.043	3.521	.5
.6	.2857	.2981	3.354	.9583	1.044	3.500	.4
.7	.2874	.3000	3.333	.9578	1.044	3.480	.3
.8	.2890	.3019	3.312	.9573	1.045	3.460	.2
.9	.2907	.3038	3.291	.9568	1.045	3.440	.1
17.0	0.2924	0.3057	3.271	0.9563	1.046	3.420	73.0
.1	.2940	.3076	3.251	.9558	1.046	3.401	.9
.2	.2957	.3096	3.230	.9553	1.047	3.382	.8
.3	.2974	.3115	3.211	.9548	1.047	3.363	.7
.4	.2990	.3134	3.191	.9542	1.048	3.344	.6
.5	.3007	.3153	3.172	.9537	1.048	3.326	.5
.6	.3024	.3172	3.152	.9532	1.049	3.307	.4
.7	.3040	.3191	3.133	.9527	1.050	3.289	.3
.8	.3057	.3211	3.115	.9521	1.050	3.271	.2
.9	.3074	.3230	3.096	.9516	1.051	3.254	.1
	Cos A	Cot A	Tan A	Sin A	Csc A	Sec A	∠ A

Table IV (*continued*)

∠ A	Sin A	Tan A	Cot A	Cos A	Sec A	Csc A	
18.0	0.3090	0.3249	3.078	0.9511	1.052	3.236	72.0
.1	.3107	.3269	3.060	.9505	1.052	3.219	.9
.2	.3123	.3288	3.042	.9500	1.053	3.202	.8
.3	.3140	.3307	3.024	.9494	1.053	3.185	.7
.4	.3156	.3327	3.006	.9489	1.054	3.168	.6
.5	.3173	.3346	2.989	.9483	1.054	3.152	.5
.6	.3190	.3365	2.971	.9478	1.055	3.135	.4
.7	.3206	.3385	2.954	.9472	1.056	3.119	.3
.8	.3223	.3404	2.937	.9466	1.056	3.103	.2
.9	.3239	.3424	2.921	.9461	1.057	3.087	.1
19.0	0.3256	0.3443	2.904	0.9455	1.058	3.072	71.0
.1	.3272	.3463	2.888	.9449	1.058	3.056	.9
.2	.3289	.3482	2.872	.9444	1.059	3.041	.8
.3	.3305	.3502	2.856	.9438	1.060	3.026	.7
.4	.3322	.3522	2.840	.9432	1.060	3.011	.6
.5	.3338	.3541	2.824	.9426	1.061	2.996	.5
.6	.3355	.3561	2.808	.9421	1.062	2.981	.4
.7	.3371	.3581	2.793	.9415	1.062	2.966	.3
.8	.3387	.3600	2.778	.9409	1.063	2.952	.2
.9	.3404	.3620	2.762	.9403	1.064	2.938	.1
20.0	0.3420	0.3640	2.747	0.9397	1.064	2.924	70.0
.1	.3437	.3659	2.733	.9391	1.065	2.910	.9
.2	.3453	.3679	2.718	.9385	1.066	2.896	.8
.3	.3469	.3699	2.703	.9379	1.066	2.882	.7
.4	.3486	.3719	2.689	.9373	1.067	2.869	.6
.5	.3502	.3739	2.675	.9367	1.068	2.856	.5
.6	.3518	.3759	2.660	.9361	1.068	2.842	.4
.7	.3535	.3779	2.646	.9354	1.069	2.829	.3
.8	.3551	.3799	2.633	.9348	1.070	2.816	.2
.9	.3567	.3819	2.619	.9342	1.070	2.803	.1
21.0	0.3584	0.3839	2.605	0.9336	1.071	2.790	69.0
.1	.3600	.3859	2.592	.9330	1.072	2.778	.9
.2	.3616	.3879	2.578	.9323	1.073	2.765	.8
.3	.3633	.3899	2.565	.9317	1.073	2.753	.7
.4	.3649	.3919	2.552	.9311	1.074	2.741	.6
.5	.3665	.3939	2.539	.9304	1.075	2.728	.5
.6	.3681	.3959	2.526	.9298	1.076	2.716	.4
.7	.3697	.3979	2.513	.9291	1.076	2.705	.3
.8	.3714	.4000	2.500	.9285	1.077	2.693	.2
.9	.3730	.4020	2.488	.9278	1.078	2.681	.1
22.0	0.3746	0.4040	2.475	0.9272	1.078	2.670	68.0
.1	.3762	.4061	2.463	.9265	1.079	2.658	.9
.2	.3778	.4081	2.450	.9259	1.080	2.647	.8
.3	.3795	.4101	2.438	.9252	1.081	2.635	.7
.4	.3811	.4122	2.426	.9245	1.082	2.624	.6
	Cos A	Cot A	Tan A	Sin A	Csc A	Sec A	∠ A

Table IV (*continued*)

∠ A	Sin A	Tan A	Cot A	Cos A	Sec A	Csc A	
.5	.3827	.4142	2.414	.9239	1.082	2.613	.5
.6	.3843	.4163	2.402	.9232	1.083	2.602	.4
.7	.3859	.4183	2.391	.9225	1.084	2.591	.3
.8	.3875	.4204	2.379	.9219	1.085	2.580	.2
.9	.3891	.4224	2.367	.9212	1.086	2.570	.1
23.0	0.3907	0.4245	2.356	0.9205	1.086	2.559	67.0
.1	.3923	.4265	2.344	.9198	1.087	2.549	.9
.2	.3939	.4286	2.333	.9191	1.088	2.538	.8
.3	.3955	.4307	2.322	.9184	1.089	2.528	.7
.4	.3971	.4327	2.311	.9178	1.090	2.518	.6
.5	.3987	.4348	2.300	.9171	1.090	2.508	.5
.6	.4003	.4369	2.289	.9164	1.091	2.498	.4
.7	.4019	.4390	2.278	.9157	1.092	2.488	.3
.8	.4035	.4411	2.267	.9150	1.093	2.478	.2
.9	.4051	.4431	2.257	.9143	1.094	2.468	.1
24.0	0.4067	0.4452	2.246	0.9135	1.095	2.459	66.0
.1	.4083	.4473	2.236	.9128	1.096	2.449	.9
.2	.4099	.4494	2.225	.9121	1.096	2.440	.8
.3	.4115	.4515	2.215	.9114	1.097	2.430	.7
.4	.4131	.4536	2.204	.9107	1.098	2.421	.6
.5	.4147	.4557	2.194	.9100	1.099	2.411	.5
.6	.4163	.4578	2.184	.9092	1.100	2.402	.4
.7	.4179	.4599	2.174	.9085	1.101	2.393	.3
.8	.4195	.4621	2.164	.9078	1.102	2.384	.2
.9	.4210	.4642	2.154	.9070	1.102	2.375	.1
25.0	0.4226	0.4663	2.145	0.9063	1.103	2.366	65.0
.1	.4242	.4684	2.135	.9056	1.104	2.357	.9
.2	.4258	.4706	2.125	.9048	1.105	2.349	.8
.3	.4274	.4727	2.116	.9041	1.106	2.340	.7
.4	.4289	.4748	2.106	.9033	1.107	2.331	.6
.5	.4305	.4770	2.097	.9026	1.108	2.323	.5
.6	.4321	.4791	2.087	.9018	1.109	2.314	.4
.7	.4337	.4813	2.078	.9011	1.110	2.306	.3
.8	.4352	.4834	2.069	.9003	1.111	2.298	.2
.9	.4368	.4856	2.059	.8996	1.112	2.289	.1
26.0	0.4384	0.4877	2.050	0.8988	1.113	2.281	64.0
.1	.4399	.4899	2.041	.8980	1.114	2.273	.9
.2	.4415	.4921	2.032	.8973	1.114	2.265	.8
.3	.4431	.4942	2.023	.8965	1.116	2.257	.7
.4	.4446	.4964	2.014	.8957	1.116	2.249	.6
.5	.4462	.4986	2.006	.8949	1.117	2.241	.5
.6	.4478	.5008	1.997	.8942	1.118	2.233	.4
.7	.4493	.5029	1.988	.8934	1.119	2.226	.3
.8	.4509	.5051	1.980	.8926	1.120	2.218	.2
.9	.4524	.5073	1.971	.8918	1.121	2.210	.1
	Cos A	Cot A	Tan A	Sin A	Csc A	Sec A	∠ A

Table IV (*continued*)

∠ A	Sin A	Tan A	Cot A	Cos A	Sec A	Csc A	
27.0	0.4540	0.5095	1.963	0.8910	1.122	2.203	63.0
.1	.4555	.5117	1.954	.8902	1.123	2.195	.9
.2	.4571	.5139	1.946	.8894	1.124	2.188	.8
.3	.4586	.5161	1.937	.8886	1.125	2.180	.7
.4	.4602	.5184	1.929	.8878	1.126	2.173	.6
.5	.4617	.5206	1.921	.8870	1.127	2.166	.5
.6	.4633	.5228	1.913	.8862	1.128	2.158	.4
.7	.4648	.5250	1.905	.8854	1.129	2.151	.3
.8	.4664	.5272	1.897	.8846	1.130	2.144	.2
.9	.4679	.5295	1.889	.8838	1.132	2.137	.1
28.0	0.4695	0.5317	1.881	0.8829	1.133	2.130	62.0
.1	.4710	.5340	1.873	.8821	1.134	2.123	.9
.2	.4726	.5362	1.865	.8813	1.135	2.116	.8
.3	.4741	.5384	1.857	.8805	1.136	2.109	.7
.4	.4756	.5407	1.849	.8796	1.137	2.102	.6
.5	.4772	.5430	1.842	.8788	1.138	2.096	.5
.6	.4787	.5452	1.834	.8780	1.139	2.089	.4
.7	.4802	.5475	1.827	.8771	1.140	2.082	.3
.8	.4818	.5498	1.819	.8763	1.141	2.076	.2
.9	.4833	.5520	1.811	.8755	1.142	2.069	.1
29.0	0.4848	0.5543	1.804	0.8746	1.143	2.063	61.0
.1	.4863	.5566	1.797	.8738	1.144	2.056	.9
.2	.4879	.5589	1.789	.8729	1.146	2.050	.8
.3	.4894	.5612	1.782	.8721	1.147	2.043	.7
.4	.4909	.5635	1.775	.8712	1.148	2.037	.6
.5	.4924	.5658	1.767	.8704	1.149	2.031	.5
.6	.4939	.5681	1.760	.8695	1.150	2.024	.4
.7	.4955	.5704	1.753	.8686	1.151	2.018	.3
.8	.4970	.5727	1.746	.8678	1.152	2.012	.2
.9	.4985	.5750	1.739	.8669	1.154	2.006	.1
30.0	0.5000	0.5774	1.732	0.8660	1.155	2.000	60.0
.1	.5015	.5797	1.725	.8652	1.156	1.994	.9
.2	.5030	.5820	1.718	.8643	1.157	1.988	.8
.3	.5045	.5844	1.711	.8634	1.158	1.982	.7
.4	.5060	.5867	1.704	.8625	1.159	1.976	.6
.5	.5075	.5890	1.698	.8616	1.161	1.970	.5
.6	.5090	.5914	1.691	.8607	1.162	1.964	.4
.7	.5105	.5938	1.684	.8599	1.163	1.959	.3
.8	.5120	.5961	1.678	.8590	1.164	1.953	.2
.9	.5135	.5985	1.671	.8581	1.165	1.947	.1
31.0	0.5150	0.6009	1.664	0.8572	1.167	1.942	59.0
.1	.5165	.6032	1.658	.8563	1.168	1.936	.9
.2	.5180	.6056	1.651	.8554	1.169	1.930	.8
.3	.5195	.6080	1.645	.8545	1.170	1.925	.7
.4	.5210	.6104	1.638	.8536	1.172	1.919	.6
	Cos A	Cot A	Tan A	Sin A	Csc A	Sec A	∠ A

Table IV (*continued*)

∠ A	Sin A	Tan A	Cot A	Cos A	Sec A	Csc A	
.5	.5225	.6128	1.632	.8526	1.173	1.914	.5
.6	.5240	.6152	1.626	.8517	1.174	1.908	.4
.7	.5255	.6176	1.619	.8508	1.175	1.903	.3
.8	.5270	.6200	1.613	.8499	1.177	1.898	.2
.9	.5284	.6224	1.607	.8490	1.178	1.892	.1
32.0	0.5299	0.6249	1.600	0.8480	1.179	1.887	58.0
.1	.5314	.6273	1.594	.8471	1.180	1.882	.9
.2	.5329	.6297	1.588	.8462	1.182	1.877	.8
.3	.5344	.6322	1.582	.8453	1.183	1.871	.7
.4	.5358	.6346	1.576	.8443	1.184	1.866	.6
.5	.5373	.6371	1.570	.8434	1.186	1.861	.5
.6	.5388	.6395	1.564	.8425	1.187	1.856	.4
.7	.5402	.6420	1.558	.8415	1.188	1.851	.3
.8	.5417	.6445	1.552	.8406	1.190	1.846	.2
.9	.5432	.6469	1.546	.8396	1.191	1.841	.1
33.0	0.5446	0.6494	1.540	0.8387	1.192	1.836	57.0
.1	.5461	.6519	1.534	.8377	1.194	1.831	.9
.2	.5476	.6544	1.528	.8368	1.195	1.826	.8
.3	.5490	.6569	1.522	.8348	1.196	1.821	.7
.4	.5505	.6594	1.517	.3848	1.198	1.817	.6
.5	.5519	.6619	1.511	.8339	1.199	1.812	.5
.6	.5534	.6644	1.505	.8329	1.201	1.807	.4
.7	.5548	.6669	1.499	.8320	1.202	1.802	.3
.8	.5563	.6694	1.494	.8310	1.203	1.798	.2
.9	.5577	.6720	1.488	.8300	1.205	1.793	.1
34.0	0.5592	0.6745	1.483	0.8290	1.206	1.788	56.0
.1	.5606	.6771	1.477	.8281	1.208	1.784	.9
.2	.5621	.6796	1.472	.8271	1.209	1.779	.8
.3	.5635	.6822	1.466	.8261	1.210	1.774	.7
.4	.5650	.6847	1.460	.8251	1.212	1.770	.6
.5	.5664	.6873	1.455	.8241	1.213	1.766	.5
.6	.5678	.6899	1.450	.8231	1.215	1.761	.4
.7	.5693	.6924	1.444	.8221	1.216	1.757	.3
.8	.5707	.6950	1.439	.8211	1.218	1.752	.2
.9	.5721	.6976	1.434	.8202	1.219	1.748	.1
35.0	0.5736	0.7002	1.428	0.8192	1.221	1.743	55.0
.1	.5750	.7028	1.423	.8181	1.222	1.739	.9
.2	.5764	.7054	1.418	.8171	1.224	1.735	.8
.3	.5779	.7080	1.412	.8161	1.225	1.730	.7
.4	.5793	.7107	1.407	.8151	1.227	1.726	.6
.5	.5807	.7133	1.402	.8141	1.228	1.722	.5
.6	.5821	.7159	1.397	.8131	1.230	1.718	.4
.7	.5835	.7186	1.392	.8121	1.231	1.714	.3
.8	.5850	.7212	1.386	.8111	1.233	1.710	.2
.9	.5864	.7239	1.381	.8100	1.234	1.705	.1
	Cos A	Cot A	Tan A	Sin A	Csc A	Sec A	∠ A

Table IV (*continued*)

∠ A	Sin A	Tan A	Cot A	Cos A	Sec A	Csc A	
36.0	0.5878	0.7265	1.376	0.8090	1.236	1.701	54.0
.1	.5892	.7292	1.371	.8080	1.238	1.697	.9
.2	.5906	.7319	1.366	.8070	1.239	1.693	.8
.3	.5920	.7346	1.361	.8059	1.241	1.689	.7
.4	.5934	.7373	1.356	.8049	1.242	1.685	.6
.5	.5948	.7400	1.351	.8039	1.244	1.681	.5
.6	.5962	.7427	1.346	.8028	1.246	1.677	.4
.7	.5976	.7454	1.342	.8018	1.247	1.673	.3
.8	.5990	.7481	1.337	.8007	1.249	1.669	.2
.9	.6004	.7508	1.332	.7997	1.250	1.666	.1
37.0	0.6018	0.7536	1.327	0.7986	1.252	1.662	53.0
.1	.6032	.7563	1.322	.7976	1.254	1.658	.9
.2	.6046	.7590	1.318	.7965	1.255	1.654	.8
.3	.6060	.7618	1.313	.7955	1.257	1.650	.7
.4	.6074	.7646	1.308	.7944	1.259	1.646	.6
.5	.6088	.7673	1.303	.7934	1.260	1.643	.5
.6	.6101	.7701	1.298	.7923	1.262	1.639	.4
.7	.6115	.7729	1.294	.7912	1.264	1.635	.3
.8	.6129	.7757	1.289	.7902	1.266	1.632	.2
.9	.6143	.7785	1.285	.7891	1.267	1.628	.1
38.0	0.6157	0.7813	1.280	0.7880	1.269	1.624	52.0
.1	.6170	.7841	1.275	.7869	1.271	1.621	.9
.2	.6184	.7869	1.271	.7859	1.272	1.617	.8
.3	.6198	.7898	1.266	.7848	1.274	1.614	.7
.4	.6211	.7926	1.262	.7837	1.276	1.610	.6
.5	.6225	.7954	1.257	.7826	1.278	1.606	.5
.6	.6239	.7983	1.253	.7815	1.280	1.603	.4
.7	.6252	.8012	1.248	.7804	1.281	1.599	.3
.8	.6266	.8040	1.244	.7793	1.283	1.596	.2
.9	.6280	.8069	1.239	.7782	1.285	1.592	.1
39.0	0.6293	0.8098	1.235	0.7771	1.287	1.589	51.0
.1	.6307	.8127	1.230	.7760	1.289	1.586	.9
.2	.6320	.8156	1.226	.7749	1.290	1.582	.8
.3	.6334	.8185	1.222	.7738	1.292	1.579	.7
.4	.6347	.8214	1.217	.7727	1.294	1.576	.6
.5	.6361	.8243	1.213	.7716	1.296	1.572	.5
.6	.6374	.8273	1.209	.7705	1.298	1.569	.4
.7	.6388	.8302	1.204	.7694	1.300	1.566	.3
.8	.6401	.8332	1.200	.7683	1.302	1.562	.2
.9	.6414	.8361	1.196	.7672	1.304	1.559	.1
40.0	0.6428	0.8391	1.192	0.7660	1.305	1.556	50.0
.1	.6441	.8421	1.188	.7649	1.307	1.552	.9
.2	.6455	.8451	1.183	.7638	1.309	1.549	.8
.3	.6468	.8481	1.179	.7627	1.311	1.546	.7
.4	.6481	.8511	1.175	.7615	1.313	1.543	.6
	Cos A	Cot A	Tan A	Sin A	Csc A	Sec A	∠ A

Table IV (*continued*)

∠ A	Sin A	Tan A	Cot A	Cos A	Sec A	Csc A	
.5	.6494	.8541	1.171	.7604	1.315	1.540	.5
.6	.6508	.8571	1.167	.7593	1.317	1.537	.4
.7	.6521	.8601	1.163	.7581	1.319	1.534	.3
.8	.6534	.8632	1.158	.7570	1.321	1.530	.2
.9	.6547	.8662	1.154	.7559	1.323	1.527	.1
41.0	0.6561	0.8693	1.150	0.7547	1.325	1.524	49.0
.1	.6574	.8724	1.146	.7536	1.327	1.521	.9
.2	.6587	.8754	1.142	.7524	1.329	1.518	.8
.3	.6600	.8785	1.138	.7513	1.331	1.515	.7
.4	.6613	.8816	1.134	.7501	1.333	1.512	.6
.5	.6626	.8847	1.130	.7490	1.335	1.509	.5
.6	.6639	.8878	1.126	.7478	1.337	1.506	.4
.7	.6652	.8910	1.122	.7466	1.339	1.503	.3
.8	.6665	.8941	1.118	.7455	1.341	1.500	.2
.9	.6678	.8972	1.114	.7443	1.344	1.497	.1
42.0	0.6691	0.9004	1.111	0.7431	1.346	1.494	48.0
.1	.6704	.9036	1.107	.7420	1.348	1.492	.9
.2	.6717	.9067	1.103	.7408	1.350	1.489	.8
.3	.6730	.9099	1.099	.7396	1.352	1.486	.7
.4	.6743	.9131	1.095	.7385	1.354	1.483	.6
.5	.6756	.9163	1.091	.7373	1.356	1.480	.5
.6	.6769	.9195	1.088	.7361	1.358	1.477	.4
.7	.6782	.9228	1.084	.7349	1.361	1.475	.3
.8	.6807	.9260	1.080	.7337	1.363	1.472	.2
.9	.6804	.9293	1.076	.7325	1.365	1.469	.1
43.0	0.6820	0.9325	1.072	0.7314	1.367	1.466	47.0
.1	.6833	.9358	1.069	.7302	1.370	1.464	.9
.2	.6845	.9391	1.065	.7290	1.372	1.461	.8
.3	.6858	.9424	1.061	.7278	1.374	1.458	.7
.4	.6871	.9457	1.058	.7266	1.376	1.455	.6
.5	.6884	.9490	1.054	.7254	1.379	1.453	.5
.6	.6896	.9523	1.050	.7242	1.381	1.450	.4
.7	.6909	.9556	1.046	.7230	1.383	1.447	.3
.8	.6921	.9590	1.043	.7218	1.386	1.445	.2
.9	.6934	.9623	1.039	.7206	1.388	1.442	.1
44.0	0.6947	0.9657	1.036	0.7193	1.390	1.440	46.0
.1	.6959	.9691	1.032	.7181	1.392	1.437	.9
.2	.6972	.9725	1.028	.7169	1.395	1.434	.8
.3	.6984	.9759	1.025	.7157	1.397	1.432	.7
.4	.6997	.9793	1.021	.7145	1.400	1.429	.6
.5	.7009	.9827	1.018	.7133	1.402	1.427	.5
.6	.7022	.9861	1.014	.7120	1.404	1.424	.4
.7	.7034	.9896	1.010	.7108	1.407	1.422	.3
.8	.7046	.9930	1.007	.7096	1.409	1.419	.2
.9	.7059	.9965	1.004	.7083	1.412	1.417	.1
45.0	0.7071	1.0000	1.000	0.7071	1.414	1.414	45.0
	Cos A	Cot A	Tan A	Sin A	Csc A	Sec A	∠ A

Odd-numbered answers

1

EXERCISE 1.1 (PAGE 3)

1. {1, 2, 3, 4} **3.** {1, 3, 5, ...} **5.** {8, 10, 12} **7.** {3, 6, 9, 12}
9. {6, −3, 0, −5} **11.** $\{\sqrt{7}, -\sqrt{3}, \sqrt{5}\}$ **13.** {6} **15.** {1, 2}, {1, 3}, {2, 3}
17. Infinite **19.** Finite **21.** Infinite **23.** Variable **25.** Constant
27. Constant **29.** Variable

EXERCISE 1.2 (PAGE 8)

1. Symmetric law **3.** Reflexive law
5. Transitive law of equality or substitution law
7. Substitution law **9.** Transitive law of inequality
11. Transitive law of equality or substitution law
13. Trichotomy law **15.** Substitution law **17.** Substitution law
19.  0 5 **21.** 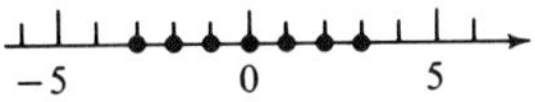 −5 0 5
23.  5 10 **25.**  0 5 **27.** $8 > 2$

29. $-4 < -3$ **31.** $x \not< y$ **33.** $3 < x < 5$ **35.** $<$ **37.** $<$ **39.** $<$
41. $=$ **43.** $2 < 5$ **45.** $x \geq y$ **47.** $x > 0$ **49.** $x \geq 0$

EXERCISE 1.3 (PAGE 13)

1. Commutative law of addition **3.** Reciprocal axiom
5. Negative axiom **7.** Closure for multiplication
9. Associative law of multiplication **11.** Distributive law
13. Commutative law of addition **15.** Associative law of addition
17. Double-negative law **19.** Commutative law of addition
21. Commutative law of multiplication **23.** Distributive law
25. Negative axiom **27.** Positive **29.** Positive **31.** 4 **33.** -2
35. 0

EXERCISE 1.4 (PAGE 17)

1. 7 **3.** -6 **5.** 3 **7.** -4 **9.** 0 **11.** 5 **13.** 5 **15.** 27
17. -6 **19.** 14 **21.** -10 **23.** 5 **25.** 1 **27.** 7 **29.** 6 **31.** 9
33. 5 **35.** 6 **37.** -63 **39.** -3 **41.** 1.05 **43.** 65.3 **45.** 132
47. 756 **49.** 0.99 **51.** 88.62

EXERCISE 1.5 (PAGE 21)

1. -15 **3.** 24 **5.** -42 **7.** 45 **9.** 0 **11.** 56 **13.** 30 **15.** -36
17. 0 **19.** -64 **21.** -8 **23.** -24 **25.** 3,523 **27.** $-16{,}598$
29. -41.15 **31.** -2.4621 **33.** 9 **35.** -4 **37.** 6 **39.** 0
41. Undefined **43.** 5 **45.** 82 **47.** -182 **49.** 2.8 **51.** -0.34
53. $-32 = 2(-16)$ **55.** $24 = (-3)(-8)$ **57.** $56 = (-7)(-8)$ **59.** 5(1/7)
61. 2(1/9) **63.** 13(1/11) **65.** 7(1/10) **67.** 3/2 **69.** 2/7 **71.** 5/8
73. 9/2

EXERCISE 1.6 (PAGE 23)

1. (2) (3) **3.** (5) (5) **5.** (3) (3) (2) **7.** Prime
9. (2) (2) (2) (7) **11.** -1(2) (19) **13.** (2) (2) (5)
15. (2) (53) **17.** -1(2) (2) (2) (5) (5) **19.** (2) (2) (2) (2) (5) (5)
21. (5) (41) **23.** -1(3) (3) (23) **25.** (2) (2) (2) (2) (3) (5)
27. (2) (2) (61)

CHAPTER REVIEW (PAGE 24)

1. {6, 12, 18} **2.** {-3, 0, 5} **3.** Transitive law of inequality
4. Substitution law **5.** Symmetric law **6.** 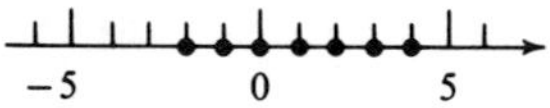

7. 

8. $x \geq 0$ **9.** Associative law of addition

10. Distributive law **11.** Commutative law of multiplication
12. Negative axiom **13.** Negative **14.** -3 **15.** -2 **16.** 5
17. -316 **18.** -144.6 **19.** $4+(-7)$ **20.** $12+(-3)$ **21.** 24
22. -130.51 **23.** 7 **24.** -2.1 **25** $3(1/5)$ **26.** $3/100$ **27.** (2)(2)(17)
28. (2)(2)(2)(3)(5) **29.** (7)(13) **30.** $-1(5)(5)(17)$

EXERCISE 2.1 (PAGE 28)

1. y^4 **3.** $2a^2b$ **5.** $2x^2y^3$ **7.** r^2+rs^3 **9.** x^2+y^4
11. $3a^2c-ab^2c^2$ **13.** $xxyyyy$ **15.** $2abbb$ **17.** $7rrrsst$
19. $-x(-x)(-x)$ **21.** $-yyx$ **23.** $-xxx(-y)(-y)$ **25.** $rrst+sst$
27. $2xxx-xyy$ **29.** $abb-abbbb$ **31.** Binomial; degree 3 in x
33. Trinomial; degree 2 in y **35.** Monomial; degree 3 in a
37. Binomial; degree 4 in x and y; degree 3 in x; degree 2 in y
39. Monomial; degree 3 in r
41. Bionomial; degree 2 in r and s; degree 2 in s; degree 2 in r
43. one **45.** two **47.** two **49.** two

2

EXERCISE 2.2 (PAGE 31)

1. $9a$ **3.** $3x$ **5.** $-2a^4b$ **7.** $-2xy^2$ **9.** $-3a$ **11.** $8c$
13. $2c+2d$ **15.** $2b-2a$ **17.** $3xy+x^2y$ **19.** $6x^3+3x^2-2x$
21. $3x+3y$ **23.** $2a^2$ **25.** $5R+S$ **27.** $2R^2+2S$ **29.** $2\pi R^2+\pi R^2h$
31. z^2-2z+2 **33.** c^2d^2+5 **35.** $4x^2-7x+8$ **37.** $4x-1$
39. $-x^2+3x-1$ **41.** $-2b$ **43.** x^2+4x-2 **45.** $a-b+2c-3$
47. $2x^2+3x+2$ **49.** $2d^2+9d+1$ **51.** $4t^3-2t^2-3t+5$
53. $-m^2+2m-14$ **55.** $-2y-1$ **57.** $2-x$ **59.** $A-1$
61. $-A-3B$

EXERCISE 2.3 (PAGE 34)

1. $6x^3$ **3.** $-20a^4$ **5.** $-2x^2y^3$ **7.** abc^2 **9.** $-x^2y^4$ **11.** $-8x^2y^3z^3$
13. $3x+3$ **15.** $2l+2w$ **17.** $-x^2-x+2$ **19.** $a^2bc-ab^2c+abc^2$
21. $a^3b-a^2b^2+ab^3$ **23.** $4rh+2\pi r^2-2r^2$ **25.** $x^2+7x+10$
27. x^2+2x-3 **29.** y^2-6y+9 **31.** $x^2-ax-2a^2$ **33.** $x^2+2xy+y^2$
35. x^2-y^2 **37.** $10x^2+17x+3$ **39.** $9a^2-6a+1$ **41.** $4x^2-25$
43. $2x^2+3ax-2a^2$ **45.** $E_1{}^2-1$ **47.** $E_1{}^2+2E_1E_2+E_2{}^2$
49. $2x^2+8x+6$ **51.** x^3+x^2-6x **53.** $3a^2+12a+12$
55. $-6x^2-9x+6$ **57.** $-2x^3-12x^2-18x$ **59.** $9x^4-33x^3+30x^2$
61. a^2-4b^2 **63.** $xy+ay+bx+ab$ **65.** a^3-b^3 **67.** $9x+9$
69. 6 **71.** $-2a^2-2a$ **73.** $4a+4$ **75.** $4a+4$ **77.** $-4x^2-11x$
79. $20x^2+4x$

EXERCISE 2.4 (PAGE 37)

1. x^2y **3.** $2a^2b$ **5.** $-7ab$ **7.** x^2y **9.** $x-2$ **11.** $2xy^2z^2$
13. $x^2(x+y)$ **15.** $-2y^3(2x+y)$ **17.** $(x+y)^2(x-y)^2$
19. $(y-z)(y+2z)^3$ **21.** $R^2(E_1+E_2)^2$ **23.** $(E_1+E_2)^2(E_1-E_2)^2$
25. $x^0=1; x \neq 0$

EXERCISE 2.5 (PAGE 39)

1. $x^4(x^2)$ **3.** $xy^2(y^2)$ **5.** $2x^3y(3y^4)$ **7.** $2(x+3)$ **9.** $4x(x+2)$
11. $3(x^2-2x-1)$ **13.** $6(4a^2+2a-1)$ **15.** $2x(x^3-2x+4)$
17. $xy(y+x)$ **19.** $I(R_1+R_2)$ **21.** $\pi R(R+2)$ **23.** $R_0(1+at)$
25. $-2(x-1)$ **27.** $-a(b+c)$ **29.** $-xy(1+x)$ **31.** $-x(1-x+x^2)$
33. $-E(2R_1-R_2)$ **35.** $-V_0(1+0.00365T)$

2

EXERCISE 2.6 (PAGE 42)

1. $(x+2)(x+6)$ **3.** $(y-5)(y-3)$ **5.** $(x-3)(x+2)$ **7.** $(x-3)(x+1)$
9. $(y+4)(y-1)$ **11.** $(x-7)(x+6)$ **13.** $(x-y)^2$ **15.** $(x+5y)(x+y)$
17. $(x-1)(x+1)$ **19.** $(a-3)(a+3)$ **21.** $(2-b)(2+b)$
23. $(ab-1)(ab+1)$ **25.** $(a-2xy)(a+2xy)$ **27.** $(3x-y)(3x+y)$
29. $(x^2-3)(x^2+3)$ **31.** $(3x+1)(x+1)$ **33.** $(3x-8)(3x+1)$
35. $(3x-a)(x-2a)$ **37.** $(3x-y)(3x+y)$ **39.** $(2x+3)^2$
41. $(2xy-5)(2xy+5)$ **43.** $3(x+2)(x+2)$ **45.** $2a(a-5)(a+1)$
47. $4(a-b)^2$ **49.** $4y(x-3)(x+3)$ **51.** $x(4+x)(3-x)$
53. $x^2y^2(x-1)(x+1)$ **55.** $(y^2+2)(y^2+1)$ **57.** $(3x^2+1)(x^2+2)$
59. $(x^2+4)(x-1)(x+1)$ **61.** $(x-2)(x+2)(x-1)(x+1)$
63. $(2a^2+1)(a-1)(a+1)$ **65.** $(x^2+2a^2)(x-a)(x+a)$
67. $x=0; x=2$ **69.** $x=2; x=1$

EXERCISE 2.7 (PAGE 44)

1. $(x+1)(ax+1)$ **3.** $(x+a)(ax+1)$ **5.** $(x+y)(x+a)$
7. $(3a-c)(b-d)$ **9.** $(3x+y)(1-2x)$ **11.** $(a^2+2b^2)(a-2b)$
13. $(x-1)(x^2+x+1)$ **15.** $(2x+y)(4x^2-2xy+y^2)$
17. $(a-2b)(a^2+2ab+4b^2)$ **19.** $(xy-1)(x^2y^2+xy+1)$
21. $(3a+4b)(9a^2-12ab+16b^2)$ **23.** $(2x-y)(x^2-xy+y^2)$
25.
$$\begin{aligned} ac-ad+bd-bc &= a(c-d)-b(c-d) \\ &= (a-b)(c-d); \\ ac-ad+bd-bc &= -a(d-c)+b(d-c) \\ &= (b-a)(d-c) \end{aligned}$$

EXERCISE 2.8 (PAGE 46)

1. 9 **3.** 7 **5.** 8 **7.** 10 **9.** 18 **11.** 2 **13.** −4 **15.** −1
17. 48 **19.** 60 **21.** 4 **23.** 15 **25.** 0 **27.** 64 **29.** 17 **31.** 17
33. 75 **35.** 24 **37.** 7 **39.** 100 **41.** 20 **43.** 13.95

CHAPTER REVIEW (PAGE 47)

1. $2xy^2 - 3x^2y$ **2.** $x - xy^2$ **3.** $x^2y + y^5$ **4.** 2 **5.** $a + b$
6. $-b - 2c$ **7.** $-2y$ **8.** $2x - 3$ **9.** $2x^2y^2$ **10.** $2x^3 - 6x^2$
11. $x^2 - 2xy - 3y^2$ **12.** $6x^2 + 15x - 18$ **13.** $-8xy^2$ **14.** $(x - 2)(x + 3)$
15. $4(x - 3)$ **16.** $2x(x^2 - 2x + 3)$ **17.** $(x - 4)(x - 3)$
18. $(2y - 1)(y + 3)$ **19.** $(x - 5y)(x + 5y)$ **20.** $x(2x - 1)(x + 2)$
21. $(2x + 1)(y + x)$ **22.** $(x - 2a)(x^2 + 2ax + 4a^2)$ **23.** 22 **24.** 45
25. −4.56

EXERCISE 3.1 (PAGE 52)

1. $\frac{-3}{5}$ **3.** $\frac{4}{7}$ **5.** $\frac{-4x}{y}$ **7.** $\frac{2}{x}$ **9.** $\frac{-3x}{2y^2}$ **11.** $\frac{-2}{x - y}$ or $\frac{2}{y - x}$

13. $\frac{-(R - S)}{3}$ or $\frac{S - R}{3}$ **15.** $\frac{L + C}{L}$ **17.** $\frac{4}{y - 3}$ **19.** $\frac{-1}{z - x}$ **21.** $\frac{2 - x}{x - 3}$

23. $\frac{2 - R}{R - S}$ **25.** $\frac{2}{R_2 - R_1}$ **27.** $\frac{x_1 - x_2}{y_1 - y_2}$

EXERCISE 3.2 (PAGE 55)

1. $\frac{3}{4}$ **3.** $\frac{5}{3}$ **5.** $\frac{-5}{11}$ **7.** $\frac{1}{4x^3}$ **9.** $\frac{y^2}{3x}$ **11.** $\frac{-a}{b}$ **13.** $\frac{1}{3a^2b^2c}$

15. $\frac{-9a^2}{2b}$ **17.** $2x - 3$ **19.** $y - 1$ **21.** $\frac{R}{2} - \frac{1}{2}$ **23.** $P + \frac{3}{2}$

25. $a^2 - 3a + 2$ **27.** $4a^2 + 2a + \frac{1}{2}$ **29.** $x^2 - 2x - \frac{3}{2}$

31. $2R_1 + 3R_2 + 4$ **33.** $RR_1 + R_2 - 1$

EXERCISE 3.3 (PAGE 58)

1. $y + 7$ **3.** $y - 2$ **5.** $x + 2$ **7.** $y - 1$ **9.** $R - 3$ **11.** $S + 7$

13. $2y + 5$ **15.** $2y - 3 + \frac{-2}{2y + 1}$ **17.** $x + 1$ **19.** $R + 2 + \frac{4}{4R - 1}$

21. $2S - 1$ **23.** $2P - 1$ **25.** $x - 6$ **27.** $2y + 3$ **29.** $5R + 4$

EXERCISE 3.4 (PAGE 60)

1. $\frac{9}{12}$ **3.** $\frac{-48}{15}$ **5.** $\frac{48}{12}$ **7.** $\frac{4}{12x}$ **9.** $\frac{-3a^2b}{3b^3}$ **11.** $\frac{xy^2}{xy}$

13. $\frac{x+y}{3(x+y)}$ **15.** $\frac{3(x+1)(x-1)}{9(x+1)}$ **17.** $\frac{27a(a+1)}{9(a+1)}$ **19.** $\frac{3(a+b)}{a^2-b^2}$

21. $\frac{3x(y-3)}{y^2-y-6}$ **23.** $\frac{-2(x+2)}{x^2+3x+2}$ **25.** $\frac{-9(a-3)}{6a^2-24a+18}$ **27.** $\frac{-2(a+b)}{b^2-a^2}$

29. $\frac{-x(x-1)}{x^2-3x+2}$

EXERCISE 3.5 (PAGE 64)

1. $\frac{9}{11}$ **3.** $\frac{10}{7}$ **5.** $\frac{x-3}{2}$ **7.** $\frac{a+b-c}{6}$ **9.** $\frac{2}{3x}$ **11.** $\frac{2x-1}{2y}$

13. $\frac{4a-b}{3x}$ **15.** $\frac{1-2x}{x+2y}$ **17.** $\frac{6-2a}{a^2-2a+1}$ **19.** 60 **21.** 120 **23.** 252

25. $6ab^2$ **27.** $24x^2y^2$ **29.** $a(a-b)^2$ **31.** $(a+b)(a-b)$

33. $(a+1)^2(a+4)$ **35.** $(x-1)^2(x+4)$ **37.** $x(x-1)^3$ **39.** $\frac{5}{8}$ **41.** $\frac{16}{35}$

43. $\frac{2+2a}{ax}$ **45.** $\frac{5y-3x^2}{yx^2}$ **47.** $\frac{R_2+R_1}{R_1R_2}$ **49.** $\frac{-a-4}{6}$

51. $\frac{2x^2+xy+2y^2}{2xy}$ **53.** $\frac{R^2+RS+S^2}{3RS}$ **55.** $\frac{5R_1+R_2}{2R_1}$ **57.** $\frac{-4}{15(x-2)}$

59. $\frac{33-x}{2(x-3)(x+3)}$ **61.** $\frac{-2ax-3a}{(3x+2)(x-1)}$ **63.** $\frac{-8x^2+2xy+6y^2}{(3x+y)(2x-y)}$

65. $\frac{-R^2-4RS+S^2}{R^2-S^2}$ **67.** $\frac{-6S-4}{(S-4)(S+4)(S-1)}$

EXERCISE 3.6 (PAGE 70)

1. $\frac{2}{3}$ **3.** $\frac{7}{10}$ **5.** $\frac{10}{3}$ **7.** $\frac{b^2}{a}$ **9.** $\frac{3c}{35ab}$ **11.** $\frac{5}{ab}$ **13.** 5

15. $\frac{a(2a-1)}{a+4}$ **17.** $\frac{R+3}{R-5}$ **19.** $\frac{-(S-3)}{S+1}$ **21.** $\frac{4}{3}$ **23.** $\frac{5}{2}$ **25.** $\frac{1}{ax^2y}$

27. $\frac{28y}{9a}$ **29.** $\frac{x}{2}$ **31.** $\frac{5}{2}$ **33.** $\frac{P-5}{P-2}$ **35.** $\frac{N-1}{N-3}$

EXERCISE 3.7 (PAGE 72)

1. $\frac{3}{2}$ **3.** $\frac{4}{7}$ **5.** $\frac{a}{cb}$ **7.** $\frac{4y}{3}$ **9.** $\frac{1}{5}$ **11.** $\frac{1}{10}$ **13.** $\frac{3r+2}{5r+1}$

15. $\frac{R_1R_2-1}{R_1R_2+1}$ **17.** $\frac{10}{7}$ **19.** $\frac{4a^2-3a}{4a+1}$ **21.** $\frac{-1}{y-3}$ **23.** $\frac{a-2b}{a+2b}$

25. $\frac{R_1+R_2}{2R_1+R_2}$ **27.** $\frac{T_1T_2}{T_2+T_1}$

CHAPTER REVIEW (PAGE 75)

1. $\frac{-3x}{2}$ **2.** $\frac{-(x-y)}{x}$ **3.** $\frac{2y^2}{3x}$ **4.** $1+y$ **5.** $2x+1$ **6.** $2T+3$

7. $\frac{5a}{3b}$ **8.** $\frac{-b^2+2b+4}{(b+1)(b-1)}$ **9.** $\frac{5}{R-S}$ **10.** $\frac{P^2}{P-Q}$ **11.** $\frac{x}{2y^2}$ **12.** $\frac{x-y}{2-y}$

13. $\frac{a}{2}$ **14.** $\frac{x-3}{(x+1)(x+3)}$ **15.** $\frac{y}{x}$ **16.** $\frac{-1}{a-1}$ or $\frac{1}{1-a}$ **17.** $\frac{S}{R^2}$

18. $\frac{1}{R^2}$ **19.** $\frac{10}{19}$ **20.** $\frac{-1}{y^2-1}$

EXERCISE 4.1 (PAGE 78)

1. no **3.** no **5.** yes **7.** no **9.** yes **11.** no **13.** no **15.** no
17. yes **19.** no

EXERCISE 4.2 (PAGE 80)

1. $x=3$ **3.** $x=1$ **5.** $x=3$ **7.** $x=4$ **9.** $x=0$ **11.** $x=\frac{-2}{3}$

13. $x=3$ **15.** $x=2$ **17.** $x=5$ **19.** $x=2$ **21.** $x=-2$

23. $x=\frac{-1}{3}$ **25.** $x=6$ **27.** $x=3$ **29.** $x=0$ **31.** $x=6$

33. $x=6$ **35.** $x=4$ **37.** $x=\frac{1}{3}$ **39.** $x=-30$ **41.** $x=-7$
43. $x=3$ **45.** $\varnothing$ **47.** $y=13$ **49.** $x=3.7$ **51.** $s=257.6$
53. $b=12$

EXERCISE 4.3 (PAGE 84)

1. $x=b+c-a$ **3.** $y=\frac{-cb}{a}$ **5.** $y=b$ **7.** $x=\frac{c}{c+1}$ **9.** $x=a+b$

4

11. $x = \dfrac{a}{a+3} + 2$ or $x = \dfrac{3a+6}{a+3}$ **13.** $x = \dfrac{a}{6a-1}$ **15.** $x = \dfrac{ab}{a+b}$

17. $x = \dfrac{4b}{2a-1}$ **19.** $k = v - gt$ **21.** $m = \dfrac{b}{a}$ **23.** $v = \dfrac{K}{p}$ **25.** $p = \dfrac{I}{rt}$

27. $g = \dfrac{2s}{t^2}$ **29.** $t = \dfrac{v-k}{g}$ **31.** $c = \dfrac{2A}{h} - b$ or $c = \dfrac{2A - hb}{h}$

33. $d = \dfrac{s - 5\pi D}{3\pi}$ **35.** $\dfrac{S-a}{S}$

EXERCISE 4.4 (PAGE 89)

1. $x < 2$ **3.** $x > 1$

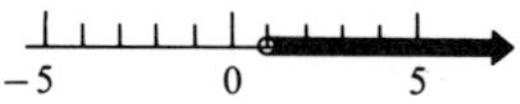

5. $x < -4$ **7.** $x < \frac{7}{2}$

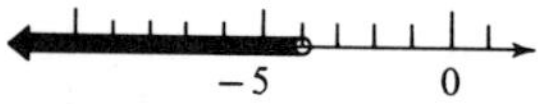

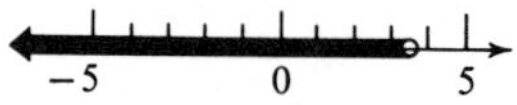

9. $x < \dfrac{-9}{2}$ **11.** $x > 5$

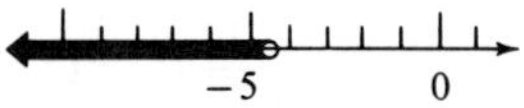

13. $x \geq 3$  **15.** $x > 3$

17. $x \leq 0$  **19.** $R < -14$

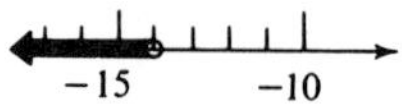

21. $T \leq -14$  **23.** $P \geq \dfrac{-6}{11}$

EXERCISE 4.5 (PAGE 91)

(Note: For any given problem there may be more than one acceptable mathematical model. To save space we have shown only one model in each case.)

1. (a) $x + (3x - 2) = 30$; (b) $x = 8$; $3x - 2 = 22$
3. (a) $l + (6l - 2) = 40$; (b) $l = 6$; $6l - 2 = 34$
5. (a) $5t + 1 = 26$; (b) $t = 5$
7. (a) $x + (x+1) + (x+2) = 42$; (b) $x = 13$; $x + 1 = 14$; $x + 2 = 15$
9. (a) $\frac{1}{3}x + \frac{1}{2}(x+1) = 33$; (b) $x = 39$; $x + 1 = 40$
11. (a) $x + (x+2) + (x+4) = 63$; (b) $x = 19$; $x + 2 = 21$; $x + 4 = 23$
13. (a) $2(x+4) = 3x + 4$; (b) Boy's age: 12 years; girl's age: 4 years
15. (a) $2(x - 10) = (x+4) - 10$; (b) Boy's age: 18 years; girl's age: 14 years
17. (a) $180d_1 = (82)(6)$; (b) $2\frac{11}{15}$ feet
19. (a) $24(x+2) = 32x$; (b) 6 feet (32 lb) and 8 feet (24 lb)
21. (a) $10(x+3) + 5x = 180$; (b) 10 nickels; 13 dimes
23. (a) $25x + 5(16 - x) = 220$; (b) 7 quarters; 9 nickels
25. (a) $10(x-1) - \frac{20}{3}x = 200$; (b) 63 ice cream bars

27. (a) $0.30n + 0.12(40) = 0.20(n + 40)$; (b) 32 gallons

29. (a) $1.00n + 0.45(12) = 0.60(n + 12)$; (b) 4.5 ounces

31. (a) $55(30 - n) + 85n = 63(30)$; (b) 8 pounds at 85¢; 22 pounds at 55¢

33. (a) $0.06A + 0.08(2{,}000 - A) = 132$;
(b) \$1400 invested at 6%; \$600 invested at 8%

35. (a) $0.05A + 0.07A = 144$; (b) \$1,200 at each rate

37. (a) $0.08A + 0.05(A + 1{,}000) = 739$;
(b) \$5300 invested at 8%; \$6300 invested at 5%

39. (a) $\dfrac{1260}{r + 120} = \dfrac{420}{r}$;
(b) Rate of airplane is 180 mph; rate of automobile is 60 mph.

41. (a) $80(t) = 40(3 + t)$; (b) 240 miles from A

43. (a) $\dfrac{d}{42} - 2 = \dfrac{d}{54}$; (b) 378 miles

45. (a) $\frac{1}{5}x + \frac{1}{15}x = 1$; (b) $3\frac{3}{4}$ days

47. (a) $\frac{1}{4}x + \frac{1}{6}x + \frac{1}{12}x = 1$; (b) 2 hours

49. (a) $\dfrac{1}{15}(6) + \dfrac{1}{x}(6) = 1$; (b) 10 hours

51. (a) $80 \leq \dfrac{78 + 64 + 88 + 76 + x}{5} < 90$;
(b) The grade must be greater than or equal to 94.

53. (a) $0.05(10{,}000 - x) + 0.07x \geq 616$; (b) \$5800 at 7%

55. (a) $0.075(20) < 0.10x + 0.05(20 - x) < 0.08(20)$;
(b) Between 12 and 10 ounces of the 10% solution

CHAPTER REVIEW (PAGE 103)

1. no **2.** yes **3.** $x = 9$ **4.** $x = \dfrac{-5}{2}$ **5.** $x = -28$ **6.** $x = \frac{5}{3}$

7. $T = \frac{9}{2}$ **8.** $r = 1$ **9.** $r = \dfrac{I}{pt}$ **10.** $g = \dfrac{r - k}{t}$ **11.** $D = \dfrac{S - 3\pi d}{5\pi}$

12. $r = \dfrac{r_1 r_2}{r_1 + r_2}$ **13.** $x \leq 4$

−5 0 5

14. $T > \dfrac{-4}{5}$

−5 0 5

15. (a) $0.04(5{,}000) + 0.03(10{,}000) + x(5{,}000) = 800$; (b) 6%

16. (a) $10d + 5(d + 2) = 160$; (b) 10 dimes, 12 nickels

17. (a) $(w + 16) \cdot 4 = w \cdot 12$; (b) 8 lb, 24 lb

18. (a) $25x = 20(x + 5)$; (b) 20 pounds

19. (a) $\dfrac{280}{r + 6} = \dfrac{210}{r}$; (b) 18 mph; 24 mph

20. (a) $\frac{1}{4}x + \frac{1}{8}x = 1$; (b) $2\frac{2}{3}$ hours

5

EXERCISE 5.1 (PAGE 107)

1. x^5 **3.** y^7 **5.** x^3 **7.** x^2y^3 **9.** x^8 **11.** x^8 **13.** x^6y^9

15. $x^6y^4z^2$ **17.** $\dfrac{x^6}{y^9}$ **19.** $\dfrac{81y^4}{x^8}$ **21.** $\dfrac{-8x^3}{125y^6}$ **23.** $72e^7$ **25.** $4\pi^4R^6$

27. $2H^4$ **29.** $\dfrac{T^4}{K}$

EXERCISE 5.2 (PAGE 110)

1. 3 **3.** 2 **5.** -3 **7.** 9 **9.** 27 **11.** 16 **13.** $\frac{1}{32}$ **15.** $\frac{8}{27}$

17. $x^{2/3}$ **19.** $x^{1/3}$ **21.** $n^{3/2}$ **23.** $m^{1/2}$ **25.** $\dfrac{a}{b^{1/4}}$ **27.** $\dfrac{a^4}{c^2}$

29. $x^{7/4}y^{3/4}$ **31.** $x^{1/2}y^{2/3}$ **33.** $x^{5/12}y^{1/6}$ **35.** $P^3V^{9/2}$ **37.** $N^{3/2}$

39. $\dfrac{81V^2}{16r}$

EXERCISE 5.3 (PAGE 112)

1. $\frac{1}{2}$ **3.** $\frac{5}{3}$ **5.** 9 **7.** $\dfrac{-1}{2}$ **9.** $\dfrac{200}{9}$ **11.** $\frac{1}{2}$ **13.** x **15.** x^7

17. $\dfrac{1}{x^6}$ **19.** $\dfrac{1}{x^{3/2}}$ **21.** $\dfrac{p^{1/2}}{m}$ **23.** $\dfrac{r^3}{R^2}$ **25.** $\dfrac{y}{x}$ **27.** $\dfrac{ay^2}{b^2x^3}$ **29.** $4x^5y^2$

31. $\dfrac{4x^2}{y^{16}}$

EXERCISE 5.4 (PAGE 114)

1. $\dfrac{1}{(x-2)^2}$ **3.** $\dfrac{(x-y)^2}{2x+y}$ **5.** $\dfrac{y(y+1)^2}{(y-2)^{1/2}}$ **7.** $\frac{3}{4}$ **9.** $\dfrac{4097}{512}$ **11.** 0

13. $\dfrac{y-x}{xy}$ **15.** $\dfrac{2s}{r}$ **17.** $\dfrac{1-S^2}{RS}$ **19.** $\dfrac{E^2+R^2}{ER}$ **21.** $R+E$

23. $\dfrac{PQ}{Q-P}$

EXERCISE 5.5 (PAGE 116)

1. 3.4×10^4 **3.** 2.1×10 **5.** 8.372×10^6 **7.** 1.4×10^{-3} **9.** 6×10^{-7}
11. 2.30×10^{-5} **13.** 1600 **15.** 600,000 **17.** 19,500 **19.** 0.00000023
21. 12,340 **23.** 400 **25.** 10^6 **27.** 0.0001 **29.** 800,000,000 **31.** 720
33. 0.4 **35.** 0.8 **37.** 3.14×10^{-4} **39.** 5×10^4

EXERCISE 5.6 (PAGE 119)

1. y^7 **3.** $x^{3/2}$ **5.** $y^{2/3}$ **7.** $x^{5/4}$ **9.** $y^{5/6}$ **11.** $x^{11/12}$ **13.** $x^{1/2}$
15. y **17.** $y^{-5/6}$ **19.** $x^{3/2}+x$ **21.** $x^{2/3}+x$ **23.** $x+x^{3/2}$
25. $R^{3/2}-R^{5/4}$ **27.** $S^{-1}+1$ **29.** $T^{-7/2}-T^{-7/6}$ **31.** $y^3(y^4)$
33. $x^{-2}(x^{-3})$ **35.** $y(y^{-3})$ **37.** $x^{1/5}(x^{2/5})$ **39.** $x^{-2/3}(x^{1/3})$
41. $y(y^{-2/3})$ **43.** $y(y^{1/2}+1)$ **45.** $R^{2/3}(R^{1/3}-1)$ **47.** $R^{3/2}(R^{-1}+1)$
49. $M^{-1/2}(M^{-1}+1)$

CHAPTER REVIEW (PAGE 120)

1. a^2b^4 **2.** $-a^6b^3$ **3.** $\frac{8x^3}{y^6}$ **4.** $\frac{r^4}{9s^4}$ **5.** 2 **6.** -2 **7.** $\frac{1}{4}$

8. $\frac{27}{8}$ **9.** y **10.** y **11.** a^3b **12.** $\frac{a^3}{b^{1/2}}$ **13.** $\frac{1}{9}$ **14.** $\frac{1}{4}$ **15.** $\frac{1}{2}$

16. 64 **17.** y **18.** $\frac{1}{y^6}$ **19.** $\frac{n}{T^{1/2}}$ **20.** R^5L **21.** $\frac{4}{9}$ **22.** $\frac{1}{xy}$

23. 2.8×10^{-4} **24.** 2.05×10^5 **25.** 10^2 **26.** 10^6 **27.** $x^{5/2}-1$
28. $R^{1/3}+R^{-7/6}$ **29.** $y^{1/5}(y^{3/5})$ **30.** $S^{-1/2}(1-S^{-1})$

EXERCISE 6.1 (PAGE 123)

1. $\sqrt{3}$ **3.** $\sqrt{x^3}$ **5.** $3(\sqrt[3]{x})$ **7.** $x\sqrt{y}$ **9.** $2(\sqrt[3]{x^2})$ **11.** $3x(\sqrt[3]{y^2})$

13. $\sqrt{x+2y}$ **15.** $\sqrt[3]{(x-y)^2}$ **17.** $\sqrt{x}-\sqrt{y}$ **19.** $\frac{1}{\sqrt[3]{4}}$ **21.** $\frac{3}{\sqrt[3]{x^2}}$

23. $\frac{1}{\sqrt{x^2-y^2}}$ **25.** $3^{1/2}$ **27.** $x^{2/3}$ **29.** $x^{1/3}y^{1/3}$ **31.** $3x^{1/2}$

33. $2^{1/3}x^{1/3}y^{2/3}$ **35.** $2xy^{2/3}$ **37.** $(x-y)^{1/2}$ **39.** $3(x^2-y)^{1/3}$

41. $x^{1/2}-2y^{1/2}$ **43.** $\frac{1}{x^{1/2}}$ **45.** $\frac{x}{y^{1/3}}$ **47.** $\frac{2}{(x+y)^{1/2}}$ **49.** 4 **51.** -5

53. 3 **55.** -4 **57.** -2 **59.** x^2 **61.** $2T^2$ **63.** S^2T^3 **65.** $\frac{2}{3}MR^4$

67. $\frac{-2}{5}M$ **69.** 8 **71.** 19 **73.** $\frac{5}{8}$

EXERCISE 6.2 (PAGE 128)

1. 3 **3.** $10\sqrt{2}$ **5.** x^2 **7.** $x\sqrt{x}$ **9.** $2x^2\sqrt{x}$ **11.** $2x^3\sqrt{2}$ **13.** 4

15. $-2x^2$ **17.** $x(\sqrt[4]{x})$ **19.** $x^2(\sqrt[5]{xy})$ **21.** 6 **23.** x^3y **25.** 2

27. x **29.** 3×10 **31.** 2×10^2 **33.** 6×10^{-2} **35.** $2\sqrt{3}\times 10^{-3}$

37. $\frac{\sqrt{5}}{5}$ 39. $\frac{\sqrt{2}}{2}$ 41. $\frac{\sqrt{2x}}{2}$ 43. $\frac{\sqrt{LC}}{C}$ 45. $\sqrt{x}$ 47. $-a\sqrt{b}$

49. $\frac{\sqrt[3]{4x^2y}}{2x}$ 51. $\frac{\sqrt[4]{2}}{2}$ 53. x^2y 55. xy^2 57. $\sqrt{3}$ 59. $\sqrt{3}$

61. $\sqrt[3]{9}$ 63. $2\sqrt{x}$ 65. $\sqrt{2x}$ 67. $y\sqrt{x}$

EXERCISE 6.3 (PAGE 130)

1. $3\sqrt{3}$ 3. $5\sqrt{3}$ 5. $12\sqrt{2x}$ 7. $4x\sqrt{y}$ 9. $3x\sqrt{xy}$ 11. $5(\sqrt[3]{5})$

13. $5(\sqrt[3]{2})$ 15. $9(\sqrt[3]{2x})$ 17. $\frac{2\sqrt{2}}{3}$ 19. $\frac{\sqrt{5}}{3}$ 21. $\frac{5\sqrt{11}}{6}$ 23. $\frac{5\sqrt{3}}{3}$

25. $\frac{-3\sqrt{5}}{5}$ 27. $\frac{11\sqrt{3}}{8}$ 29. $\frac{5\sqrt{R}}{12}$ 31. $\frac{\sqrt{L}}{2}$ 33. $\frac{-3a\sqrt{b}}{2}$

EXERCISE 6.4 (PAGE 132)

1. $6-2\sqrt{5}$ 3. $3\sqrt{2}+\sqrt{6}$ 5. $2\sqrt{3}+3\sqrt{2}$ 7. $1-\sqrt{5}$

9. $x-2\sqrt{x}-3$ 11. $x-9$ 13. $x-6\sqrt{x}+9$ 15. $-4+\sqrt{6}$ 17. 3

19. $7-2\sqrt{10}$ 21. $x-\sqrt{3x}-6$ 23. $x+\sqrt{xy}-6y$ 25. $2(1+\sqrt{3})$

27. $3(\sqrt{7}-1)$ 29. $2(2+\sqrt{3})$ 31. $6(\sqrt{3}+1)$ 33. $4(1+\sqrt{y})$

35. $\sqrt{3}(y-x)$ 37. $\sqrt{2}(1-\sqrt{3})$ 39. $\sqrt{x}(\sqrt{3}+\sqrt{5})$ 41. $1+\sqrt{3}$

43. $1-\sqrt{3}$ 45. $3+\sqrt{3}$ 47. $1+\sqrt{3}$ 49. $1-\sqrt{x}$ 51. $x-y$

EXERCISE 6.5 (PAGE 135)

1. $\frac{\sqrt{2}}{2}$ 3. $\frac{x\sqrt{5}}{5}$ 5. $\frac{2\sqrt{2y}}{y}$ 7. $\frac{\sqrt{3}-1}{2}$ 9. $\frac{2\sqrt{7}+4}{3}$ 11. $\frac{4-4\sqrt{x}}{1-x}$

13. $\frac{x\sqrt{x}+3x}{x-9}$ 15. $\frac{x+\sqrt{xy}}{x-y}$ 17. $\frac{\sqrt{y-1}-y+1}{2-y}$ 19. $\frac{1}{\sqrt{3}}$ 21. $\frac{a}{3\sqrt{a}}$

23. $\frac{2}{\sqrt{2T}}$ 25. $\frac{-1}{2+2\sqrt{2}}$ 27. $\frac{x-1}{3\sqrt{x}+3}$ 29. $\frac{x-y}{x\sqrt{x}+x\sqrt{y}}$

EXERCISE 6.6 (PAGE 137)

1. 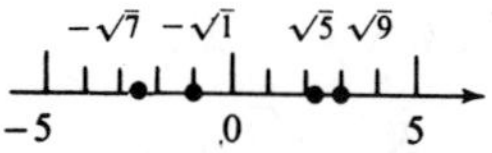

3. 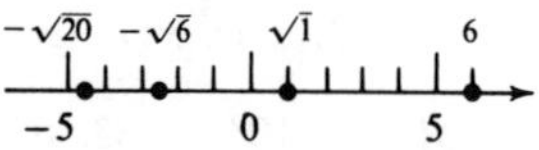

5. 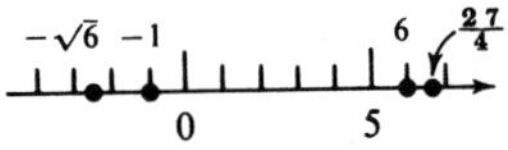

7. 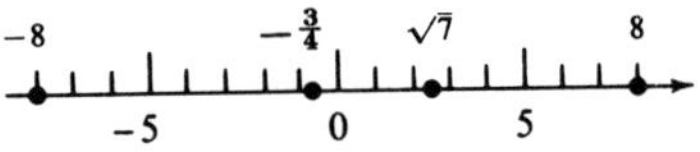

9. 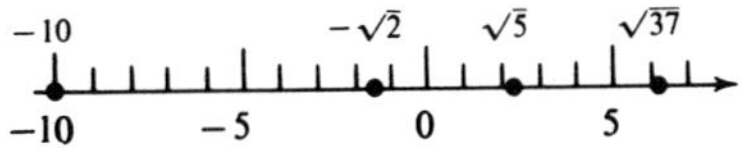

11. 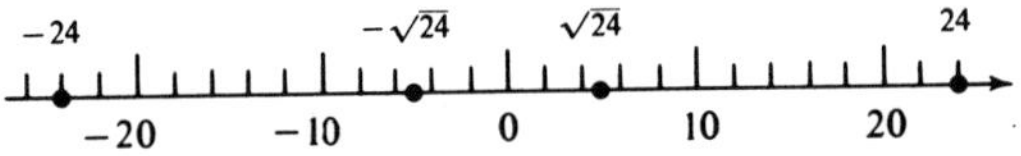

13. 27 **15.** 0.4796 **17.** 6.8 **19.** 0.28 **21.** 35.4 **23.** 6.91

25. 22.7 **27.** 4.484 **29.** 111.8 **31.** 2.006

EXERCISE 6.7 (PAGE 140)

1. $3i$ **3.** $6i$ **5.** $16i\sqrt{2}$ **7.** $2i$ **9.** $1-i$ **11.** $-1-i\sqrt{2}$

13. $5-i$ **15.** $-5-5i$ **17.** $5-5i$ **19.** $3+2i$ **21.** $2+3i$

23. $3-2i$ **25.** 8 **27.** -6 **29.** 7 **31.** $4+5i\sqrt{2}$ **33.** $1-7i\sqrt{5}$

35. $\frac{3}{4}+\frac{7i\sqrt{3}}{6}$

EXERCISE 6.8 (PAGE 143)

1. -2 **3.** -6 **5.** 3 **7.** $6+3i$ **9.** $12-6i$ **11.** $6-10i$

13. $10+2i$ **15.** $7-i$ **17.** $-1+5i$ **19.** $4i$ **21.** -6 **23.** $-4\sqrt{3}$

25. $-2+i\sqrt{2}$ **27.** $22+4i\sqrt{3}$ **29.** 14 **31.** $-i$ **33.** $\frac{3i}{2}$

35. $-1-i$ **37.** $\frac{9}{13}-\frac{6}{13}i$ **39.** $\frac{3}{13}+\frac{2}{13}i$ **41.** i

43. a. -1 b. 1 c. $-i$

CHAPTER REVIEW (PAGE 145)

1. $\sqrt{x^3}$ **2.** $\sqrt[3]{x}-\sqrt[3]{y}$ **3.** $\sqrt[3]{x-y}$ **4.** $\frac{5}{\sqrt[3]{x}}$ **5.** 8 **6.** -2

7. SR^2 **8.** $\frac{2}{3}T^2$ **9.** $2\sqrt{3}$ **10.** $xy(\sqrt[3]{xy^2})$ **11.** $a\sqrt{b}$

12. $4\sqrt{2}\times 10^{-2}$ **13.** $\frac{\sqrt{15}}{5}$ **14.** $\frac{3\sqrt{2x}}{x}$ **15.** $\frac{3(\sqrt[3]{4x^2})}{x}$ **16.** $\sqrt{3y}$

17. $13\sqrt{2}$ **18.** $(4a+2)\sqrt{3a}$ **19.** 0 **20.** $\frac{5\sqrt{x}}{3}$ **21.** -7

22. $2x + 5\sqrt{x} - 3$ **23.** $2(1 - \sqrt{2})$ **24.** $\sqrt{y}(1 + \sqrt{5})$ **25.** $2 + \sqrt{3}$

26. $\sqrt{a} - \sqrt{b}$ **27.** $\dfrac{3}{2\sqrt{3}}$ **28.** $\dfrac{-1}{1 - \sqrt{3}}$

29. Number line with points marked at $-\sqrt{25}$, $-\sqrt{2}$, $\sqrt{9}$, $\sqrt{29}$; scale labels -5, 0, 5 **30.** 2.94 **31.** $6i$ **32.** $1 + i\sqrt{2}$

33. $2 - i$ **34.** $6 + 2i$ **35.** -10 **36.** $13 - i$ **37.** $-3\sqrt{2}$

38. $12 - i\sqrt{3}$ **39.** $2 - 2i$ **40.** $\dfrac{1 - i\sqrt{3}}{2}$

EXERCISE 7.1 (PAGE 149)

1. $\{2, 3\}$ **3.** $\{0, \frac{3}{2}, -2\}$ **5.** $\left\{1, \dfrac{-5}{2}\right\}$ **7.** $\{1, -2, 3\}$ **9.** $\left\{\dfrac{-2}{5}, \dfrac{1}{3}, \dfrac{3}{4}\right\}$

11. $\{3, -2\}$ **13.** $\{\frac{3}{2}, -1\}$ **15.** $\left\{0, \dfrac{-1}{2}, 3\right\}$ **17.** $\{\frac{5}{2}, -3\}$ **19.** $\{0, -2\}$

21. $\{-3, 3\}$ **23.** $\left\{\dfrac{2}{3}, \dfrac{-2}{3}\right\}$ **25.** $\{4, -1\}$ **27.** $\{-7, 2\}$ **29.** $\{1\}$

31. $\{1, \frac{1}{2}\}$ **33.** $\{-2, 3\}$ **35.** $\{\frac{3}{2}, -2\}$ **37.** $\{-3, 1\}$ **39.** $\left\{\dfrac{-10}{3}, 1\right\}$

41. $\{13, 2\}$ **43.** $x^2 - 5x + 6 = 0$ **45.** $2x^2 - 5x + 3 = 0$

47. $8x^2 - 10x + 3 = 0$ **49.** $9x^2 - 6x + 1 = 0$ **51.** $x^2 + 1 = 0$

53. $x^2 - 2x + 2 = 0$

EXERCISE 7.2 (PAGE 153)

1. $\{2, -2\}$ **3.** $\{2, -2\}$ **5.** $\left\{\dfrac{10}{3}, \dfrac{-10}{3}\right\}$ **7.** $\{\sqrt{5}, -\sqrt{5}\}$

9. $\{2\sqrt{3}, -2\sqrt{3}\}$ **11.** $\{\sqrt{P}, -\sqrt{P}\}$ **13.** $\left\{\sqrt{\dfrac{P}{Q}}, -\sqrt{\dfrac{P}{Q}}\right\}$

15. $\left\{\sqrt{\dfrac{N}{2M}}, -\sqrt{\dfrac{N}{2M}}\right\}$ **17.** $\{\sqrt{RT}, -\sqrt{RT}\}$ **19.** $\{3, -1\}$

21. $\{-1, -4\}$ **23.** $\{6 + \sqrt{5}, 6 - \sqrt{5}\}$ **25.** $\{2 + \sqrt{3}, 2 - \sqrt{3}\}$

27. $\{R + 2, R - 2\}$ **29.** $\left\{\dfrac{4 - Q}{P}, \dfrac{-4 - Q}{P}\right\}$ **31.** $\left\{\dfrac{T + \sqrt{5}}{S}, \dfrac{T - \sqrt{5}}{S}\right\}$

33. $\{-B + \sqrt{C}, -B - \sqrt{C}\}$ **35.** $\left\{\dfrac{C + \sqrt{L}}{R}, \dfrac{C - \sqrt{L}}{R}\right\}$

EXERCISE 7.3 (PAGE 156)

1. (a) 1; (b) $(x+1)^2$ **3.** (a) 9; (b) $(x-3)^2$ **5.** (a) $\frac{9}{4}$; (b) $(x+\frac{3}{2})^2$

7. (a) $\frac{49}{4}$; (b) $(x-\frac{7}{2})^2$ **9.** (a) $\frac{1}{4}$; (b) $(x-\frac{1}{2})^2$

11. (a) $\frac{1}{16}$; (b) $(x+\frac{1}{4})^2$ **13.** $\{2, -6\}$ **15.** $\{1\}$ **17.** $\{-4, -5\}$

19. $\{1+\sqrt{2}, 1-\sqrt{2}\}$ **21.** $\{\frac{1}{2}, -2\}$ **23.** $\left\{\dfrac{-2+\sqrt{2}}{2}, \dfrac{-2-\sqrt{2}}{2}\right\}$

25. $y=(x+1)^2-4$ **27.** $y=(x+\frac{1}{2})^2+\frac{3}{4}$ **29.** $y=(x-1)^2+4$

31. $y=(x+1)^2+\frac{2}{3}$ **33.** $(x-2)^2+(y-2)^2=25$

35. $(x+3)^2+(y-1)^2=4$ **37.** $(x-1)^2+(y+4)^2=2$

39. $(x-\frac{1}{2})^2+(y+1)^2=4$ **41.** $x=\dfrac{-b\pm\sqrt{b^2-4ac}}{2a}$

7

EXERCISE 7.4 (PAGE 160)

1. $\{2, 1\}$ **3.** $\{1\}$ **5.** $\left\{\dfrac{-1+\sqrt{5}}{2}, \dfrac{-1-\sqrt{5}}{2}\right\}$ **7.** $\{2, \frac{3}{2}\}$

9. $\left\{\dfrac{1+i\sqrt{7}}{4}, \dfrac{1-i\sqrt{7}}{4}\right\}$ **11.** $\left\{\dfrac{3}{2}, \dfrac{-5}{2}\right\}$ **13.** $\left\{3, \dfrac{-3}{2}\right\}$ **15.** $\{2k, -k\}$

17. $\left\{\dfrac{1+\sqrt{1-4ac}}{2a}, \dfrac{1-\sqrt{1-4ac}}{2a}\right\}$ **19.** 1; two unequal real solutions

21. 24; two unequal real solutions **23.** 0; one real solution

25. -543; two imaginary solutions **27.** -8; two imaginary solutions

29. $k=4$ **31.** $k\leq -2$

EXERCISE 7.5 (PAGE 162)

1. $\{64\}$ **3.** $\{4\}$ **5.** No solution **7.** $\{\frac{15}{2}\}$ **9.** $\{4\}$ **11.** $\{13\}$

13. $\{5\}$ **15.** $\{0\}$ **17.** $\{4\}$ **19.** $A=\pi r^2$ **21.** $L=\dfrac{1}{R^3}$

23. $R=\pm\sqrt{Q^2-P^2}$

EXERCISE 7.6 (PAGE 164)

1. 4, 9 **3.** 7, 8; $-7, -8$ **5.** 6, 7 **7.** 4 or $\frac{1}{4}$ **9.** -12 **11.** 80 and 40

13. $5\sqrt{2}$ seconds or approximately 7.07 seconds **15.** $\frac{5}{2}$ seconds, $\frac{5}{4}\sqrt{6}$ seconds

17. 3 mph on bicycle and 1 mph on foot **19.** 36 people

CHAPTER REVIEW (PAGE 166)

1. $\{2, -5\}$ **2.** $\{0, 3\}$ **3.** $\{5, -3\}$ **4.** $6x^2 + x - 2 = 0$ **5.** $\left\{\frac{3}{2}, \frac{-3}{2}\right\}$

6. $\{c + \sqrt{b}, c - \sqrt{b}\}$ **7.** $\left\{2, \frac{-1}{2}\right\}$ **8.** $\{2 + \sqrt{6}, 2 - \sqrt{6}\}$

9. $y = (x - 2)^2 + 5$ **10.** $(x + 2)^2 + (y - 3)^2 = 4^2$ **11.** $\{\frac{1}{2}, -3\}$

12. $\left\{\frac{-3 + \sqrt{93}}{6}, \frac{-3 - \sqrt{93}}{6}\right\}$ **13.** $\left\{k, \frac{-k}{2}\right\}$ **14.** $k = 6$ **15.** $\{64\}$

16. $\{8\}$ **17.** 12 **18.** 8 and 17 **19.** 40 mph

20. 3 hours; 6 hours

8

EXERCISE 8.1 (PAGE 170)

1. 2 **3.** -1 **5.** $2\frac{1}{2}$ **7.** $a + 2$ **9.** 9 **11.** 1 **13.** 1

15. $a^2 - 2a + 1$ **17.** $x = -1$ **19.** $x = -4$ **21.** $x = -6$

23. $x = -1, 1$ **25.** $x = 3, -3$ **27.** a. $a^2 - b^2$ b. $2x + 1$

29. $C = 2\pi r$; $\{r \mid r > 0\}$ **31.** $P = 4s$; $\{s \mid s > 0\}$ **33.** $d = 40t$; $\{t \mid t > 0\}$

35. $h^2 = b^2 + 36$; $\{b \mid b > 0\}$ **37.** $A = \frac{s^2(\sqrt{3})}{4}$; $\{s \mid s > 0\}$

39. $P = 20 + \frac{A}{5}$; $\{A \mid A > 0\}$

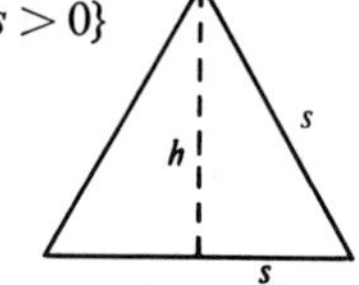

EXERCISE 8.2 (PAGE 174)

1. a. (0, 7) b. (2, 9) c. $(-2, 5)$ **3.** a. (2, 5) b. $(\frac{1}{2}, 2)$ c. (0, 1)

5. a. $(0, \frac{3}{2})$ b. $(\frac{3}{2}, 0)$ c. $(\frac{1}{2}, 1)$ **7.** $\{(-1, 1), (0, 0), (1, -1)\}$

9. $\{(-2, 4), (-1, 1), (0, 0), (1, 1), (2, 4)\}$ **11.** $\{(1, 1), (3, \frac{1}{3}), (5, \frac{1}{5})\}$

13. $\{(x, y) \mid y = x + 1\}$ **15.** $\{(2, 12.56), (4, 25.12), (6, 37.68)\}$

17. $\{(1, 3.14), (3, 28.26), (5, 78.50)\}$ **19.** $\{(5, 1200), (10, 1400), (20, 1800)\}$

21. $\{(1, \frac{1}{4}), (1, \frac{1}{2}), (9, \frac{3}{4})\}$ **23.** $\{(0, 32), (100, 212), (150, 302)\}$

EXERCISE 8.3 (PAGE 179)

1. Quadrant I **3.** Quadrant IV **5.** Quadrant II **7.** Quadrant III

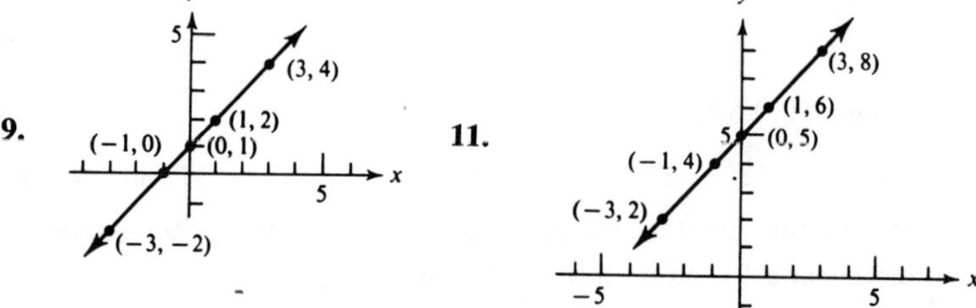

13.

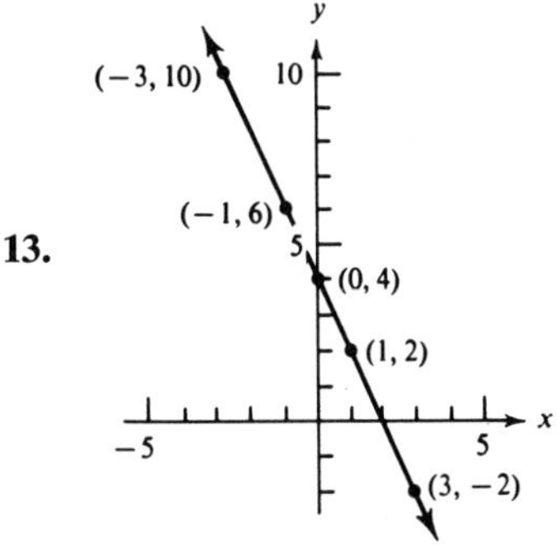

15.

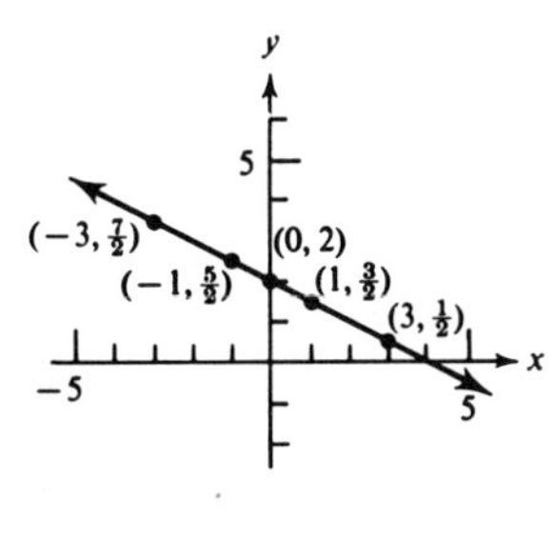

17.

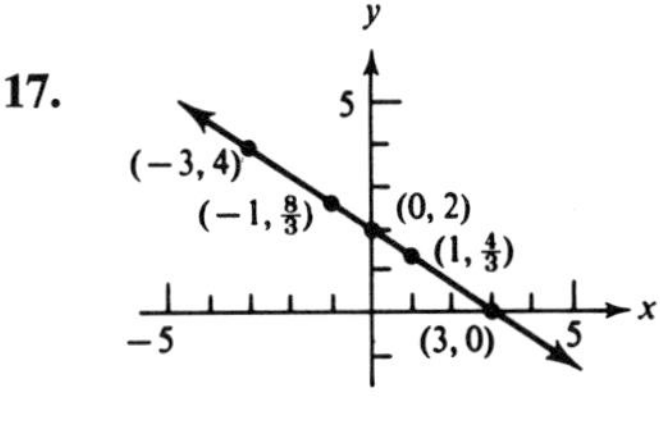

19.

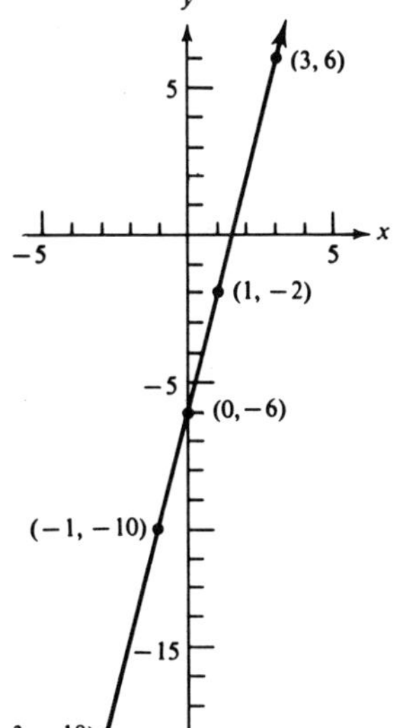

21.

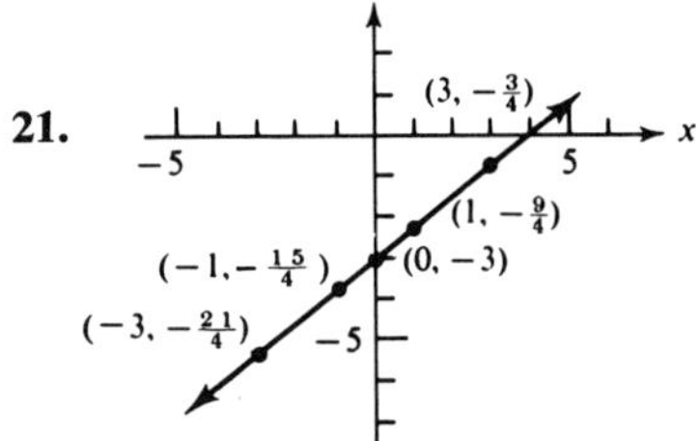

23.

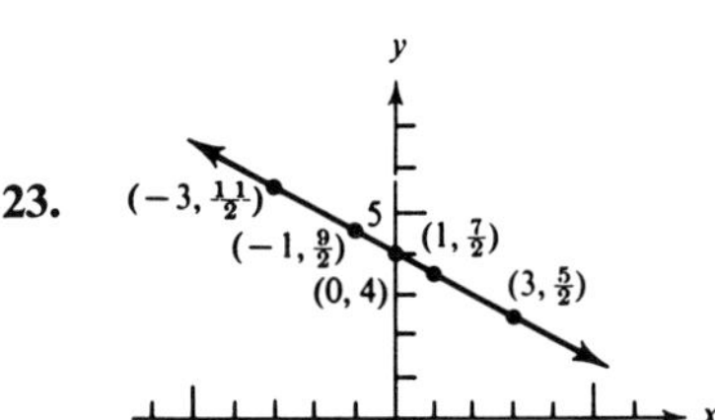

25.

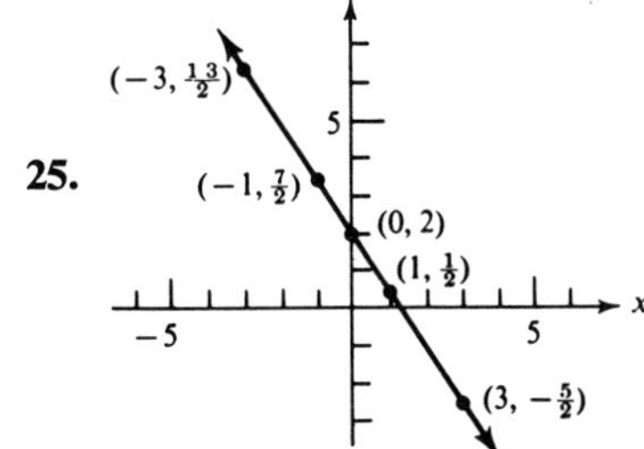

8

27.

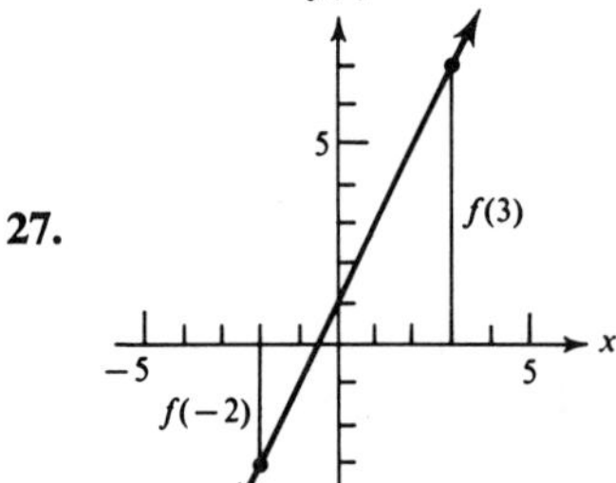

29.

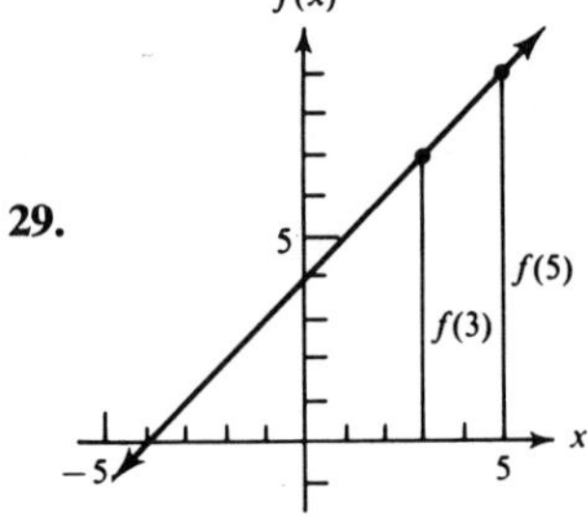

31. Range: {1, 2, 3, 4}

EXERCISE 8.4 (PAGE 180)

1.

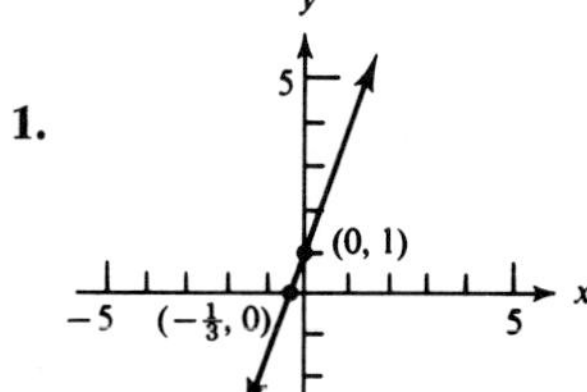

1 is the y-intercept; $-\frac{1}{3}$ is the x-intercept

3.

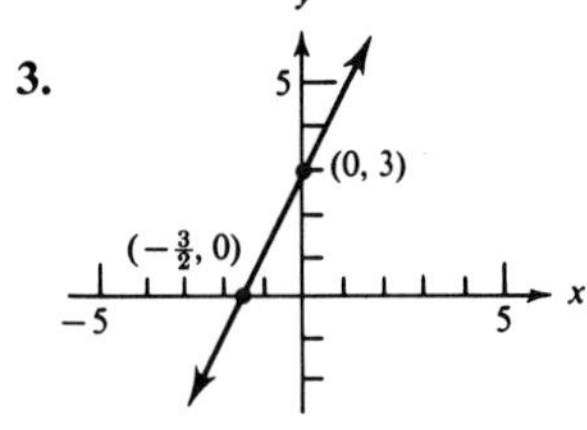

3 is the y-intercept; $-\frac{3}{2}$ is the x-intercept

5.

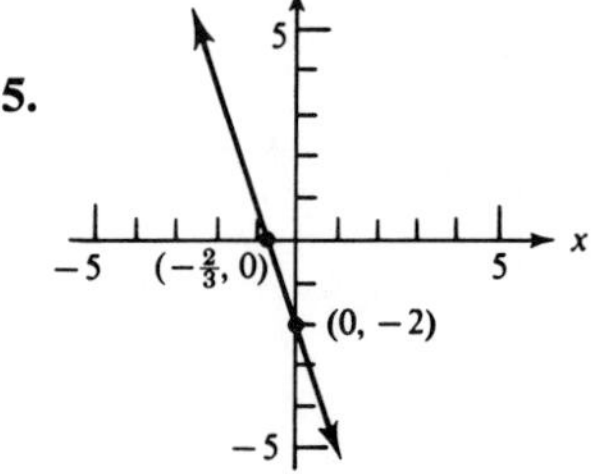

-2 is the y-intercept; $-\frac{2}{3}$ is the x intercept

7. 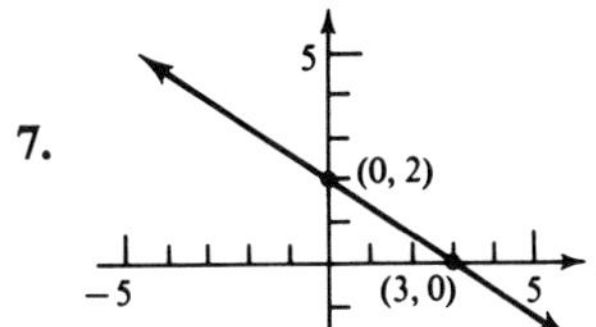

2 is the y-intercept; 3 is the x-intercept

9. 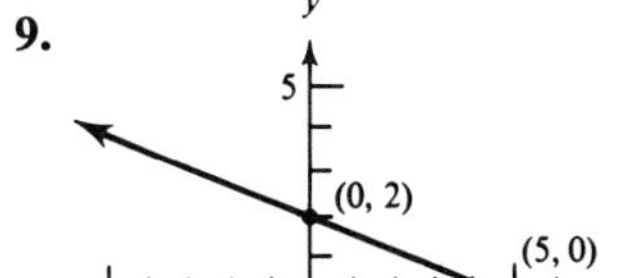

2 is the y-intercept; 5 is the x intercept

11. 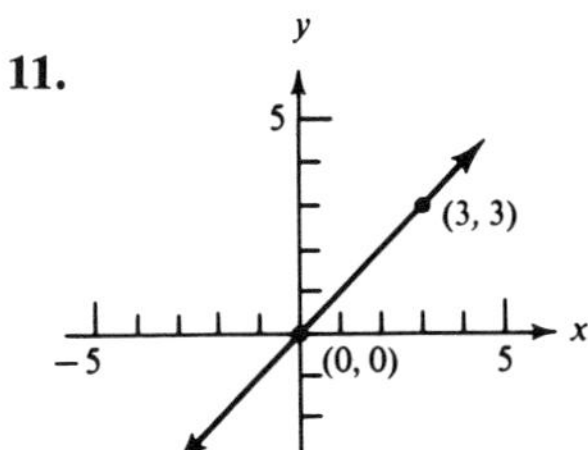

0 is the y-intercept; 0 is the x-intercept

13. 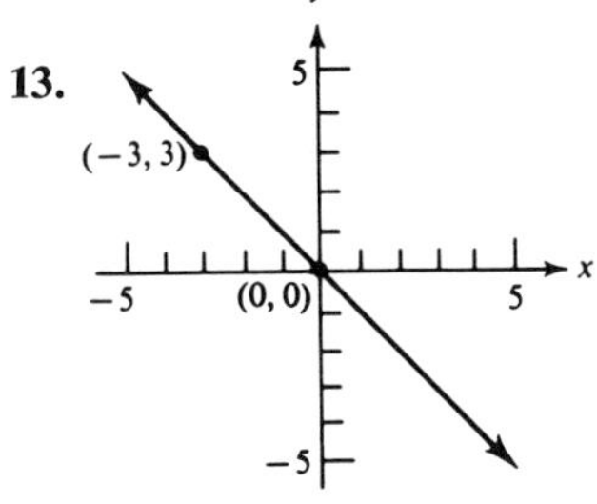

0 is the y-intercept; 0 is the x-intercept

15. 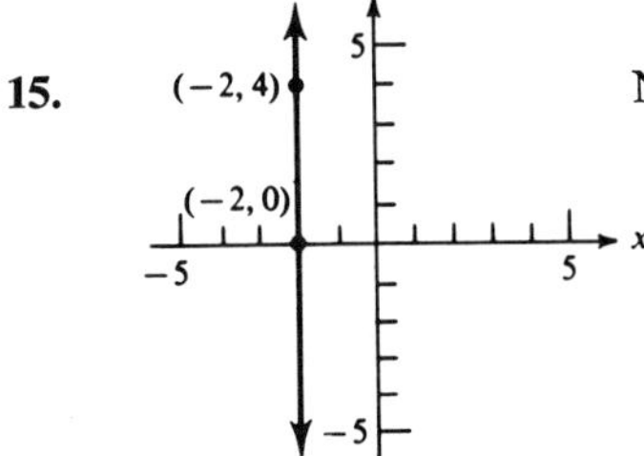

No y-intercept; -2 is the x-intercept

17. 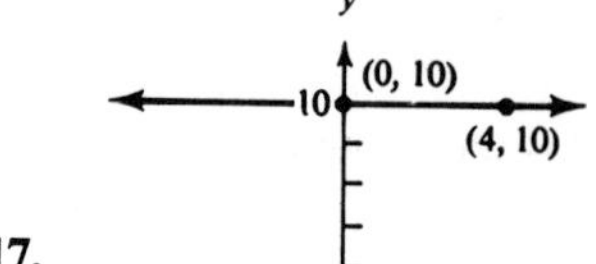
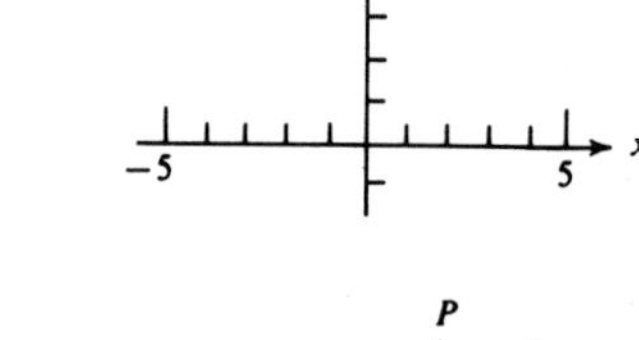

10 is the y-intercept; no x-intercept

19. 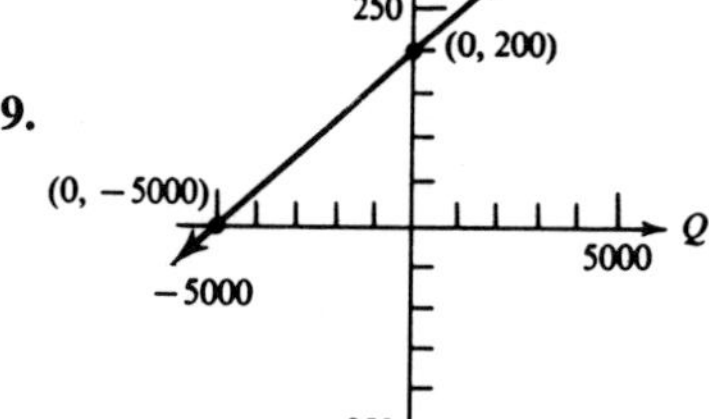

21. 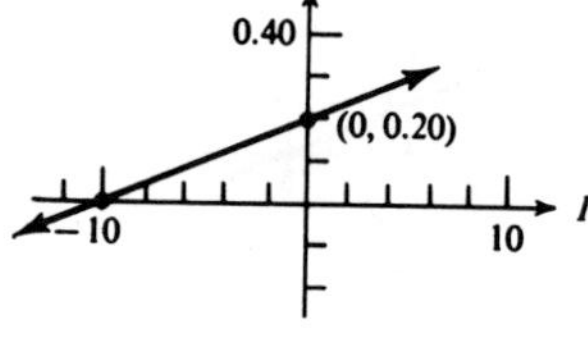

23. 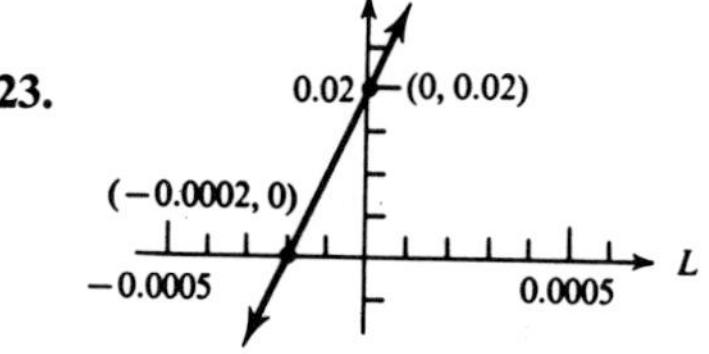

25. 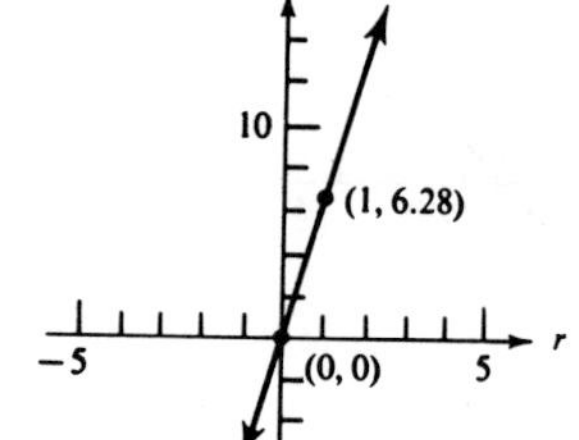

EXERCISE 8.5 (PAGE 188)

1. 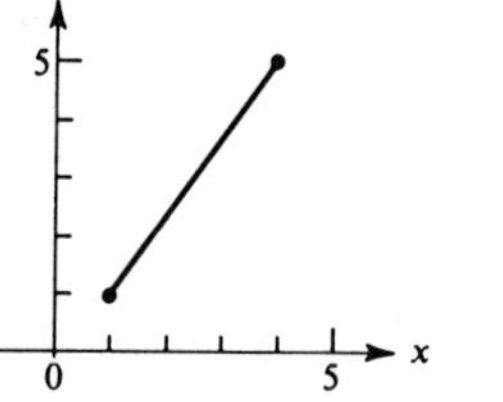

3. 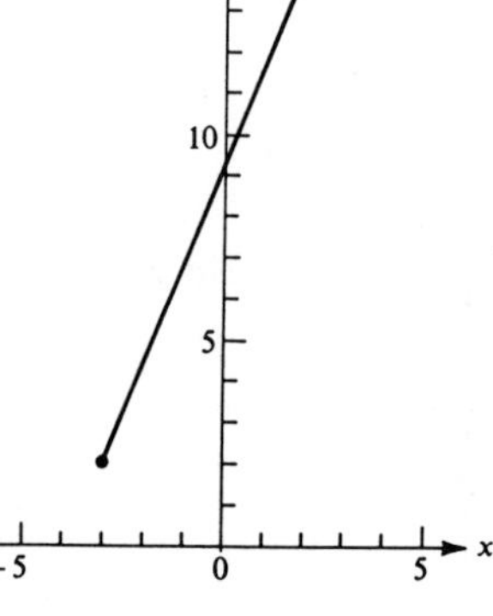

5. 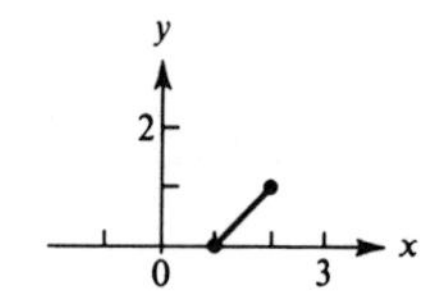

7.

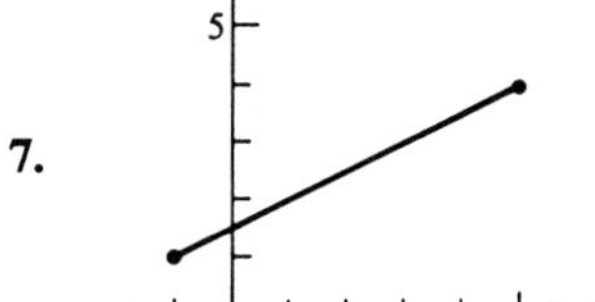

9.

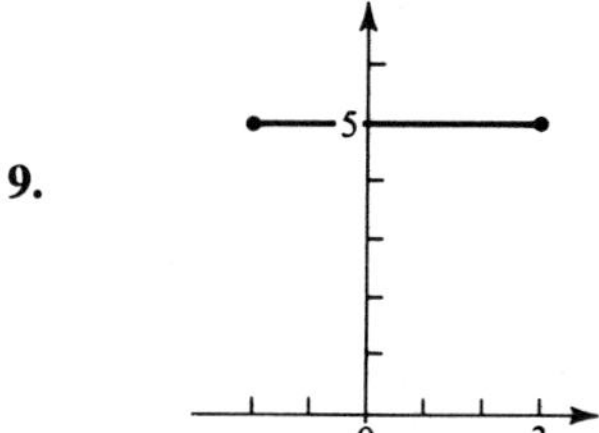

11.

13.

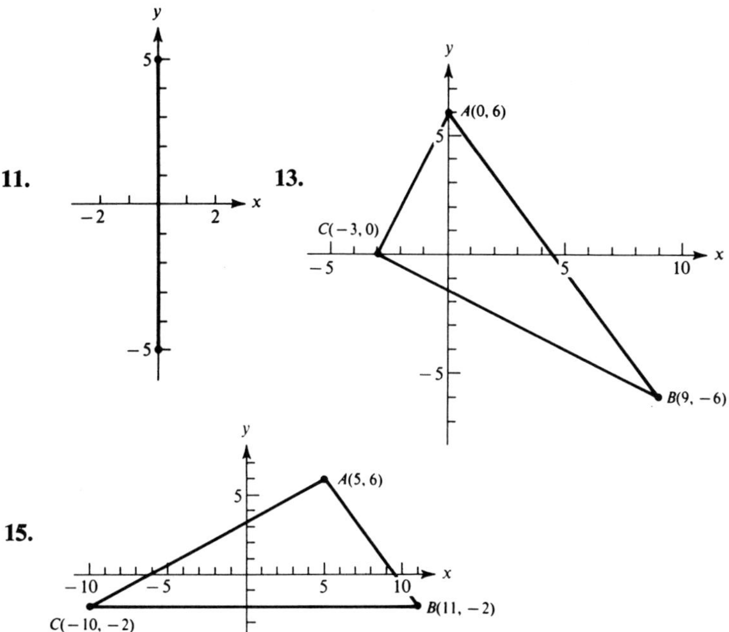

15.

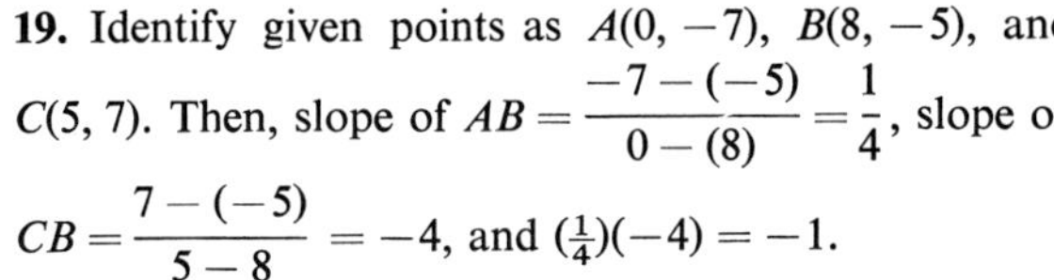

17. Identify vertices as $A(0, 6)$, $B(9, -6)$, and $C(-3, 0)$. Then by the distance formula, $(AB)^2 = 225$ and $(BC)^2 + (AC)^2 = 180 + 45 = 225$; the triangle is a right triangle.

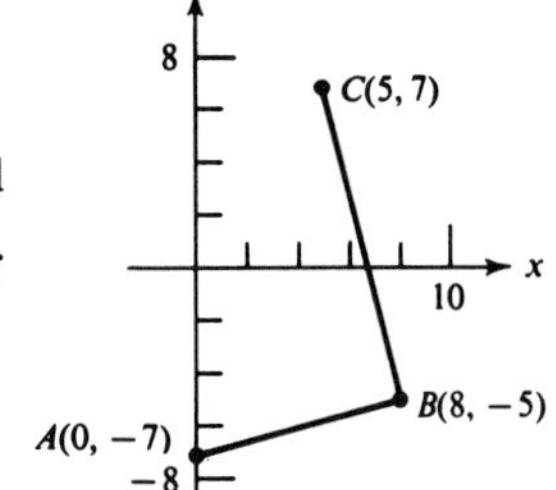

19. Identify given points as $A(0, -7)$, $B(8, -5)$, and $C(5, 7)$. Then, slope of $AB = \dfrac{-7-(-5)}{0-(8)} = \dfrac{1}{4}$, slope of $CB = \dfrac{7-(-5)}{5-8} = -4$, and $(\frac{1}{4})(-4) = -1$.

21. Identify given points as $A(2, 4)$, $B(3, 8)$, $C(5, 1)$, and $D(4, -3)$. Then the slopes are: AB, $m_1 = \frac{-5}{4}$; BC, $m_2 = \frac{36}{5}$; CD, $m_3 = \frac{-5}{4}$; DA, $m_4 = \frac{36}{5}$. Therefore, since $m_1 = m_3$ and $m_2 = m_4$, it follows that AB is parallel to CD, DA is parallel to BC, and $ABCD$ is a parallelogram.

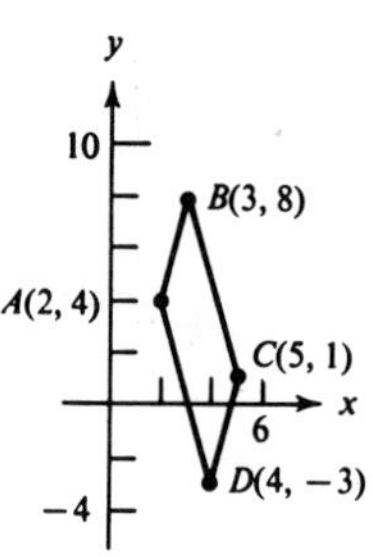

EXERCISE 8.6 (PAGE 191)

8

1.

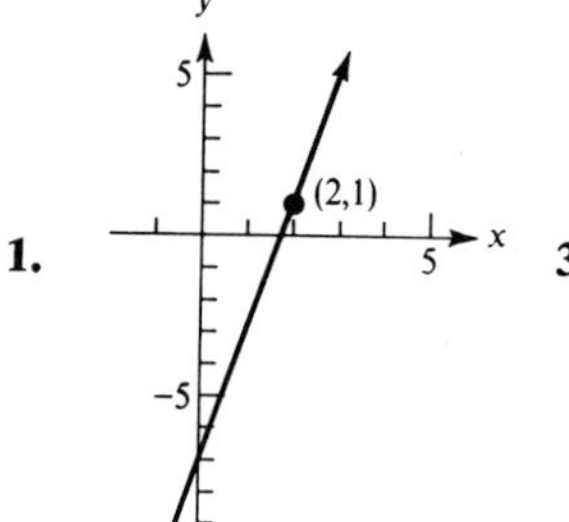

3.

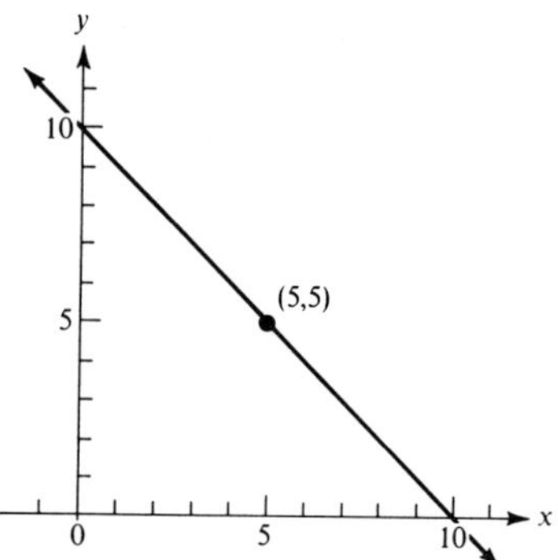

5.

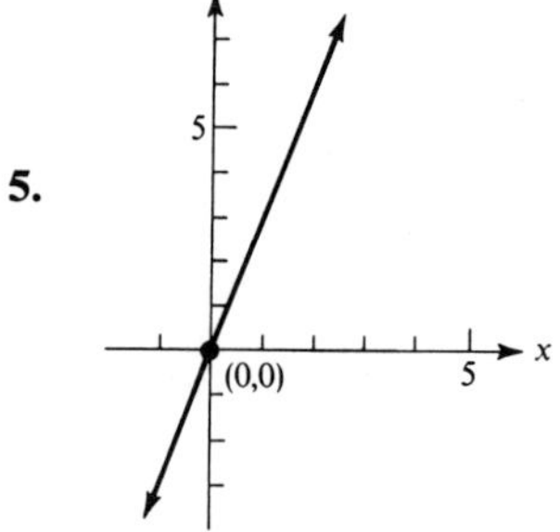

7.

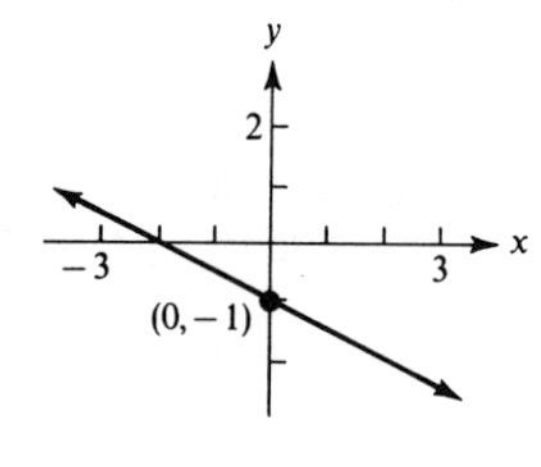

9.

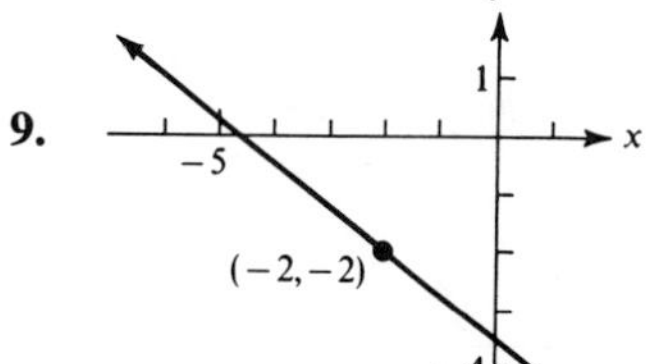

11.

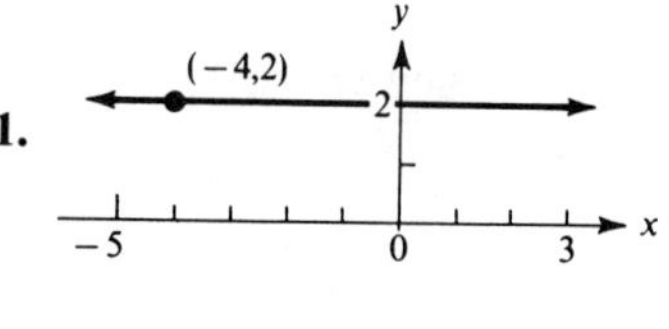

13. $y = -x + 3$, slope: -1, y-intercept: 3

15. $y = \frac{-3}{2}x + \frac{1}{2}$, slope: $\frac{-3}{2}$, y-intercept: $\frac{1}{2}$

17. $y = \frac{1}{3}x - \frac{2}{3}$, slope: $\frac{1}{3}$, y-intercept: $\frac{-2}{3}$

19. $y = \frac{8}{3}x$, slope: $\frac{8}{3}$, y-intercept: 0 **21.** $y = 0x - 2$, slope: 0, y-intercept: -2

23. Since graph is a line perpendicular to the x-axis, there is no slope and no y-intercept.

25.

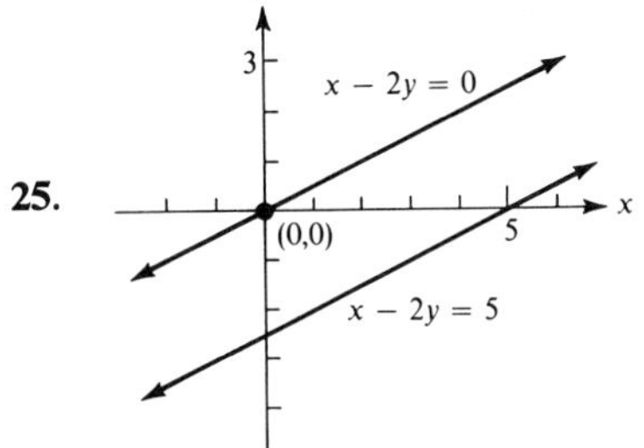

27.

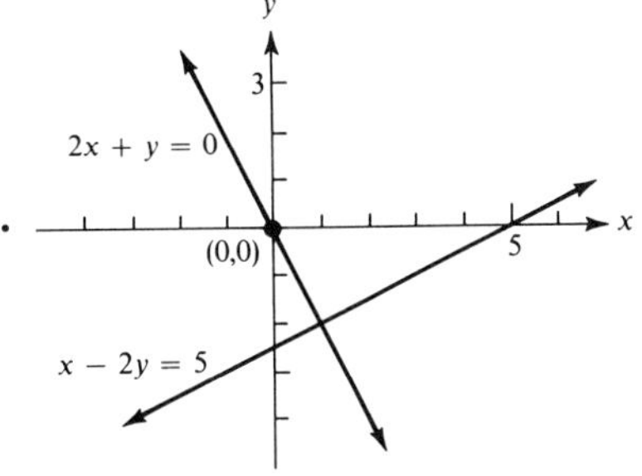

29. Since $y - y_1 = m(x - x_1)$ and $m = \frac{y_2 - y_1}{x_2 - x_1}$, by substitution, it follows that

$$y - y_1 = \left(\frac{y_2 - y_1}{x_2 - x_1}\right)(x - x_1)$$

EXERCISE 8.7 (PAGE 194)

1.

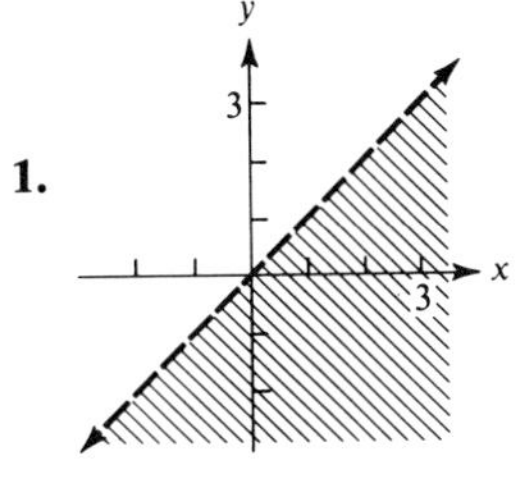

3.

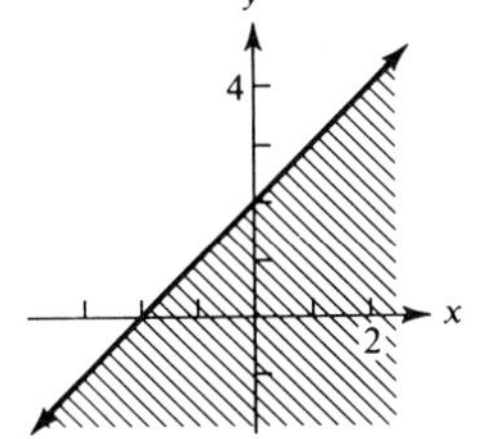

5.

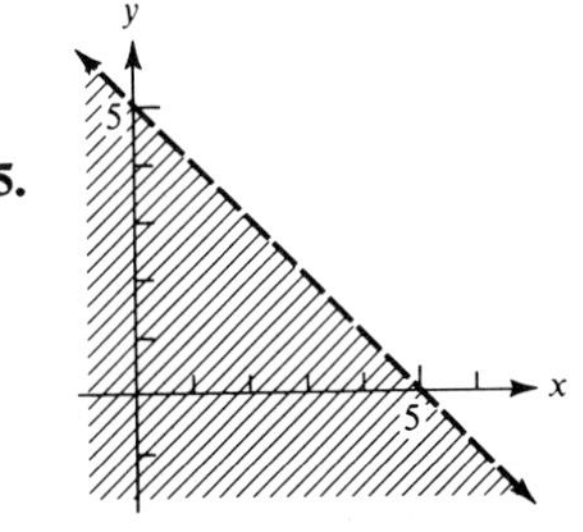

7.

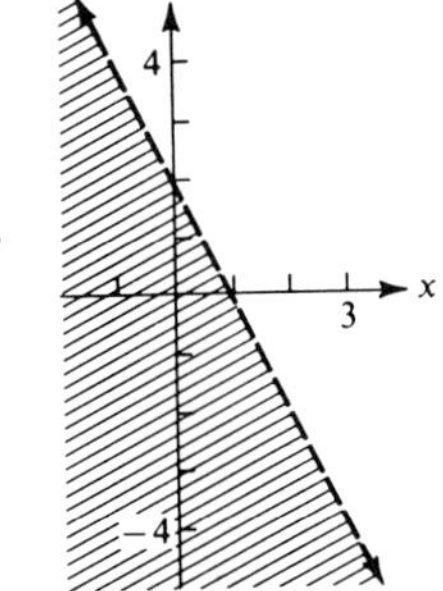

9.

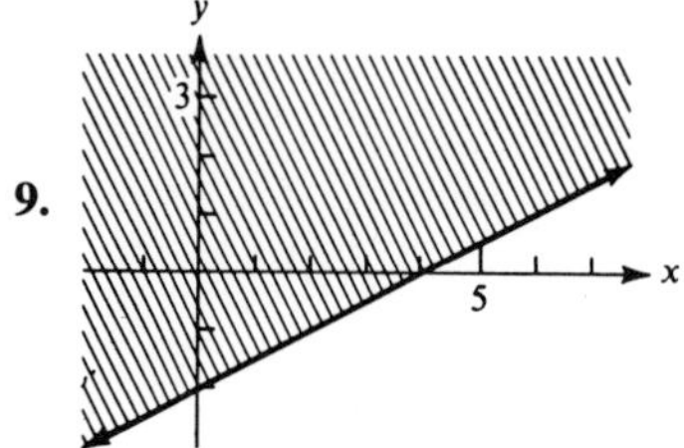

11.

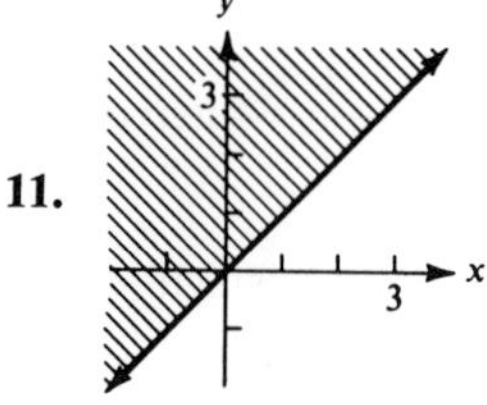

13.

15.

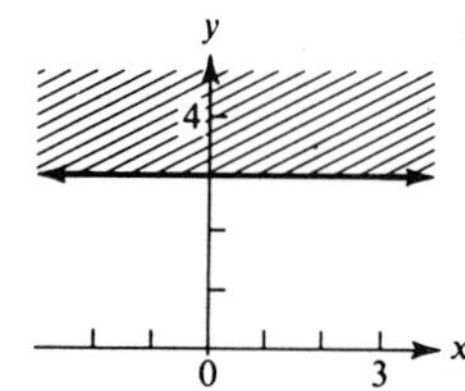

17.

19.

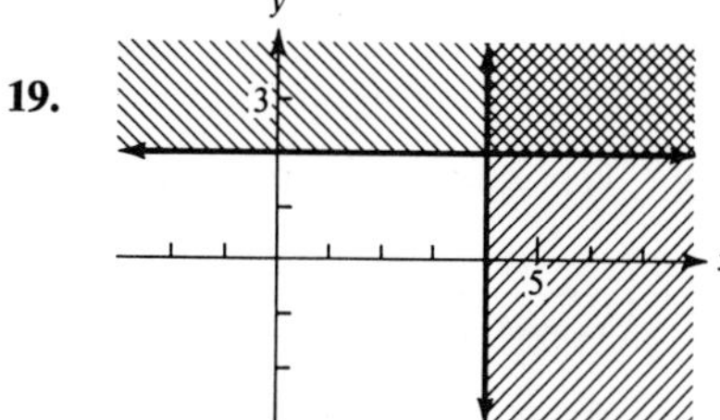

21.

23.

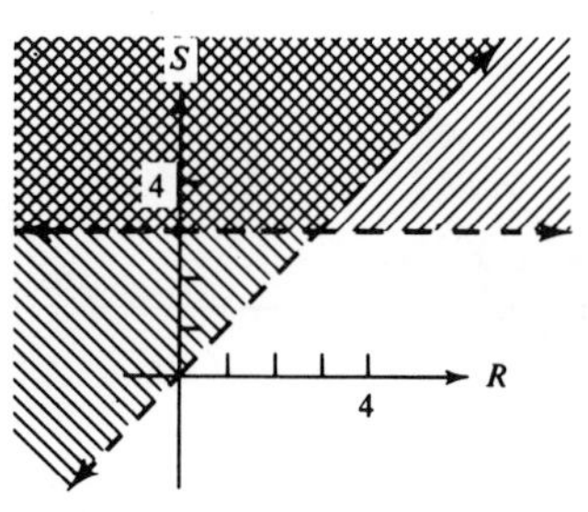

25.

CHAPTER REVIEW (PAGE 196)

1. 11 **2.** 18 **3.** $2x+3$ **4.** $P=2l+14$

5. a. $(0, -\frac{1}{6})$ b. $(\frac{1}{2}, 0)$ c. $(3, \frac{5}{6})$ **6.** $\{(-10, 14), (20, 68), (80, 176)\}$

7.

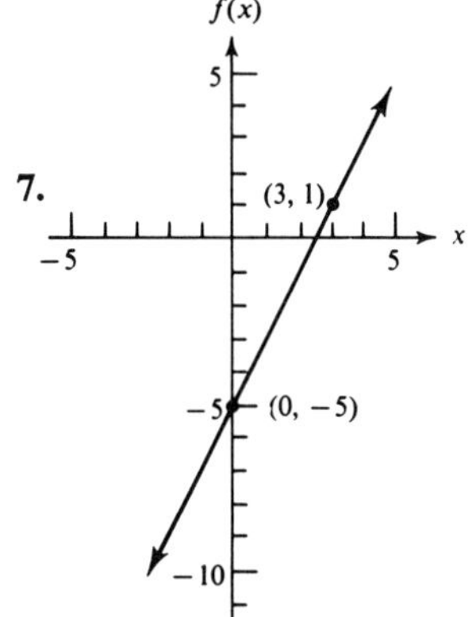

8.

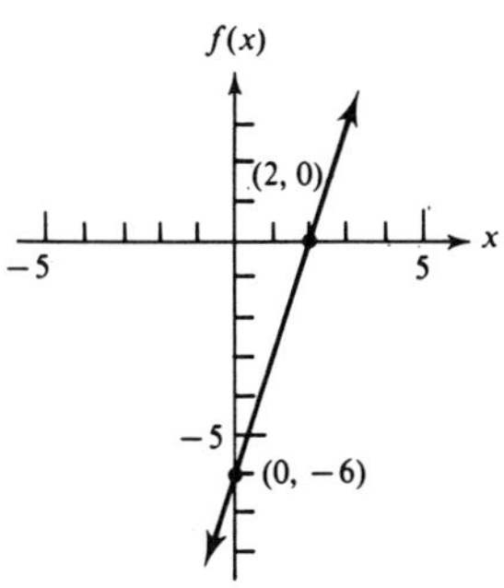

9.

10.

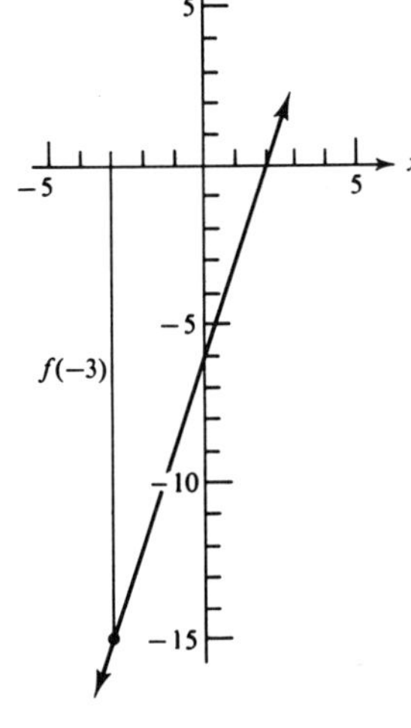

11.

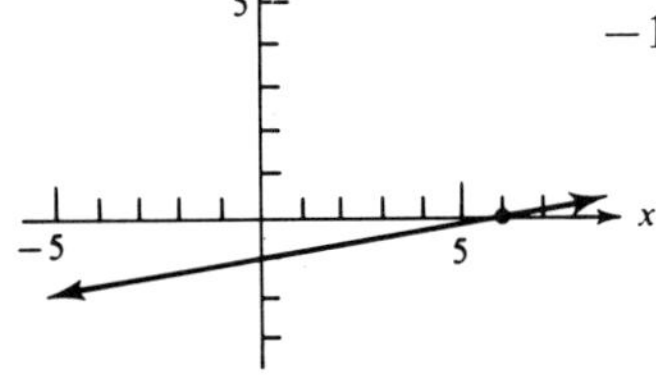

-1 is the y-intercept, 6 is the x-intercept.

12. 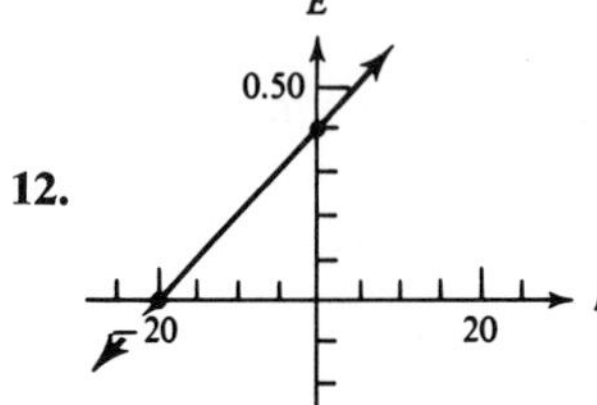

0.40 is the E-intercept, -20 is the I-intercept.

13. $\sqrt{178}$ **14.** $m = \frac{13}{3}$ **15.** $2x - y - 1 = 0$

16. $y = \dfrac{-2}{3}x + 2$; slope $= \dfrac{-2}{3}$; y-intercept $= 2$ **17.** $2x + 3y - 1 = 0$

18. $2x + 5y - 31 = 0$

19. 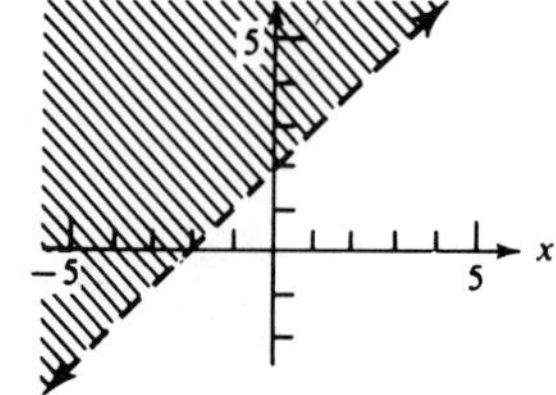

20. 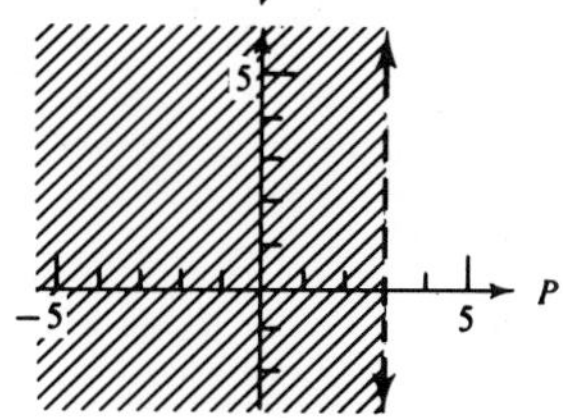

EXERCISE 9.1 (PAGE 200)

1. a. $(-4, 17), (-3, 10), (-2, 5), (-1, 2), (0, 1), (1, 2), (2, 5), (3, 10), (4, 17)$

b. 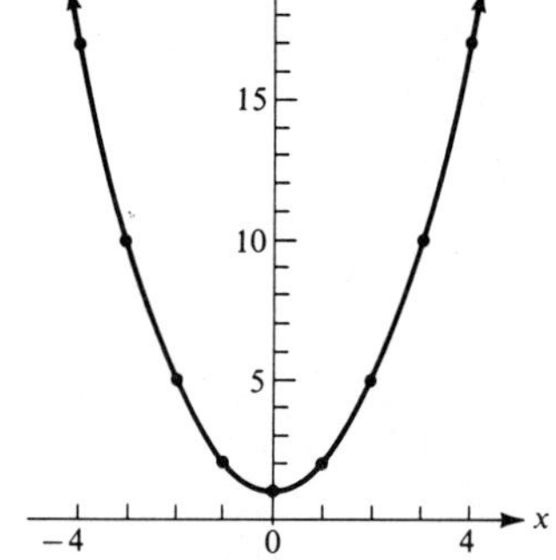

3. a. (−4, 13), (−3, 6), (−2, 1), (−1, −2), (0, −3), (1, −2), (2, 1), (3, 6), (4, 13)

b.

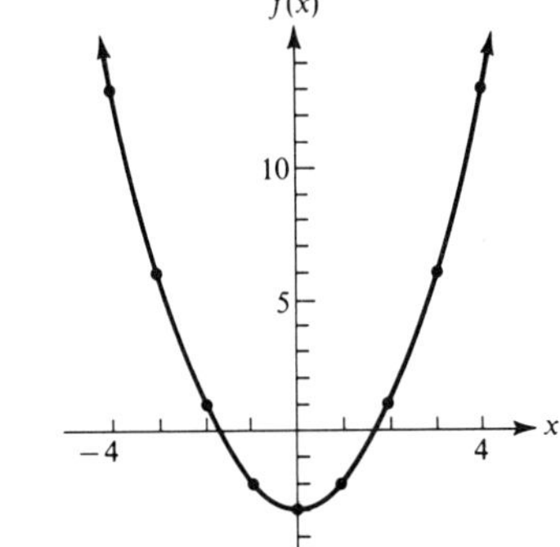

5. a. (−4, −12), (−3, −5), (−2, 0), (−1, 3), (0, 4), (1, 3), (2, 0), (3, −5), (4, −12)

b.

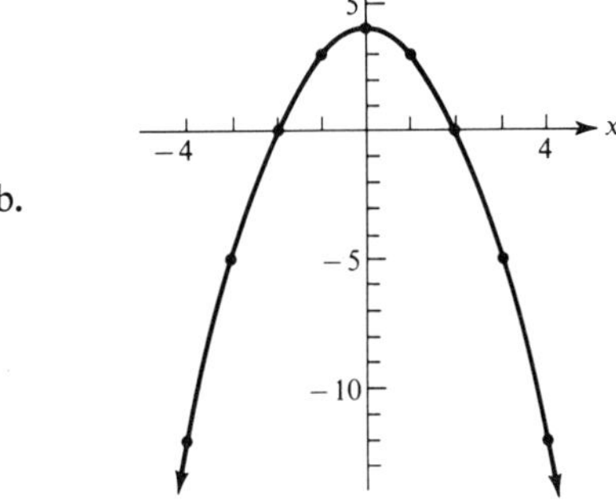

7. a. (−4, 44), (−3, 24), (−2, 10), (−1, 2), (0, 0), (1, 4), (2, 14), (3, 30), (4, 52)

b.

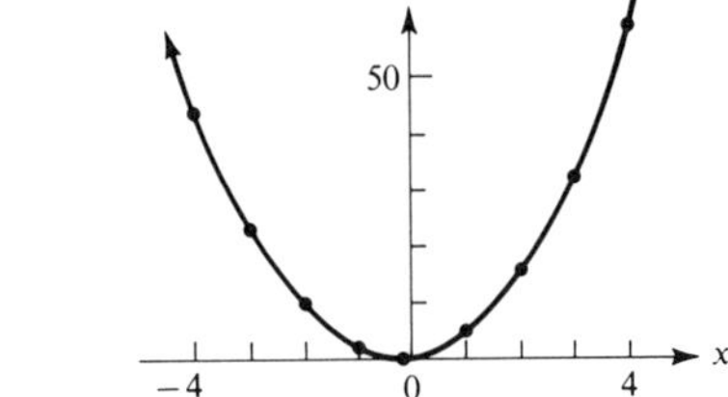

9. a. $(-4, 9), (-3, 4), (-2, 1), (-1, 0), (0, 1), (1, 4), (2, 9), (3, 16), (4, 25)$

b.

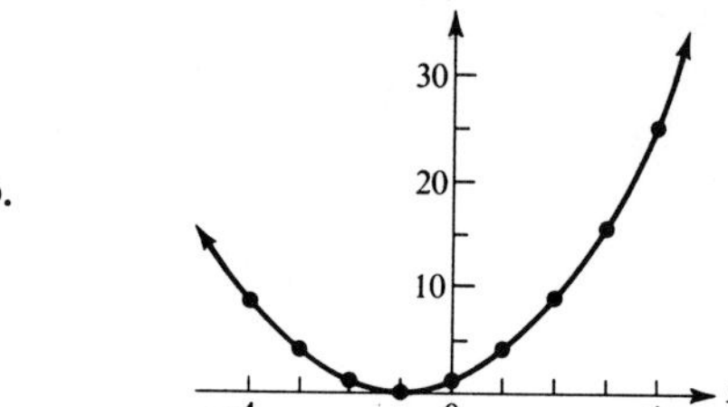

11. a. $(-4, -39), (-3, -24), (-2, -13), (-1, -6), (0, -3), (1, -4), (2, -9),$
$(3, -18), (4, -31)$

b.

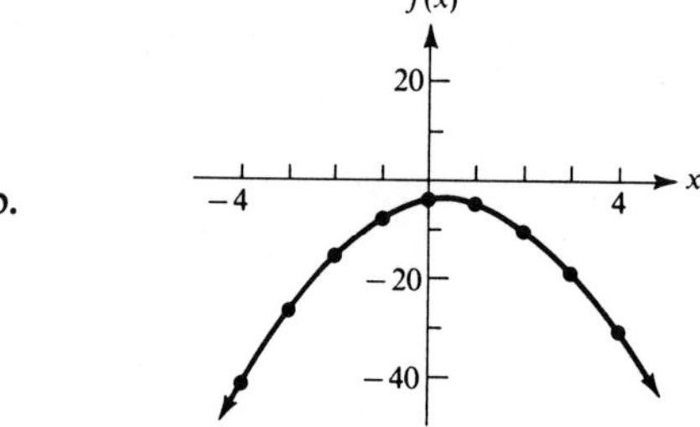

13. a. $(-4, -19), (-3, -11), (-2, -5), (-1, -1), (0, 1), (1, 1), (2, -1), (3, -5),$
$(4, -11)$

b.

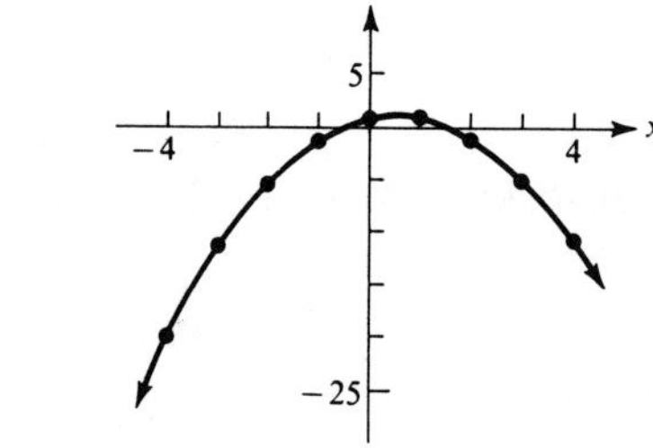

15.

EXERCISE 9.2 (PAGE 204)

1.

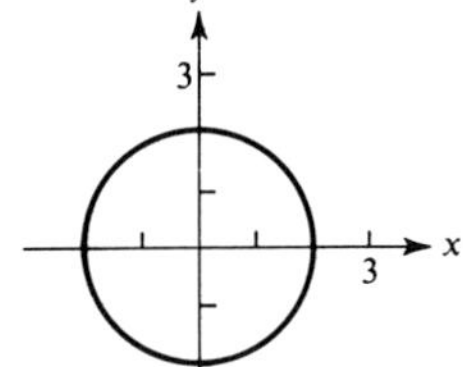

3.

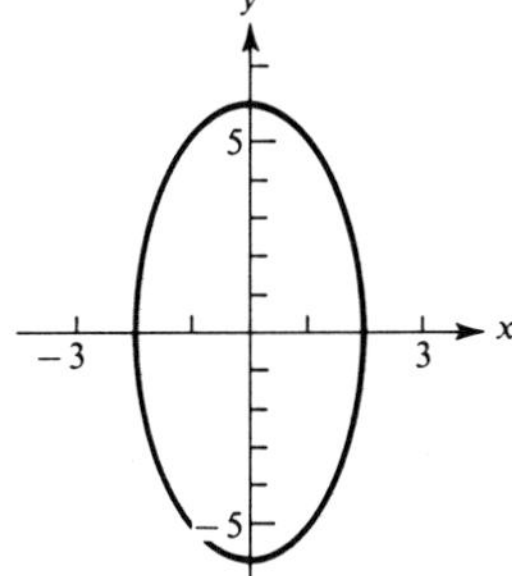

5.

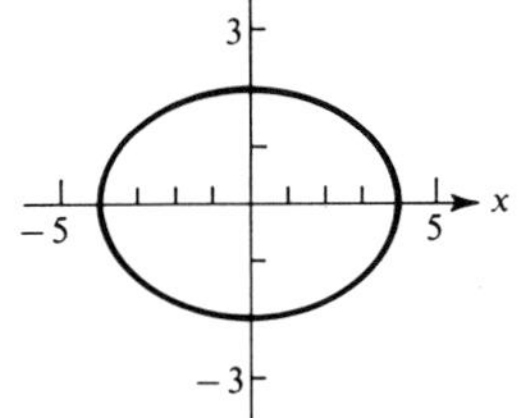

7.

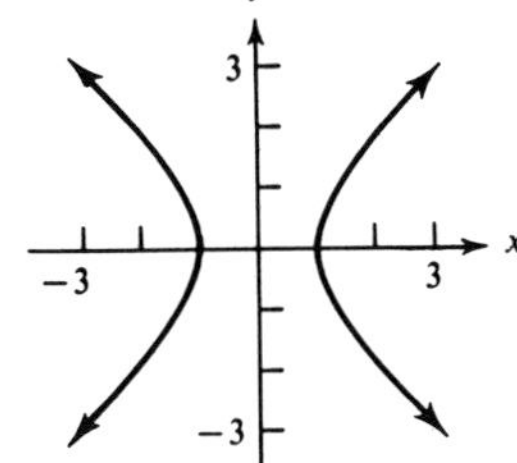

9.

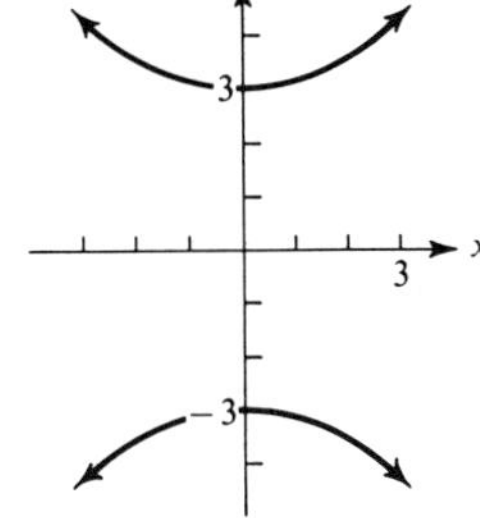

11.

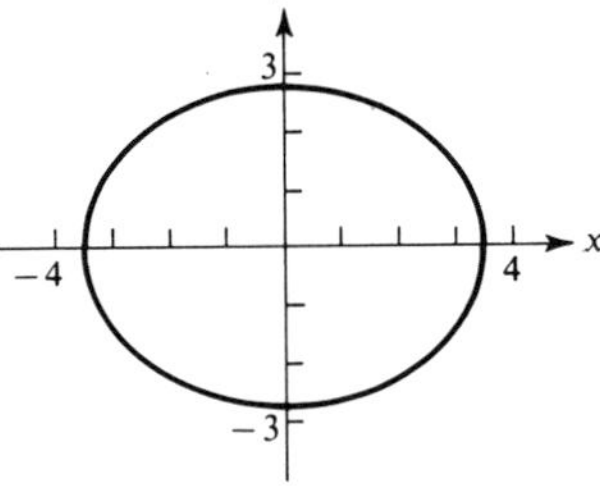

13.

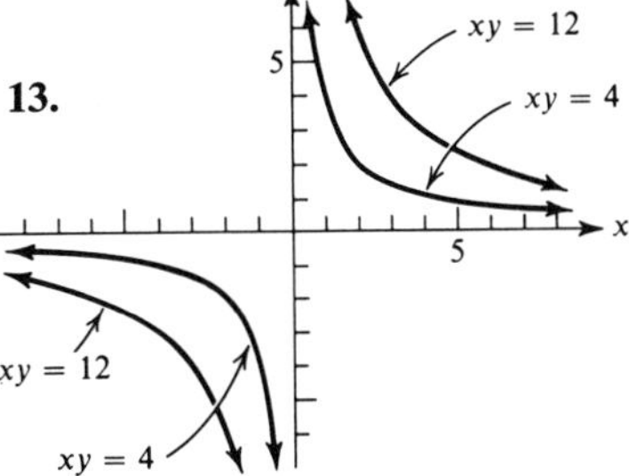

They belong to the solution sets of both equations.

15.

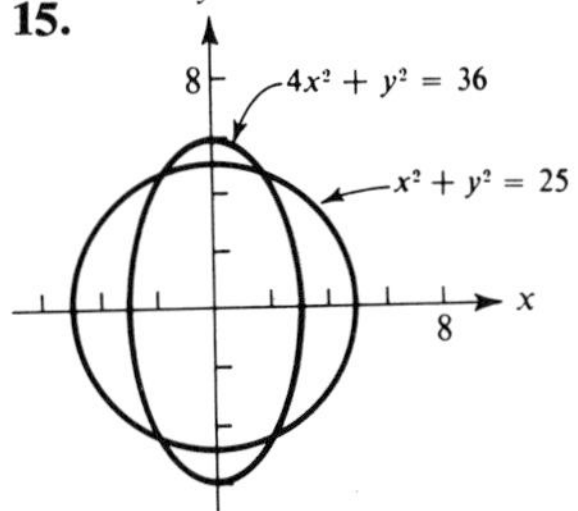

17. 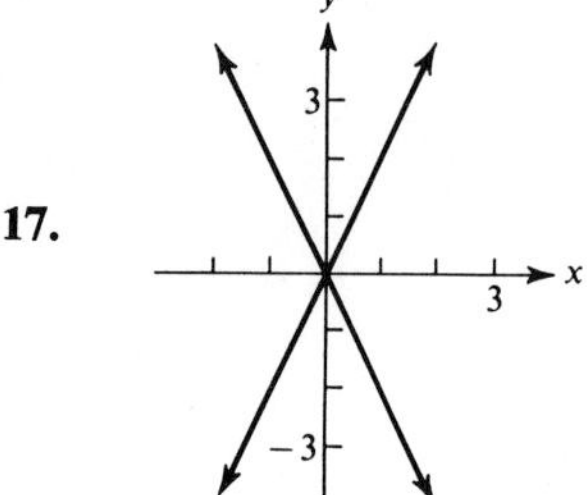

The graph is always two intersecting straight lines.

EXERCISE 9.3 (PAGE 207)

1. Circle

3. Parabola

5. Ellipse

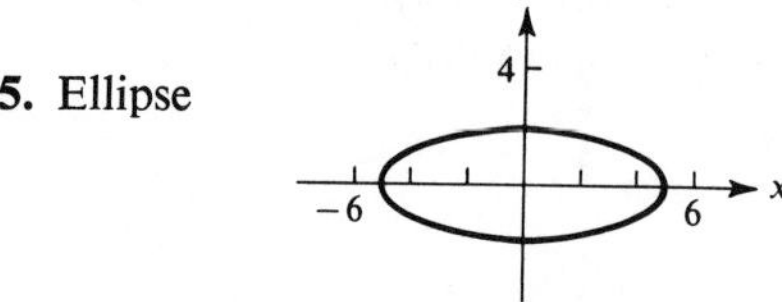

7. Hyperbola

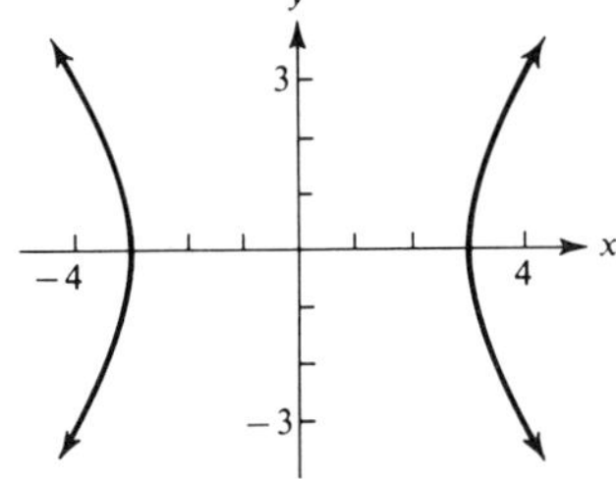

9. Two intersecting lines

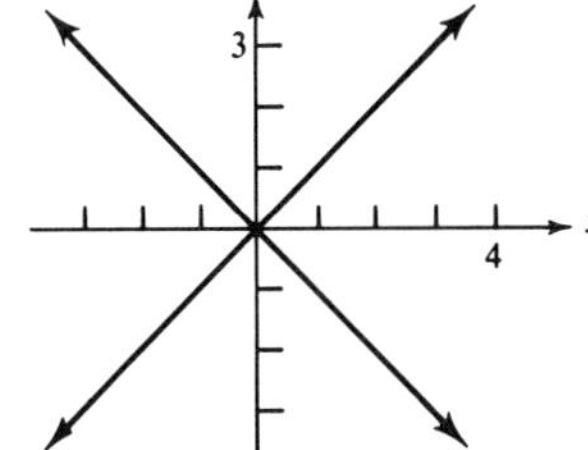

11. Parabola

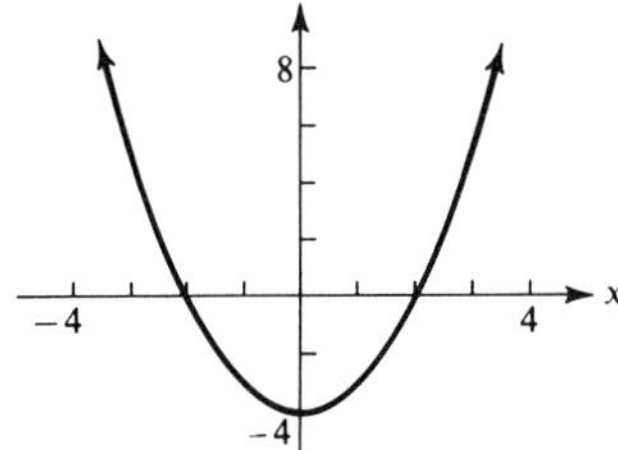

13. Circle

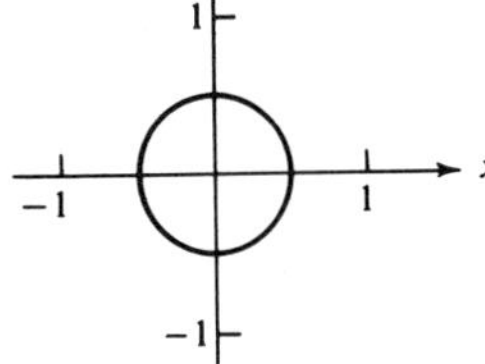

15. Ellipse

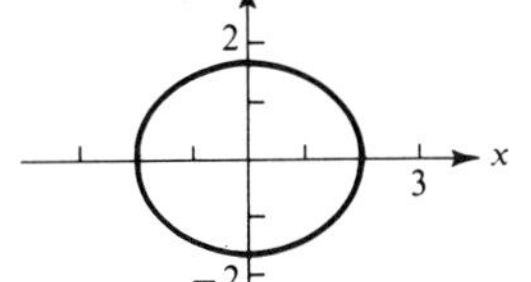

17. Parabola

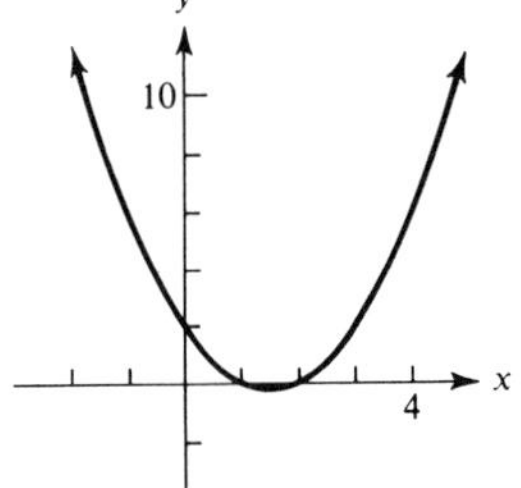

19. 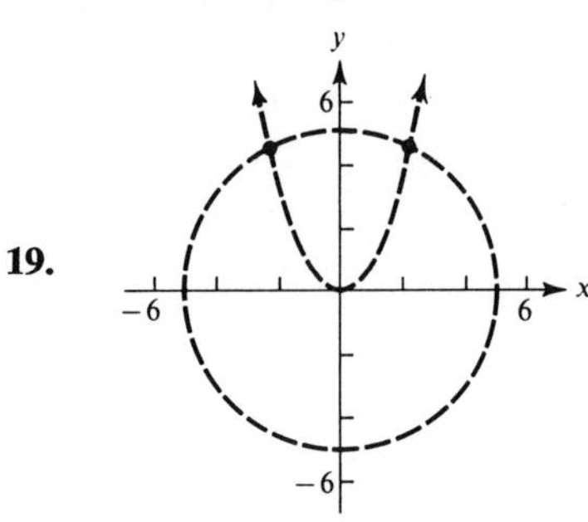

EXERCISE 9.4 (PAGE 210)

1. $d = kt$ **3.** $I = \frac{k}{R}$ **5.** $V = klw$ **7.** 3 **9.** 200 **11.** 2 **13.** $\frac{32}{9}$
15. 264 **17.** 16 **19.** 400 ft **21.** 160 lb/sq ft **23.** 1665 lb

9

EXERCISE 9.5 (PAGE 214)

9. 3872 yards **11.** 3.4 pounds **13.** 6912 cubic inches **15.** 7.6 quarts
17. 22.2 yards **19.** 11.1 inches **21.** 5.52 meters **23.** 290.63 square feet
25. 7.10 liters

CHAPTER REVIEW (PAGE 215)

1. hyperbola **2.** circle **3.** parabola **4.** ellipse

5. 6. 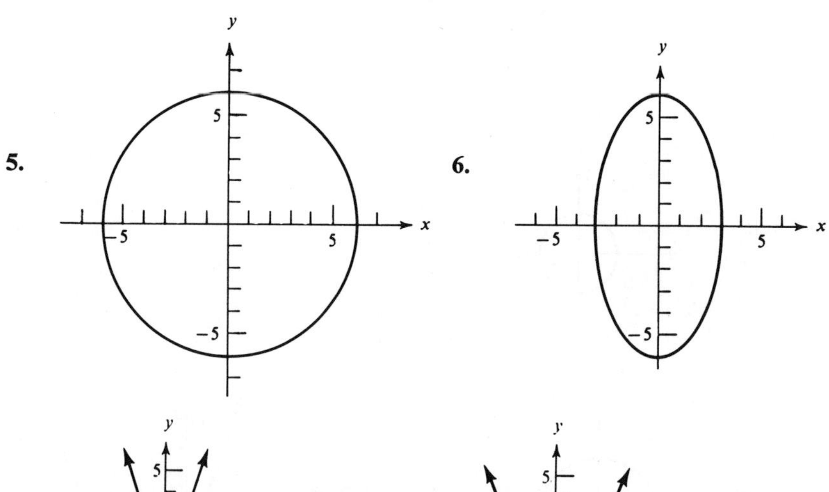

7. 8. 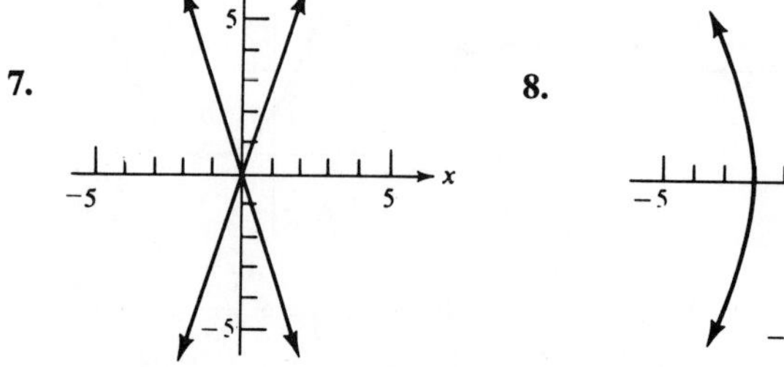

9. 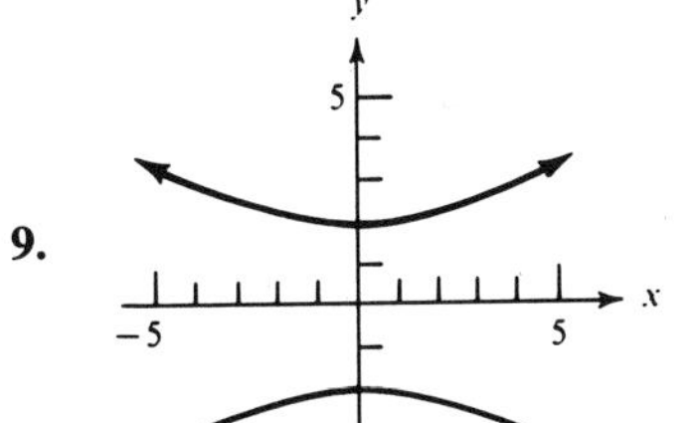

10. 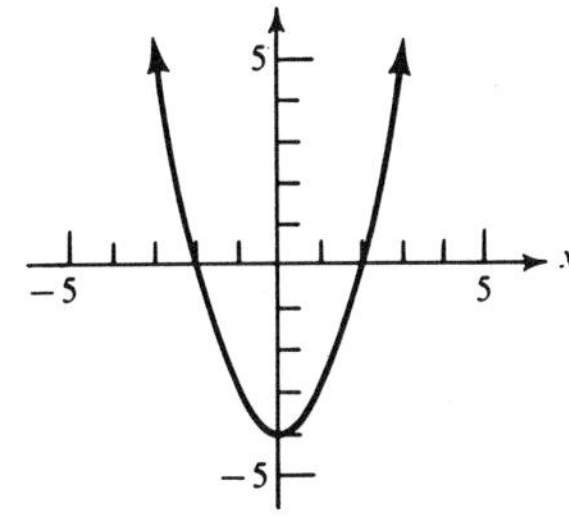

11.

12. 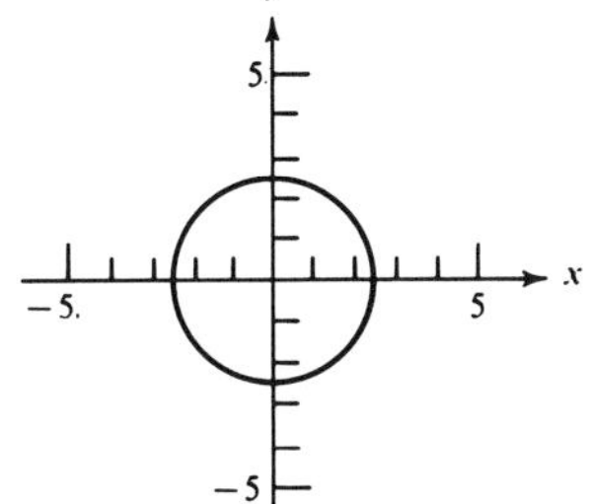

13. $y = 9$ **14.** 158 pounds **15.** 432 rpm **16.** 64 posts **19.** 2.12 miles
20. 2112 ounces

EXERCISE 10.1 (PAGE 221)

1. 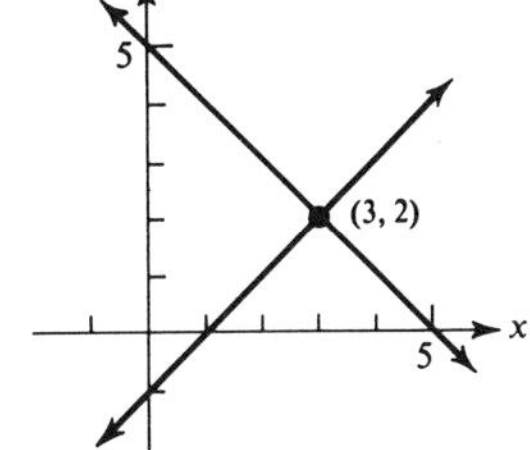

3. 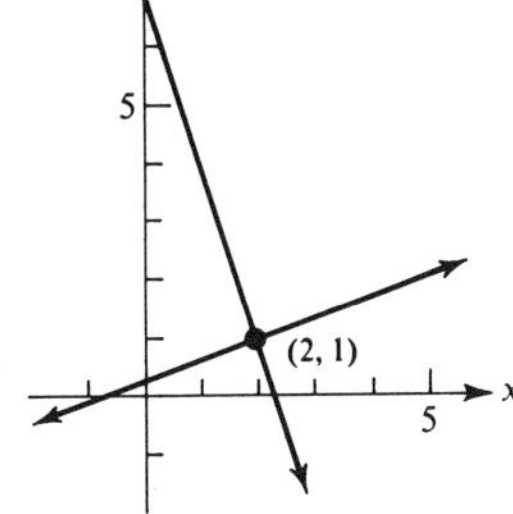

5. 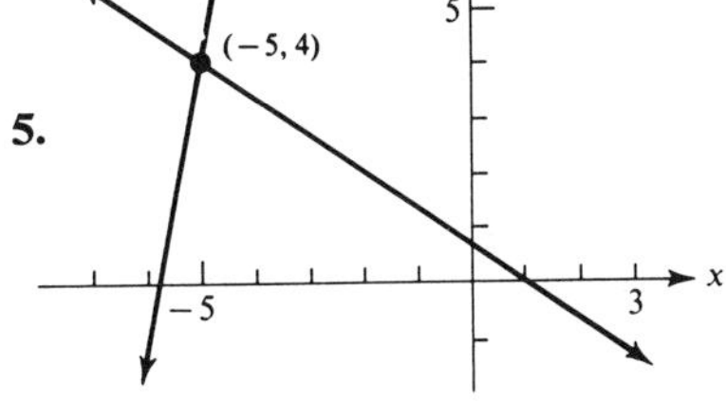

7. 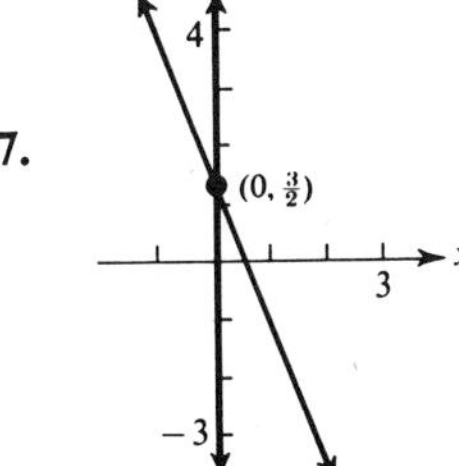

9. 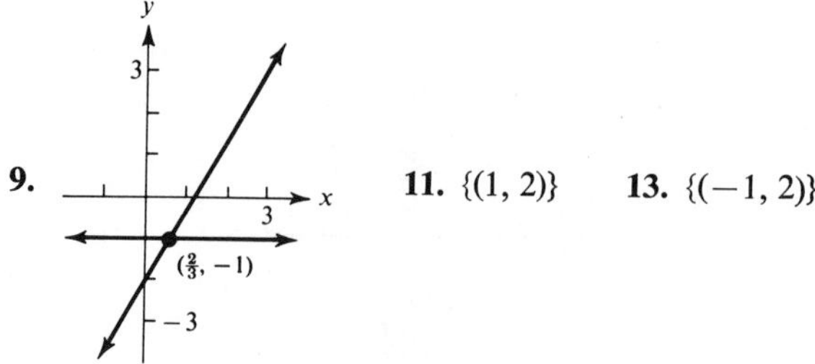

11. $\{(1, 2)\}$ **13.** $\{(-1, 2)\}$

15. Infinitely many solutions; dependent **17.** $\varnothing$; inconsistent

19. 16 and 22 **21.** 7 lb, 21 lb **23.** \$1000 at 6%, and \$800 at 5%

25. Speed of boat is 3 mph in still water; current moves at 1 mph

27. $33\frac{1}{3}$ ounces 75% alcohol; $16\frac{2}{3}$ ounces of 90% alcohol **29.** $a = 1$, $b = -1$

EXERCISE 10.2 (PAGE 224)

1. $\{(1, 2, -1)\}$ **3.** $\{(2, -2, 0)\}$ **5.** $\{(2, 2, 1)\}$

7. Not consistent and independent **9.** $\{(3, -1, 1)\}$ **11.** $\{(4, -2, 2)\}$

13. 3, 6, 6 **15.** 60 nickels, 20 dimes, 5 quarters

17. 40 inches, 60 inches, 55 inches **19.** $a = \dfrac{1}{3}$, $b = \dfrac{-25}{3}$, and $c = \dfrac{38}{3}$

EXERCISE 10.3 (PAGE 229)

1. 1 **3.** -12 **5.** 2 **7.** $-\dfrac{254}{225}$ **9.** $\{(1, 1)\}$ **11.** $\{(2, 2)\}$ **13.** $\{(6, 4)\}$

15. Inconsistent, $\varnothing$ **17.** $\{(4, 1)\}$ **19.** $\left\{\left(\dfrac{1}{a+b}, \dfrac{1}{a+b}\right)\right\}$

21. $\begin{vmatrix} a & a \\ b & b \end{vmatrix} = ab - ab = 0$

23. $\begin{vmatrix} a_1 & b_1 \\ a_2 & b_2 \end{vmatrix} = a_1b_2 - a_2b_1; \quad -\begin{vmatrix} b_1 & a_1 \\ b_2 & a_2 \end{vmatrix} = -(b_1a_2 - b_2a_1) = -b_1a_2 + b_2a_1$

$= a_1b_2 - a_2b_1$

25. $\begin{vmatrix} ka & a \\ kb & b \end{vmatrix} = kab - kba = 0$

EXERCISE 10.4 (PAGE 234)

1. 3 **3.** 9 **5.** -4 **7.** -1 **9.** -5 **11.** 17 **13.** 0 **15.** 0

17. -206 **19.** x^3 **21.** 0 **23.** $-2ab^2$ **25.** $\{3\}$ **27.** $\left\{2, -\frac{17}{7}\right\}$

29. 5 **31.** 6

EXERCISE 10.5 (PAGE 237)

1. Prop. 1 **3.** Prop. 3 **5.** Prop. 4 and Prop. 3 **7.** Prop. 2 **9.** Prop. 4

11. Prop. 4 **13.** Prop. 4 (twice) **15.** Prop. 5 **17.** Prop. 6

19. Prop. 6 **21.** Prop. 6 **23.** $\begin{vmatrix} 1 & 3 \\ 0 & -4 \end{vmatrix}$

25. $\begin{vmatrix} 1 & -2 & 1 \\ 0 & 7 & 1 \\ 0 & 2 & 1 \end{vmatrix}$ **27.** $\begin{vmatrix} 0 & 1 & -2 \\ 0 & 2 & 2 \\ 1 & 1 & 3 \end{vmatrix}$ **29.** -5 **31.** 12 **33.** 16

35. -35 **37.** -98 **39.** 6

EXERCISE 10.6 (PAGE 241)

1. $\{(1, 1, 1)\}$ **3.** $\{(1, 1, 0)\}$ **5.** $\{(1, -2, 3)\}$ **7.** $\{(3, -1, -2)\}$

9. The solution set is either $\varnothing$ or has an infinite number of members.

11. $\left\{\left(1, -\frac{1}{3}, \frac{1}{2}\right)\right\}$ **13.** $\{(-5, 3, 2)\}$

EXERCISE 10.7 (PAGE 244)

1.

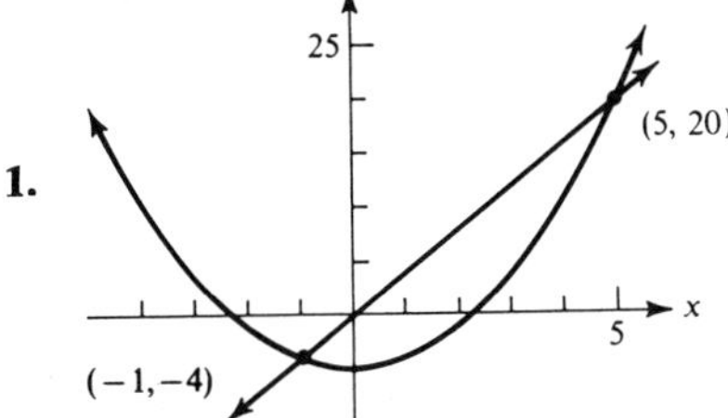

3.

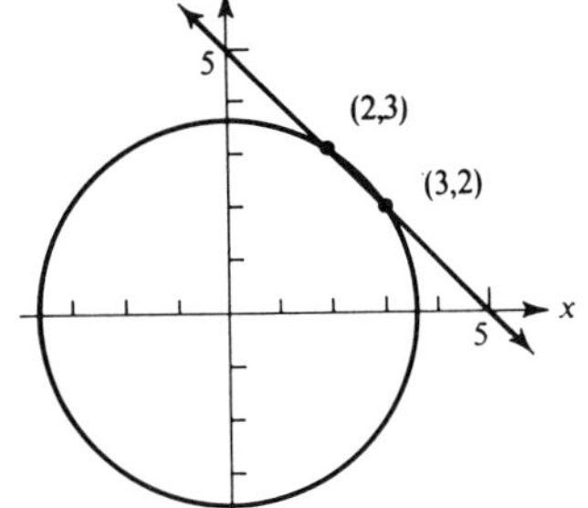

5.

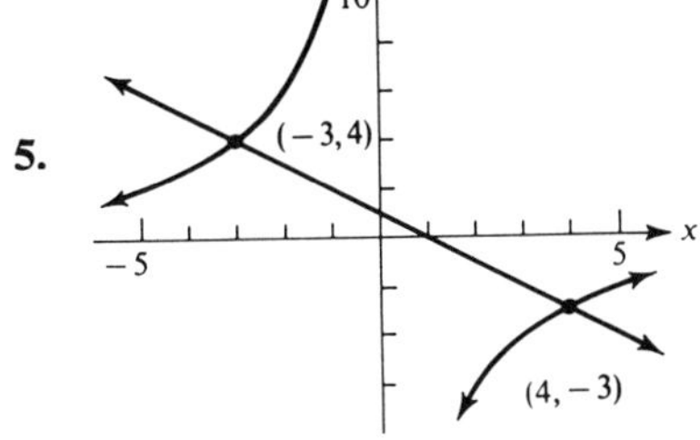

7.

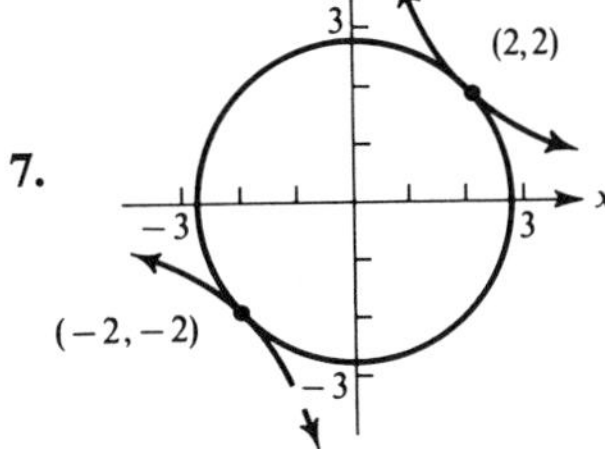

10

9. $\{(i\sqrt{7}, 4), (-i\sqrt{7}, 4)\}$

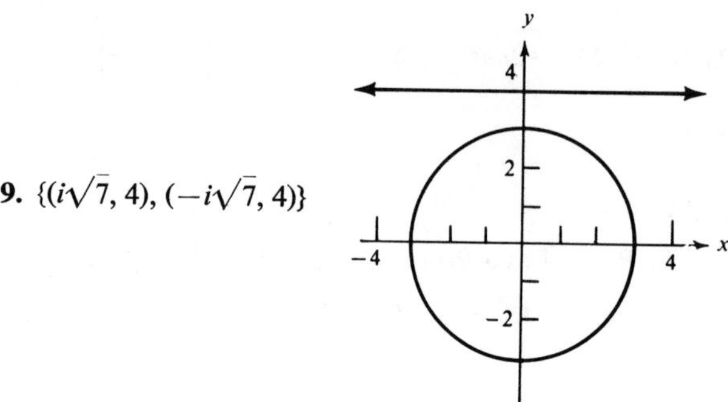

11. $\{(3, 1), (2, 0)\}$ 13. $\{(-1, -3)\}$ 15. $\{(0, -1), (\frac{30}{13}, \frac{7}{13})\}$ 17. 2, 3
19. 1, 12 21. 7 and 2; -15 and 24

EXERCISE 10.8 (PAGE 248)

1. $\{(1, 3), (-1, 3), (1, -3), (-1, -3)\}$ 3. $\{(1, 2), (-1, 2), (1, -2), (-1, -2)\}$
5. $\{(3, \sqrt{2}), (-3, \sqrt{2}), (3, -\sqrt{2}), (-3, -\sqrt{2})\}$
7. $\{(2, 1), (-2, 1), (2, -1), (-2, -1)\}$
9. $\{(\sqrt{3}, 4), (-\sqrt{3}, 4), (\sqrt{3}, -4), (-\sqrt{3}, -4)\}$
11. $\{(1, -2), (-1, 2), (2, -1), (-2, 1)\}$
13. $\{(2, -2), (-2, 2), (2i\sqrt{2}, i\sqrt{2}), (2i\sqrt{2}, -i\sqrt{2})\}$

CHAPTER REVIEW (PAGE 249)

1. $\{(\frac{1}{2}, \frac{7}{2})\}$ 2. $\{(1, 2)\}$ 3. a. Consistent and independent b. Dependent
4. a. Inconsistent b. Consistent and independent 5. $\{(2, 0, -1)\}$
6. $\{(2, 1, -1)\}$ 7. -13 8. $\{(-7, 4)\}$ 9. -2 10. 12 11. Prop. 1
12. Prop. 6 13. Prop. 4 and 3 14. Prop. 4 15. 15 16. $\{(2, -1, 3)\}$
17. The solution set is either $\emptyset$ or has an infinite number of members.
18. $\{(4, -13), (1, 2)\}$ 19. $\{(\frac{6}{7}, -\frac{37}{7}), (2, -3)\}$
20. $\{(2, 3), (2, -3), (-2, 3), (-2, -3)\}$

EXERCISE 11.1 (PAGE 253)

1. $(0, 1), (1, 3), (2, 9)$ 3. $(-2, \frac{1}{4}), (0, 1), (2, 4)$ 5. $(-3, 8), (0, 1), (3, \frac{1}{8})$
7. $(-2, \frac{1}{100}), (-1, \frac{1}{10}), (0, 1)$

9.
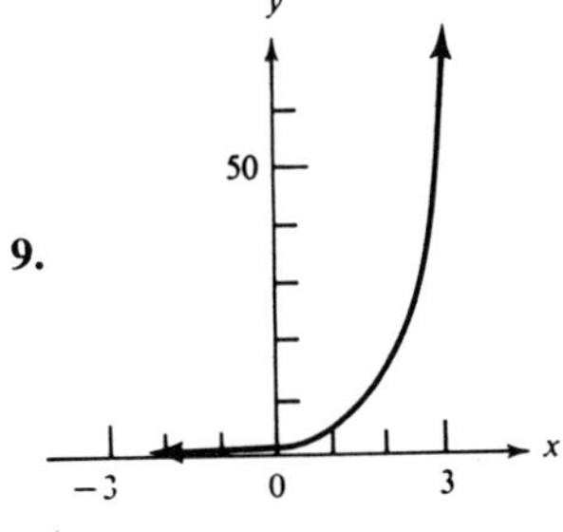

11.
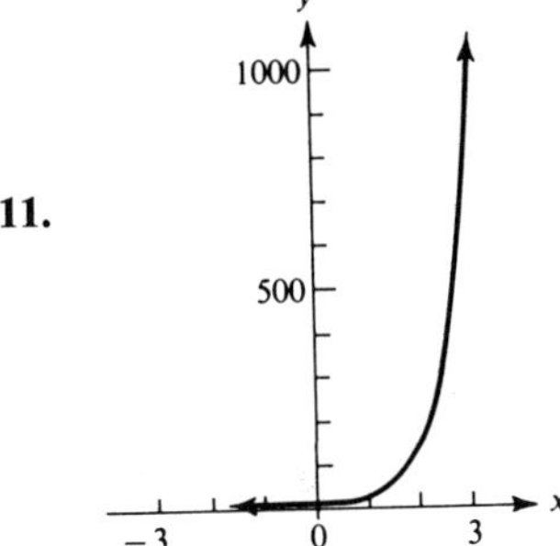

13.
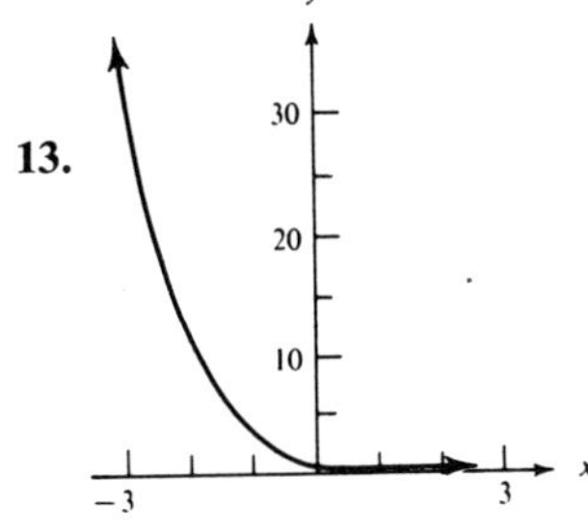

15.
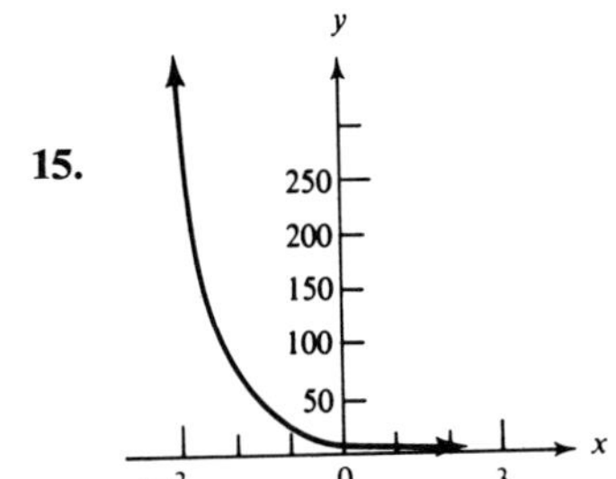

EXERCISE 11.2 (PAGE 256)

1. $\log_4 16 = 2$ **3.** $\log_3 27 = 3$ **5.** $\log_{1/2} \frac{1}{4} = 2$ **7.** $\log_8 \frac{1}{2} = -\frac{1}{3}$

9. $\log_{10} 100 = 2$ **11.** $\log_{10} 0.1 = -1$ **13.** $2^6 = 64$ **15.** $3^2 = 9$

17. $(\frac{1}{3})^{-2} = 9$ **19.** $(10)^3 = 1000$ **21.** $(10)^{-2} = 0.01$ **23.** 2 **25.** 3

27. $\frac{1}{2}$ **29.** -1 **31.** 1 **33.** 2 **35.** -1 **37.** 2 **39.** 2 **41.** 64

43. -3 **45.** 100 **47.** 4 **49.** 0, 0, 0

EXERCISE 11.3 (PAGE 258)

1. $\log_b x + \log_b y$ **3.** $\log_b x - \log_b y$ **5.** $5 \log_b x$ **7.** $\frac{1}{3} \log_b x$

9. $2 \log_b x + 3 \log_b y$ **11.** $\frac{1}{2} (\log_b x - \log_b z)$

13. $\frac{1}{3} \log_{10} x + \frac{2}{3} \log_{10} y - \frac{1}{3} \log_{10} z$ **15.** $\frac{1}{2} \log_{10} x + \frac{2}{3} \log_{10} y$

17. $\frac{1}{2} \log_{10} x + \frac{1}{2} \log_{10}(x - y)$ **19.** $\log_{10} 2 + \log_{10} \pi + \frac{1}{2} \log_{10} l - \frac{1}{2} \log_{10} g$

21. $\log_b(xy)$ **23.** $\log_b\left(\frac{1}{x}\right)$ **25.** $\log_b(x^2y^2)$ **27.** $\log_b\left(\frac{x^3y}{z^2}\right)$

29. $\log_{10} \frac{x(x - 2)}{z^2}$

31. $\frac{1}{4}\log_{10} 8 + \frac{1}{4}\log_{10} 2 = \frac{1}{4}\log_{10} 2^3 + \frac{1}{4}\log_{10} 2 = \frac{3}{4}\log_{10} 2 + \frac{1}{4}\log_{10} 2$

$$= (\tfrac{3}{4} + \tfrac{1}{4})\log_{10} 2 = \log_{10} 2$$

33. By Law III, $10^{2\log_{10}x} = 10^{\log_{10}x^2}$; by relationship on page 255, $10^{\log_{10}x^2} = x^2$. Hence, $10^{2\log_{10}x} = 10^{\log_{10}x^2} = x^2$.

35. $\log_{10}[\log_3(\log_5 125)] = \log_{10}[\log_3(\log_5 5^3)] = \log_{10}[\log_3(3\log_5 5)]$

$$= \log_{10}[\log_3 3] = \log_{10} 1 = 0$$

37. $\log_b b^2 + \log_b b^3 = 2\log_b b + 3\log_b b = 5\log_b b = \log_b b^5$

39. Show that the left and right members of the equality reduce to the same quantity. For the left side: $\log_b 4 + \log_b 8 = \log_b 2^2 + \log_b 2^3 = 2\log_b 2 + 3\log_b 2 = 5\log_b 2$; for the right side: $\log_b 64 - \log_b 2 = \log_b 2^6 - \log_b 2 = 6\log_b 2 - \log_b 2 = 5\log_b 2$.

11

EXERCISE 11.4 (PAGE 263)

1. 2 **3.** 3 **5.** -2 or $8-10$ **7.** 0 **9.** -4 or $6-10$ **11.** 4
13. 0.8280 **15.** 1.9227 **17.** 2.5011 **19.** $9.9101-10$ **21.** $8.9031-10$
23. 2.3945 **25.** 4.10 **27.** 36.7 **29.** 0.0642 **31.** 16.0 **33.** 5480
35. 0.000718 **37.** 9.10 **39.** 5000 **41.** 113

EXERCISE 11.5 (PAGE 267)

1. 0.6246 **3.** 0.7937 **5.** 3.1824 **7.** 4.5695 **9.** $9.7095-10$
11. $7.9218-10$ **13.** 3.225 **15.** 89.38 **17.** 10.52 **19.** 0.05076
21. 0.7485 **23.** 0.7495

EXERCISE 11.6 (PAGE 270)

1. 4.014 **3.** 2.299 **5.** 64.34 **7.** 2.010 **9.** 3.436×10^{-10}
11. 0.04582 **13.** 9.872 **15.** 4.746 **17.** 1.394 **19.** 3.483 **21.** 57.81
23. 2.117 **25.** 1.11 sec

EXERCISE 11.7 (PAGE 272)

1. {2.81} **3.** {0.89} **5.** {0.78} **7.** {−0.53} **9.** {−2.10} **11.** {−1.47}

Certain of the following answers are approximate values. They are consistent with the given data and the use of four-place tables.

13. 1.34 **15.** 5% **17.** 20 yrs **19.** 2.4% **21.** 12 yrs
23. \$7396, \$7430 **25.** 7 **27.** 7.7 **29.** 6.2 **31.** 1.0×10^{-3}
33. 2.5×10^{-6} **35.** 6.3×10^{-8} **37.** 12.3 grams
39. 30 in. of mercury; 16.1 in. of mercury

CHAPTER REVIEW (PAGE 276)

1. 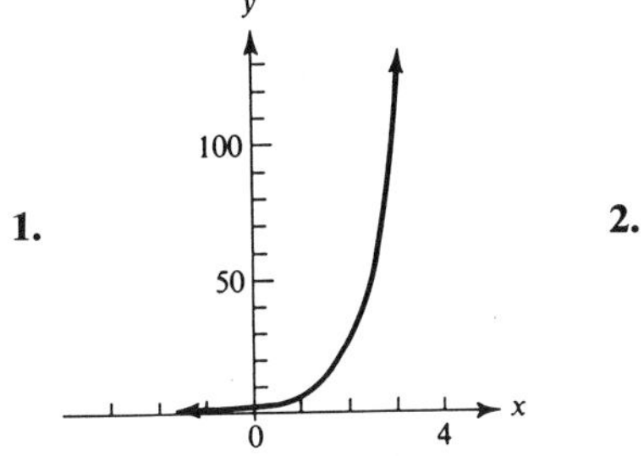

2.

3. a. $\log_9 27 = \frac{3}{2}$ b. $\log_{4/9} \frac{2}{3} = \frac{1}{2}$ 4. a. $5^4 = 625$ b. $10^{-4} = 0.0001$

5. 27 6. $\sqrt[3]{3}$ 7. a. $\log_b 3 + 2 \log_b x + \log_b y$ b. $\frac{1}{2} \log_b x + \log_b y - 2 \log_b z$

8. $6(\log_2 8 - \log_2 4) = 6(\log_2 2^3 - \log_2 2^2) = 6(3 \log_2 2 - 2 \log_2 2)$
$= 6 \log_2 2 = 6 \cdot 1 = 6$

9. $\log_2[\log_4(\log_2 16)] = \log_2[\log_4(\log_2 2^4)] = \log_2[\log_4(4 \log_2 2)] = \log_2[\log_4 4] = \log_2 1 = 0$

10. Show that the left and right members reduce to the same number.
For the left side: $\frac{1}{2} \log_2 16 + \log_4 16 = \frac{1}{2} \log_2 2^4 + \log_4 4^2 = 2 \log_2 2 + 2 \log_4 4$
$= 2(1) + 2(1) = 4;$

for the right side: $8 \log_{16} 4 = 8(\frac{1}{2}) = 4$

11. a. 9.8531 − 10 b. 3.2577 12. a. 0.0150 b. 5290 13. a. 7.52 b. 354

14. 1.4461 15. 0.03162 16. 26.09 17. 1.387 18. 3.129 19. 2.465

20. 0.9226

EXERCISE 12.1 (PAGE 282)

1. 34° 3. 90° 5. ∠1, ∠4 7. ∠1, ∠3, ∠4 9. (a) 67° (b) 157°

11. (a) 75° 50′ (b) 165° 50′ 13. (a) 27° 56′ (b) 117° 56′

15. (a) 74.3° (b) 164.3° 17. (a) 65.43° (b) 155.43°

19. (a) 70.18° (b) 160.18° 21. 15.5° 23. 88.2° 25. 120.75°

27. 14° 12′ 29. 54° 20′ 31. 12° $22\frac{1}{2}$′ 33. 15° 30′, 88° 12′, 54° 48′

35. 30°, 150° 37. 14°, 76°

EXERCISE 12.2 (PAGE 285)

1. ∠5 = 72°, ∠6 = 118° 3. ∠13 = 47°, ∠14 = 133°

5. ∠7 = 102°, ∠8 = 78° 7. ∠17 = 72°, ∠18 = 118° 9. 142° 50′

11. 93° 30′ 13. $\frac{14}{3}$ 15. 3.2 17. 4.2 19. 9

EXERCISE 12.3 (PAGE 288)

1. 32° 3. 124° 5. 56° 7. 60° 9. 45° 11. 45° 13. 32°

15. △*FCE* 17. △*DEB*, △*DEC*, △*BDC* 19. *DE*

21. $b = 4$, $c = 5.7$ **23.** $a = b = 6.0$ **25.** $e = 2.5$, $f = 5.1$
27. $e = 0.2$, $d = 0.4$ **29.** Equilateral **31.** Scalene **33.** 60° **35.** 45°
37. 2.4 **39.** 6.6 **41.** 3.4 **43.** 13 **45.** $\sqrt{41}$ **47.** $9 + \sqrt{41}$ **49.** 10
51. 42.5 **53.** $h^2 + \left(\frac{s}{2}\right)^2 = s^2$, from which $h^2 = \frac{3}{4}s^2$, or $h = \frac{\sqrt{3}}{2}s$.

EXERCISE 12.4 (PAGE 293)

1. (a) Trapezoid (b) Quadrilateral (c) $\angle A = 150°$, $\angle B = 90°$
3. (a) Parallelogram (b) Quadrilateral (c) $\angle B = \angle C = 150°$, $\angle D = 30°$
5. 5.8 **7.** 3.4 **9.** 4.3 **11.** 11.6 **13.** 1.6 **15.** 2.8 **17.** 60°

19. 120° **21.** 1.8 **23.** 10.0 **25.** 9.4 **27.** $s\sqrt{2}$
29. 320 tiles, 48 feet of moulding **31.** 2.1 gallons

33.

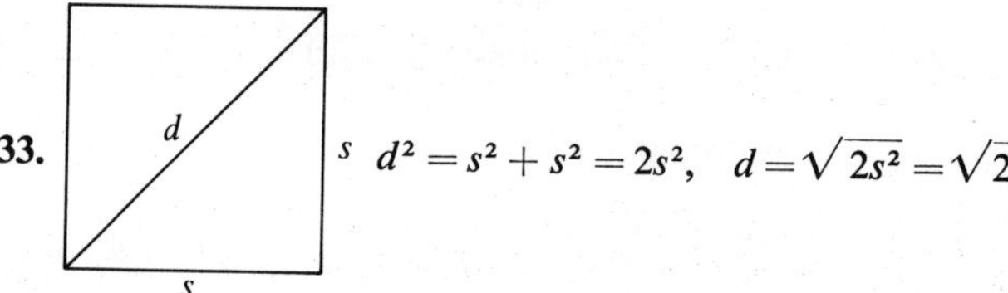

$d^2 = s^2 + s^2 = 2s^2$, $d = \sqrt{2s^2} = \sqrt{2}s$

EXERCISE 12.5 (PAGE 297)

1. Radius **3.** Chord **5.** Tangent **7.** Central angle **9.** 60°
11. 120° **13.** 90° **15.** 60° **17.** 2π **19.** $\frac{2\pi}{3}$ **21.** 4π **23.** 22.1
25. 89.3 **27.** 7.2 **29.** 18.5 **31.** 24.5 **33.** 55.3 **35.** 11.3
37. 727.3 (727 complete plates) **39.** 672.2 **41.** 7.3 inches

EXERCISE 12.6 (PAGE 300)

1. $V = 36$ cubic inches, $S = 66$ square inches
3. $V = 6.1$ cubic inches, $S = 25.0$ square inches
5. $V = 94.7$ cubic inches, $S = 115.6$ square inches
7. $V = 6.5$ cubic feet, $S = 13.8$ square feet
9. $V = 3368.9$ cubic inches, $S = 1086.8$ square inches
11. $V = 9.0$ cubic inches, $S = 28.6$ square inches
13. 129.0 cubic inches **15.** 0.2 cubic inches **17.** 38.2 cubic inches
19. 371.9 cubic inches **21.** 395.7 **23.** $781.83 **25.** 1053.9 pounds

CHAPTER REVIEW (PAGE 303)

1. a. (a) $61° 46'$ (b) $151° 46'$ b. (a) $55.4°$ (b) $145.4°$
2. a. $114.4°$ b. $67.25°$ **3.** a. $17° 24'$ b. $18° 36'$ **4.** $112° 30'$, $67° 30'$
5. $80°$ **6.** $100°$ **7.** $118°$ **8.** $\frac{9}{4}$ **9.** 7.1 **10.** 6.0 **11.** 5.2
12. 3.5 **13.** Perimeter, 14.2; area, 7.8 **14.** Perimeter, 9.5; area, 3.5
15. (a) 24 (b) 8.5 **16.** 47 feet $\times$ 37 feet
17. (a) 17.8 inches (b) 249.6 square inches **18.** 18.8 square inches
19. Volume, 376.8 cubic inches; area, 290.3 square inches **20.** 13 inches

EXERCISE 13.1 (PAGE 306)

13

1. $17°$, 15 **3.** $22°$, 35 **5.** $\angle B = 40°$, $\angle C = 120°$, $c = 204$
7. $\angle A = 60°$, $\angle B = 20°$, $b' = 12$, $c' = 35$ **9.** 6 **11.** $\frac{21}{2}$ **13.** 5.4
15. 19.4 **17.** 157 feet **19.** 117 feet

EXERCISE 13.2 (PAGE 312)

1. $\sin A = \frac{4}{5}$, $\cos A = \frac{3}{5}$, $\tan A = \frac{4}{3}$, $\csc A = \frac{5}{4}$, $\sec A = \frac{5}{3}$, $\cot A = \frac{3}{4}$

3. $\sin A = \dfrac{3}{\sqrt{10}}$, $\cos A = \dfrac{-1}{\sqrt{10}}$, $\tan A = -3$,
$\csc A = \dfrac{\sqrt{10}}{3}$, $\sec A = -\sqrt{10}$, $\cot A = \dfrac{-1}{3}$

5. $\sin A = \dfrac{1}{\sqrt{2}}$, $\cos A = \dfrac{1}{\sqrt{2}}$, $\tan A = 1$, $\csc A = \sqrt{2}$, $\sec A = \sqrt{2}$, $\cot A = 1$

7. $\sin A = \dfrac{\sqrt{3}}{2}$, $\cos A = \dfrac{1}{2}$, $\tan A = \sqrt{3}$, $\csc A = \dfrac{2}{\sqrt{3}}$, $\sec A = 2$, $\cot A = \dfrac{1}{\sqrt{3}}$

9. $\sin A = \dfrac{-4}{5}$, $\cos A = \dfrac{3}{5}$, $\tan A = \dfrac{-4}{3}$, $\csc A = \dfrac{-5}{4}$, $\sec A = \dfrac{5}{3}$, $\cot A = \dfrac{-3}{4}$

11. $\sin A = \dfrac{\sqrt{7}}{4}$, $\cos A = \dfrac{3}{4}$, $\tan A = \dfrac{\sqrt{7}}{3}$, $\csc A = \dfrac{4}{\sqrt{7}}$, $\sec A = \dfrac{4}{3}$, $\cot A = \dfrac{3}{\sqrt{7}}$

13. $\sin A = 1$, $\cos A = 0$, $\tan A$ (undefined),
$\csc A = 1$, $\sec A$ (undefined), $\cot A = 0$

15. $\sin A = 0$, $\cos A = -1$, $\tan A = 0$,
$\csc A$ (undefined), $\sec A = -1$, $\cot A$ (undefined)

17. $\cos A = \dfrac{\sqrt{3}}{2}$, $\tan A = \dfrac{1}{\sqrt{3}}$, $\csc A = 2$, $\sec A = \dfrac{2}{\sqrt{3}}$, $\cot A = \sqrt{3}$

19. $\sin A = \frac{3}{5}$, $\cos A = \frac{4}{5}$, $\csc A = \frac{5}{3}$, $\sec A = \frac{5}{4}$, $\cot A = \frac{4}{3}$

13

21. $\sin A = \dfrac{1}{\sqrt{2}}$, $\cos A = \dfrac{1}{\sqrt{2}}$, $\tan A = 1$, $\csc A = \sqrt{2}$, $\cot A = 1$

23. $\sin A = \dfrac{3}{\sqrt{10}}$, $\tan A = 3$, $\csc A = \dfrac{\sqrt{10}}{3}$, $\sec A = \sqrt{10}$, $\cot A = \dfrac{1}{3}$

25. $\sin A = \dfrac{2}{\sqrt{13}}$, $\cos A = \dfrac{3}{\sqrt{13}}$, $\csc A = \dfrac{\sqrt{13}}{2}$, $\sec A = \dfrac{\sqrt{13}}{3}$, $\cot A = \dfrac{3}{2}$

27. $\sin A = \dfrac{\sqrt{3}}{2}$, $\cos A = \dfrac{1}{2}$, $\tan A = \sqrt{3}$, $\sec A = 2$, $\cot A = \dfrac{1}{\sqrt{3}}$

29. $\sin A = 1$, $\tan A$ (undefined), $\csc A = 1$, $\sec A$ (undefined), $\cot A = 0$

31. $\sin A = 0$, $\cos A = 1$, $\csc A$ (undefined), $\sec A = 1$, $\cot A$ (undefined)

33. Impossible, since $\sec A \geq 1$ for these angles.

35. $\sec A = \dfrac{d}{x} = \dfrac{1}{\frac{x}{d}} = \dfrac{1}{\cos A}$

37. $\sin^2 A + \cos^2 A = \dfrac{y^2}{d^2} + \dfrac{x^2}{d^2} = \dfrac{y^2 + x^2}{d^2} = \dfrac{d^2}{d^2} = 1$

39. $\cot^2 A + 1 = \dfrac{x^2}{y^2} + 1 = \dfrac{x^2 + y^2}{y^2} = \dfrac{d^2}{y^2} = \csc^2 A$

41. II **43.** IV **45.** III **47.** III

EXERCISE 13.3 (PAGE 318)

1. $\dfrac{\sqrt{3}}{2} \approx 0.87$ **3.** $\dfrac{\sqrt{2}}{2} \approx 0.71$ **5.** 1 **7.** $-\frac{1}{2} = -0.50$ **9.** 2 **11.** 0

13. $-\dfrac{2\sqrt{3}}{3} \approx -1.15$ **15.** 1 **17.** 30°, 150° **19.** 45° **21.** 30°

23. 150° **25.** 0°, 180° **27.** 45°, 135° **29.** 150° **31.** 150° **33.** 2.3

35. 2 **37.** 2.4 **39.** 2.3 **41.** 0 **43.** 1.3

EXERCISE 13.4 (PAGE 321)

1. 0.6905 **3.** 0.7192 **5.** 0.1016 **7.** 0.7431 **9.** 5.576 **11.** 0.9391
13. 1.0005 **15.** 1.4388 **17.** 0.9896 **19.** 1.0000 **21.** 12° 40′
23. 39° 30′ **25.** 73° 20′ **27.** 20° 40′, 159° 20′ **29.** 31.7° **31.** 29.7°
33. 10.3° **35.** 45.1° **37.** 0.3795 **39.** −2.073 **41.** 2.280
43. 0.8892 **45.** 2.697 **47.** 1.0118 **49.** 72° 56′ **51.** 22° 46′
53. 42° 45′ **55.** 54.59°, 125.41° **57.** 70.71° **59.** 17.85°

EXERCISE 13.5 (PAGE 327)

1. $c = 27.5$, $\angle A = 33.1°$, $\angle B = 56.9°$
3. $b = 12.4$, $\angle A = 27.3°$, $\angle B = 62.7°$
5. $a = 19.1$, $b = 20.5$, $\angle B = 47°$
7. $b = 96.3$, $c = 99.2$, $\angle B = 76°$
9. $b = 110.5$, $c = 240.8$, $\angle A = 62.7°$
11. $b = 25.2$, $\angle A = 48.3°$, $\angle B = 41.7°$
13. $a = 43.3$, $b = 83.2$, $\angle B = 62.5°$
15. $a = 74.8$, $c = 110.4$, $\angle A = 42.7°$
17. $b = 107.48$, $\angle A = 19° 28'$, $\angle B = 70° 32'$
19. $a = 140$, $\angle A = 36° 52'$, $\angle B = 53° 8'$
21. $a = 50.80$, $b = 99.79$, $\angle B = 63°$
23. $a = 9.64$, $c = 37.26$, $\angle B = 75°$
25. $a = 99.07$, $b = 299.00$, $\angle B = 71° 40'$
27. $c = 135.45$, $\angle A = 38° 10'$, $\angle B = 51° 50'$
29. $a = 39.97$, $b = 74.23$, $\angle B = 61° 42'$
31. $a = 19.54$, $c = 33.98$, $\angle A = 35° 6'$

33. $\sin A = \frac{a}{c} = \cos B$ **35.** $\tan A = \frac{a}{b} = \cot B$ **37.** $\sec A = \frac{c}{b} = \csc B$

13

EXERCISE 13.6 (PAGE 329)

1. 64.3 feet **3.** 8° 30′ **5.** 62° 00′ **7.** 75° 30′ **9.** 9.1 feet
11. 44.0° **13.** 49.4 feet **15.** 25.6 feet **17.** 259.4 feet

EXERCISE 13.7 (PAGE 334)

1. $\angle C = 70° 00'$, $b = c = 14.6$ **3.** $\angle C = 75° 00'$, $a = 26.9$, $b = 22.0$
5. $\angle C = 99° 00'$, $a = 8.5$, $b = 10.0$
7. $\angle A = 24° 20'$, $\angle C = 112° 30'$, $c = 54.0$
9. $\angle A = 45° 40'$, $\angle B = 63° 30'$, $b = 23.0$
11. $\angle C = 44.6°$, $b = 10.9$, $c = 8.0$
13. $\angle B = 117.1°$, $a = 84.6$, $b = 102.6$
15. $\angle A = 26.3°$, $\angle C = 91.2°$, $c = 22.5$
17. $\angle A = 113.5°$, $\angle C = 20.0°$, $a = 84.7$

EXERCISE 13.8 (PAGE 335)

1. 42.2 feet **3.** 9.9 feet **5.** 162.1 miles **7.** $x = 50.1$ miles, $y = 74.4$ miles
9. 1.0 mile **11.** 15,152.4 square feet

EXERCISE 13.9 (PAGE 340)

1. $\angle A = 34.3°$, $\angle B = 115.7°$, $c = 4.4$
3. $\angle B = 26.3°$, $\angle C = 33.7°$, $a = 7.8$
5. $\angle A = 83.2°$, $\angle B = 52.5°$, $c = 6.9$ **7.** $\angle B = \angle C = 76.1°$, $a = 1.1$
9. $\angle A = 35° \, 30'$, $\angle B = 48° \, 10'$, $\angle C = 96° \, 20'$
11. $\angle A = 27° \, 50'$, $\angle B = 32° \, 10'$, $\angle C = 120° \, 00'$
13. $\angle A = 38° \, 20'$, $\angle B = 36° \, 50'$, $\angle C = 104° \, 50'$
15. $\angle A = 18° \, 40'$, $\angle B = 19° \, 30'$, $\angle C = 141° \, 50'$
17. If $\angle C = 90°$, then $c^2 = a^2 + b^2 - 2ab \cdot \cos 90° = a^2 + b^2$.

EXERCISE 13.10 (PAGE 342)

1. 35.4 feet **3.** 79.8 miles **5.** 62° 50′ **7.** 55° 10′ **9.** 31° 10′
11. 137° 30′ **13.** 20.5 pounds

CHAPTER REVIEW (PAGE 345)

1. 19.8 **2.** 3.5

3. $\sin A = \dfrac{3}{\sqrt{13}}$, $\cos A = \dfrac{-2}{\sqrt{13}}$, $\tan A = \dfrac{-3}{2}$,

$\csc A = \dfrac{\sqrt{13}}{3}$, $\sec A = \dfrac{-\sqrt{13}}{2}$, $\cot A = \dfrac{-2}{3}$

4. $\sin A = \dfrac{\sqrt{33}}{7}$, $\tan A = \dfrac{\sqrt{33}}{4}$, $\csc A = \dfrac{7}{\sqrt{33}}$, $\sec A = \dfrac{7}{4}$, $\cot A = \dfrac{4}{\sqrt{33}}$

5. a. $-\sqrt{2}$ b. -1 **6.** a. 60° b. 135° **7.** a. 0.9283 b. -1.5818
8. a. 142° 30′ b. 45° 50′, 134° 10′ **9.** a. 0.6792 b. 1.5727
10. a. 41° 14′ b. 137° 57′ **11.** $\angle B = 75° \, 50'$, $a = 1.6$, $b = 6.2$ **12.** 39° 31′
13. $\angle A = 71° \, 00'$, $a = 15.0$, $c = 11.6$ **14.** $\angle A = 27.0°$, $\angle C = 48.3°$, $c = 11.3$
15. \$3425.45 **16.** \$24,605.04 **17.** $\angle B = 42.6°$, $\angle C = 88.8°$, $a = 5.1$
18. $\angle A = 38° \, 00'$, $\angle B = 44° \, 10'$, $\angle C = 97° 50'$ **19.** 14.3 pounds
20. 14° 17′ with 9 pound force, 21° 43′ with 6 pound force

EXERCISE A.1 (PAGE 353)

(A) **1.** 1100 **3.** 1250 **5.** 1700 **7.** 1080 **9.** 1180 **11.** 1350
13. 1550 **15.** 1930

(B) **1.** 200 **3.** 220 **5.** 250 **7.** 382 **9.** 386 **11.** 202 **13.** 241
15. 287 **17.** 302 **19.** 304

(C) **1.** 500 **3.** 580 **5.** 690 **7.** 890 **9.** 945 **11.** 515 **13.** 525 **15.** 650 **17.** 800 **19.** 942

EXERCISE A.2 (PAGE 360)

1. 8.00 **3.** 36.0 **5.** 180.0 **7.** 126.0 **9.** 1240 **11.** 7.50 **13.** 62.0 **15.** 3.25 **17.** 14.10 **19.** 26.0 **21.** 9.64 **23.** 55.4 **25.** 601 **27.** 19.80 **29.** 2840 **31.** 770 **33.** 63.6 **35.** 84.5 **37.** 884 **39.** 10,590 **41.** 284 **43.** 773 **45.** 1190 **47.** 206 **49.** 232 **51.** 1551 **53.** 3850 **55.** 200

EXERCISE A.3 (PAGE 364)

1. 2.00 **3.** 6.00 **5.** 2.00 **7.** 6.75 **9.** 100.0 **11.** 7.75 **13.** 22.8 **15.** 2.08 **17.** 1.095 **19.** 45.3 **21.** 0.0941 **23.** 0.757 **25.** 0.927 **27.** 4.92 **29.** 60.4 **31.** 0.500 **33.** 0.1428 **35.** 0.0312 **37.** 1.25 **39.** 0.00425

A

EXERCISE A.4 (PAGE 370)

1. 3.00 **3.** 0.900 **5.** 1.150 **7.** 2.24 **9.** 0.0328 **11.** 6.00 **13.** 1.233 **15.** 0.0229 **17.** 51.9 **19.** 3540 **21.** 31.0 **23.** 9440 **25.** 16.80 **27.** 0.303 **29.** 0.622 **31.** 62.1 **33.** 2.19 **35.** 42.7 **37.** 0.0028 **39.** 3.20

EXERCISE A.5 (PAGE 375)

1. 4.00 **3.** 4.00 **5.** 2.00 **7.** 4.00 **9.** 6.00 **11.** 8.00 **13.** 5.00 **15.** 8.53 **17.** 175.4 **19.** 20.4 **21.** 103.1 **23.** 8.87 **25.** 4.66 **27.** 0.1300 **29.** 0.572 **31.** 5.55; 9.80 **33.** 37.7; 0.0630 **35.** 0.1428 **37.** 0.0548

EXERCISE A.6 (PAGE 378)

1. 12.0 **3.** 6.00 **5.** 25.8 **7.** 243 **9.** 14.25 **11.** 250 **13.** 28.0 **15.** 25.0% **17.** 85.7% **19.** 4.57% **21.** 0.222% **23.** 172.0% **25.** 97.1% **27.** 4.54% **29.** 75.0 **31.** 4140 **33.** 6800 **35.** 26,700 **37.** 21,800 **39.** 287

EXERCISE A.7 (PAGE 384)

1. 9.00 **3.** 0.250 **5.** 44.9 **7.** 1274 **9.** 5.58 **11.** 60.0 **13.** 0.0434 **15.** 2.00 **17.** 2.83 **19.** 28.3 **21.** 1.732 **23.** 1.80

25. 0.0362 **27.** 0.409 **29.** 5.24 **31.** 4.34 **33.** 0.553 **35.** 2.62
37. 0.0743 **39.** 0.74 **41.** 59.5 **43.** 0.712 **45.** 1.285

EXERCISE A.8 (PAGE 389)

1. 8.00 **3.** 17.6 **5.** 0.0000176 **7.** 11,900 **9.** 314 **11.** 64.0
13. 59,200 **15.** 0.00120 **17.** 2.00 **19.** 6.00 **21.** 3.24 **23.** 0.698
25. 1.91 **27.** 2.29 **29.** 0.327 **31.** 6.81 **33.** 305 **35.** 2.29
37. 43.3 **39.** 3.22 **41.** 0.0833 **43.** 0.299 **45.** 0.139

EXERCISE A.9 (PAGE 392)

A

1. 0.477 **3.** 0.161 **5.** 0.964 **7.** 2.36 **9.** 1.268 **11.** 1.477
13. 2.477 **15.** 3.147 **17.** 23.6 **19.** 127 **21.** 9.477 − 10 **23.** 8.477 − 10
25. 7.806 − 10 **27.** 0.236 **29.** 0.0127 **31.** 0.673 **33.** 1.511 **35.** 313
37. 8.924 − 10 **39.** 0.428 **41.** 3.732 **43.** 8.973 − 10 **45.** 0.0001396
47. 0.316 **49.** 0.0000513

EXERCISE A.10 (PAGE 397)

1. 0.500 **3.** 0.375 **5.** 0.956 **7.** 0.1822 **9.** 0.269 **11.** 0.0698
13. 0.0419 **15.** 0.0218 **17.** 38.9° **19.** 25.5° **21.** 30.0° **23.** 70.2°
25. 13.5° **27.** 3.57° **29.** 0.72° **31.** 0.500 **33.** 0.866 **35.** 0.891
37. 0.250 **39.** 0.165 **41.** 0.0349 **43.** 0.0384 **45.** 0.0471 **47.** 36°
49. 63.1° **51.** 68.4° **53.** 83.4° **55.** 60°

EXERCISE A.11 (PAGE 402)

1. 0.577 **3.** 0.277 **5.** 0.325 **7.** 0.0524 **9.** 0.0384 **11.** 1.732
13. 7.59 **15.** 38.2 **17.** 0.264 **19.** 0.0454 **21.** 1.921 **23.** 3.51
25. 0.879 **27.** 10.98° **29.** 3.90° **31.** 1.33° **33.** 26.57° **35.** 3.57°
37. 77.95° **39.** 80.8° **41.** 72.65° **43.** 82.83° **45.** 45.85° **47.** 85.33°
49. 88.17° **51.** 88.9°

EXERCISE A.12 (PAGE 407)

1. 10.00 **3.** 3.24 **5.** 0.494 **7.** 30.0° **9.** 15.9° **11.** 21.9°
13. $\angle B = 59°$, $b = 28.6$, $c = 33.4$ **15.** $\angle A = 30°$, $\angle B = 60°$, $b = 3.98$
17. $\angle B = 48.2°$, $\angle A = 41.8$, $c = 25.5$ **19.** $\angle B = 57.6°$, $a = 14.7$, $b = 23.1$
21. $b = 23.6$, $c = 22.6$, $\angle C = 70°$ **23.** $\angle A = 100.6°$, $c = 30.6$, $a = 62.8$
25. $\angle A = 26.7°$, $b = 1003$, $\angle B = 123.3°$
27. $\angle A = 25.7°$, $\angle B = 116.6°$, $b = 35.7$
29. $\angle C = 25.6°$, $\angle A = 100.1°$, $a = 3.89$

Index

Symbols

The page where the symbol is first used is shown in parentheses.

$\{a, b\}$	the set whose elements (or members) are a and b (p. 2)
A, B, C, etc.	names of sets (p. 2)
$\emptyset$	the null set or the empty set (p. 2)
R	the set of real numbers (p. 3)
$=$	is equal to or equals (p. 7)
$<$	is less than (p. 7)
$\leq$	is less than or equal to (p. 7)
$>$	is greater than (p. 7)
$\geq$	is greater than or equal to (p. 7)
$\neq$, $\nless$, etc.	is not equal to, is not less than, etc. (p. 7)
$\|a\|$	the absolute value of a (p. 13)
a^m	the mth power of a or a to the mth power (p. 27)
$\{x \mid \ldots\}$	the set of all x such that . . . (p. 89)
$a^{1/n}$	the nth root of a (p. 108)
$\sqrt[n]{a}$	the nth root of a (p. 123)
$\approx$	is approximately equal to (p. 136)
C	the set of complex numbers (p. 140)
i	imaginary unit (p. 139)
$a + bi$	complex number (p. 139)
$\pm$	plus or minus (p. 159)